The World of Galaxies

Gérard and Antoinette de Vaucouleurs in Paris, 1962.

Harold G. Corwin, Jr. Lucette Bottinelli
Editors

The World of Galaxies

Proceedings of the Conference
"Le Monde des Galaxies"
Held 12–14 April 1988 at the
Institut d'Astrophysique de Paris
in Honor of
Gérard and Antoinette de Vaucouleurs
on the Occasion of His 70th Birthday

With 183 Illustrations

Springer-Verlag
New York Berlin Heidelberg
London Paris Tokyo Hong Kong

Harold G. Corwin, Jr.
Department of Astronomy
University of Texas
Austin, TX 78712
U.S.A.

Lucette Bottinelli
Observatoire de Paris,
Section de Meudon
92190 Meudon
France

Library of Congress Cataloging-in-Publications Data
Symposium, Le monde des galaxies (1988 : Paris, France)
The world of galaxies : proceedings of the Symposium, Le monde des galaxies, in honor of Gerard and Antoinette de Vaucouleurs on the occasion of his seventieth birthday / Harold G. Corwin, Lucette Bottinelli, editors.
p. cm.
1. Galaxies—Congresses. 2. Vaucouleurs, Gérard de, 1918- 3. Vaucouleurs, Antoinette de. I. Vaucouleurs, Gérard Henri de, 1918- . II. Vaucouleurs, Antionette de. III. Corwin, Harold G. IV. Bottinelli, Lucette. V. Title.
QB851.S94 1988
523.1 ′ 12—dc20 89-10084

Softcover reprint of the hardcover 1st edition 1989

Camera-ready copy supplied by editors.

9 8 7 6 5 4 3 2 1

ISBN-13: 978-1-4613-9358-0 e-ISBN-13: 978-1-4613-9356-6
DOI: 10.1007/978-1-4613-9356-6

This book is dedicated to the memory of

Antoinette de Vaucouleurs

(1921 – 1987)

who helped to show so many of us the way through

The World of Galaxies.

Comité d'Honneur

Mesdames et Messieurs:

H. Alfvén (Royal Institute of Technology, Sweden)
V. A. Ambartsumian (Byurakan Astrophysical Observatory, Armenia, U.S.S.R.)
A. Berroir (INSU, France)
F. Bertola (Osservatorio Astronomico di Padova, Italy)
E. M. Burbidge (University of California, San Diego, U. S. A.)
G. Burbidge (University of California, San Diego, U. S. A.)
G. Courtés (LAS, Marseille, France)
W. Cunningham (University of Texas at Austin, U. S. A.)
R. D. Davies (University of Manchester, England, U. K.)
C. Dewitt-Morette (University of Texas at Austin, U. S. A.)
H. Elsässer (Max-Planck-Institut für Astronomie, Heidelberg, F. R. G.)
S. Feneuille (CNRS, France)
C. Fréjacques (CNRS, France)
H. van der Laan (ESO, Garching bei München, F. R. G.)
N. U. Mayall (Tucson, Arizona, U. S. A.)
G. Monnet (Canada-France-Hawaii Telescope, Hawaii, U. S. A.)
J.-C. Pecker (Collège de France, Institut d'Astrophysique, Paris, France)
N. G. Roman (Chevy Chase, Maryland, U. S. A.)
E. Schatzman (Observatoire de Nice, France)
B. Takase (Tokyo Astronomical Observatory, Japan)
G. Wlérick (Observatoire de Meudon, France)
L. Woltjer (ESO, Garching bei München, F. R. G.)

Scientific Organizing Committee

J. Audouze
J. Bahcall
F. N. Bash (co-chair)
M. Capaccioli (co-chair)
K. C. Freeman
L. Gouguenheim
J. Heidmann
E. Khachikian
P. van der Kruit
D. Sciama
H. J. Smith
R. B. Tully

Local Organizing Committee

L. Bottinelli
Y. Bousquet
S. Collin
F. Delmas
P. Fouqué
M. Gros
J.-C. Pecker (chair)

Introduction

From 12 April to 14 April 1988, 120 of Gérard and Antoinette de Vaucouleurs's friends and colleagues gathered at the Institut d'Astrophysique in Paris to celebrate Gérard's 70th birthday and his remarkable career in Astronomy. The gathering also honored the memory of Antoinette (who died 29 August 1987 after a long illness) and her own no less remarkable career. This volume collects the 24 invited review papers and the 60 contributed poster papers presented at the meeting.

Gérard de Vaucouleurs

Gérard de Vaucouleurs was born on 25 April 1918 in Paris, where he spent his boyhood. He became an active amateur astronomer in the early 1930's, making extensive observations of Mars, Jupiter, and variable stars (including the bright supernova of 1937 in IC 4182). He also began life-long interests in astronomical photography and galaxy cataloguing during this period. In 1939, he met the director of the Paris transport system and an equally avid amateur astronomer, Julien Péridier. De Vaucouleurs worked at Péridier's private observatory at Le Houga in southwestern France on and off throughout the next decade. His undergraduate work was in mathematics, astronomy, and experimental physics; this, combined with his interest in observational astronomy, formed his life-long empirical approach to science.

After spending 18 months in the French army early in 1939 – 41, Gérard returned to the Péridier observatory and then, in 1943, to his studies at the Sorbonne, where he met Antoinette. They were married in October 1944, and both eventually became graduate students at the Institut d'Astrophysique (1945 – 49). It was there that they were fellow students of J.-C. Pecker, and were influenced by Jean Cabannes (at the Sorbonne), Paul Couderc (Observatoire de Paris), and Daniel Chalonge (Institut d'Astrophysique), among others.

De Vaucouleurs's intensive studies of photography led to the publication of several books on photography including *Manuel de Photographie Scientifique* with J. Dragesco and P. Selme, perhaps the most thorough exploration of practical scientific photography ever to see print. He then applied this knowledge of photography to the problem of the distribution of light in nebulae: the $r^{1/4}$ law was first published in 1948, a year before he received his (first) doctorate.

Always a prolific writer, he already had half a dozen books to his credit by this time, and his 1949 thesis (on Rayleigh scattering of light) is the 79th in his list of over 500 published books, papers, articles, reviews, and reports (a nearly complete list of these is given in *Gérard and Antoinette de Vaucouleurs — A Life for Astronomy*, published in 1989 by World Scientific).

In 1950, the de Vaucouleurs emigrated to London where he produced a weekly radio science program for the French Section of the BBC. The next move was to the Commonwealth Observatory at Mt. Stromlo in Canberra, Australia in 1951. This marked a return to active observational astronomy for de Vaucouleurs. It was here that he called attention in 1953 to the belt of galaxies stretching across the northern sky, and a similar flattened structure in the south. He was the first to interpret these as superclusters of galaxies. In doing so, he pioneered modern studies of the distribution of galaxies throughout the universe. The Australian years also saw the completion of a first revision of the Shapley-Ames catalogue of bright galaxies, a survey of the southern Shapley-Ames objects with the 30-inch Reynolds Reflector, extensive work on the Magellanic Clouds, and continued observations of Mars and variable stars. This work culminated in 1957 with his earning a D. Sc. degree from the Australian National University.

Short stays at Lowell Observatory in Flagstaff, Arizona (1957 – 58), and at Harvard College Observatory in Cambridge, Massachusetts (1958 – 60) preceded the de Vaucouleurs's move in 1960 to the newly-formed Astronomy Department at the University of Texas in Austin, where they finally settled. Research at Lowell — where de Vaucouleurs met and was strongly influenced by Harold Johnson — centered on photoelectric observations of galaxies. In charge of Harvard's planetary research program, de Vaucouleurs began a project to map the surface of Mars, and was among the first to apply computers to the determination of precise positions of Martian surface features. These studies yielded the rotation rate of Mars to a precision not surpassed until the reduction of Viking spacecraft data in the 1970's.

At Texas, de Vaucouleurs continued his studies of the photometric properties of individual galaxies, superclusters and the distribution of galaxies, and mapping the surface of Mars. He developed an interest in kinematics of galaxies, and built the "Galaxymeter," a device that successfully combined a photoelectric photometer, an image tube spectrograph, a Fabry-Perot interferometer, and a photographic reducing camera. Though a few simple changes in the optical path of the instrument switched it from one mode to another, the de Vaucouleurs used it mostly for obtaining interferograms of late-type galaxies rich in Hα emission. The *First Reference Catalogue of Bright Galaxies*, co-authored with Antoinette, appeared in 1964, and found immediate application in the first all-sky survey of galaxy groups, de Vaucouleurs's contribution to the classic *Galaxies and the Universe*. The series of photoelectric observations of galaxies in the Johnson *UBV* system begun by de Vaucouleurs at Lowell in 1957 is still in progress today at McDonald; this may be the longest-running extragalactic observing project in the history of Astronomy. The de Vaucouleurs produced a *Second Reference Catalogue of Bright Galaxies* in 1976, helped in this endeavor by Harold G. Corwin, Jr.

Though he realized very early in his career that determining distances to galaxies would be vital to understanding their properties, de Vaucouleurs's intense concentration on the problem of the distance scale really began in the mid-1970's. Drawing on his extensive background in physics and mathematics, he developed a broad-based approach to the distance scale problem that relies not on just two or three distance indicators, but over a dozen. His series of seven papers "The Extragalactic Distance Scale" in *The Astrophysical Journal* stand as models of his way of doing proper observational astronomy: collect all the data one can, analyze them carefully by looking for and correcting systematic errors, and only then use the data for their intended purpose.

Always keenly aware of the value of collaboration, he has developed an international following of colleagues and co-authors. Astronomers in France, Italy, Great Britian, Norway, Australia, Argentina, and half a dozen other countries have worked with him on his many projects. A current example is the *Third Reference Catalogue of Bright Galaxies* being readied for publication by not only the de Vaucouleurs and Corwin, but by Ronald Buta, George Paturel, and Pascal Fouqué, the latter two in France. L. Bottinelli and L. Gouguenheim have enjoyed a collaboration with de Vaucouleurs for many years, studying the H I properties of galaxies, particularly as they apply to the distance scale work. De Vaucouleurs's many students are similarly spread far and wide, and his papers have appeared in virtually all of the world's major astronomical publications. Throughout his career, he has also devoted a surprisingly large percentage of his output to writing for the public or for amateur astronomers.

Of the many awards and honors that de Vaucouleurs has received for his work and writing (including the Herschel Medal of the Royal Astronomical Society, the Henry Norris Russell Lectureship of the American Astronomical Society, election to the U. S. National Academy of Sciences, and lately, election to the Academy of Sciences of Argentina), perhaps none has pleased him more than the Janssen Prize of the Société Astronomique de France, awarded to him and Antoinette for their lifetime of work together devoted to Astronomy.

Antoinette de Vaucouleurs

Antoinette de Vaucouleurs, born in Paris on 14 November 1921, studied mathematics, physics, and astronomy at the Sorbonne from 1944 to 1948. Beginning her career in research in 1949 at the Institut d'Astrophysique, she worked in spectroscopy, finding new doublets and perturbations in the infrared spectrum of potassium. She also assisted her husband in many of his early observational studies of galaxies in France. While in England in the early 1950's, she volunteered her time as a research assistant at the University of London's Mill Hill

Observatory, measuring parallax plates.

In Australia, Antoinette not only continued with her help to her husband's work, but carried on her own work in spectroscopy. This culminated in her publication in 1957 of the first *quantitative* spectral and luminosity classifications of 366 southern early-type stars. While at Lowell Observatory, reducing her husband's first *UBV* photoelectric photometry of galaxies, she noticed that the *U*-band data for several bright Seyfert galaxies fluctuated by several tenths of a magnitude, much more than would normally be allowed by observational error alone. She pointed this out to Gérard who told her, "Of *course* galaxies aren't variable." He suggested that night sky fluctuations were responsible, and was very contrite a decade later when the variations of the Seyfert nucluei were firmly established by other observers. The episode serves to illustrate Antoinette's acumen and attention to detail that served her well as she continued the enormous efforts involved in producing the *Reference Catalogues* at Texas. Her last work was the literature search for the data that will appear in the *Third Reference Catalogue*.

She continued her work in spectroscopy after the move to Texas, but her attention shifted more and more to galaxies. Her name appears on several long lists of redshifts of galaxies, and her work with her husband on the bar of the LMC was the first quantitative analysis of the stellar population of a galaxy from its spectrum. She took over the chore of handling most of the data reduction from the many observing runs at McDonald that she shared with her husband. Again, her meticulous attention to detail caught many errors in the literature and in the data that would otherwise have gone undetected.

She was passionately dedicated to astronomy, and authored or co-authored over 60 research papers and books during her career. She contributed to countless other papers, often turning down a co-authorship when she felt that her often vital work deserved no more than an acknowledgment. She was just as dedicated to her husband, and it may be that her greatest contribution was not her own remarkable astronomical skill, but the fact that she used it to fully support Gérard and his work during their long journey together to study the galaxies.

Acknowledgments

Le Monde des Galaxies was a wonderful conference. Helping us to make it so was our gracious host Prof. Jean Audouze, Director of the Institut d'Astrophysique de Paris. He was ably assisted by the Local Organizing Committee and by Jean Begot, Jacques Fagot, Nicole Hallet, Jean Heidmann, Helena Hedreul, Laurence Lericque, Jean Mouette, and Francoise Warin. These wonderful colleagues together made the meeting run so smoothly that we hardly noticed them doing their jobs.

During the assembly of the Proceedings, the editors received invaluable help from Fritz Benedict, Susan Cady, Charles Geiger, Lucienne Gouguenheim, Connie Mahaffey, Cecil Martinez, and Annalisa Palacios. We also appreciate the help of our professional editor at Springer-Verlag, Jeffery Robbins, who guided us through the publication process with skill, humor, and remarkable patience.

Table of Contents*

*Invited review papers are listed in bold type.

The Magellanic Clouds

Photometric Properties of the Galaxy and of Other Galaxies

Kinematics and Dynamics of Galaxies

The Local Supercluster

The Cosmic Distance Scale

List of Participants

Ables, H. D.	U. S. Naval Observatory, Flagstaff, Arizona, U. S. A.
Alcaino, G.	Instituto Isaac Newton, Santiago, Chile
Arsenault, R.	European Southern Observatory, Garching bei München, F. R. G.
Audouze, J.	Institut d'Astrophysique de Paris, France
Bahcall, J. N.	Institute for Advanced Study, Princeton, New Jersey, U. S. A.
Balkowski, C.	Observatoire de Meudon, France
Barbaro, G.	Dipartimento di Astronomia, Università di Padova, Italy
Bash, F. N.	Astronomy Department, University of Texas, Austin, Texas, U. S. A.
Baxter, D. A.	University College, Mathematics Institute, Cardiff, Wales, U. K.
Bender, R.	Landessternwarte Königstuhl, Heidelberg, F. R. G.
Bertin, G.	Classe di Scienze, Scuola Normale Superiore, Pisa, Italy
Bertola, F.	Osservatorio Astronomico di Padova, Italy
Binney, J. J.	Department of Theoretical Physics, University of Oxford, U. K.
Bonoli, F.	Osservatorio Astronomico di Bologna, Italy
Bonoli, C.	Osservatorio Astronomico di Padova, Italy
Bottinelli, L.	Observatoire de Meudon, France
Brosche, P.	Observatorium Hoher-List der Universitäts-Sternwarte Bonn, F. R. G.
Buchert, T.	Max Planck Institut, Garching bei München, F. R. G.
Burbidge, E. M.	Center for Astrophysics and Space Science, University of California, San Diego, La Jolla, California, U. S. A.
Burstein, D.	Department of Physics, Arizona State University, Tempe, Arizona, U. S. A.
Buson, L.	Osservatorio Astrofisico, Asiago, Italy
Buta, R.	Astronomy Department, University of Texas, Austin, Texas, U. S. A.
Calvani, M.	Istituto di Padova, Italy

Capaccioli, M.	Osservatorio Astronomico di Padova, Italy
Caulet, A.	Department of Physics and Astronomy, University of Alabama, Tuscaloosa, Alabama, U. S. A.
Christian, C. A.	Canada-France-Hawaii Telescope, Kamuela, Hawaii, U. S. A.
Collin, S.	Institut d'Astrophysique de Paris, France
Corwin, H. G.	McDonald Observatory and Astronomy Department, University of Texas, Austin, Texas, U. S. A.
Couch, W.	Anglo-Australian Observatory, Epping, N. S. W., Australia
Courtès, G.	Laboratoire d'Astronomie Spatiale du CNRS et Observatoire de Marseille, France
D'Onofrio, M.	Osservatorio Astronomico, Padova, Italy
Davies, R. L.	Kitt Peak National Observatory, Tucson, Arizona, U. S. A.
Davies, R. D.	University of Manchester, Nuffield Radio Astronomy Laboratories, Jodrell Bank, England, U. K.
Davoust, E.	Observatoire de Toulouse, France
Dettmar, R.-J.	Radioastronomisches Institut der Universität Bonn, F. R. G.
Feast, M. W.	South African Astronomical Observatory, Cape Province, South Africa
Federici, L.	Osservatorio Astronomico di Bologna, Italy
Feretti, L.	Istituto Radioastronomia, Bologna, Italy
Focardi, P.	Dipartimento di Astronomia, Università di Bologna, Italy
Fosbury, R. A. E.	Space Telescope — European Coordinating Facility, European Southern Observatory, Garching bei München, F. R. G.
Fouqué, P.	Observatoire de Meudon, France
Freeman, K. C.	Mt. Stromlo Observatory, Canberra, A. C. T., Australia
Galetta, G.	Dipartimento di Astronomico, Università di Padova, Italy
Garcia-Barreto, J.	Instituto de Astronomia, Universidad Nacional de Mexico
Gioia, I. M.	Center for Astrophysics, Cambridge, Massachusetts, U. S. A.
Giuricin, G.	Dipartimento di Astronomia, Trieste, Italy
Gouguenheim, L.	Observatoire de Meudon, France
Gros, M.	Institut d'Astrophysique de Paris, France
Grosbol, P.	European Southern Observatory, Garching bei München, F. R. G.

Hanes, D. A.	Astronomy Group, Physics Department, Queen's University, Kingston, Ontario, Canada
Hatzidimitriou, D.	Department of Astronomy, University of Edinburgh, Scotland, U. K.
Hayli, A.	Observatoire de Lyon, France
Heidmann, J.	Observatoire de Meudon, France
Helou, G.	IPAC, California Institute of Technology, Pasadena, California, U. S. A.
Henning, P. A.	Astronomy Program, University of Maryland, College Park, Maryland, U. S. A.
van der Hulst, J. M.	Netherlands Foundation of Radioastronomie, Dwingeloo, The Netherlands
Kerr, F.	Astronomy Program, University of Maryland, College Park, Maryland, U. S. A.
van der Kruit, P. C.	Kapteyn Laboratorium, Groningen, The Netherlands
van der Laan, H.	European Southern Observatory, Garching bei München, F. R. G.
Lafon, J.-P.	Observatoire de Meudon, France
Lewis, J. R.	Institute of Astronomy, Cambridge, England, U. K.
MacGillivray, H. T.	Royal Observatory, Edinburgh, Scotland, U. K.
Madore, B. F.	California Institute of Technology, Pasadena, California, U. S. A.
Malin, D.	Anglo-Australian Observatory, Epping, N. S. W., Australia
Martin, P.	Département de Physique, Université Laval, Québec, Canada
Marziani, P.	Osservatorio Astronomico di Padova, Italy
Masnou, J.-L.	Observatoire de Meudon, France
Michard, R.	Observatoire de Nice, France
Mitra, S.	Astronomy Department, University of Texas, Austin, Texas, U. S. A.
Moss, C.	Institute of Astronomy, Cambridge, England, U. K.
Nicholson, R.	Royal Greenwich Observatory, East Sussex, England, U. K.
Nieto, J.-L.	Observatoire de Toulouse, France
Nottale, L.	Observatoire de Meudon, France
Odewahn, S.	Astronomy Department, University of Texas, Austin, Texas, U. S. A.
Okamura, S.	Kiso Observatory, Japan
Palumbo, G. G. C.	Dipartimento di Astronomia, Università di Bologna, Italy

Patsis, P. A.	Department of Astronomy, University of Athens, Greece
Paturel, G.	Observatoire de Lyon, France
Pecker, J.-C.	Institut d'Astrophysique de Paris, France
Pence, W. D.	Space Telescope Science Institute, Baltimore, Maryland, U. S. A.
Poulain, P.	Observatoire de Toulouse, France
Prieto, M.	Instituto de Astrofisica de Canaries, Universidad de La Laguna, Tenerife, Spain
Prieur, J.-L.	Observatoire de Toulouse, France
Prugniel, P.	Observatoire de Toulouse, France
Pucacco, J.	Istituto di Fisica, Università "La Sapienza," Roma, Italy
Rampazzo, R.	Osservatorio Astronomico di Brera, Milano, Italy
Rifatto, A.	Osservatorio Astronomico di Padova, Italy
Roberts, M. S.	National Radio Astronomy Observatory, Charlottesville, Virginia, U. S. A.
Rosino, L.	Osservatorio Astronomico di Padova, Italy
Rubin, V. C.	Department of Terrestrial Magnetism, Carnegie Institution of Washington, Washington, D. C., U. S. A.
Rudnicki, K.	Jagiellonian University Observatory, Krakow, Poland
Ruffini, R.	International Center for Relativistic Astrophysics, Dipartimento di Fisica, Università di Roma, Italy
Sancisi, R.	Kapteyn Laboratorium, Groningen, The Netherlands
Secco, L.	Dipartimento di Astronomia, Università di Padova, Italy
Seiden, P. E.	IBM Research Center, Yorktown Heights, New York, U. S. A.
Shaw, M. A.	Department of Astronomy, The University of Manchester, England, U. K.
Simien, F.	Observatoire de Lyon, France
Smith, H. J.	McDonald Observatory and Astronomy Department, University of Texas, Austin, Texas, U. S. A.
Sparks, W. B.	Space Telescope Science Institute, Baltimore, Maryland, U. S. A.
Sulentic, J. W.	Department of Physics and Astronomy, University of Alabama, Tuscaloosa, Alabama, U. S. A.
Sullivan, W. T.	Department of Astronomy, University of Washington, Seattle, Washington, U. S. A.
Tammann, G.	Astronomy Institute, University of Basel, Binningen, Switzerland
Taylor, K.	Anglo-Australian Observatory, Epping, N. S. W., Australia
Teague, P. F.	Mt. Stromlo Observatory, Canberra, A. C. T., Australia

Thonnat, M.	I. N. R. I. A., Valbonne, France
Trinchieri, G.	Osservatorio Astrofisico di Arcetri, Firenze, Italy
Tully, R. B.	Institute for Astronomy, University of Hawaii, Honolulu, Hawaii, U. S. A.
Vallenari, A.	Osservatorio Astronomico di Padova, Italy
de Vaucouleurs, G.	McDonald Observatory and Astronomy Department, University of Texas, Austin, Texas, U. S. A.
Vettolani, G.	Istituto di Radioastronomia, Bologna, Italy
Wainscoat, R. J.	NASA Ames Research Center, Moffett Field, California, U. S. A.
Wayte, S.	Mt. Stromlo Observatory, Canberra, A. C. T., Australia
Westerlund, B. E.	Astronomical Observatory, Uppsala, Sweden
Whitmore, B. C.	Space Telescope Science Institute, Baltimore, Maryland, U. S. A.
Wlerick, G.	Observatoire de Meudon, France

Gérard de Vaucouleurs and His Friends at the Institut d'Astrophysique de Paris, 14 April 1988.

Welcome and Opening Address

Jean Audouze

Le Directeur, Institut d'Astrophysique, Paris, France

Honneur, émotion, et tristesse, sont les trois mots française par lesquels je voudrais ouvrir ce colloque international "Le Monde des Galaxies," organisé à la fois en l'honneur du Professeur Gérard de Vaucouleurs et à la mémoire de Madame Antoinette de Vaucouleurs. Yes, we all have been very sad to learn, at the end of the summer of 1987, of the death of Antoinette de Vaucouleurs who is so much and so closely associated with the scientific work we are going to celebrate during these three days. Professor Gérard de Vaucouleurs, we share with you this immense loss and we of course dedicate this meeting to the memory of your wife. When Jean–Claude Pecker asked me to host this meeting, and when you, Gérard de Vaucouleurs, wrote to me that you feel as if you are a member of Institut d'Astrophysique de Paris, I felt that the choice made by the organising committee to have this meeting here, and the very kind appreciation of Gérard, is indeed one of the highest honours which has been bestowed on this Institute during the decade I have worked here.

Gérard and Antoinette de Vaucouleurs were among the very first scientists who have belonged to this Institute. After having been chosen by Professor Jean Cabannes as one of his graduate students, you Gérard, completed here your thesis work on Rayleigh diffusion in gas and liquids. Here you have been Attaché de Recherche au Centre National de la Recherche Scientifique from 1945 to 1950, and you have benefitted from the support of Henri Mineur and Daniel Chalonge who foresaw your dedication to extragalactic astrophysics, and your exceptional scientific talents. It is in this Institute that the most famous galactic luminosity relation in $r^{1/4}$, now known as the de Vaucouleurs law, was established. Jean–Claude Pecker will evoke more precisely and completely than I can the main steps of your magnificent scientific career, and will list your many achievements. I will just end up these few welcoming remarks by saying that the French astronomical community is very proud to have exported to the U. S., and especially to one of the most powerful states of the Union, two of the best representatives of our community.

Je terminerai en vous redisant, très cher Gérard de Vaucouleurs, toute notre respectueuse amitié et affection et en réaffirmant que cette maison est vraiment la vôtre. Au nom de tous les membres de l'IAP, je suis très heureux d'accueillir ici les participants au symposium "Le Monde des Galaxies" qui se tient en votre honneur et d'ouvrir les travaux de ce colloque.

Galaxy Catalogues and Surveys

Harold G. Corwin, Jr.

McDonald Observatory and Astronomy Department, University of Texas at Austin, U. S. A.

Introduction

I'd like to begin this survey of surveys and catalogues of galaxies with a story about Gérard de Vaucouleurs and myself. When I first went to work for him in the summer of 1965, I was the proud owner of a copy of the then–new *Reference Catalogue of Bright Galaxies* (RC1; 1964). I don't often seek autographs for my books, but this was a special occasion and a special book, so I asked Gérard and Antoinette for their autographs. After signing the title page, Gérard began leafing through the catalogue and noticed some of my penciled "updates" in the B(0) column. Learning that these magnitudes had not (yet!) been reduced to the B(0) system, he scolded me for putting the wrong numbers in the catalogue. Properly chastized, I immediately wrote a short computer program to do the reduction, and spent the necessary time to get the correct data into the correct columns.

The point of that short lesson in cataloguing has stuck with me over the years: one of the major tasks of the astronomical cataloguer is to *systematize* the data that goes into the book. What has become clear to me much more recently is that that idea was first put into practice for galaxy catalogues by Gérard de Vaucouleurs. With one major exception that I shall mention later, the *First Reference Catalogue* was the first galaxy catalogue to offer researchers astrophysically useful data which had been corrected for systematic error and then reduced to well–defined standard systems.

In this survey, I shall put de Vaucouleurs's cataloguing work into historical perspective, show why it has been so important for extragalactic research over the past quarter century, and attempt to devine the directions that galaxy cataloguing might take in the future.

History

Galaxy cataloguing is not as old as star cataloguing because of the unfortunate accident of the location of the Magellanic Clouds deep in the southern sky. The Asian and European cultures from which we derive our scientific heritage did not record the two Clouds until the 16th century. By that time, however, the Andromeda Galaxy had been in the catalogues for a full seven centuries. It was, of course, listed with all other nebulae and star clusters — the apparently motionless non–stellar astronomical objects.

By the time Charles Messier published his famous catalogue in 1781, there were 140 of these non–stellar objects known (41 are galaxies). Messier included 98 of these in his list, omitting seven additional objects known to him or his colleague Pierre Méchain[1] for which they did not have good positions at the time of publication. Messier's goal was to establish a fairly complete list of the brighter known non–stellar objects so that comet hunters would not be misled into discovering false comets. Nevertheless, he realized that to be useful, his list would have to be accurate, so he made careful observations of all of the objects that he could see from Paris, recording their positions and descriptions as accurately as he could. While a few discrepancies exist in his list, Messier's catalogue is still one of the most relatively error–free that has ever been published.

Then came William Herschel. Finally freed from the drudgery of teaching music and composition by his 1781 discovery of Uranus, and his subsequent royal stipend, he was able to devote himself fully to his first love, astronomy. After making more than 200 generally unsatisfactory telescope mirrors, Herschel finally mastered the art, and by October of 1783, was making regular "sweeps" of the heavens with a Newtonian reflector with an aperture of 18.7 inches. Within a remarkable single observing season, he recorded 500 new nebulae, most of them galaxies, and was able to announce in the summer of 1784 the discovery of what we now call the Local Supercluster. He, of course, did not know *what* his great "stratum" of nebulae represented, but there is little doubt that he visualized it as the nebular analogue of the Milky Way: a belt of celestial objects circling the sky.[2]

Herschel eventually discovered 2500 nebulae and star clusters (Herschel 1786, 1789, 1802). His son John carried on the task of charting the heavens, eventually discovering about 1770 more non–stellar objects, all but 500 of these in the southern hemisphere (Herschel 1833, 1847). Though preceded by Lacaille (1755) and Dunlop (1828) in the south, John Herschel — during a four–year stay at the Cape of Good Hope (1834 – 1838) — was the first to provide accurate positions

[1] These have since been given "Messier numbers" by Camille Flammarion, Helen Saywer Hogg, Owen Gingerich, and Kenneth Glyn–Jones.

[2] Herschel has left us to wonder whether he indeed thought of the belt of nebulae as a flattened disk–like system — as he eventually came to visualize the Milky Way — or simply as a belt. My temptation to attribute a modern interpretation to Herschel is considerable, but one must "read between the lines" of Herschel's papers to find *any* interpretation of what he saw.

and descriptions for the brighter southern nebulae.

By the mid-1800's, it became apparent to several astronomers that a complete catalogue of all known non-stellar objects needed to be assembled and published. Thus was born John Herschel's so-called "General Catalogue" of 1864.[3] Since most of the objects were found by the Herschel's with 18.7-inch telescopes, the GC has acquired a reputation of being more "homogeneous" than most later catalogues, and has occasionally been used as a source list of galaxies "complete" to the 14th magnitude.

Through the latter two-thirds of the nineteenth century, visual micrometric measurements of the nebulae were just about the only aspect of their study that could be easily quantified. This work was carried on at several observatories, but reached its peak right here in Paris. Now little known, the massive quarter-century of work by Guillaume Bigourdan (1917, and references therein) resulted in the precise measurement of *all* the known nebulae visible to him with the 310-cm "Great West Equatorial" refractor. The impetus behind his work was to determine the proper motions, if any, of the nebulae, and thus to provide at least a statistical estimate of their average distances. Bigourdan, and the many others involved in this work, of course found no nebula with an unambigous proper motion. This null result led to indeterminant "statistical parallaxes" for the nebulae, and a general — if largely unspoken — concensus that the laborious micrometric work was of little value at best, or totally wasted time at worst. [4]

So, unfortunately, most of this work did not find its way into J. L. E. Dreyer's *A New General Catalogue of Nebulae and Clusters of Stars*... published by the Royal Astronomical Society as the first part of its *Memoirs*, Volume **49** (NGC; 1888). This work was intended to update the GC, gathering positions and brief descriptions of all known nebulae and star clusters into one convenient list. Since it lists almost all the bright nearby galaxies, it is still common to refer to the NGC for galaxy nomenclature. Only in recent years have the more modern catalogues and finding lists with their variety of names[5] been anywhere nearly as popular as the NGC.

The NGC is far from homogeneous in its coverage of the sky. Though more than half its 7840 entries are due to the Herschel's efforts, many of the remain-

[3]Curiously, Sir John's "General Catalogue" does not bear the title by which it has always been known. It is simply called *Catalogue of Nebulae and Clusters of Stars* on the title page. Nevertheless, it was the first widely-available listing of all known nebulae and star clusters — it is in deed, if not in name, a *general* catalogue.

[4]However, the work has given us precise *relative* positions for the nuclei of most of the brighter galaxies. These positions, if properly reduced and corrected to our present astrometric reference frames, have the potential of providing far more accurate positions than are possible using the over-exposed galaxy images on modern sky survey copy plates, prints, or films. Comparitive studies have shown that the best of the visual astrometry has mean errors of less than a second of arc. This is could be accurate enough to help in better understanding the fundamental reference frames in use at the time, and thus to a better knowledge of the motions of the reference stars.

[5]*e.g.* DDO, VV, UGC, ESO, Zwicky, Markarian, Arakelian, and that annoyingly ubiquitous fellow Anonymous!

der of the objects were found with telescopes as large as 1.8 meters. Dreyer's *Index Catalogues...* (IC) of 1895 and 1908 are even less homogeneous than the NGC: nearly two–thirds of the entries in the *Second Index Catalogue* were discovered by photography. Thus, in a few small areas of the sky, the census of galaxies at the turn of the century is complete to nearly 17th magnitude, while neighboring areas might contain unlisted 13th magnitude objects. The IC, however, does bring us into the modern era of photographic galaxy surveys. The early pioneers in astronomical photography — Roberts, Barnard, Keeler, Wolf and his colleagues at Heidelberg, the several Harvard observers working under Pickering's direction — are all represented with entries in the IC.

Dreyer realized after the *Second Index Catalogue* that there was little point in trying to keep up with the flood of new discoveries by periodically publishing updated catalogues of all known nebulae. After all, observers were finding on single plates as many galaxies as were contained in William or John Herschel's entire catalogues. In any event, most of these new objects were faint and relatively difficult to study. It was clear that the new science of astrophysics was going to concentrate on the bright nebulae, at least for developing an understanding of the distances and properties of individual objects.

So, while detailed lists of newly discovered nebulae continued to be published occasionally, the era of the all–sky general catalogue of all known non–stellar objects had passed. At least, so it seemed at the time. One other such catalogue appeared in 1980, Dixon and Sonneborn's massive *A Master List of Nonstellar Optical Astronomical Objects*, and it seems likely that at least electronic versions of this or other compendia will continue to be available. But this is getting ahead of the story a bit.

I hope I've made it clear that up to this point, the catalogues of non–stellar objects have been primarily used as finding lists. There is little astrophysically useful information in them, and even that has to be treated with some caution. For example, attempts have been made to quantify the magnitude and diameter data in the NGC, and the distribution of the NGC objects has also received some attention. However, the selection effects in this and earlier catalogues were poorly understood, and generally defeated attempts at deriving useful information from the catalogues themselves. Instead, as I've said, the main usefulness of the early cataloguing work has been to provide finding lists of objects for further study.

The Early Twentieth Century

With the rapid acceptance in the 1920's of most of the nebulae as stellar systems external to our own Milky Way, there came the need for a catalogue of just the brightest of them. This was provided by Harlow Shapley and Adelaide Ames of Harvard College Observatory in 1932. The open clusters and planetaries, the globulars and diffuse nebulae that "clutter up" the NGC are (mostly!) weeded out of the Shapley–Ames catalogue. I can at last drop the word "nebula" and begin to speak of "galaxies" and "extragalactic astronomy" in our own 20th century jargon.

Yet the Shapley–Ames catalogue, like its predecessors, is little more than a finding list of galaxies with descriptive entries. Its descriptions are in terms that we recognize as "modern" in the sense of being quantified (diameters given in arc minutes and mean magnitudes estimated from several plates). But the small scale survey plates from the 1–, 2–, and 3–inch aperture refractors did not allow precise magnitude measurements to be made, and the diameters and types are from a heterogeneous list of sources. In addition, studies of the "completeness" of the Shapley–Ames catalogue show that it is only 50% complete at blue magnitude $\sim$ 12.5, in spite of its authors' hope to achieve virtual completeness at magnitude 13.0.

Steps to the Reference Catalogues

It was against this background that Gérard de Vaucouleurs began his first catalogue of deep sky objects in 1934. This was a listing of the objects by NGC number with their position, a notation of the nature of the object when known, and references to published material. The Second World War interrupted this effort as it did so many others in Europe. Thus, it was not until 1949 that the de Vaucouleurs (Gérard and Antoinette were married during the War), with Paul Griboval, began their updating of the Shapley–Ames catalogue. Ideally, this updating should contain the data needed to compare galaxies in different parts of the sky with one another, and the catalogue should also contain data that would be useful in the study of the global properties of galaxies.

In other words, the data for each object should be *systematized* in such a way that they would be useful for astrophysical studies of the galaxies. This is the de Vaucouleurs's great contribution to galaxy cataloguing. They raised it from the level of simply compiling finding lists to a fully realized scientific endeavor in its own right.

This ideal was not achieved overnight. Intermediate steps include the card file to which de Vaucouleurs even now often turns when he wants basic information about a galaxy. Most of the data here — painstakingly entered by hand, first by Griboval, then by one or the other of the de Vaucouleurs — are from the early 20th century listings from Mt. Wilson, Harvard, Lund, Heidelberg, and the few other observatories then active in nebular and extragalactic astronomy. This card file led to an interim revision of the Shapley–Ames catalogue, published as an Australian National University Mimeogram (de Vaucouleurs1953). Unfortunately not widely circulated,[6] this is nonetheless de Vaucouleurs's first published galaxy catalogue. It is also the one systematized catalogue that I

[6] I, for example, had never seen a copy until de Vaucouleurs, reading through a draft of this review, dug out a copy to show me.

mentioned earlier as pre–dating RC1. The goal that de Vaucouleurs set himself in this work was to provide *total* magnitudes for the Shapley–Ames galaxies on the photometric system defined by the North Polar Sequence. In doing this, he also found that the errors in the Shapley–Ames magnitudes are primarily dependent on surface brightness, and provided a first estimate of the correction necessary to bring them into systematic agreement with the photoelectric data.

The de Vaucouleurs continued work through the 1950's and early 1960's at various observatories — first Mt. Stromlo, then Lowell, Harvard, and finally McDonald — on various aspects of cataloguing. Systems of magnitudes, diameters, morphological types, color indices, and radial velocities were set up in a series of publications that culminated in 1964 with the first *Reference Catalogue of Bright Galaxies.* This large book (its size was the most common complaint about RC1) contained not only 100,000 information elements on about 2600 galaxies, but detailed notes and references for most of these objects. It nicely summed up the first half–century of extragalactic research.

Recent Catalogues and Surveys

In the meantime, the production and wide distribution of the Palomar Sky Survey as photographic prints had provided the impetus for many other surveys of galaxies, some of which led to major catalogues; for example, the *Morphological Catalogue of Galaxies* by Vorontsov–Velyaminov and his associates (MCG; 1962, 1963, 1964, 1968, 1974) the *Catalogue of Galaxies and of Clusters of Galaxies* by Zwicky and his colleagues (CGCG; 1961, 1963, 1965, 1966, 1968a,b), and the Abell catalogue of rich galaxy clusters (1958). Other shorter lists also appeared: a list of low surface brightness galaxies, nominally dwarves (DDO; van den Bergh 1959, 1966; see also Fisher and Tully 1975), a reclassification of the Shapley–Ames galaxies including luminosity classes (van den Bergh 1960), Vorontsov's *Atlas of Interacting Galaxies* (VV; 1959), Arp's atlas of peculiar galaxies (1966), Zwicky's several lists and catalogue of compact galaxies (CSCG; 1971 and references therein; 1975), the fifteen lists of Markarian galaxies with ultraviolet excesses (see Mazzarella and Balzano 1986 and references therein), and so on. Most of these share at least one characteristic with the GC and NGC: they are still finding lists — surveys of *what's there* — rather than completely systematized catalogues. Of course, most of them were constructed in a systematic fashion. Had they not been, we would not use them now.

More recently, we have had other specialized surveys of extragalactic objects:

- southern galaxies [by Lauberts (1982 and references therein), and the de Vaucouleurs's and myself (1985)],
- southern *peculiar* galaxies (Arp and Madore 1987),
- northern galaxies (Nilson 1973),

- high surface brightness galaxies (Arakelian 1975),
- more low surface brightness galaxies (Karachentseva 1968, 1973a),
- isolated galaxies (Karachentseva 1973b),
- double galaxies (Karachentsev 1972, Turner 1976),
- triple galaxies (Karachentseva *et al.* 1979)
- ultraviolet–excess galaxies [Markarian *et al.* 1982 and references therein (see also Mazzarella and Balzano 1986); Markarian *et al.* 1986 and references therein; Takase and Miyauchi–Isobe 1988 and references therein (see also Takase and Miyauchi–Isobe, this conference); and Pesch and Sanduleak 1988 and references therein],
- nearby galaxies (Kraan-Korteweg 1986, Tully and Fisher 1987, Tully 1988),
- groups of galaxies (de Vaucouleurs 1975, Turner and Gott 1976, Geller and Huchra 1983, Tully 1988, Maia *et al.* 1989),
- compact groups of compact galaxies (Baier and Tiersch 1979, and references therein)

— just about any sort of galaxy or association of galaxies that you care to name now has its own catalogue. For a few of these, the systematic errors in the data have been investigated and corrected, but most still serve primarily as finding lists.

Older catalogues are being revised, too. We now have a *Second Reference Catalogue* (RC2; de Vaucouleurset al. 1976), a revised NGC (Sulentic and Tifft 1973), a revised NGC *and* IC (Sinnott 1988), *two* Revised Shapley–Ames Catalogues (Sandage and Tammann 1981, 1987), and a revised Abell catalogue (Abell *et al.* 1989). A third *Reference Catalogue* will appear in about a year, and there is no reason to doubt that continual revisions of these and other catalogues will continue to be published in the future.

Another type of survey that deserves mention is the data survey. Here, a specific data element or elements for a specified sample of galaxies is collected. Examples include the famous CfA redshift survey (Huchra *et al.* 1985), the collections of visual and infrared photometric data that have been assembled at Texas under the de Vaucouleurs's direction (Longo and de Vaucouleurs 1983, 1985, 1988), the lists of precise positions for the UGC galaxies (Dressel and Condon 1976) and the CGCG galaxies (Santagata *et al.* 1987 and references therein), the several lists of neutral hydrogen line widths (*e.g.* Bottinelli *et al.* 1981, Huchtmeier *et al.* 1983), and the similar lists of central velocity dispersions (Whitmore *et al.* 1985 and references therein). Sometimes, these lists have been systematized, sometimes not. But all are important and necessary aids to our common goal of understanding the galaxies.

Finally, a recent development is the extragalactic database, a large collection of data for galaxies accessible "on-line" or via magnetic media of some sort. These are exemplified by the database at Lyons (Paturel 198?), and that currently being set up by Helou, Madore, and their colleagues at JPL/IPAC in California (Helou, private communication). The goal of both of these databases is to collect *all* data available for galaxies, making them available in both their published form and — eventually — in systematized form.

I have so far covered in more or less detail the history of galaxy catalogues and surveys, sorting out I hope, the several different types that we deal with on a daily basis. I have also tried to make the de Vaucouleurs's contribution to the topic clear: their study of the data looking for systematic errors in them, and the correction of the data for those errors, then the statistical reduction of the corrected data to well-understood systems.

The Future

I'd like to speculate a little bit about the near future of galaxy cataloguing and surveying. First, it seems obvious that before the end of this century, we shall be able to produce from the two IIIa–J sky surveys[7] detailed lists of galaxies with reasonably accurate photometric parameters derived by scanning machines from the survey plates. (A preliminary version of such a galaxy catalogue could already be assembled at the Space Telescope Science Institute in Baltimore from their 25–μm scans of the southern IIIa-J survey and their so-called "quick V" survey from Palomar [M. Shara, private communication].) There have been suggestions that the photometric data actually be kept on the survey plates themselves in the form of silver granules rather than as magnetic bits on disk or tape (Kibblewhite *et al.* 1975). However, I don't see this happening. As optical compact discs and parallel processors operating at gigaflop speeds become the rule over the next few years, digital records will be much easier to handle, and probably even more compact to store — bit for bit — than the original photographic data.

Spectroscopic surveys along the lines of the CfA surveys are continuing, of course. Multiobject spectrographs will speed the acquisition of data in galaxy clusters, and more use of Fabry–Perot interferometry for kinematic studies of late–type galaxies will increase (see *e.g.* the review by Pence in this volume).

Optical spectrophotometry, however, will probably remain at the mercy of the vagueries of the weather, and will probably not be widely applied to galaxies until orbital or lunar observatories become a reality. The same is probably true for CCD photometric surveys, since — in the near future at least — surface photometry will still be more or less dependent on aperture photometry for absolute calibration.

[7] Presently almost finished at Siding Spring and just begun at Palomar.

Infrared observers have made great progress in developing area detectors in the past few years, and the extragalactic catalogue drawn from the IRAS survey has recently become available (Lonsdale *et al.* 1985). Radio mapping and surveying is now relatively easy at many frequencies. Extragalactic astronomy at wavelengths shorter than optical is still in its infancy, at least in relative terms, and also awaits permanent space observatories for the widespread collection of survey data.

The catalogues themselves will certainly develop further along lines that are already clear. The computerized databases mentioned above will collect raw data of all types. The specialized databases, exemplified by the extended CfA zone surveys (Huchra *et al.* 1989), the IRAS database in Pasadena (see *e.g.* Beichman *et al.* 1988), the Bologna redshift collection (Palumbo *et al.* 1983), and the neutral hydrogen lists maintained here in France (Bottinelli *et al.* private communication) will proliferate as the need for these data continues to grow.

The systematized catalogues of reduced data such as the *Reference Catalogues* may have to be updated piecemeal in the future because of the enormous amount of data now available, and the effort necessary to search for and correct the systematic effects in the raw data. For example, simply deriving global photometric parameters for 3500 of the galaxies which will be listed in RC3 took about two man *years* of work (Buta, private communication). Much of this time was spent in studying and removing the systematic errors in the data, though individual review of the data for each galaxy — still a necessary step — also consumed a considerable amount of time. Thus, periodic updating of sections of the computerized versions of the systematized catalogues will probably be necessary.

However, until access to these computerized data bases is as easy as taking a book off the shelf and turning pages to the right entry, astronomers are also going to want books and lists printed on paper for ready reference. Money is still a factor, too. While the cost for a data bit in a book continues to increase, the cost for that same data bit in electronic form continues to decrease. Even amortized across the cost of many users and projects, however, the present costs of computing equipment, software development, and access to data via telecommunications networks are still much greater than for books bearing the same data. While this will probably change in the future, references to previous studies of individual galaxies will remain most cheaply and easily done from printed books and catalogues.

At the moment, we are obviously in a transition period between seeing data distributed purely on paper, or purely electronically. Gérard and Antoinette de Vaucouleurs have been pioneers during this transition period, not just showing us how to collect, reduce, and systematize the data, but also showing us how to easily distribute it in both printed and electronic form. They have set us an example — and high standards — by their work, and we should be foolish to ignor that example — and those standards.

References

Abell, G. O. 1958, *Astrophys. J. Suppl.* **3**, 211.
Abell, G. O., Corwin, H. G., and Olowin, R. P. 1989, *Astrophys. J. Suppl.* in press.
Arakelian, M. A. 1975, *Soob. Byurakan Obs.* **47**, 1.
Arp, H. C. 1966, *Astrophys. J. Suppl.* **14**, 1.
Arp, H. C. and Madore, B. F. 1987, *A Catalogue of Southern Peculiar Galaxies and Associations* (Cambridge: Cambridge Univ. Press).
Baier, F. W. and Tiersch, H. 1979, *Astrofiz.* **15**, 33.
Beichman, C. A., Neugebauer, G., Habing, H. J., Clegg, P. E., and Chester, T. J. 1988, *Infrared Astronomical Satellite (IRAS) Catalogs and Atlases*, Vol. **1**, Explanatory Supplement (Washington, D. C.: NASA).
van den Bergh, S. 1959, *Publ. David Dunlap Obs.* **II**, No. 5 (DDO).
van den Bergh, S. 1966, *Astron. J.* **71**, 922.
van den Bergh, S. 1960, *Publ. David Dunlap Obs.* **II**, No. 6.
Bigourdan, G. 1917, *Observations de nébuleuses et d'amas stellaires*, in five volumes (Paris: Gauthier-Villars).
Bottinelli, L., Gouguenheim, L., and Paturel, G. 1982, *Astron. Astrophys. Suppl.* **47**, 171.
Corwin, H. G., de Vaucouleurs, A., and de Vaucouleurs, G. 1985, *Southern Galaxy Catalogue*, Univ. Texas Monograph in Astron. No. **4** (Austin: Univ. Texas Astron. Dept.) (SGC).
Dixon, R. S. and Sonneborn, G. 1980 *A Master List of Nonstellar Optical Astronomical Objects* (Columbus: Ohio State Univ. Press) (MOL).
Dressel, L. L. and Condon, J. J. 1976, *Astrophys. J. Suppl.* **31**, 187.
Dreyer, J. L. E. 1888, *Mem. Roy. Astron. Soc.* **49**, 1 (NGC).
Dreyer, J. L. E. 1895, *Mem. Roy. Astron. Soc.* **51**, 185 (IC 1).
Dreyer, J. L. E. 1908, *Mem. Roy. Astron. Soc.* **59**, 105 (IC 2).
Dunlop, J. 1828, *Phil. Trans. Roy. Soc. London* **118**, 113.
Fairall, A. P. and Jones, A. 1988, *Publ. Astron. Dept. Univ. Cape Town* No. **10**.
Fisher, J. R. and Tully, R. B. 1975, *Astron. Astrophys.* **44**, 151.
Geller, M. J. and Huchra, J. P. 1983, *Astrophys. J. Suppl.* **52**, 61.
Haynes, R. F., Huchtmeier, W. K. H., Siegman, B. C., and Wright, A. E. 1975, *A Compendium of Radio Measurements of Bright Galaxies* (Melbourne: CSIRO).
Herschel, W. 1784, *Phil. Trans. Roy. Soc. London* **74**, 437.
Herschel, W. 1786, *Phil. Trans. Roy. Soc. London* **76**, 457.
Herschel, W. 1789, *Phil. Trans. Roy. Soc. London* **79**, 212.
Herschel, W. 1802, *Phil. Trans. Roy. Soc. London* **92**, 477.
Herschel, J. 1833, *Phil. Trans. Roy. Soc. London* **123**, 359.
Herschel, J. 1847, *Results of Astronomical Observations Made ... at the Cape of Good Hope*; ... (London: Smith, Elder, and Co.).

Herschel, J. 1864, *Phil. Trans. Roy. Soc. London* **154**, 1 (GC).
Huchra, J., Davis, M., Latham, D., and Tonry, J. 1983, *Astrophys. J. Suppl.* **52**, 89.
Huchra, J., Geller, M., de Lapparent, V., and Corwin, H. G. 1989, *Astrophys. J. Suppl.*, in press.
Huchtmeier, W. K., Richter, O.-G., Bohnenstengel, H.-D., and Hauschildt, M. 1983, *A General Catalog of HI Observations of External Galaxies*, ESO Scientific Preprint No. **250**.
Karachentsev, I. D. 1972, *Soob. Spets. Astrofiz. Obs.* **7**, 3.
Karachentseva, V. E., Karachentsev, I. D., and Shcherbanovsky, A. L. 1979, *Astrofiz. Iss.-Izv. Spets. Astrofiz. Obs.* **11**, 3 (English translation: 1979, Bul. Spec. Astrophys. Obs. — N. Caucasus) **11**, 1).
Karachentseva, V. E. 1968, *Publ. Byurakan Obs.* **39**, 61.
Karachentseva, V. E. 1973a, *Astrofiz. Iss.-Izv. Spets. Astrofiz. Obs.* **5**, 10.
Karachentseva, V. E. 1973b, *Soob. Spets. Astrofiz. Obs.* **8**, 3.
Kibblewhite, E. J., Bridgeland, M. T., Hooley, T., and Horne, D. 1975, in *Image Processing Techniques in Astronomy*, ed. C. de Jager and H. Nieuwenhuijzen (Dordrecht: Reidel), p. 245.
Kraan-Korteweg, R. C. 1986, *Astron. Astrophys. Suppl.* **66**, 255.
Lacaille, N.-L. 1755, *Mém. Acad. Roy. Sci.*, p. 286.
Lauberts, A. 1982, *The ESO/Uppsala Survey of the ESO (B) Atlas* (Garching bei München: ESO) (ESO/Upps).
Longo, G. and de Vaucouleurs, A. 1983, *A General Catalogue of Photoelectric Magnitudes and Colors in the* U,B,V *System of 3,578 Galaxies Brighter than the 16-th* V*-Magnitude (1936–1982)*, Univ. Texas Monograph in Astron. No. **3** (Austin: Univ. Texas Astron. Dept.).
Longo, G. and de Vaucouleurs, A. 1985, *Supplement to the General Catalogue of Photoelectric Magnitudes and Colors in the* U,B,V *System* Univ. Texas Monograph in Astron. No. **3a** (Austin: Univ. Texas Astron. Dept.).
Lonsdale, C. J., Helou, G., Good, J. C., and Rice, W. 1985, *Catalogued Galaxies and Quasars Observed in the IRAS Survey* (Pasadena: JPL).
Maia, M. A. G., da Costa, L. N., and Latham, D. W. 1989, *Astrophys. J. Suppl.*, in press.
Markarian, B. E., Lipovetskii, V. A., and Stepanyan, D. 1982, *Astrofizika*, **17**, 321.
Markarian, B. E., Stepanyan, D., and Erastova, L. K. 1986, *Astrofizika*, **25**, 551.
Mazzarella, J. M. and Balzano, V. A. 1986, *Astrophys. J. Suppl.* **62**, 751.
Messier, C. 1781, *Conn. des Temps pour 1784*, 227 (M).
Nilson, P. 1973, *Uppsala General Catalogue of Galaxies*, (Uppsala: Roy. Soc. Sci. Uppsala) (UGC).
Palumbo, G. G. C., Tanzella-Nitti, G., and Vettolani, G. 1983, *Catalogue of Radial Velocities of Galaxies* (New York: Gordon and Breach).
Pesch, P. and Sanduleak, N. 1988, *Astrophys. J. Suppl.* **66**, 309.

Sandage, A. and Tammann, G. A. 1981, *A Revised Shapley–Ames Catalogue of Bright Galaxies* (Washington, D. C.: Carnegie Inst. of Washington).

Sandage, A. and Tammann, G. A. 1987, *A Revised Shapley–Ames Catalogue of Bright Galaxies*, Second Edition (Washington, D. C.: Carnegie Inst. of Washington).

Santagata, N., Basso, L., Gottardi, M., Palumbo, G. G. C., Vettolani, G., and Vigotti, M. 1987, *Astron. Astrophys. Suppl.* **70**, 191.

Shapley, H. and Ames, A. 1932, *Ann. Astron. Obs. Harvard Col.* **88**, No. 2.

Sinnott, R. 1988, *NGC 2000.0* (Cambridge, Mass.: Sky Publ. Corp.).

Sulentic, J. W. and Tifft, W. G. 1973, *The Revised New General Catalogue of Nonstellar Astronomical Objects* (Tucson: Univ. of Arizona Press).

Takase, B. and Miyauchi–Isobe, N. 1988, *Ann. Tokyo Astron. Obs., Ser. 2*, **22**, 41.

Tully, R. B. and Fisher, J. R. 1987, *Nearby Galaxies Atlas* (Cambridge: Cambridge Univ. Press).

Tully, R. B. 1988, *Nearby Galaxies Catalog* (Cambridge: Cambridge Univ. Press).

Turner, E. L. 1976, *Astrophys. J.* **208**, 20.

Turner, E. L. and Gott, J. R. 1976, *Astrophys. J. Suppl.* **32**, 409.

de Vaucouleurs, A. and Longo, G. 1985, *Catalogue of Visual and Infrared Photometry of Galaxies from 0.5 μm to 10 μm (1961–1985)* Univ. Texas Monograph in Astron. No. **5** (Austin: Astronomy Department).

de Vaucouleurs, G. 1953, "A Revision of the Harvard Survey of Bright Galaxies," Australian National Univ. Mimeogram.

de Vaucouleurs, G. and de Vaucouleurs, A. 1964, *Reference Catalogue of Bright Galaxies* (Austin: Univ. of Texas Press) (RC1).

de Vaucouleurs, G., de Vaucouleurs, A., and Corwin, H. G. 1976, *Second Reference Catalogue of Bright Galaxies* (Austin: Univ. of Texas Press) (RC2).

Vorontsov–Velyaminov, B. A. 1959, *Atlas and Catalog of Interacting Galaxies* (Moscow: Sternberg Astron. Inst., Moscow State Univ.) (VV).

Vorontsov–Velyaminov, B. A. 1977, *Astron. Astrophys. Suppl.* **28**, 1 (VV).

Vorontsov–Velyaminov, B. A., Archipova, V. P., and Krasnogorskaja, A. A. 1962, 1963, 1964, 1968, 1974, *Morphological Catalogue of Galaxies*, in five volumes (Moscow: Moscow State Univ.) (MCG).

Whitmore, B. C., McElroy, D. B., and Tonry, J. L. 1985, *Astrophys. J. Suppl.* **59**, 1.

Zwicky, F., Herzog, E., Kowal, C. T., Wild, P., and Karpowicz, M. 1961, 1963, 1965, 1966, 1968a, 1968b, *Catalogue of Galaxies and of Clusters of Galaxies*, in six volumes (Pasadena: California Inst. of Technology) (CGCG)

Zwicky, F. 1971, *Catalogue of Selected Compact Galaxies and of Post–Eruptive Galaxies* (Berne: Zwicky) (CSCG).

Zwicky, F., Sargent, W. L. W., and Kowal, C. T. 1975, *Astron. J.* **80**, 545.

Discussion

Binney: We are surely at the point at which we should be planning the dissemination of catalogues in machine–readable form — possibly in addition to publication in book form. Clearly, publication in machine–readable form is cheaper than in book form, and for many purposes superior, as well as offering the opportunity for regular updates.

Should we not be considering what software system we should employ for this purpose? No proprietory system is likely to provide the flexibility and transportablity which the scientific community requires. Can we devise a simple database system based on ASCII files for which we can publish the source code in a high–level language such as FORTRAN or C?

Corwin: I agree completely that proprietory database systems are not likely to be adequate for all the kinds of studies for which we use machine–readable catalogues. For the same reason, I'm not sure that the simple database system that you envision would be widely used. I am in favor of distributing the catalogued data in ASCII (or some other widely accepted) format — perhaps packed to save tape or disk space, as long as unpacking software is included. This would leave each individual or group free to use or to develop a database system suitable to their own needs.

Hanes: You remarked that early in his work [William] Herschel recognized the non–uniform distribution of galaxies and incipiently recognized the "Local Supercluster." Did his opinion on its nature disappear as he added more nebulae, or was he still convinced of its reality? More interestingly, over the intervening years and decades, did other astronomers fully appreciate this phenomenon, or did it escape rediscovery until Gérard's work in the 1940's and 1950's?

Corwin: Herschel remained convinced of the reality of the "stratum" of nebulae that he had found — many of the additional objects beyond his first 500 are of course located in it — but did not speculate much further as to its physical nature. Every other astronomer who has studied the distribution of the nearby nebulae has noted the existence of Herschel's stratum, and most have remarked on other chains and filaments in other parts of the sky — the Perseus–Pisces chain and the Southern Supercluster, for example, were noted by John Herschel and his contemporaries. The reasons that the Local Supercluster had to be "rediscovered" by de Vaucouleurs are based in Hubble's observational work on the distribution of the *faint* ($m_{pg} \simeq 19$) galaxies, and with the general acceptance that these observations supported a relatively simple homogeneous and isotropic expanding universe, *á la* Einstein, Lemaître, and Friedmann. Large scale structures such as the Local Supercluster would violate the so–called "Cosmological Principle," so were generally ignored by cosmologists (observational as well as theoretical) until redshift–based distances became available for large numbers of galaxies.

De Vaucouleurs: We should remember that Reynolds and the Scandinavian astronomers (Holmberg and Reiz in particular) led by Lundmark commented on the flattened distribution of the nearby galaxies right through the 1920's, '30's, and '40's.

Kerr: Galaxies have usually been thought of as independent and self–contained objects. Now there is great interest in interacting galaxies. How do you think future cataloguers will handle them?

Corwin: With difficulty, I'm afraid. Consider, for example, the problems in photometrically separating the components of an interacting system, or in resolving them with broad beam 21–cm observations (unless they are well–separated in velocity space, of course). There is no *easy* solution to these cases, and I suspect that each one will have to be handled on an individual basis.

De Vaucouleurs: For RC2, we indicated data for an entire interacting system by putting a dot after the name. Where we could give data for each of the objects separately, we included additional listings with no dots after the name.

Burstein: I think that the current status of our knowledge of galaxies is comparable to that which was known of stars in the 1910's and 1920's. At that time, there were a few comprehensive catalogues (*e.g.* the HD, the GC of double stars, etc.). (Indeed, I think a good analogy can be drawn between the *Henry Draper Catalogue* and the *First Reference Catalogue of Bright Galaxies* concerning their influences on their respective fields.) Have we now arrived at a time in our field at which no one person, or group, can systematically catalog everything about galaxies? Might we not be better off having those specialists in specific fields handle the cataloguing of those specific data?

Corwin: Your suggestion about splitting the cataloguing tasks makes a lot of sense for at least one other reason besides the one (specialization) that you mentioned: time. There are simply too many extragalactic data now available for even a group to handle in a reasonable amount of time. For example, I mentioned in the review that the derivation of total magnitudes and colors for 3500 of the galaxies to be included in RC3 took about two man *years* of work, primarily by Ron Buta. Rather than publishing an all–inclusive "RC4," as useful as it might be, I suspect that a sample of galaxies complete to some limit will form a continuously updated "Bright Galaxy Catalogue" analogous to the Yale *Bright Star Catalogue*. We already see the specialization in catalogues such as the Texas collections of UBV (Longo and de Vaucouleurs 1983, 1985) and infrared photometry (de Vaucouleurs and Longo 1988), the Bologna (Palumbo *et al.* 1983), Cape Town (Fairall and Jones 1988), and CfA (Huchra *et al.* 1983) redshift lists, and the radio continuum compendium (Haynes *et al.* 1975). This trend will obviously continue.

De Vaucouleurs: There will still remain a need for a general "reference catalogue," however.

Pecker: It may be appropriate to say that Messier came to Paris in his wooden shoes as a young peasant attracted both by the "big" city, and by the stars. He was hired by Delisle (at the Collège de France) and observed until his death at the same observatory (where the Museum of Cluny is now located). He never earned a degree — but was still elected a member of the Académie des Sciences!

Corwin: He also earned the respect of astronomers throughout the world. William Herschel, for example, did not include any of Messier's nebulae or clusters in his own catalogue, so that these "Messier objects" would continue to be referred to by their Messier numbers in honor of their first cataloguer.

The Updated Version of the Catalogue of Radial Velocities of Galaxies

G. G. C. Palumbo[1,2], G. Vettolani[3], G. Baiesi-Pillastrini[1], and A. P. Fairall[4]

[1]Dipartimento di Astronomia, Università di Bologna, Italy
[2]Istituto T.E.S.R.E./CNR, Bologna, Italy
[3]Istituto di Radioastronomia/CNR, Bologna, Italy
[4]Department of Astronomy, University of Capetown, South Africa

Since the *Catalogue of Radial Velocities of Galaxies* (CRVG) (Palumbo *et al.* 1983) was published, an increasing number of redshifts are published every year. To provide a comprehensive list of data to the astronomical community which so well received the first edition of the CRVG, we have undertaken the task of updating the catalogue with the aim of publishing a new edition in early 1989 complete with all redshifts that have appeared in the literature up to the end of 1987. What follows is a concise description of the present state of development of the project.

New redshifts collected since 1981: about 10,000 contained in approximately 300 references. References containing redshifts from 1981 to 1985: about 600. Estimated number of new references since the publication of CRVG: about 850. Estimated number of redshifts in the complete updated CRVG: about 30,000. The new edition will appear both as a book and as a tape, both available from Gordon and Breach Scientific Publisher (New York).

New features planned for the new edition: additional names such as UGC, MCG, Zwicky; cluster membership; numerous tables for particular lists such as Markarian, Vorontsov-Velyaminov, IC, NGC, Arp, etc.; and plots in coordinate and redshift space.

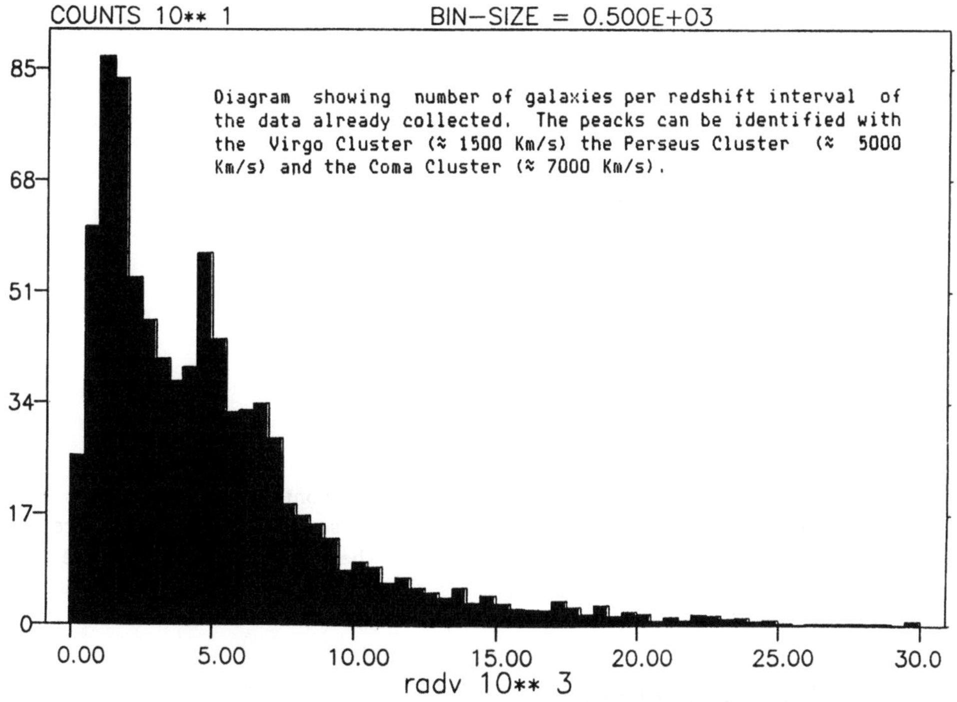

Figure 1: Number of galaxies per redshift interval for the data already collected. The peaks in the distribution can be identified with the Virgo Cluster ($V \approx 1500\,\mathrm{km\,sec^{-1}}$), the Perseus Cluster ($V \approx 5000\,\mathrm{km\,sec^{-1}}$), and the Coma Cluster ($V \approx 7000\,\mathrm{km\,sec^{-1}}$).

Reference

Palumbo, G. G. C., Tanzella–Nitti, G., and Vettolani, G. 1983, *Catalogue of Radial Velocities of Galaxies* (New York: Gordon and Breach).

A Survey Catalogue of Ultraviolet–Excess Galaxies

B. Takase and N. Miyauche–Isobe

Tokyo Astronomical Observatory, Japan

Abstract

Eight survey catalogues have been compiled which contain 4,409 ultraviolet–excess galaxies in an area of the sky of some 2,400 square degress. They were detected on three–image photographs taken with the 105–cm Kiso Schmidt telescope. A description of the catalogues is given, and a comparison with the Byurakan and Case surveys for similar objects is shown.

1. The Kiso Survey

Our survey for ultraviolet–excess galaxies is based on *UGR* three–color photographs taken with the 105–cm Kiso Schmidt telescope. Exposure times are so set that the three images of A0 stars are equally bright on the plate. Therefore, any galaxy whose *U* image is brighter than others is regarded as a UV–excess galaxy having bluer color than A0 stars.

The detection of Kiso UV–excess galaxies (hereafter abbreviated as KUG's) has so far been made on 80 plates covering some 2,400 square degrees, and eight catalogues have been compiled (Takase and Miyauchi–Isobe 1984, 1985a,b, 1986a,b, 1987a,b, and 1988). The number of listed KUGs totals 4,409, and so the number density is 1.8 per sq. degree.

2. The KUG Catalogue

The format of the catalogue is given in Table 1. KUG names are composed of their values of α and δ. Apparent size and magnitude are measured and estimated, respectively, on the blue Palomar Sky Survey prints.

Table 1: Format of the KUG Catalogue

No.	KUG Name	R. A. (1950.0)	Dec.	Mor. Type	App. Size	App. Mag.	UVX Deg.	Other Name
31*	0742+625	7 42 38.8	62 30 20	C	0.3x0.3	15.5	L	MK82
32	0743+614	7 43 05.2	61 27 31	?	0.2x0.1	17.5:	L	
33	0743+610	7 43 07.5	61 03 25	Sk	1.7x0.8	14.5	M	MK10
34	0743+624	7 43 20.6	62 26 57	Sp	1.1x1.0	15.4	L	U4015
35	0743+616	7 43 49.3	61 37 07	Ic:	0.6x0.3	17:	M	
36	0745+590	7 45 04.0	59 02 42	?	0.4x0.2	16.5:	L	
37	0745+587	7 45 24.8	58 47 09	Ig:	0.3x0.1	17:	M	
38	0745+611	7 45 37.0	61 09 37	C:	0.2x0.2	18:	L	
39	0745+620	7 45 50.1	62 03 54	Pd:	0.4x0.2	17:	M	
40	0746+611	7 46 42.0	61 06 27	?	0.2x0.2	17.5:	L	

The degree of UV–excess is estimated from the Kiso UGR plate, and is divided into three categories, high (H), medium (M), and low (L). According to the calibration made by Noguchi *et al.* (1980), they correspond to $U - B$ colors of ≤ -0.5, ~ -0.3, and ~ 0, respectively.

The morphological type is assigned according to the Kiso classification scheme (Takase *et al.* 1983). The meaning of the designations is **Ic**: irregular with clumpy HII regions; **Ig**: irregular with giant HII regions; **Sk**: spiral with knotty HII regions; **Sp**: spiral with peculiar bar and/or nucleus; **Pi**: pair of interacting components; **Pd**: pair of detached components; and **C**: compact.

3. Comparison with Other Surveys

Among 1,500 Markarian galaxies (MKG's), 381 are included in the Kiso survey area, of which 60 or 16% are too red to be KUG's. The detection of MKG's (Markarian *et al.* 1982) are based on an objective prism spectral survey with the Byurakan 1–m Schmidt. The second Byurakan survey (Markarian *et al.* 1986) now being carried out, is deeper than the first one by ≈ 2 magnitudes, and intends to find not only Markarian–type blue galaxies, but also all of the emission line galaxies — even if they are red in color — together with all blue stellar objects (BSO's) to the survey limit.

The Case survey (Pesch and Sanduleak 1988) is also a spectral survey, but with the 61–cm Case/Burrell Schmidt. The selection criteria for the Case Galaxies (CG's) is either the presence of a continuum much bluer than that of average galaxies, or the appearance of such emission features as $N_1 + N_2$ and/or $[O\,II]_{3727}$.

Table 2 compares the four surveys.

Table 2: Comparison of Surveys

Survey	Area Covered (deg^2)	Number of Objects	Number Density (deg^{-2})	Limiting Magnitude
Kiso	2,400	4,409	1.8	16 – 18.5*
Byurakan 1	17,000	1,500	0.1	17.5
Byurakan 2	80	521†	6.5†	19.5
Case	960	895	0.9	> 18

*Different from plate to plate depending on the observational conditions.
†BSO's are included.

References

Markarian, B. E., Lipovetskii, V. A., and Stepanyan, D. 1982, *Astrofizika*, **17**, 321 (last list of MKG's).

Markarian, B. E., Stepanyan, D., and Erastova, L. K. 1986, *Astrofizika*, **25**, 551 (latest list of the second Byurakan Survey).

Noguchi, T. Maehara, H., and Kondo, M. 1980, *Ann. Tokyo Astron. Obs., Ser. 2*, **18**, 55.

Pesch, P. and Sanduleak, N. 1988, *Astrophys. J. Suppl.* **66**, 309 (latest list of the Case Survey).

Takase, B., Noguchi, T., and Maehara, H. 1983, *Ann. Tokyo Astron. Obs., Ser. 2*, **19**, 440.

Takase, B. and Miyauchi–Isobe, N. 1984, 1985a,b, 1986a,b, 1987a,b, and 1988, Ann. Tokyo Astron. Obs., Ser 2, **19**, 595 (List I); **20**, 237 (II); **20**, 335 (III); **21**, 127 (IV); **21**, 181 (V); **21**, 251 (VI); **21**, 363 (VII); and **22**, 41 (VIII).

The Edinburgh/Durham Southern Sky Galaxy Survey

C. A. Collins, N. H. Heydon–Dumbleton, and H. T. MacGillivray

Department of Astronomy, University of Edinburgh
and Royal Observatory, Edinburgh, Scotland, U. K.

Introduction

We are currently completing the Edinburgh/Durham galaxy catalogue of the southern sky using COSMOS measurements of glass copy plates of the IIIa–J Southern Sky survey taken with the U. K. 1.2–m Schmidt Telescope (UKST). We have constructed a mosaic of 60 fields around the SGP ($\sim$ 1500 square degrees) and intend to add a similar number soon to complete the catalogue. Since each COSMOS scan of a UKST plate covers a region of $5.3^o \times 5.3^o$, our catalogue will cover $\sim$ 1 steradian of the southern sky. The Edinburgh/Durham galaxy catalogue is one of the largest and most complete surveys of its kind, containing more than one million galaxies down to a limiting magnitude of $b_j = 20$.

Image Classification and Magnitude Calibration

In order to acheive homogeneity, we have carried out an extensive investigation of position and magnitude dependent systematic errors in image classification, *i.e.* star/galaxy discrimination (see Heydon–Dumbleton *et al.* 1988a). Our classification procedure involves correcting for the variation in the measured parameters of saturated images across each plate, and uses an automated classification algorithm to ensure uniform and consistent classification over the whole of each plate and from plate to plate. The automated algorithm objectively classifies images using their distribution in various COSMOS parameter–magnitude classification planes, and uses image structure information to assign, for each plate, optimum magnitude ranges over which each classification parameter is used. We

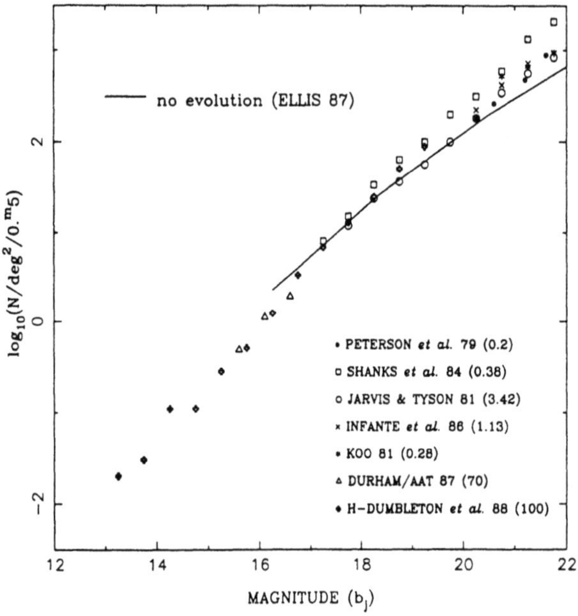

Figure 1: Differential number–magnitude galaxy counts to $b_j \sim 19.5$.

have estimated the completeness of our catalogue to be at least 95% for galaxies and to have < 10% contamination by stars.

For unsaturated images on the same plate, we have shown (MacGillivray and Stobie 1984; Heydon–Dumbleton *et al.* 1988b) that the COSMOS image magnitude is linearly related to the apparent magnitude (b_j). The accuracy of galaxy photometry is therefore set by the accuracy to which sky surface brightness can be estimated for each plate. To reduce these estimates to within the limit set by Geller *et al.* (1984), we have obtained at least two galaxy calibration sequences (in B and V) on every other field in our survey. With this configuration, each field without sequences has four surrounding fields which are calibrated. We can reduce errors in calibration to less than 0.04 mag r. m. s. without recourse to complicated global density fitting techniques which tend to propagate errors from plate to plate (Geller *et al.* 1984).

Number Counts and Angular Correlation Function

In Figure 1, we present differential galaxy number–magnitude counts down to $b_j \sim 19.5$ together with counts from other authors. Our counts are from a six–plate, well–calibrated region of the survey ($\sim$ 100 square degrees of sky, and

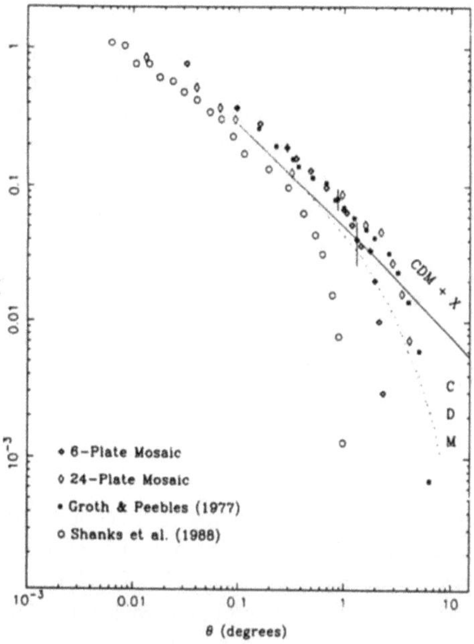

Figure 2: Angular correlation functions scaled to the depth of the Lick Survey.

photometric accuracy better than 0.04 mag). Because of its size, this region is the first data set to accurately sample galaxy counts at bright magnitudes ($b_j <$ 20). Over these magnitudes, our counts do not show significant deviation from a $10^{0.6 \cdot m}$ slope. This result is in general agreement with the deeper photographic counts of Shanks *et al.* (1984) and shows deviation from a "No–Evolution" model fainter than $b_j \sim 18$ (*cf.* Ellis 1987).

We have calculated the angular correlation function, $\omega(\Theta)$, for two different subsets of the catalogue. The first was obtained from the 100 square degree well-calibrated region, while the second is for a partially calibrated 24–plate region (~ 500 square degrees), with a limiting magnitude of $b_j = 20.0 \pm 0.2$ mag r. m. s. These correlation functions are scaled to the depth of the Lick Galaxy Survey ($b_j \sim 18.6$) and are shown in Figure 2 together with the correlation function obtained by Groth and Peebles (1977) for the Lick survey.

Our results for the accurately calibrated six–plate mosaic indicate a break from a -0.71 power law on scales $\sim 2^o$ ($7h^{-1}$ Mpc). Our partially–calibrated 24–plate region gives a break at $\sim 2.5^o$ (10h^{-1} Mpc) as do the results of Groth and Peebles from the Lick data. There is, however, no evidence in our data for break-scales as small as $3h^{-1}$ Mpc as found by Shanks *et al.* (1988). The difference in amplitude on large scales between our six–plate and 24–plate mosaics is probably due to a $+0.20$ mag calibration error in the latter introducing excess power on large scales.

Our observation of a break in $\omega(\Theta)$ on scales of $\sim 7h^{-1}$ Mpc lends some support to the predictions of a cold dark matter (CDM) model (the dotted line in Figure 2, from the calculations of Bond and Couchman 1987), which shows a significant turn–down on these scales. Our confirmation of the break point limits the amount of large–scale power in the mass fluctuation spectrum in any theory of galaxy formation. Such large–scale power is required to explain other observations of large–scale structure. Bond and Couchman have calculated the effect of a hybrid model (CDM + extra large–scale power), shown as the curve CDM + X in Figure 2. Since the main effect of extra power on large–scales is to remove the break in $\omega(\Theta)$, its confirmation here contradicts models with a standard fluctuation spectrum and extra large–scale power.

References

Bond, J.R. and Couchman, H. 1987, *Proceedings* of Second Canadian Conference on General Relativitity and Relativistic Astrophysics, in press.

Ellis, R.S. 1987, in *Observational Cosmology*, I.A.U. Symp. No. **124**, eds. A. Hewitt, G. Burbidge, and L. Z. Fang (Dordrecht: Reidel), p. 367.

Geller, M. J., de Lapparent, V. and Kurtz, M. J. 1984, *Astrophys. J. Letters* **287**, L55.

Groth, E. J. and Peebles, P. J. E. 1977, *Astrophys. J.* **217**, 385.

Heydon–Dumbleton, N. H., Collins, C. A. and MacGillivray, H. T. 1988a, *Mon. Not. Roy. Astr. Soc.*, submitted.

Heydon–Dumbleton, N. H., Collins, C. A. and MacGillivray, H. T. 1988b, in *Large-Scale Structure in the Universe — Observational and Analytic Methods* (Heidelberg: Springer–Verlag) in press.

MacGillivray, H. T. and Stobie, R. S. 1984, *Vistas in Astron.* **27**, 433.

Shanks, T., Stevenson, P. R. F., Fong, R. and MacGillivray, H. T. 1984, *Mon. Not. Roy. Astr. Soc.* **206**, 767.

Shanks, T., Hale–Sutton, D., Fong, R. and MacGillivray, H. T. 1988, *Mon. Not. Roy. Astr. Soc.*, submitted.

The AAO Distant Galaxy Cluster Survey

Warrick J. Couch[1], Richard S. Ellis[2], and David F. Malin[1]

[1]Anglo–Australian Observatory, Epping, N.S.W., Australia.
[2]Physics Department, University of Durham, England, U.K.

Introduction

Galaxy cluster catalogues such as those of Abell (1958) and Zwicky *et al.* (1961–68) have served as invaluable resources in the study of clusters and their constituent galaxies. They have provided astronomers with a consistently selected collection of objects with a range of properties and at close to moderate distances ($0.0 \leq z \leq 0.3$). However, those interested in studying clusters at distances beyond the limits of these Schmidt plate based catalogues have not been so well provided for. No deep large–scale systematic cataloguing of clusters has been undertaken, thus obliging workers to use whatever clusters have been reported in the literature — a dangerous practice given the heterogeneity of the methods used in selecting such objects.

After at least a decade of photography on fine-grain plates at the prime–focus of 4–m (and larger) telescopes, a substantial area of sky has now been covered to a depth and detail well beyond that reached by the Schmidt sky surveys. Motivated by the need for a systematic survey of clusters at these distances, together with our interest in using clusters as probes of galaxy evolution, we have over the last 4 years been constructing a deep cluster catalogue using copies of the large number of very high quality sky–limited plates that have now been taken on the Anglo–Australian Telescope (AAT).

In this paper we give a summary description of the survey and its status; a more detailed discussion is to be published elsewhere (Couch *et al.* 1988).

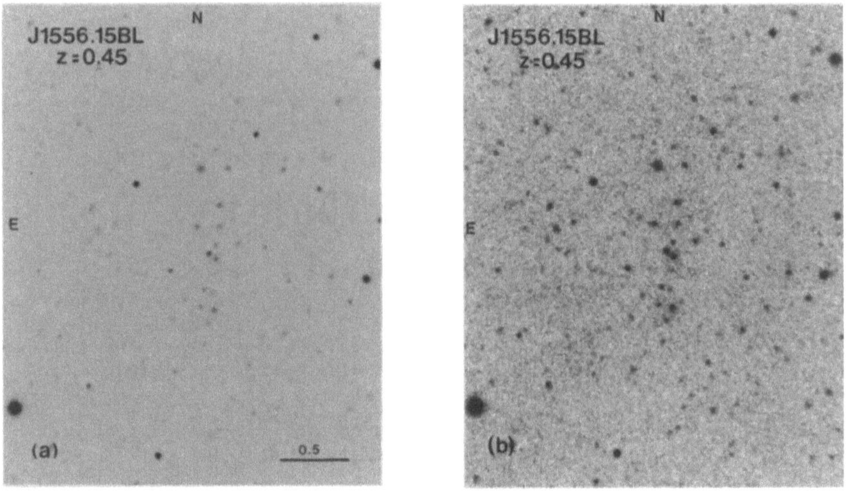

Figure 1: (a). A print of one of our clusters as seen on the original AAT plate; (b) a print of the same cluster as seen on a high contrast film derivative made from the original plate.

Methods and Selection Criteria

Crucial to our search has been the production by Malin of ***high contrast*** (HC) film derivatives which are routinely made from all deep AAT plates. This technique, which is better known for revealing extended features of very low surface brightness (*e.g.* the shells around E galaxies; Malin and Carter 1980) also greatly enhances the visibility of compact images at the plate limit. This can be seen in the "before" and "after" prints shown in Figure 1. Quantitative evaluation of the depths achieved when eyeball scanning such films indicates that image detection is complete to $B_J = 24.5$ and, in comparison with plate measuring machines, the method has the ability to reveal many more fainter images (Couch *et al.* 1984).

Using the HC films we have each independently identified clusters on both blue (IIIaJ + GG385) and red (IIIaF + RG630) plates. To maintain uniformity, only those plates with uniform emulsions and taken in seeing better than 2 arcsec have been searched. Our method of detection involves visually identifying overdensities and quantifying them in terms of a "sigma-excess" given by the relation :

$$\sigma = (N_c - N_f)/\sigma_f .$$

Here, N_c is the number of galaxies counted within a 1.5 Mpc = 3 arc-minute ($H_0 = 50$, $q_0 = 0.1$, $< z > \sim 0.5$) diameter circle centred on the overdensity; N_f and σ_f are the mean and standard deviation of the background "field" count as

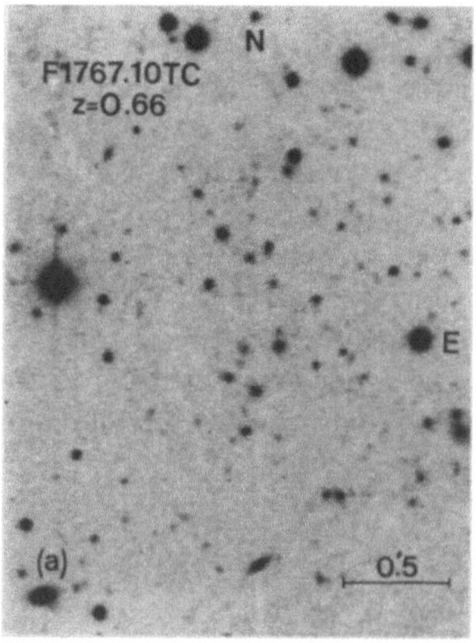

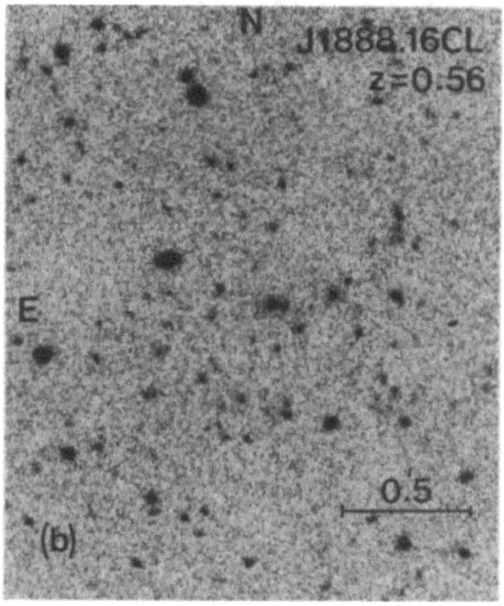

Figure 2: (a). A deep CCD I frame of the currently most distant confirmed cluster in our catalogue, F1767.10TC at $z = 0.66$; (b) the $z = 0.56$ cluster J1888.16CL as seen on the blue high contrast film on which it was discovered.

measured in ~20 such circles at random locations across the plate.

To date 56 fields (covering ~56 square degrees) have been surveyed yielding a total of 68 clusters in the range $3 \leq \sigma \leq 10$. In comparison, model calculations show that at the survey limit, an Abell richness class 3 cluster at z~0.5 will appear as a 3–7σ enhancement depending upon the passband used and the amount of luminosity evolution it has undergone. Pictures of two representative clusters are shown in Figure 2.

Cluster Redshifts

In order to check the reality of the clusters and to eliminate from the catalogue any candidates which are due to line-of-sight projection effects, a programme of obtaining spectroscopic redshifts for a number of galaxies in each cluster has been carried out in parallel with our survey. Redshift information has been secured for 30 of our clusters so far of which <10% are due to projection effects. Importantly, the confirmed clusters have redshifts in the range $0.28 \leq z \leq 0.66$ ($< z >= 0.44$) indicating that our catalogue overlaps but *extends well beyond* the redshift limit of the Abell catalogue.

Related Work

Our catalogue has already been used for a source of targets in a number of programmes undertaken by ourselves or in collaboration with others. These together with projects planned in the future include: detailed investigation of the evolutionary status of the clusters via CCD imaging and multi-object spectroscopy, use of the clusters as targets in a distant supernova search programme, use of the clusters as targets in a comparison of the fluxes seen by IRAS towards high and low redshift clusters, high resolution imaging of the clusters at optical and far-uv wavelengths using the Hubble Space telescope, and extension of the catalogue to northern declinations using HC derivatives made from the 4-m plates available in the Kitt Peak archive.

References

Abell, G. O. 1958, *Astrophys. J. Suppl.* **3**, 211.
Couch, W. J., Ellis, R. S., Kibblewhite, E. J., Malin, D. F., and Godwin, J. 1984, *Mon. Not. Roy. Astron. Soc.* **209**, 307.
Couch, W. J., Ellis, R. S., MacLaren, I., and Malin, D. F. 1988, in preparation.
Malin, D. F. and Carter, D. 1980, *Nature* **285**, 643.
Zwicky, F., Herzog, E., Wild, P., Karpowicz, M., and Kowal, C. T. 1961–1968, *Catalogue of Galaxies and Clusters of Galaxies*, 6 volumes (Pasadena: California Inst. of Technology).

Galaxy Morphology and Classification

Ronald Buta

Department of Astronomy, University of Texas, Austin, Texas, U.S.A.

Abstract

A brief discussion of the de Vaucouleurs revised Hubble classification system is given, with special emphasis on ring and lens morphologies. The possible physical interpretations of each dimension of the system are summarized.

1. Introduction

Of the many topics in astronomy Gérard de Vaucouleurs has worked in over the past 40 years, I think the one which brought him some of the greatest enjoyment and personal satisfaction was his research on galaxy morphology and classification. Not only is the beauty of galaxies enjoyable from an aesthetic viewpoint, but galaxy morphology is rich with the enigmas and puzzles that qualify it as one of the major research topics of today. Although for many years, there was so little firm theoretical understanding of galaxy morphology that the topic was little more than descriptive, this is certainly no longer true. Galaxy morphology today is, in fact, undergoing a major revolution. Not only has the Hubble Sequence been revitalized as a tool for understanding galaxy structure and evolution, but the many additional phenomena, such as bars and rings, that used to be regarded as secondary, have also been brought to the forefront. Never before, I believe, have the many types of structures visible in galaxy images been understood as well as today, although much remains to be learned.

It is not hard to summarize the main problems of galaxy morphology. Why is the Hubble sequence a continuous sequence, and what physical parameters underlie this continuity? What is the role of angular momentum, dissipation, and galaxy–galaxy interactions? How do the various patterns originate, and how do they change with time? What types of objects define the patterns? What role do the patterns have on the star formation processes in the disk? What is the role of dark matter? These questions have all been asked but definitive answers have been elusive. Ultimately, the answers to these questions will hinge on an understanding of galaxy formation, a topic which is currently a subject of considerable uncertainty and debate.

In honor of Gérard de Vaucouleurs, I should like to briefly describe his 1959 galaxy classification system, and then focus on a few aspects of galaxy morphology that have received much attention over the last few years. Comprehensive reviews on this subject have already been given by Sandage (1975) and Kormendy (1982). I will concentrate mainly on ring and lens morphologies in normal disk galaxies (S0 to Sbc), since these features were of great interest to Gérard and exemplify how much our understanding has progressed about galaxy morphology as a whole. Very comprehensive discussions of elliptical galaxies can be found in *I. A. U. Symposium No.* **127** (de Zeeuw 1987) and other articles in the present volume, while a thorough review of irregulars can be found in Gallagher and Hunter (1984, 1987).

2. The Revised Hubble Classification System (RHS)

Gérard de Vaucouleurs's revised Hubble system is described in a lengthy *Handbuch der Physik* article published in 1959. It was not the only revision of the Hubble system available at the time: Hubble himself had been working on a revision since the 1930's, but died before completing it. In 1961, Allan Sandage published Hubble's revision in the form of an atlas. Both systems are still in active use today, and there are good reasons for this: the original Hubble system provided the first ordering of galaxies that represented more than just smooth changes in the properties of optical morphology. As data began to be accumulated for large numbers of galaxies, many physical parameters (*e.g.* colors, HI content, concentration indices, surface brightnesses) were found to correlate with these changes, implying that the Hubble sequence contains information on the basic physics of galaxies.

Although de Vaucouleurs's revision followed Hubble's framework in placing galaxies along a sequence of "stages" and in recognizing bar and ring morphologies, his view of the spread in morphologies at a given stage was clearly different from Hubble's: instead of a multi–pronged fork, de Vaucouleurs proposed to describe galaxy structure with a three–dimensional classification *volume* whose principal axis was the stage and whose secondary axes were the "family" and the "variety." These latter characteristics refer, respectively, to the presence or absence of a bar and of an *inner* ring. Continuity was implied along each axis. To highlight this, de Vaucouleurs introduced special notation to allow for transition cases. For family, SB would still denote barred spirals while SA would denote non-barred ("ordinary") spirals. Intermediate barred cases are denoted S$\underline{\mathrm{A}}$B, SAB, or SA$\underline{\mathrm{B}}$, depending on the degree of development of a bar. The same kind of notation was used for variety: pure ringed systems were denoted (r) while pure spiral (s–shaped) systems were denoted (s), with objects of intermediate characteristics highlighted by a compound notation such as ($\underline{\mathrm{r}}$s), (rs), and (r$\underline{\mathrm{s}}$).

For stage, de Vaucouleurs adopted two refinements. First, transition stages, Sab and Sbc, between the original Hubble spiral categories Sa, Sb, and Sc were recognized to eliminate the need for notation such as Sa^+, Sb^-, etc. Second, more stages were added along the Hubble sequence beyond Sc to allow for intermediate spiral stages between Sc and Irr I. De Vaucouleurs adopted Shapley and Paraskevopoulos's (1940) suggestion of the need for an Sd classification (very late, small–bulge systems like NGC 7793), and identified the Magellanic Clouds as an even later stage along the spiral sequence, Sm. Transition stages were denoted as Scd and Sdm. The true Irr I galaxies were denoted Im. S0 galaxies were divided into three categories, $\mathrm{S0}^-$, $\mathrm{S0}^\circ$, and $\mathrm{S0}^+$, but these do not correlate exactly with the three S0 categories in Hubble's revision (Sandage 1961). As in Hubble's revision, S0 galaxies are placed between E and Sa galaxies on the stage sequence, the transition stage to spirals being denoted S0/a. The various combinations and permutations of symbols possible gave de Vaucouleurs's classification system a complexity unprecedented in galaxy morphology studies, but as he noted, symbols can be dropped according to the amount of detail visible in an image or the desired level of complexity.

The availability of two revised Hubble systems means that care must be exercised in interpretation of published Hubble types, particularly the stage parameter where the stage "Sc" in the Hubble revision encompasses galaxies classified over the wide range of stages, Sbc–Sm, in de Vaucouleurs's revision. The arrangement of the late spiral stages and the validity of de Vaucouleurs's extensions have been repeatedly demonstrated from the correlations found between type and a wide variety of measured parameters (de Vaucouleurs 1977; see also Figure 1). The extensions are therefore fully justified and well–founded, and form one of the major advantages of de Vaucouleurs's revision over Hubble's revision.

Care must also be exercised in the interpretation of the part of the classifications involving bars and rings. Recently, much research has focussed on these secondary structures; on their classifications, independent observers are not very consistent. For example, some galaxies classified as (r)–variety by de Vaucouleurs (1963) are classified as (s)–variety by Sandage and Tammann (1981;ST) because the inner ring appears to be made of very tightly wrapped spiral arms. Hubble's revised classification system, as applied by ST, has very broad categories along the spiral sequence that encompass a wide range of forms, yet the ring category they envision is extremely conservative (as is also the bar category). I believe that overconservatism leads to unrealistically sharp edges

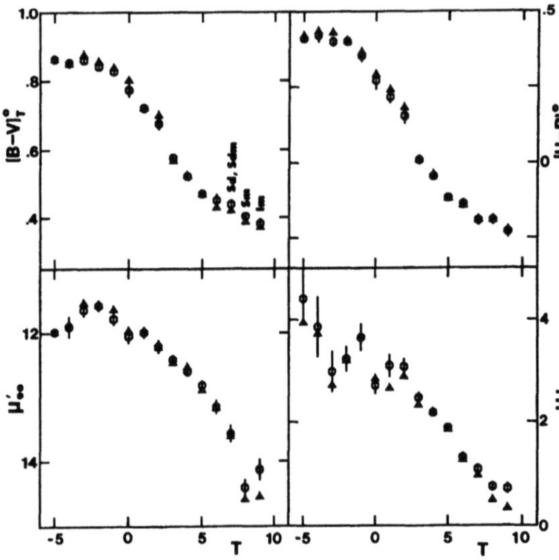

Figure 1: Dependence of several photometric parameters on the stage in de Vaucouleurs's system. Here HI refers to the hydrogen index and μ'_{eo} is the mean surface brightness within the effective isophote. Based on new data from the *Third Reference Catalogue of Bright Galaxies.* Triangles are median values, while circles are means. Stages Sd and Sdm are combined here.

to the morphological "cells" that may be misleading; the cell boundaries are in reality rather fuzzy. The advantage of de Vaucouleurs's revision is that it views galaxy structure in the broadest possible way, that is, as a continuum of forms in all possible dimensions.

Surprisingly, many astronomers today still persist in adhering to the original Hubble revision, perhaps because it is simpler to understand and apply in practice, and because it serves their needs. However, modern extragalactic astronomy demands a system with greater versatility and applicability, which can serve as a basis for the type of research that can now be done with the sophisticated telescopes and instruments currently in use.

3. Extensions and Additions

Over the past 10–15 years, the imaging capabilities for galaxies have been greatly improved, especially with the advent of CCD's. Not only this, but the great Sky Surveys produced with Schmidt telescopes since the 1950's have greatly increased the number of galaxies for which detailed morphology can be clearly studied. This exposive growth in the number of available images has in recent years led to great curiosity about the apparently secondary and more subtle aspects of galaxy structure, and to a few possible revisions to the classification systems. Some of these are described here.

a. Rings and Pseudo–Rings

It is well–known that in addition to the inner rings used for the variety dimension in the RHS, two other types of ring are occasionally found: *nuclear* rings, sometimes seen in the nuclei of barred galaxies, and *outer* rings, occasionally seen enveloping the main bodies of early–type spirals and lenticulars. Although these features may not be as widely present or as easily discernible as family, variety, and stage, their characteristics and their physical interest make it worthwhile to recognize them with their own type symbols. In the revised Hubble classification notation, outer rings and pseudo–rings do have type symbols, (R) and (R$'$), respectively, to be used preceding the main part of the classification. However, for nuclear rings and pseudo–rings, there is currently no notation, and the first possible extension or addition to the RHS that I propose is the use of (nr) and (nr$'$) for such structures. Note that this classification is not entirely straightforward: some nuclear rings appear to be pseudo–rings only because of dust (*e.g.* NGC 1512, Lindblad and Jorsater 1981; NGC 4314, Benedict 1980). Table 1 illustrates the possible use of this notation and some examples.

The second possible extension or addition to the RHS is perhaps more subtle. In a recent paper (Buta 1986a), I have suggested a possible refinement to the classification of outer rings which is motivated by the recent theoretical work of Schwarz (1979, 1981). From the appearance of over 300 outer rings and pseudo–rings in SB and SAB galaxies on the SRC IIIa–J Southern Sky Survey films, it appears that at least two distinct subtypes exist: the first type (R_1') shows the characteristics that the arms which emerge from each end of the bar seem to wind 180° and return to the opposite side, dipping in to form a broad figure eight pattern (Figure 2a, NGC 3504); in the second type (R_2'), the arms go an extra 90° before intersecting each other, and no figure eight characteristic is made (Figure 2c, ESO 294-16). Between these extremes, intermediate cases are found which appear to show aspects of both types (Figure 2b, ESO 509-98).

These distinctive morphologies were discovered by Schwarz (1979) in simulations of barred galaxy gas dynamics, and it is a small triumph for the theory that they were found in real galaxies *after* they had been predicted to exist. The predictions came at an especially opportune time because the availability of the deep IIIa–J SRC Southern Sky Survey films made it possible to detect outer rings and pseudo–rings in galaxies that otherwise would have been very difficult to study on the ESO Quick–Blue Survey or the Palomar Sky Survey plates. For reasons described in section 4, the mixed morphology (R_1R_2') was not predicted to exist, but its apparent existence in real galaxies underscores once again the "fuzziness" characteristic of galaxy morphological cells.

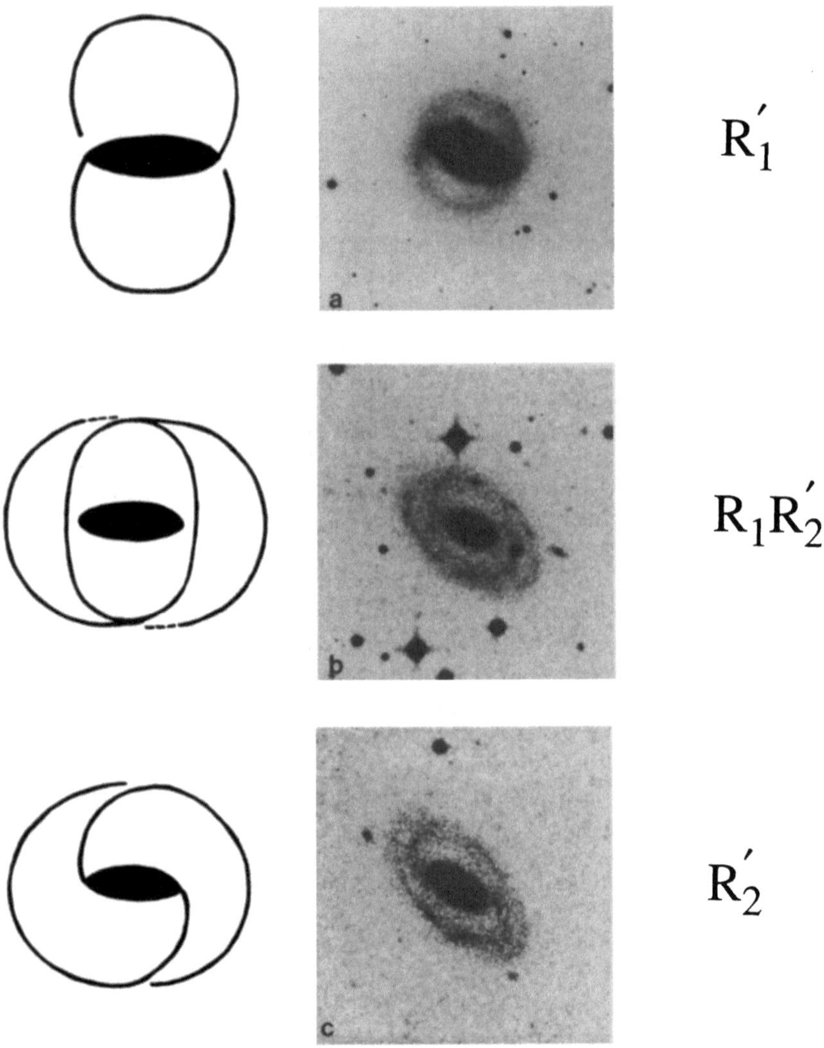

Figure 2: New Outer Pseudo–Ring Morphologies and Notation

Table 1
Summary of Additional Notation to Revised Hubble System

Structure	Notation	Examples	Type
Nuclear ring	(nr),(nr′)	NGC 1512	SB(r,nr)ab
		NGC 4314	(R′)SB(l,nr′)a
		NGC 3081	(R)SAB(r,nr)0/a
Nuclear lens	(nl)	NGC 1433	(R_1')SB(r,nl)ab
Inner lens	(l)	NGC 1553	SA(rl)0°
Outer Lens	(L)	NGC 1440	(L)SB(s)0^+
		NGC 4596	(L)SB(r)0^+
Type I Schwarz Outer Ring/ Pseudo–Ring	(R_1),(R_1')	NGC 3504	(R_1')SAB(s)ab
		ESO 508-78	(R_1')SB(r)a
		NGC 1808	(R_1')SAB(s)a
Type II Schwarz Outer Ring/ Pseudo–Ring	(R_2),(R_2')	NGC 688	(R_2')SAB(s)b
		NGC 841	(R_2')SAB(s)b
		IC 1439	(R_2')SAB(s)a
		ESO 294-16	(R_2')SB(s)a
Mixed–mode Outer Ring/ Pseudo–Ring	(R_1R_2')	ESO 509-98	(R_1R_2')SB(s)a
		NGC 1291	(R_1R_2')SB(l,nl)0/a
		IC 1438	(R_1R_2')SAB(r)0/a

The relative frequencies of the different ring patterns have not been firmly established. De Vaucouleurs and Buta (1980) selected carefully defined samples from the *Second Reference Catalogue of Bright Galaxies* (de Vaucouleurs, *et al.* 1976, hereafter RC2) to judge the frequencies of inner rings. These occur in about 25% of spirals, but if inner pseudo–rings are considered, the total frequency would exceed 50%. In a study exclusively dealing with barred spirals, Kormendy (1979) found that 76% of early–to–intermediate barred spirals possess inner rings. Outer rings appear to be most frequent near stage S0/a (de Vaucouleurs 1975). The tentative results of Buta (1986a) suggested that the R_1' subtype is perhaps three times as frequent as the R_2' subtype, although only 55% of recognizeable outer pseudo–rings could be categorized in either way. No definitive statistics exist yet for nuclear rings, whose detection is strongly influenced by selection effects.

b. Lenses

A lens is a relatively uniform or shallow zone in the luminosity distribution of disk galaxies which, on direct photographs, appears to have a fairly sharp edge. Several are illustrated in the Hubble Atlas. They were first recognized by Hubble in S0 galaxies, but were never formally given a type symbol. Instead, they were regarded as characteristic of S0 galaxies in general. They also do not have a type symbol in de Vaucouleurs's revision, although de Vaucouleurs has proposed using the code "L" for S0 galaxies, to imply "lenticular," and has presented information on lens sizes in the notes to the *Reference Catalogue of Bright Galaxies* (de Vaucouleurs and de Vaucouleurs 1964). Kormendy (1979, 1982) has done detailed studies of the frequency and other properties of lenses, and believes them to be an important "distinct component" of galaxies whose true frequency cannot be judged from RC2 types. He has criticized RC2 types because he believes these misclassify lenses as inner rings. He has also pointed out that even though all S0 galaxies might be denoted "L," many do not have lenses.

These disagreements reflect some of the subjectivity of galaxy classification, and I wish to make two points about them. First, since the RC2 does not give quantitative information on the strength of luminosity enhancements associated with rings, one cannot determine whether a ring is strong or weak from RC2 types. However, if a ring is very weak, then in less than optimum seeing the enhancement may be washed out and one will see a "plateau" (or lens) in the light distribution. The distinction between calling a galaxy "inner-ringed" or "inner-lensed" then reduces to semantics in such circumstances, unless rings and lenses are genuinely distinct phenomena. A second point is that rings tend to be blue in color and fairly narrow, while lenses are red and broad. When both are present in the same galaxy, they overlap and have the same shape, implying a close causal connection.

This suggests at least two interpretations: either rings and lenses are variations on the same phenomenon, differing only in the mean age of the stellar population and the amplitude of the enhancement which is superposed on the usual decreasing background of the bulge plus disk; or they are distinct phenomena which happen to coincide in position and shape. The first interpretation would view lenses as being a direct result of ring formation, in the sense that the lens would represent the aging component of a ring whose star formation rate has been fairly uniform over a Hubble time, such that a more diffuse, broader component has been built up. This view finds some some support in the integrated colors of rings (Buta 1986b, 1987a, 1988b), the existence of non-axisymmetric color gradients (Buta 1986b), the existence of lens analogues of each type of ring (de Vaucouleurs 1974; Kormendy 1979; Sandage and Brucato 1979; Buta 1986b), and the existence of rings lacking an old component (*e.g.* NGC 1398, Buta 1988b) or a young component (Gallagher and Wirth 1980). How diffuse an old ring will be will most likely depend on its age and velocity dispersion.

Only for several lenses have velocity dispersions so far been measured, so that a definitive comparison cannot yet be made. As shown by Kormendy (1984), lenses tend to be fairly hot.

However, the nature of lenses is probably more complicated than might be inferred from the above arguments, because lenses usually contain a lot of light and at the same time have oval intrinsic shapes which make them characteristically bar-like. It is quite interesting how some galaxies show inner, outer, and nuclear rings just like those in SB galaxies, yet no obvious bar crosses the inner ring. In such a circumstance, we might conclude that the lens cannot be simply an "old" ring, but is the near-axisymmetric analogue of a bar; it is therefore distinct from rings, but capable of appearing to be associated with them in the same manner as an SB-type bar. Several interpretations have been proposed in this vein. Kormendy (1979, 1981, 1982) believes that lenses originate secularly from bar dissolution in SB galaxies. Although bars are believed to be fairly robust (Miller and Smith 1979b), Contopoulos (1983) has stated that strong bars can generate enough stochasticity inside corotation that stars on stochastic orbits will populate a less elongated, oval zone. On the other hand, Athanassoula (1983) believes that lenses represent disk instabilities just like conventional SB-type bars, the main difference concerning the degree of random motions.

Regardless of the physical interpretation of lenses, I agree with Kormendy (1979) that perhaps they should be given their own type symbols in the classification. He has proposed the use of (l) for inner lenses and (L) for outer lenses, with (nl) an obvious extension for nuclear lenses.

The relative frequencies of lenses are as poorly known as those of rings. Statistics for inner lenses (l) are most secure. Kormendy (1979) found using a carefully defined sample that inner lenses are present in 54% of early-type SB galaxies. The frequencies of nuclear and outer lenses remain to be determined.

4. Physical Interpretations of Revised Hubble Classifications

The establishment of an ordering of galaxies into taxonomic categories is clearly only one step towards understanding the physical nature of galaxy structure. It is important first to re-state what Hubble himself emphasized: that the classification system is entirely empirical and has no basis in theory. Nevertheless, we wish to understand the physical parameters which may underlie each dimension of the RHS, and here I wish to summarize a few of the current ideas.

Table 2 lists some of the parameters which may provide the physical basis of each dimension of the RHS. The stage and family have been thoroughly discussed in reviews by Sandage (1986) and Kormendy (1981, 1982), and I refer the reader to those papers for more details. Variety was the least-known dimension of the system until recently, and here I would like to focus mainly on this dimension as it provides an opportunity for a beautiful confrontation between theory and observation.

It is probably fairly safe to say that the main characteristic galaxies were "born" with is their bulge–to–disk ratio. Whether these two components formed together, or if one preceded the other, is still a point of controversy. Nevertheless, once the bulge and disk formed non–axisymmetric disturbances could have developed in the stellar distribution, and bars or bar–like distortions could have grown rapidly. Once these structures formed, they could stir up any remaining gas in the disk; *e.g.* bars could drive trailing spiral structures in the outer disk. Any patterns would evolve slowly in form, owing to gravity torques and shocks.

The ring–like patterns defining the (r)–variety as well as nuclear and outer rings are believed to be related to orbit resonances with a bar, oval distortion, or density wave (de Vaucouleurs and Freeman 1972; Duus and Freeman 1975; Schwarz 1979; Schempp 1982; Kormendy 1982 and further references therein; Athanassoula 1983). The link with resonances was most firmly established by Schwarz (1979), who used a particle dynamical scheme to simulate the flow of gas in response to a static bar potential. Using a fairly simple method to simulate dissipation via cloud–cloud collisions, he found that gas in a barred galaxy can naturally collect near the main orbit resonances: inner Lindblad resonance (ILR), outer Lindblad resonance (OLR), and the inner 4:1 ultraharmonic resonance (UHR).

The rings form because large secular effects can occur at these resonances (Lynden–Bell and Kalnajs 1972), where periodic orbits achieve their maximum local eccentricity (Contopoulos 1979). If the bar forcing is strong enough, orbits slightly within a resonance can become sufficiently elongated that they will cross orbits slightly outside a resonance which are either shaped differently (as at UHR; Schwarz 1984c) or elongated in the orthogonal sense (as at ILR and OLR; Sanders and Huntley 1976). If gas clouds are present, they will not be able to settle into such orbits without experiencing dissipation. Spiral shock fronts develop in the gas, but these slowly change their form owing to gravity torques exerted by the bar. The net effect is that gas in the spirals tends to slowly drift and settle into ring–like concentrations near one or several of the major resonances.

Schwarz identified ILR, UHR, and OLR with the nuclear, inner, and outer rings, respectively, of barred galaxies. A considerable body of observational evidence has provided support for these interpretations. This has come from studies of the distributions of apparent shapes, relative sizes, and apparent orientations of rings with respect to bars (Kormendy 1979; Athanassoula *et al.* 1982; Schwarz 1984a; Buta 1984, 1986a) and from surface photometry and spectroscopy of individual examples (Buta 1984 and references therein; 1986b,1987a,b,c, 1988a,b). One of the great successes of Schwarz's work is the prediction of the distinct outer ring morphologies discussed in section 2. The morphologies arise because at OLR, spiral structure and evolution are influenced by two major families of periodic orbits, a perpendicular–aligned family slightly inside the resonance and

Table 2
Possible Physical Bases of Stage, Family, and Variety

Dimension	Basis	Selected References
Stage	(a) Initial M(h) distribution?	Sandage *et al.* 1970
	(b) Initial density?	Gott and Thuan 1976
	(c) Halo mass fraction?	Tinsley 1981
	(d) Total mass?	Tully *et al.* 1982
	(e) Initial σ_v of dark matter?	Lake and Carlberg 1988
Family	(a) $T_{mean}/\|W\|$?	Ostriker and Peebles 1973
	(b) Rotation curve, mass dist.?	Lynden–Bell 1979
	(c) Angular momentum transfer?	Lynden–Bell and Kalnajs 1972
Variety	(a) Bar pattern speed?	de Vaucouleurs and Freeman 1972; Schwarz 1979, 1984b
	(b) Rotation curve, mass dist.?	Schwarz 1979
	(c) Gas fraction, total mass?	Huntley 1980; de Vaucouleurs and Buta 1980
	(d) Bar strength?	Schwarz 1984c
	(e) Evolutionary state of pattern?	Schwarz 1979

a parallel–aligned family slightly outside the resonance. Owing to dissipation, a pseudo–ring can develop near OLR whose shape and orientation are similar to one of these families; which one dominates depends on the gas distribution, the strength of the bar, and the ability of the dissipation to deplete the other family. A natural interpretation of the mixed morphology in Figure 2b is that both of these orbits are populated, although such a circumstance ought to be disallowed as the orbits would cross. Indeed, Schwarz did not predict the existence of a mixed outer ring morphology. However, there is clearly more work to be done, and further insight into the nature of these structures may be found in models with growing bars (Elmegreen and Elmegreen 1985 and references therein).

These findings allow us to make some evaluations of the differences that might underlie the family and variety dimensions in the RHS more clearly than could be done before. First, an SB(s) spiral could be a case where the pattern speed Ω_p is high enough to preclude the existence of the main ring–forming resonances within corotation (that is, ILR and UHR). OLR could still exist, so that the spiral pattern could extend to this resonance and still evolve into an outer ring or pseudo–ring. On the other hand, if Ω_p is low such that the

ILR and the UHR exist, then the galaxy could become an SB(r,nr). If all three resonances exist, then the galaxy could become an (R)SB(r,nr). A good example where all three probably do exist is NGC 3081.

In SA galaxies, the interpretation of inner rings seems to be different. De Vaucouleurs and Buta (1980) showed that inner ring sizes depend strongly on both family and variety, in the sense that linear and relative ring diameters decrease from SB to SA and from S0/a to Sd. The first correlation was discovered by de Vaucouleurs, and is depicted in his famous cross-sectional drawing of the RHS (de Vaucouleurs and de Vaucouleurs 1964). It can be interpreted in terms of a correlation between ring size and apparent bar strength, or in terms of different resonance associations (that is, "(r)" in SA galaxies is associated with a different resonance from "(r)" in SB galaxies). I believe the observations favor the latter interpretation. Most of the inner rings of SA galaxies are probably linked to ILR, not the UHR, although features analogous to SB inner rings and lenses are identifiable quite frequently in SA systems (Kormendy 1979; Buta 1984, 1986a). Some SA inner rings lie almost exactly at the "turnover radius" of the rotation curve (Buta 1987a), again implying a possible link with ILR rather than UHR.

This brings me to the final question I wish to discuss in this conference, the possibility that galaxies could change "cell" positions over time scales equal to or less than a Hubble time. Sandage (1986) has asked how long, at present day star formation rates, it would take for a galaxy of a given stage to use up enough of its remaining gas to change by one step in the classification. He finds, using integrated birth rate functions, that this time is a function of increasing gas fraction and decreasing luminosity, and that, depending on the mode of star formation, it could be equal to or longer than a Hubble time. In fact, Sandage believes that the evidence supports the possibility that galaxies may have acquired their final position on the Hubble Sequence not long after the initial collapse, and most have not changed by one step since that time. Of course, this would apply only in the case of isolated evolution. Mergers and collisions can modify galaxy morphology on very short time scales, as much recent research continues to establish (e.g. Hernquist and Quinn 1987; Schweizer and Seitzer 1988; Whitmore and Bell 1988; Higdon 1988 and references therein).

In the case of family and variety, and independent of any external influence, the time scales for significant change could be much less than a Hubble time. For example, in n–body numerical models of galaxy collapse, bars can develop in a few disk rotations (Hohl 1975; Miller and Smith 1979a,b), and in the n–body simulations of gas flow in bar potentials (Schwarz 1979; Combes and Gerin 1985), rings can develop after only 7–10 bar rotations. Thus, it would seem that if secular evolution is occurring in galaxy structure, the most significant changes may be among family and especially variety, since rings evolve from more open spiral patterns. If bars can grow or weaken via angular momentum transfer processes (Lynden–Bell and Kalnajs 1972), or via interactions with the spheroid (Kormendy 1979), then family could change from SB to SAB, or from

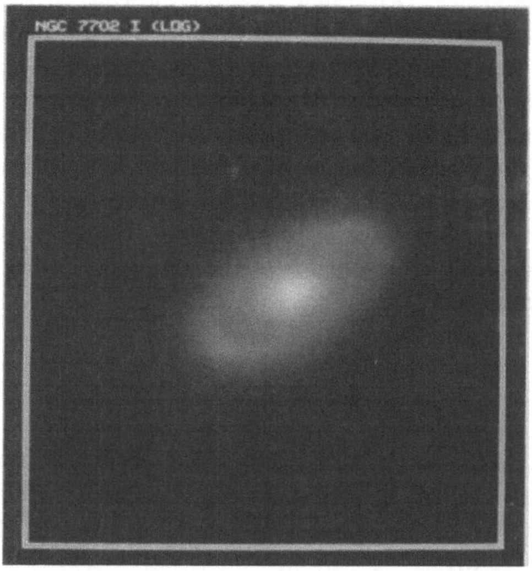

Figure 3: CCD image of NGC 7702 (I–band, AAT), a "non–barred" ringed S0 (note possible small nuclear bar; field stars are removed).

SA to SAB to SB. That such evolution may be taking place is provided by non–barred, ringed S0 galaxies like NGC 7702 (Figure 3). That a galaxy with little or no obvious bar could have such a bright inner ring made purely of stars is a clear contradiction to theoretical models which rely on bar and gas dynamics for ring formation. Either NGC 7702 once had a strong bar that caused a ring to form before it vanished, or ring formation does not require the presence of a strong bar or a significant amount of gas. Clearly, such galaxies need to be studied in greater detail before we can consider our understanding of family and variety complete.

Note that resonances may not be able to explain all rings observed in disk galaxies. Some possibly form by accretion of gas–rich satellites. A particularly striking case may be Hoag's Object (Schweizer *et al.* 1987) and many "polar–ringed" galaxies (Schweizer, Whitmore, and Rubin 1983).

5. Conclusions

In this brief review, I have by no means attempted to cover all of the papers concerning galaxy morphology published since the excellent Kormendy (1982) Saas Fee Lectures. However, in agreement with Kormendy, I hope I have shown that galaxy morphology continues to be an exciting topic of research that today is making rapid advances owing to the improvements in both theoretical and observational capabilities. Although most of the cells of the revised Hubble classification system have not yet been thoroughly studied, major breakthroughs have been made (e.g., in understanding rings), and it is likely that further progress will be made as galaxies are studied in ever greater detail. There is still, however, much to be learned and more observations to be made, specifically within the family and variety dimensions of the RHS, as I believe further research on those characteristics should be very fruitful. I have highlighted the usefulness and versatility of Gérard de Vaucouleurs's revised Hubble classification system, which is one of the finest achievements of the man whose work we are honoring with this symposium.

References

Athanassoula, E., 1983, in *I. A. U. Symp. 100, Internal Kinematics and Dynamics of Galaxies*, ed. E. Athanassoula (Dordrecht: Reidel), p. 243.

Athanassoula, E., Bosma, A., Creze, M., and Schwarz, M. P. 1982, *Astron. Astrophys.* **107**, 101.

Benedict, G. F. 1980, *Astron. J.* **85**, 513.

Buta, R. 1984, "The Structure and Dynamics of Ringed Galaxies," *Univ. Texas Publ. in Astron.* No. **23**.

Buta, R. 1986a, *Astrophys. J. Suppl.* **61**, 609.

Buta, R. 1986b, *Astrophys. J. Suppl.* **61**, 631.

Buta, R. 1987a, *Astrophys. J. Suppl.* **64**, 1.

Buta, R. 1987b, *Astrophys. J. Suppl.* **64**, 383.

Buta, R. 1987c, *Bul. Am. Astron. Soc.* **19**, 1063.

Buta, R. 1988a, *Astrophys. J. Suppl.* **66**, 233.

Buta, R. 1988b, in preparation.

Combes, F. and Gerin, M. 1985, *Astron. Astrophys.* **150**, 327.

Contopoulos, G. 1979, in *Photometry, Kinematics, and Dynamics of Galaxies*, ed. D. S. Evans (Austin: Astron. Dept. Univ. Texas), p. 425.

Contopoulos, G. 1983, *Astrophys. J.* **275**, 511.

Duus, A. and Freeman, K. C. 1975, in *La Dynamique des Galaxies Spirales*, ed. L. Weliachew (Paris: CNRS), p. 419.

Elmegreen, B. G. and Elmegreen, D. M. 1985, *Astrophys. J.* **288**, 438.

Gallagher, J. and Hunter, D. A. 1984, *Ann. Rev. Astron. Astrophys.* **22**, 37.

Gallagher, J. and Hunter, D. A. 1987, *Astron. J.* **94**, 43.

Gallagher, J. S. and Wirth, A. 1980, *Astrophys. J.* **241**, 567.

Gott, J. R. and Thuan, T. X. 1976, *Astrophys. J.* **204**, 649.

Hernquist, L. and Quinn, P. 1987 in *I. A. U. Symp. 127, The Structure and Dynamics of Elliptical Galaxies*, ed. T. de Zeeuw (Dordrecht: Reidel), p. 467.

Higdon, J. L. 1988, *Astrophys. J.* **326**, 146.

Hohl, F. 1975, in *I. A. U. Symp. 69, Dynamics of Stellar Systems*, ed. A. Hayli (Dordrecht: Reidel), p. 349.
Huntley, J. M. 1980, *Astrophys. J.* **238**, 524.
Kormendy, J. 1979, *Astrophys. J.* **227**, 714.
Kormendy, J. 1981, in *The Structure and Evolution of Normal Galaxies*, ed. S. M. Fall and D. Lynden–Bell (Cambridge: Cambridge University Press), p. 85.
Kormendy, J. 1982, in *Morphology and Dynamics of Galaxies* (Geneva: Geneva Observatory), p. 115.
Kormendy, J. 1984, *Astrophys. J.* **286**, 116.
Lake, G. and Carlberg, R. G. 1988, *Astron. J.* **96**, 1581.
Lindblad, P. O. and Jorsater, S. 1981, *Astron. Astrophys.* **97**, 56.
Lynden–Bell, D. 1979, *Mon. Not. Roy. Astron. Soc.* **187**, 101.
Lynden–Bell, D. and Kalnajs, A. J. 1972, *Mon. Not. Roy. Astron. Soc.* **157**, 1.
Miller, R. H. and Smith, B. F. 1979a, *Astrophys. J.* **227**, 407.
Miller, R. H. and Smith, B. F. 1979b, *Astrophys. J.* **227**, 785.
Ostriker, J. P. and Peebles, P. J. E. 1973, *Astrophys. J.* **186**, 467.
Sandage, A. 1961, *The Hubble Atlas of Galaxies*, Publ. No. **618** (Washington, D.C.: Carnegie Institution of Washington).
Sandage, A. 1975, in *Galaxies and the Universe*, ed. A. Sandage, M. Sandage, and J. Kristian (Chicago: Univ. of Chicago Press), p. 1.
Sandage, A. 1986, *Astron. Astrophys.* **161**, 89.
Sandage, A. and Brucato, R. 1979, *Astron. J.* **84**, 472.
Sandage, A. and Tammann, G. 1981, *A Revised Shapley-Ames Catalogue*, Publ. No. **635** (Washington, D.C.: Carnegie Institution of Washington).
Sandage, A., Freeman, K. C., and Stokes, N. R. 1970, *Astrophys. J.* **160**, 831.
Sanders, R. H. and Huntley, J. M. 1976, *Astrophys. J.* **209**, 53.
Schempp, W. V. 1982, *Astrophys. J.* **258**, 96.
Schwarz, M. P. 1979, Ph.D. Thesis, Australian National University.
Schwarz, M. P. 1981, *Astrophys. J.* **247**, 77.
Schwarz, M. P. 1984a, *Astron. Astrophys.* **133**, 222.
Schwarz, M. P. 1984b, *Mon. Not. Roy. Astron. Soc.* **209**, 93.
Schwarz, M. P. 1984c, *Proc. Astron. Soc. Aus.* **5**, 458.
Schweizer, F. and Seitzer, P. 1988, *Astrophys. J.* **328**, 88.
Schweizer, F., Whitmore, B., and Rubin, V. 1983, *Astron. J.* **88**, 909.
Schweizer, F., Ford, W. K., Jedrzejewski, R., and Giovanelli, R. 1987, *Astrophys. J.* **320**, 454.
Shapley, H. and Paraskevopoulos, J. 1940, *Proc. Nat. Acad. Sci.* **26**, 31.
Tinsley, B. M. 1981, *Mon. Not. Roy. Astron. Soc.* **194**, 63.
Tully, R. B., Mould, J. R., and Aaronson, M. 1982, *Astrophys. J.* **257**, 527.
de Vaucouleurs, G. 1959, *Handbuch der Physik* **53**, 275.
de Vaucouleurs, G. 1963, *Astrophys. J. Suppl.* **8**, 31.

de Vaucouleurs, G. 1974, in *I. A. U. Symp. 58, The Formation and Dynamics of Galaxies*, ed. J. R. Shakeshaft (Dordrecht: Reidel), p. 335.
de Vaucouleurs, G. 1975, *Astrophys. J. Suppl.* **29**, 193.
de Vaucouleurs, G. 1977, in *The Evolution of Galaxies and Stellar Populations*, ed. B. Tinsley and R. Larson (New Haven: Yale University Observatory), p. 43.
de Vaucouleurs, G. and de Vaucouleurs, A. 1964, *Reference Catalogue of Bright Galaxies*, (Austin: Univ. of Texas Press).
de Vaucouleurs, G., de Vaucouleurs, A., and Corwin, H. G. 1976, *Second Reference Catalogue of Bright Galaxies*, (Austin: Univ. of Texas Press).
de Vaucouleurs, G. and Freeman, K. C. 1972, *Vistas in Astron.* **14**, 163.
de Vaucouleurs, G. and Buta, R. 1980, *Astrophys. J. Suppl.* **44**, 451.
de Zeeuw, T. 1987, ed. *I. A. U. Symp. 127, The Structure and Dynamics of Elliptical Galaxies*, (Dordrecht: Reidel).
Whitmore, B. and Bell, M. 1988, *Astrophys. J.* **324**, 741.

Discussion

M. Burbidge: In the galaxies with very well separated outer rings, what information have we from the colors of the rings, as compared with the inner part of the galaxies, that would indicate age or population type of the stars in the rings?

Buta: Very few pure outer-ringed galaxies have so far been studied, but it appears that the outer rings have a younger population than bars or bulges. They are bluer in color than those structures, though normally not as blue as inner rings. They also are concentrations of HI gas (e.g. NGC 1291). However, outer rings (and also inner rings) are largely stellar rather than gaseous. In (R)SB0 galaxies, outer rings are sufficiently red to indicate an old stellar population.

G. Helou: The standards used to define the classification system are always face-on galaxies. It must be harder to classify edge-on systems. Are there systematic trends of misclassification as a function of inclination?

Buta: Yes. Inclination can have a serious effect on the recognition of bars and rings, estimation of luminosity class, and other subtle details, but is perhaps less of a problem for stage since we can still detect the relative prominence of the bulge and disk. However, bulge-to-disk ratio is only one of Hubble's three criteria for assignment of stage. In the case of family and variety, it would be expected that bars and rings would be lost as the inclination approaches 90°, and indeed in RC2 there is a deficiency of inner rings among edge-on spiral galaxies. In S0 galaxies, however, it is possible to infer an edge-on view of a ring or lens, or even a bar, with detailed surface photometry (*e.g.* NGC 4762).

J.-L. Nieto: You rightly mentioned the physical phenomena responsible for the structures of disk galaxies in the Hubble sequence. These galaxies certainly evolve by themselves. This may be quite different for the earliest types (E's and S0's, and maybe Sa's) where the disk structure has probably been affected by many small mergings. Another point is that the continuity between morphological types that the whole Hubble sequence suggests may be broken in the E class, if some E galaxies result from strong catastrophic mergings.

Buta: Yes, it is true that E galaxies may be more influenced by mergers than later types, but since the Hubble classification system was formulated independent of any theory, we cannot argue against the continuity it implies simply because more than one mechanism may place unrelated beasts in the same morphological cell. The continuity is an observed fact, regardless of formation mechanisms. A problem arises only when we attempt to *interpret* the physical meaning of the Hubble sequence or its secondary dimensions. Since no "cell" has been so thoroughly studied that it can be regarded as "well–understood," I would not be surprised if more puzzles are in store.

Roger Davies: Morphological classification of disk galaxies has been successful in drawing inferences about the kinematics of disk galaxies. Are there any proposals to provide a finer morphological classification for diskless galaxies (which is becoming increasingly complex — presence of disks, dust, non–elliptical isophotes) that might eventually lead to a better understanding of these systems?

Buta: None yet that I am aware of. Currently, the diskless, or "E" galaxy cell has been the subject of so much detailed research that we have only recently realized how complicated it is. The effects you mention are usually very subtle and sometimes not even obvious just by looking at photographs; they, therefore, would not be easy to incorporate into a classification system based purely on a visual inspection of a photograph or a CCD image. However, diskless galaxies with dust need to be considered more carefully than before, because these have sometimes been misclassified as S0's because of the dust.

J. Binney: You have emphasized classical Lindblad resonances in the plane. I think you should not forget resonances between the planar frequencies and the frequency of motion perpendicular to the plane (for example, Pfenniger's work).

Buta: Yes, I agree that the planar Lindblad resonances are not likely to be the only ones of importance, but in dissipational models of barred spirals (such as those of Schwarz), they are probably most relevant for the ring phenomena. However, in some barred S0 galaxies, the perpendicular resonances may also be important in influencing the morphology. I can think of one particularly striking case, NGC 7020, where random motions in the bar may not be washing out some of the finer details of the stellar orbits in three dimensions.

R. D. Davies: Within a given morphological (T) stage there is a wide range of masses (for example, 100:1 in $T = 4$ galaxies). Are there any morphological indicators of the mass (or size) of a galaxy of a given T value? I know luminosity class correlates with mass, but what about the other parameters you have used — AB, rs, etc? It would be surprising if the mass did not leave its mark on the morphology (at a given T).

Buta: Concerning family, the presence or absence of a bar does not seem strongly correlated with mass or luminosity. It is well known that bars are observed throughout the entire disk galaxy sequence, over a range of absolute magnitudes $M_B^o = -16$ to -21 at least, and are especially prevalent among the magellanic spirals and irregulars. However, for variety, the situation is different. Rings are most prominent among massive, early-type galaxies, but among later type, lower luminosity systems, either weak pseudo–rings or no rings at all are observed. There is therefore an indication that mass has an influence on variety. This is a problem that needs to be further explored.

D. Hanes: You commented that Hubble's original classification predated any theoretical interpretation: indeed he emphasized the need to avoid such interpretations. More recently, you have been encouraged by Schwarz's gas dynamical models to seek, and find, subtle differences in ring morphology. Do you see a danger of reading into a difficult and subjective exercise differences which another observer might not see, given no predisposition or expectation? How do you check yourself?

Buta: Yes, it is certainly possible that other observers, with no predisposition, might not see the subtle differences in ring morphology well enough to recognize them as important or significant. However, the fact that models of barred galaxies have reached a point where they can even make predictions concerning morphology (as opposed to kinematics, for example) is a significant advance in itself. The Schwarz predictions were well–defined, even simplistic, and came at an opportune time. I see no danger in attempting to confront those models as thoroughly as possible. The real danger lies not in recognizing the subtle differences in ring morphologies, but in interpreting them. Not all rings or pseudo–rings can be placed into the context of the models, and I check myself by not attempting to force–fit all rings into my preconceptions. There is much more to be addressed in this field.

J.–C. Pecker: Some years ago, van den Bergh introduced a sequence of so–called "anemic" galaxies. Is this compatible with de Vaucouleurs's classification? If not, why?

Buta: Van den Bergh's system is really not compatible with de Vaucouleurs's system, nor even with Hubble's own revision. The only similarity is in the recognition of a sequence of spiral disk galaxies dependent on bulge–to–disk ratio. The main difference is in the placement of S0 galaxies in an analogous sequence parallel to spirals, with "anemics" lying on an intermediate sequence between them. Surface photometry and color studies have not provided any support for this interpretation, and the system has not found general use.

M. S. Roberts: A high fraction of classical Seyfert galaxies have outer ring structures although this is not the case for the Seyfert–type galaxies discovered in the Markarian lists. (i) Are there new statistics on this correlation? and (ii) Are there explanations of these differences?

Buta: I am not aware of any new statistics on this correlation, although it is well worth exploring. The differences, however, would seem to imply that the Seyfert phenomenon is owed to more than one mechanism. Orbit resonances, galaxy–galaxy interactions, or accretion could all induce the phenomenon over some period, the latter two mechanisms leading to non–ringed Seyferts.

B. C. Whitmore: In a recent paper (*Astrophys. J.* **278**, 61), I used Principle Component Analysis to identify the fundamental physical parameters of spiral galaxies. The data base consisted of rotational velocities, velocity dispersions, $B - H$ colors, total luminosities, etc. for 60 Sa, Sb, and Sc galaxies in the Rubin *et al.* sample. Although we included bar and ring morphologies in the analysis, the sample was not very sensitive to these characteristics since galaxies with strong bars and rings were not included in the original sample. Could you comment on any strong correlations you are aware of between physical properties and the presence of bars or rings.

Buta: Yes, I can make a few comments based on a preliminary analysis of RC3 data. First, at a given stage (or narrow range of stages), integrated colors do not depend very significantly on family or variety, implying that stage is indeed the fundamental parameter. However, the data suggest one important correlation, that (r)–variety spirals are poorer in HI than (s)– variety spirals at a given stage, based on the hydrogen index. Some support for this is given by the relative frequency of the (r)–variety as a function of stage: among late-type galaxies, the (r)–variety is rare, and if a ring is present, it is usually broken and indistinct (a pseudo–ring). It appears that abundant gas may inhibit ring formation, among other things. I believe that a worthy project would be to attempt to measure basic parameters for as many galaxies as possible for each (family, variety)–cell over a narrow range of stages, say Sab–Sbc, in a balanced way, to better establish any basic differences.

CCD Observations of a Sample of Emission Line Galaxies

P. Focardi[1] and R. Merighi[2]

[1]Dipartimento di Astronomia, Università di Bologna, Italy
[2]Osservatorio Astronomico di Bologna, Italy

We are undertaking a morphological and (eventually) photometric study of a sample of emission–line galaxies from UGC (Nilson 1973).

The total sample consists of 120 galaxies; at present, we have B and R CCD images for about 60 objects. The sample has been built up by inspecting our collection of about 600 photographic spectra. The spectra are low dispersion (~120Å mm^{-1}), and have been obtained at the Loiano 152–cm telescope equipped with a Boller and Chivens spectrograph, and an EMI intensifier tube.

All galaxies with spectra showing emission features have been included in our sample; no selection has been made based on the presence, the strength, or the width of any particular line. This implies that different excitation mechanisms (such as photoionization by hot stars, power–law continuum sources, or shock–wave heating) can be at work on our sample.

The spectra have been obtained in the general context of detecting large–scale structures such as filaments, cells, voids, and the so–called "field galaxy" population. A wide range of local galaxy density is covered by our sample. Emission lines have been found to be much more common in field than in cluster galaxies (Osterbrock 1960, Gisler 1978, and Dressler *et al.* 1985). With our sample, we could probably look for this kind of correlation in regions of intermediate density. The CCD images will allow us to look for other kinds of environmental effects such as mergers, tidal encounters, etc.

The images have been obtained with the Loiano telescope equipped with an RCA CCD (Bregoli *et al.* 1986). The B and R bands will give us an indication of the contribution from the two major stellar populations in galaxies.

A first glance, the data show a complex morphology for almost all the galaxies. Strong isophotal twisting in central regions, double nuclei, wrapped filaments, and bright knots are common features.

The data are currently being analyzed. Two of the CCD images are shown here; the frames have been bias–subtracted and flat field corrected.

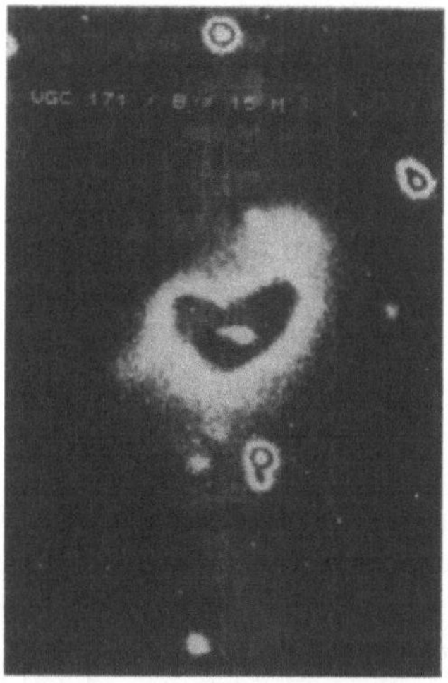

Figure 1. 15 minute *B* CCD image of UGC 171, taken with the Loiano 1.52–cm telescope.

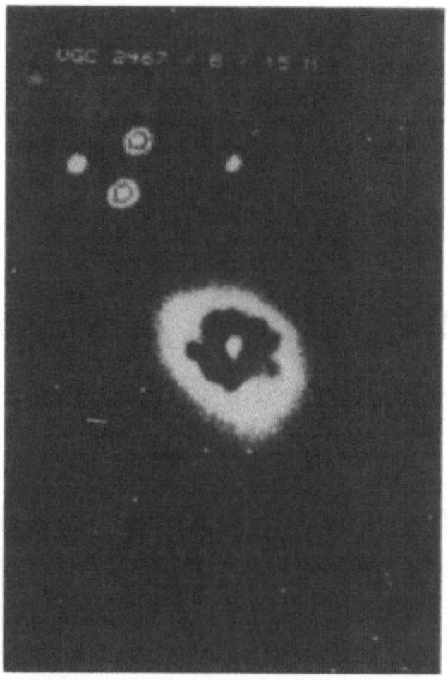

Figure 2. Same for UGC 2467.

References

Bregoli, G., Federici, L., Focardi, P., Merighi, R., Oculi, L., Piccioni, A., Volta, O., and Zitelli, V. 1986, "The Loiano CCD System" in *The Optimization of the Use of CCD Detectors in Astronomy*, ed. J. P. Balateau and S. D'Odorico (Garching bei München: ESO), p. 177.

Dressler, A., Thompson, I. B., and Shectman, S. A. 1985, *Astrophys. J.* **288**, 481.

Gisler, G. R. 1978, *Mon. Not. Roy. Astron. Soc.* **183**, 633.

Osterbrock, D. E. 1960, *Astrophys. J.* **132**, 325.

Nilson, P. 1973, *Uppsala General Catalogue of Galaxies*, Nova Acta R. Soc. Scient. Uppsaliensis, Ser. **V: A.**, Vol. 1.

The Manifold of Galaxian Parameters

P. Brosche and F.–Th. Lentes

Sternwarte der Universität Bonn, F.R.G.

There is no doubt that galaxies exist in a manifold of versions in the colloquial sense of *manifold*. The accumulated results of research on galaxies allow us, however, to also ascribe the mathematical sense to this term. This implies the existence of some parameters which vary *continuously* within the observed manifold. It is clear that this demand, as many others, must be translated into an approximation if one applies it to the real and finite body of observations. The main quantitative questions are (in this sequence): what is the true dimension p of the manifold and what are the functional dependencies between the m observed parameters.

The statistical method of factor analysis (or, in practice about the same, principal component analysis) is the method of choice. In astronomy, there seems to be a psychological barrier against the application to data of different physical dimensions. In fact, the first astronomical use by Deeming (1963) for spectral classification concerned various line strengths at different wavelengths, that is, quantities of the same nature. For our present problem, the first use was made by the first author (Brosche 1973). It was related to spiral galaxies. Here, and in later papers on these galaxies (Bujarrabal *et al.* 1981; Mebold and Reif 1981; Reif 1982; Whitmore 1984; Watanabe *et al.* 1985) and on ellipticals (Brosche and Lentes 1982, 1983, 1985; Lentes 1983; Efstathiou and Fall 1984; Watanabe *et al.* 1985; Lentes 1985; Vader 1986; Djorgovski and Davis 1987; and references in these papers) the most important and agreeing result was the dimension number

$$p = 2$$

The coefficients which connect every observable with the two eigenvectors also allow us to express every observable in terms of two other independent observables. Here we are starting from *empirical* data, and we are concentrating and condensing them to a maximum compactness (which may require a new term *eucriny* as proposed by Brosche 1985). At this level, the often posed question "What are the two variables?" is illegitimate. We remind the reader of a linear vector space: if it has p dimensions, *any* p independent vectors can serve as a basis. Orthogonal p–tupels have a certain advantage; in our galaxian case ($p = 2$), orthogonal *pairs* have this property. Only to the degree that the sample properties agree with the total manifold can one ascribe a physical meaning to the statistical independence of any two parameters. On the other hand, one certainly should avoid using a highly correlated pair for the description of the manifold.

The coefficients mentioned above enable us to exhibit the results (because $p = 2$) in one figure showing all the *gradients* of the different variables as functions of two arguments: nearly parallel or antiparallel gradients belong to highly correlated variables and vice versa. Among the many mutual correlations and relations contained in the matrix of coefficients (and in the figure), the correlation between absolute luminosity and rotational velocity later has become the subject of special interest and named accordingly the Tully–Fisher relation. It is only a correlation and the influence of the second dimension could be expressed, *e.g.*, by a radius. This has been rediscovered by Vera Rubin and her co-workers (1985).

In our view, the general result $p = 2$ is a hint towards the fact that all other characteristica of protogalaxies besides mass and relative angular momentum are forgotten because of virialisation during the early evolution. A simple model of this kind contains cloud collisions as a driving process and star formation as a retarding process (Brosche 1970). It has been found that a star formation rate not only proportional to the square of the gaseous density, but as well to the r. m. s. velocity of the interstellar clouds, is compatible with the empirical relations (Brosche and Lentes 1985). This and other modifications have been added to the original model (Brosche, Caimmi and Secco 1988). In such a way, the model is able to explain (a) a continuous Hubble sequence from ellipticals towards spirals, and (b) within the spirals, the appearance of a minimal axis ratio at an intermediate type (Sd, according to de Vaucouleurs, 1974).

Our studies owe much of their observational basis to the work of G. and A. de Vaucouleurs. Moreover, they were the first to recognize the timeliness of these studies. The first author gratefully remembers the invitation for a colloquium in Austin in 1972 and further support (de Vaucouleurs 1974).

References

Brosche, P. 1970, *Astron. Astrophys.* **6**, 240.
Brosche, P. 1973, *Astron. Astrophys.* **23**, 234.
Brosche, P. 1985, *Naturwissenschaften* **72**, 668.
Brosche, P. and Lentes, F.–Th. 1982, *Mitt. Astron. Ges.* **55**, 116.
Brosche, P. and Lentes, F.–Th. 1983, in *Internal Kinematics and Dynamics of*

Galaxies, I.A.U. Symp. No. **100**, ed. E. Athanassoula (Dordrecht: Reidel), p. 377.

Brosche, P. and Lentes, F.-Th. 1985, *Astron. Astrophys.* **153**, 157.

Brosche, P., Caimmi, R., and Secco, L. 1988, *Attie Memorie Accademia Patavina*, in press.

Bujarrabal, V., Guibert, J., and Balkowski, C. 1981, *Astron. Astrophys.* **104**, 1.

Deeming, T. J. 1963, *Mon. Not. Roy. Astron. Soc.* **127**, 493.

Djorgovski, S. and Davis, M. 1987, *Astrophys. J.* **313**, 59.

Efstathiou, G. and Fall, S. M. 1984, *Mon. Not. Roy. Astron. Soc.* **206**, 453.

Lentes, F.-Th. 1983, in *Statistical Methods in Astronomy*, ed. E. J. Rolfe (Noordwijk: ESA SP-201), p. 73.

Lentes, F.-Th. 1985, Ph. D. dissertation, Math.-Nat. Fakultät Universität Bonn.

Mebold, U. and Reif, K. 1981, *Mitt. Astron. Ges.* **51**, 143.

Reif, K. 1982, Ph. D. dissertation, Math.-Nat. Fakultät Universität Bonn.

Rubin, V. C., Burstein, D., Ford, W. K., and Thonnard, N. 1985, *Astrophys. J.* **189**, 81.

Vader, J. P. 1986, *Astrophys. J.* **306**, 390.

de Vaucouleurs, G. 1974, in *The Formation and Dynamics of Galaxies*, I.A.U. Symp. No. **58**, ed. J. R. Shakeshaft (Dordrecht: Reidel), p. 1 (see especially pp. 13ff, 24ff, 39ff).

Watanabe, M., Kodaira, K. and Okamura, S. 1985, *Astrophys. J.* **292**, 72.

Whitmore, B. C. 1984, *Astrophys. J.* **178**, 61.

Toward an Automatic Classification of Galaxies

Monique Thonnat

Institut National de Récherche en Informatique et en Automatique, Valbonne, France

Introduction

The classification of galaxies into various morphological types is a very difficult problem that only a few specialists attempt (*e.g.* de Vaucouleurs 1959, Sandage 1961). The major difficulty lies in the uniqueness of the direction of sight under which we observe the three-dimensional galaxy. Until now, galaxy classification has been achieved only by visual inspection of photographical plates. However, the increasing quantity of data provided by new instruments (large field Schmidt telescopes, space telescopes, area detectors, high speed digitizers) and the need for objective and explicit reasoning, have at last put the automation of galaxy classification within reach (Lauberts and Valentijn 1984, Thonnat 1985). We present here an automatic method for classifying galaxies, which can be integrated into a data processing program of photographic plates (see Figure 1). The complete processing of a photographic plate is made in three stages:

- The plate measuring the brightness of an astronomical field is digitized and the galaxies in it are detected.
- For each galaxy detected, specific parameters are extracted.
- Finally, these parameters are used to classify the processed galaxy image; the classification is made with an expert system rather than merely a standard pattern recognition algorithm (Ballard and Brown 1982).

The first stage consists of two processes: (1) digitization and (2) discrimination of galaxies from stars and other astronomical objects or artifacts. These are well-known in astronomical image processing (*e.g.* Bijaoui 1981 and Jarvis and Tyson 1981).

The second and third stages are developed in the following sections.

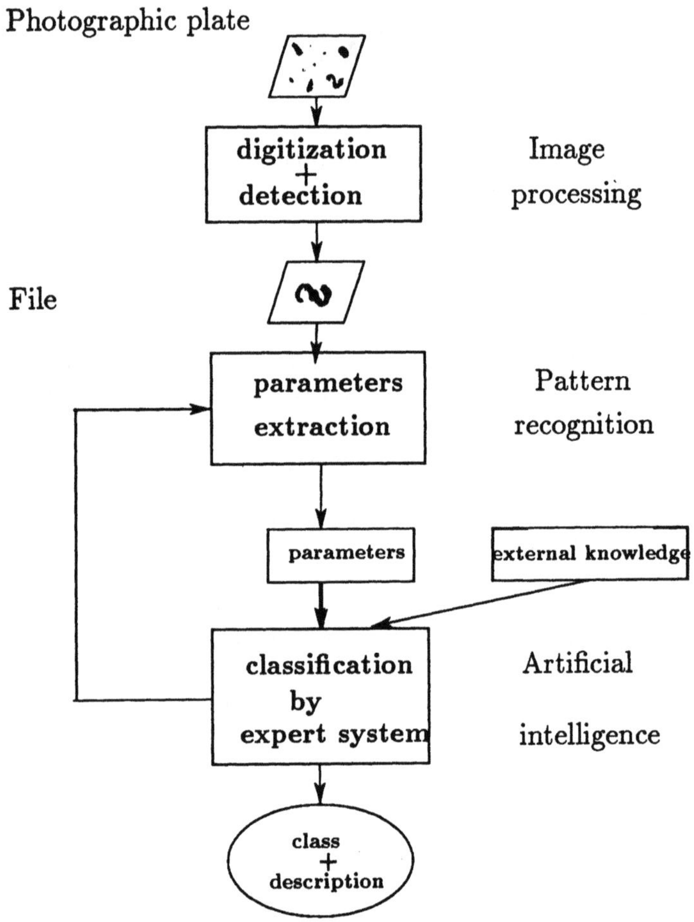

Figure 1: Synopsis of the processing.

Description of the Objects

The Classification Systems

The earliest classification system still in use today was defined by Hubble in 1926. Since then, some modifications (by *e.g.* Hubble 1936, de Vaucouleurs 1959, and Sandage 1975) have been made to obtain today's most commonly used system, the Hubble revised system. The classification of galaxies may be described with a continuous label coding **T** or a hierarchical label coding **H**.

The morphological type coding **T** is an integer value varying between -5 and $+10$, which characterizes the degree of inhomogeneity of the morphology. The lowest value ($\mathbf{T} = -5$) corresponds to a perfect ellipsoid with an eccentricity below 0.7. As **T** increases, the intrinsic shape of the galaxy becomes flatter and may be described as the superposition of a small sphere and a surrounding flat disk ($\mathbf{T} \leq -1$). Positive values of **T** correspond to the presence of spiral structure inside the disk. The extreme value ($\mathbf{T} = 10$) represents a very irregular shape.

The hierarchical label coding **H** represents the different patterns present in the galaxy. This code is an alphanumerical chain, where each character represents a special pattern. The main character defines the general class of the galaxy: "E" stands for elliptical, "L" for lenticular and "S" for spiral. The second character defines the presence of a transverse bar: "A" stands for a missing bar, "B" for a present bar, and "X" for the intermediate case. The third character specifies the shape of the structure in the disk: "R" stands for a ring, "S" for a spiral with no ring (or S-shape) and "T" for the intermediate case. Other characters specify the continuous morphological type **T**, or some peculiarity "P" in the morphology.

For example, "LA" represents a lenticular galaxy (L), with no bar (A); and "SBR3" represents a spiral galaxy (S) with a bar (B), an internal ring (R), and a type **T** equal to (3).

Examples of Images

Figure 2 (left) shows some examples of typical images of galaxies provided by the digitization and detection stage of the processing of a photographical plate. These galaxies in the Virgo cluster have been observed by J.–D. Strich with the Schmidt telescope at CERGA. [1] The sampling rate (pixel size) is (20×20) microns, and the image size is (512×512) pixels.

Among the properties of such images, we must notice that:

1. These images have a very low signal to noise ratio, which becomes critical for faint objects;

2. There is no clear limit discriminating the galaxy from the background;

[1] CERGA: Centre d'Études et de Recherches Geodynamiques et Astronomiques.

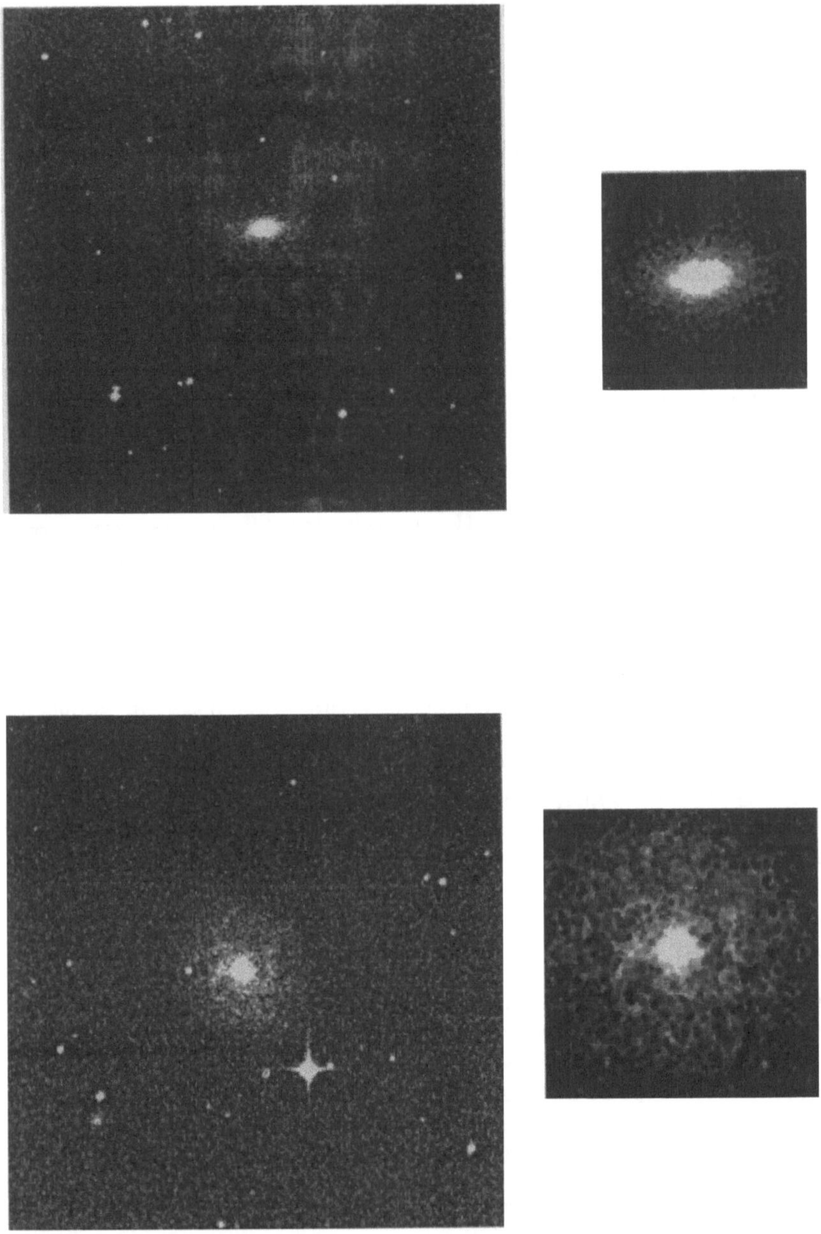

Figure 2: Galaxy images before and after preprocessing: top, an elliptical galaxy (NGC 4473); and bottom, a spiral galaxy (NGC 4571).

3. Other objects may be close, or even partially overlapping the galaxy.

Preprocessing

These three properties show that a preprocessing phase is necessary to enhance the quality of the image before classification parameters can be easily extracted from the isolated galaxy. During this preprocessing, we must perform the following functions:

- locate the exact position of the galaxy;
- define the limits of the galaxy;
- build a map of the background;
- subtract extraneous objects in the neighborhood;
- and eliminate the noise.

The data processing techniques needed for these functions are strongly dependent on the quality of the images. Since we want the system of classification to be completly automatic, even in this preprocessing phase, we must avoid all interactive methods, (in particular, manual selection of thresholds) and choose the most adaptive techniques.

Figure 2 (right) shows the results of the preprocessing of the two galaxies previously described.

Parameter Extraction

In this section, we present the various parameters extracted from the galaxies to describe their shape.

Principal Axis

The first parameter computed is the orientation of the principal (major) axis of the galaxy. The value of its direction is given by the angle θ which minimizes the moment of inertia J:

$$J = \sum_{x,y} d(x,y)(y\cos\theta - x\sin\theta)^2$$

Let θ_1 and θ_2 be the two solutions of

$$\tan(2\theta) = 2\frac{\sum_x \sum_y d(x,y)xy}{\sum_x \sum_y d(x,y)(x^2 - y^2)}$$

with $J(\theta_1) < J(\theta_2)$.

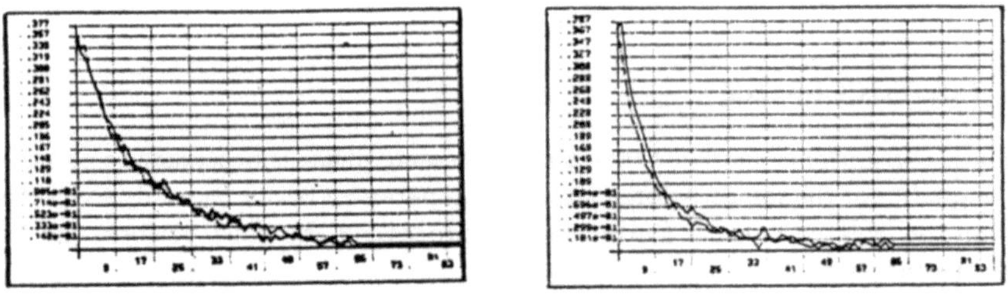

Figure 3: Distribution of the luminosity from the center of NGC 4473 in the (left) θ_1 and the $\theta_{1+\pi}$ directions, and (right) in the θ_2 and the $\theta_{2+\pi}$ directions.

- θ_1 is the orientation of the principal axis;
- θ_2 is the orientation of the small (minor) axis (axis orthogonal to the principal axis).

A set of curves representing the intensity of the galaxy in different directions from the center are built. These directions are along the two principal half-axes and the two small half-axes. Figure 3 displays the shape of these curves for the elliptical galaxy NGC 4473.

Ellipticity

The shape of the galaxy is very dependent on its angle of inclination to our line of sight. Therefore, an important measure is the apparent ellipticity of the galaxy $e = 1 - \frac{b}{a}$, where a and b are the length of the principal axis and the length of the orthogonal axis, respectively.

This value is provided by the previous curves indicating the galaxy's length along the principal and small axes. As these curves are noisy and have a smooth slope due to diffusion in the photographic emulsion, we use an integrated length value.

Let $F(x)$ be the average distribution of the intensity along the principal axis, and $f(x)$ the average distribution of the intensity along the small axis; let w_1 and w_2 be the width of the galaxy in the two directions. The widths are then deduced from

$$\int_0^{w_1} F(x)dx = \frac{3}{4}\int_0^{\infty} F(x)dx$$

and

$$\int_0^{w_2} f(x)dx = \frac{3}{4}\int_0^{\infty} f(x)dx.$$

Finally, the apparent ellipticity is given by $e = 1 - \frac{w_2}{w_1}$.

Size

In order to estimate the reliability of the measured parameters, we compute the size of the galaxy S assuming an elliptical shape: $S = \pi w_1 w_2$.

Average Profile

Using the measure of the apparent ellipticity e, we build the curve of the average intensity distribution from the center $g(x)$. The curve $g(x)$ is the radial distribution of the intensity in the new reference frame $(X, Y = \frac{w_1}{w_2 y})$, with X being the coordinate along the principal axis and Y the coordinate along the small axis .

Projected Profile

Theoretical studies (de Vaucouleurs 1976, Watanabe *et al.* 1982) have shown that the radial variation of the luminosity from the centers of galaxies is the sum of two functions. The first is the spheroidal component characterizing elliptical galaxies.

- The spheroid intensity I_1 is given in terms of the radius r by $I_1(r) = a_1 e^{-b_1 r^{\frac{1}{4}}}$.
- The density $d = \log(\frac{I}{I_0})$ is given in terms of the radius by $d_1(r) = \alpha_1 - \beta_1 r^{\frac{1}{4}}$.

The second function is the flat component characterizing the disks of lenticular and spiral galaxies.

- The disk intensity I_2 is given in terms of the radius r by $I_2(r) = a_2 e^{-b_2 r}$.
- The density $d = \log(\frac{I}{I_0})$ is given in terms of the radius by $d_2(r) = \alpha_2 - \beta_2 r$.

In order to remove the effect of the apparent tilt of the galaxy, we build a new curve: the projected profile d_p. This curve is obtained by orthogonal projection of the densities d of the galaxy along the principal axis.

$$d_p(u) = \sum_{x,y,u=x\cos\theta+y\sin\theta} d(x,y)$$

where θ is the direction of the principal axis, and x and y are the coordinates of the pixels comprising the galaxy image. From this curve we extract two parameters:

- The parameter (profile) which is the ratio of the root mean square errors found by approximating the projected profile with d_1 and d_2 for r greater than the radius of the bulge. The estimate of the radius of the bulge is obtained from the average profile previously found:

$$profile = \frac{\sum_{r=r_{bulge}}^{r_{limit}} (d_p(r) - d_2(r))^2}{\sum_{r=r_{bulge}}^{r_{limit}} (d_p(r) - d_1(r))^2}.$$

- A parameter (*linear_err*) measures directly the root mean square error found by approximating the complete projected profile for $r = 0$ to $r = r_{max}$ with a linear function:

$$linear\_err = \sum_{r=0}^{r_{limit}} (d_p(r) - d_1(r))^2$$

Contours

Contour Building

In order to describe the variation of the structure in the different regions of the galaxy, we need to extract several isophotes from the image. The isophotes must be completely representative of each region in the galaxy and they must be found with no interactive processes. First, we compute five thresholds from the distribution of the density along the principal axis $F(x)$, then we use these thresholds to obtain five binary images to which we apply an edge detector algorithm.

The five thresholds $t_i, i\epsilon[1, \ldots, 5]$ are given by

$$\int_0^{t_i} F(x)dx = \frac{1}{n_i} \int_0^{\infty} F(x)dx$$

with n_i respectively equal to 0.20, 0.50, 0.75, 0.85, and 0.90.

For each thresholded image, we use a Sobel edge detector, then we exhibit the maximal chain from the contour chaining algorithm in the INRIMAGE library (Cipiere 1984).

Figure 4 shows the contours associated with the two galaxies from the previous figures.

Parameters Extracted from these Contours

From each of these contours, we extract five parameters characterizing its shape: ellipticity, angle of the principal axis, relative position of the center, compactness, and distance of the closest ellipse.

- *The angle of the principal axis*: In the same way as for the image, we compute the angle θ_i which minimizes the inertial moment J_i and the orthogonal angle ψ_i.

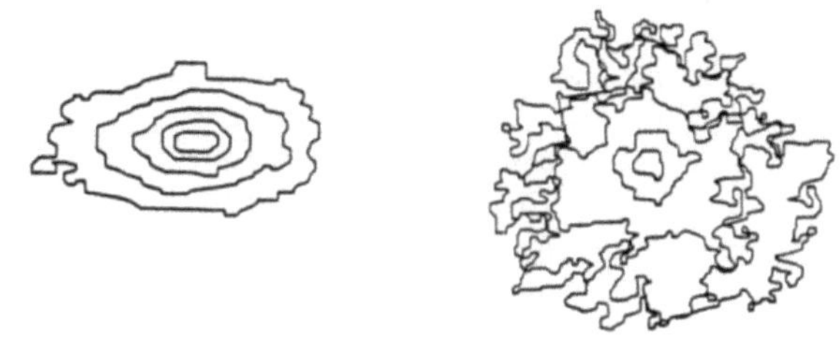

Figure 4: Contours extracted from the images of NGC 4473 (left) and NGC 4571 (right).

- *The ellipticity*: For each contour we compute the ellipticity (the description has been given above).
- *The relative position of the center*: For each contour we compute the euclidian distance ($center\_err$) between the center of the bulge and the center of gravity of the contour g_i.
- *The compactness*: The compactness C measures the roundness of a shape and is a minimum for a circle ($c \leq 4\pi$). $c = \frac{Perimeter^2}{area}$. In fact, we take into account the value of the estimated ellipticity $e = 1 - \frac{b}{a}$, so we measure the quantity, $C = \frac{b}{a}\frac{c}{e\pi}$.
- *The distance of the closest ellipse*: The last parameter ($ellipse\_err$) is the distance between each contour and the ellipse which has its main axis in the θ_i direction, an ellipticity e, and is centered on the center of gravity g_i.

Normalization is performed as follows: Let S_E be the area of the ellipse, S_C the area of the contour and, S_{dif} the sum of the areas between the two closed curves

$$ellipse\_err = \frac{S_{dif}}{S_E + S_C}$$

S_{dif} is approximated by dividing the area into small triangles, the vertices of which are 1) the points of the contour C_i, 2) the projections of these points on the ellipse p_i, and 3) the intersections of the contour and the ellipse P_j. Let x_{p_i} and y_{p_i} be the coordinates of the projections; and let p_i, x_i, and y_i be the coordinates of the points C_i belonging to the contour. Then

$$x_{p_i} = a\cos(\arctan(\frac{ay_i}{bx_i}))$$

and

$$y_{p_i} = b\sin(\arctan(\frac{ay_i}{bx_i})).$$

Example of Measured Parameters

We display here the parameters extracted from an image of NGC 4569.

orientation :	79.33
ellipticity :	0.50
linear_err :	0.039
profile :	2.77
area :	38013.3

Contours :	*Center_err*	*Ellipse_err*	*Compactness*	*Angle*	*Ellipticity*
*c*1 :	1	0.06	1.6	−6.3	0.09
*c*2 :	2	0.21	5.4	70.8	0.36
*c*3 :	11	0.16	7.1	83.6	0.58
*c*4 :	9	0.21	11.8	76.8	0.47
*c*5 :	7	0.14	8.6	76.1	0.50

Automatic Classification

Choice of the Method

In the second section (Description of the Objects), we saw that the various three–dimensional models of galaxy classes are not isotropic. Now, the images of galaxies that we wish to classify are the two–dimensional projections on photographic plates of the real objects. So, for each three–dimensional model, we observe a great diversity in the shape of the digitized images.

The main problem of this classification is matching a two–dimensional image with a three–dimensional model. This problem cannot be expressed as a graph isomorphism problem, and standard pattern matching algorithms (Pavlidis 1977) are not applicable in this case. Actually, we have a valid hypothesis with criteria similar to those used by visual classifiers (for example, if the apparent eccentricity of a galaxy is large, the galaxy is probably seen edge–on, so the type T is probably greater than −5). So, since we need a tool which works on both the descriptions of the classes and on these criteria, we have developed an expert system.

Artificial intelligence (AI) methods provide easy symbolic manipulation by means of languages as LISP or PROLOG (or other object–oriented languages), as well as the possibility of using heuristics to decrease the complexity of the problem.

Expert systems stand among the most attractive tools recently developed in the AI community (*e.g.* Nilsson 1982 and Hayes–Roth *et al.* 1983). These systems have often been used to process huge amounts of data (for example, MYCIN, a computerized medical consultation system, Shortliffe 1976), In addition to the properties mentioned above, expert systems also offer modularity and extensibility. Indeed, they represent a complete separation of the knowledge base from the control structure, and the addition of new rules in the knowledge base does not necessitate other modifications in the system. Several computerized pattern recognition systems (*e.g.* Hanson and Riseman 1978, Brooks 1981) using this approach have been proposed.

The expert system we present in the following section has been specially designed to solve the problem of object classification.

The Knowledge Base

Previous studies (Granger *et al.* 1984, Thonnat *et al.* 1985) have shown that a classical rule interpreter is not suitable for classifying objects into well-known classes. Therefore, we need to introduce into the knowledge base explicit descriptions of the various models to allow an estimate of the difference between the object and the possible models. Moreover, the knowledge base must contain rules; more precisely, the formalism of the production rules expresses in an explicit way the subjective criteria used by the experts, and allows the translation of the quantified parameters into symbolic descriptors.

So, a system using both production rules and frame-like objects (prototypes) to describe models has been developed. It has been implemented in an object-oriented extension of LeLisp (Chaillot *et al.* 1984), an efficient language in which to build data structures like production rules or frames. An extensive description of this system can be found in Granger and de Mongareuil (1987).

The Symbolic Parameters

The description of a class is made by the astronomer in terms of symbolic parameters which represent the different structural patterns of the galaxies. We have introduced these symbolic parameters into the knowledge base in addition to the quantified parameters previously described.

The symbolic parameter H : This symbolic parameter corresponds to the hierarchical label coding; it takes the values of the possible galaxy classes, *i.e.* H: E, L, S, E0, E1, ..., E6, LA, LB, LA-, LB-, LA+, LB+, SA, SB, SA0, ..., SA7, SB0, ..., and SB7. Because irregular galaxies can not yet be classified automatically, the associated values of H are not assigned.

The symbolic parameter T : This parameter corresponds to the continuous label coding. Though it takes numerical values, it is not a measured parameter; each value represents a label for the Hubble stage. T: -5, -3, -2, -1, 0, 1, ..., and 7. Values greater than 7 are not assigned, as they would correspond to irregular galaxies.

The symbolic parameter bar : This parameter specifies knowledge about the possible presence of a transverse bar. Therefore, the symbolic values are *bar*: "present," "absent," and "unknown."

The symbolic parameter shape : This parameter specifies the global structure of the galaxy; it takes the values *shape*: "elliptical," "average," and "spiral."

The symbolic parameter isophotes : This parameter indicates the degree of pertubation of the isophotal contours; the possible values are: *isophotes*: "smooth," "normal," and "distorted."

The symbolic parameter arms : This parameter specifies the development of arms in the galaxy; the values are: *arms*: "absent," "incipient," "evident," and "branched."

The symbolic parameter bulge : This parameter indicates whether the central bulge can be detected in the projected profile. The values are: *bulge*: "visible" or "invisible."

The symbolic parameter flatness : This parameter indicates the apparent ellipticity of the galaxy. It can have the values: *flatness*: "null," "negligible," "very faint," "faint," "light," or "average."

The symbolic parameter centering : The parameter *centering* measures the difference between the center of the galaxy and the center of the contours: *centering*: "good," "average," "mediocre," or "indifferent."

The symbolic parameter profile concavity : This parameter indicates if the degree of concavity of the curve of the projected profile. It can assume the values *profile concavity*: "great," "average," or "null."

The symbolic parameter validity : This important parameter specifies the quality of the observation and thus the quality of the measured parameters; it is function of the size of the image of the galaxy. The values are: *validity*: "good" or "bad."

We have seen in the previous section that for each processed galaxy five contours are constructed; these contours are described with both symbolic and quantified parameters. The quantified parameters are those which have been extracted from each contour: *center_err, ellipse_err, compactness, angle, ellipticity.*

Three symbolic parameters describe also the contours in the different regions of the galaxy.

The symbolic parameter contour c_i*, shape* : This parameter indicates the shape of each contour in the same way as the symbolic parameter *shape*; it takes the same values as the global parameter: *contour* c_i*, shape*: "elliptical," "average," and "spiral."

The symbolic parameter contour c_i*, isophotes* : This parameter indicates the degree of pertubation of each contour; like the global parameter associated with the whole galaxy, it can take the values: *contour* c_i*, isophotes*: "smooth," "normal," or "distorded."

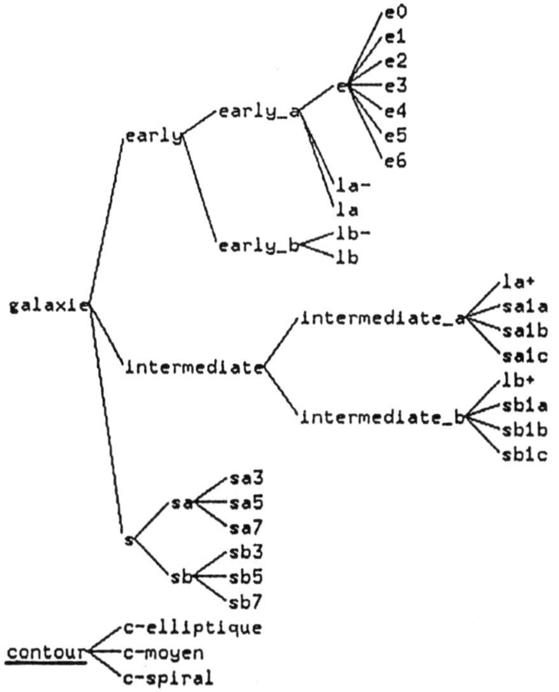

Figure 5: The heirarchy of the classes.

The symbolic parameter contour c_i, flatness : This parameter indicates if the ellipticity of each contour is not too great: *contour c_i, flatness*: "valid."

The Prototypes

The prototypes are frame–like objects, and are used to define the various classes. They are organized in a hierarchy which strictly reflects the hierarchy of the classes (see Figure 5). The descriptors of a prototype characterize the objects belonging to the corresponding class. Symbolic, numeric, and complex structures are represented by different data structures. Some descriptors may have complex structures as shown in the following example.

Class Contour	
compactness :	[0, 10000];
ellipticity :	[−0.9, 1];
angle :	[−90, 90];
ellipse_err :	[0, 10000];
center_err :	[0, 10000];
shape :	*elliptical, average, spiral*;
isophotes :	*unknown*;
flatness :	*unknown*.

A branch of the tree of the prototypes is displayed below (the prototype Galaxy is the root node of the tree):

Class Galaxy	
H :	*unknown*;
T :	[−5, 10];
shape :	*unknown*;
bar :	*unknown*;
profile :	[0, 10000];
ellipticity :	[−0.9, 1];
orientation :	[−90, 90];
isophotes :	*unknown*;
linear_err :	[0, 100000];
area :	[0, 900000];
flatness :	*unknown*;
profile concavity :	*unknown*;
centering :	*unknown*;
arms :	*unknown*;
bulge :	*unknown*;
validity :	*unknown*;
c1 :	*Class contour*;
c2 :	*Class contour*;
c3 :	*Class contour*;
c4 :	*Class contour*;
c5 :	*Class contour*.
Class S	*upperclass Galaxy*
H :	*S*;
T :	[3, 7];
shape :	*spiral*;
validity :	*good*.
Class SB	*upperclass S*
H :	*SB*;
bar :	*present*.

Class SB7	*upperclass SB*
H :	$SB7$;
T :	$[7, 7]$;
centering :	*indifferent*;
arms :	*late*.

The Facts

The facts correspond to data particular for the current case and are temporarily added to the knowledge base. For our application, the facts represent the information associated with one galaxy. The base of facts consists only of the object to be classified. This object has the same structure as the root prototype. At the beginning, all the values of its descriptors are unknown, except those corresponding to the measured parameters. At the end of the procedure, the symbolic descriptors of the object have the same values as the prototype representing the class of the object.

The Rules

The rules are composed of three parts: the conditions, the actions, and the comments. They represent the operating knowledge of the expert. The rules are used to attach a symbolic value to a descriptor of the object. In order to decrease the number of rules needed, the knowledge base is automatically structured. A procedure of initialization attaches a few rules to each prototype of the tree. The attachment is performed if the field action of the rule operates on a descriptor of the prototype.

Example of a rule:

Rule 8:

IF ellipticity of contour3 $\gg$ ellipticity 0.1
ellipticity of contour3 $>$ 0.4
THEN bar is present

"As the ellipticity of the third contour is greater than the global ellipticity of the galaxy and is important, a bar is present."

The Control Structure

Control is guided only by the prototypes. The object, which is *a priori* described only by the measured parameters, is associated with the root prototype galaxy. If no descriptor is incompatible with this prototype, then the root prototype is considered as the current prototype. According to the context, the current prototype moves down the prototype tree.

All the successors of a current prototype which are consistent with the present values of the object descriptors are taken as hypotheses. We want to saturate the base of facts, which means building a complete description of the object that we want to classify; so, all the hypotheses are considered. But,

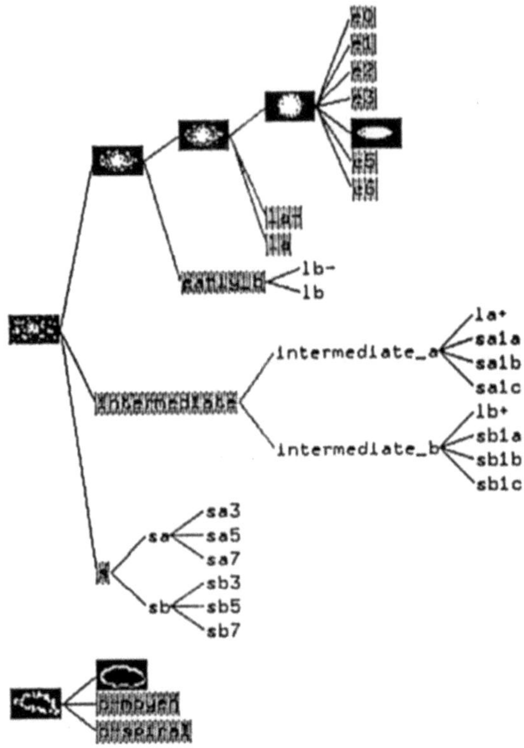

Figure 6: The path during the classification.

since we want natural reasoning, the prototype tree is scanned first in depth (all the inferences of an hypothesis are considered before studying another one). Once a prototype is taken as a possible hypothesis, the associated base of rules is scanned; the action fields of the activated rules increase the description of the object. Then, the prototype is validated if the distance between the object and the prototype is negligible. This distance is defined by the sum of the intermediate distances between the corresponding descriptors.

Figure 6 shows the path in the hierarchy of classes during the classification of NGC 4473.

The initial description of NGC 4473 is (measured parameters):

orientation :	−3.18
ellipticity :	0.4
linear_err :	0.061
profile :	0.35
area :	10935

Contours :	*Center_err*	*Ellipse_err*	*Compactness*	*Angle*	*Ellipticity*
*c*1 :	0	0.07	1.6	−1.8	0.41
*c*2 :	1	0.04	1.9	−1.5	0.35
*c*3 :	0	0.05	3.2	−2.3	0.41
*c*4 :	0	0.05	3.8	−1.4	0.43
*c*5 :	5	0.09	6	−3	0.42

Twenty four production rules (among the 106 rules) have been activated in order to complete the description of the measured galaxy. The final symbolic description of the galaxy is:

validity :	*good(certain)*
arms :	*absent(certain)*
shape :	*elliptical(certain)*
bulge :	*visible(certain)*
flatness :	*faint(certain)*
profile concavity :	*important(certain)*
centering :	*good(certain)*
bar :	*absent(certain)*
isophotes :	*smooth(certain)*
class :	*E4(compatibility : total; incompatibility : null)*

Results

Reliability Tests

A knowledge base of 37 prototypes and 106 rules has been built. The system has been tested on a set of 21 galaxies in the Virgo cluster digitized from the same Schmidt plate. The results of the automatic classification of these galaxies, is presented in the following table:

Name	Real class	Result	Comment
NGC 4459	LAR+	LA+	correct
NGC 4569	SXT2	SX3	correct
NGC 4474	L...P*	LA+/LB+	correct
NGC 4473	E.5..	E4	correct
NGC 4477	LBS.*$	L	correct
NGC 4571	SAR7	SA7	correct
NGC 4468	LA...*	bad validity	small area
NGC 4531	S..1	LA+	slight difference
NGC 4419	SBS0	SB1/SA1	almost correct
IC 3392	SA.3	SB1	almost correct (dust lanes)
NGC 4548	SBT3	SB1/SB3	correct
NGC 4461	LBS+*	LA/LB	almost correct
NGC 4438	SAS0	LB/SB1	correct (dust lanes)
NGC 4421	SBS0	LB/SB1	correct
IC 800	SBT4P$	SB5	correct
NGC 4523	SBS8	SB	irregular
NGC 4595	SXT3$	SA1/SA3	correct
NGC 4639	SXT4	SB3	correct
NGC 4540	SXT6	SB7	correct
NGC 4406	.E.3..	E3	correct
NGC 4374	.E.1..	E0	correct

Although the results are quite good, false detection of a bar in the presence of transverse dust lanes remains a problem.

Tests have shown that the method needs a minimum size for the image of the galaxy (about 50 × 50). So, the rule:

if area < 2500
then validity bad

is activated to prevent a misclassification. In order to decrease the computing time, the parameter validity is set to *good* for each successor node of the root node galaxy; so, if the previous rule is activated, no successor may be validated and the classification returns class *unknown* regardless of the other values of the measured parameters.

The knowledge base does not yet contain information about irregular or peculiar galaxies, or about ring structures. This lack of information is mainly due to the absence of such galaxies in the sampling set, so some study remains in this domain.

Results obtained on a set of galaxies observed with a different telescope (the Schmidt telescope of Observatoire de Haute–Provence) and with different exposure times, show that the knowledge base must contain information about these symbolic parameters: *telescope* and *exposure time*. Different branches of the tree of the prototypes must be scanned as functions of the values of these parameters.

Conclusions

We have presented a complete automatic method for classifying complex objects (galaxies), using an expert systems approach.

Although the knowledge base, presently composed of 37 prototypes and 106 rules, must be completed to enhance the precision, the Hubble classes obtained by this method are quiet good.

In its current state, the system can classify a galaxy in approximately 122 seconds: 120 seconds is needed for the extraction of the parameters (which is of course highly dependent on the size of the image) and 2 seconds for the inference phase.

We are developing another system which will classify images coming from various sources (*e.g.* high or low resolution images). The image processing phase will use a new expert system generator specialized for the monitoring of visual procedures (Thonnat and Clement 1988). The knowledge base of the classification phase will be extended to the irregular Magellanic galaxies and ring–shaped galaxies.

Among the advantages of this methodology, we especially note:

1. Objectivity: we always obtain the same classification for the same data, which is not the case for the human experts.

2. Explicitness of the knowledge: it is very important to be conscious of the knowledge we use and to structure it to fit the problem at hand.

3. Multi–expertise: with this method, it is possible to group several kinds of knowledge or skills (*e.g.* astronomy, optical instruments, and image processing).

4. Explanations: it is possible to check the reasoning step by step. This feature allows easy learning, easy modification, development, and refining of existing systems of classification.

References

Ballard, D. H. and Brown, C. M. 1982, *Computer Vision* (New York: Prentice–Hall).

Bijaoui, A. 1981, *Image et Information* (Paris: Masson).

Brooks, R. A. 1981, *AI J.* **16**, 285.

Chailloux, J., Devin, M., and Hullot, J. M. 1984, "LeLisp, a portable and efficient lisp system," in 5TH *ACM Conference on Lisp and Functional Programming.*

Cipiere, P. 1984, "Manuel de reference du logiciel INRIMAGE," (Valbonne: INRIA).

Granger, C. and de Mongareuil, A. 1987, "The reference manual," *Ilog.*

Granger, C., Thonnat, M., and Vignard, P. 1984, "Etude d'un système expert pour la classification de galaxies: SYGAL," *Journées d'étude sur les systèmes experts et leurs applications.*

Hanson, A. and Riseman, E. 1978, in *Computer Vision Systems*, ed. A. Hanson and E. Riseman (New York: Academic Press), p. 303.

Hayes–Roth, F., Watermanand, D. A., and Lenat, D. B. 1983, in *Building Expert Systems* (Reading, MA: Addison–Wesley).

Hubble, E. P. 1936, *The Realm of Nebulae* (Oxford: Oxford Univ. Press).

Jarvis, J. F. and Tyson, J. A. 1981, *Astron. J.* **86**, 476.

Lauberts, A. and Valentijn, E. A. 1983, *The Messenger*, No. **34**, 10.

Nilsson, N. J. 1982, in *Principles of Artificial Intelligence* (Berlin: Springer Verlag).

Pavlidis, T. 1977, in *Structural pattern recognition* (Berlin: Springer Verlag).

Sandage, A. R. 1961, *The Hubble Atlas of Galaxies* (Washington, D. C.: Carnegie Inst. of Washington).

Sandage, A. R. 1975, in *Galaxies and the Universe*, ed. A. Sandage, M. Sandage, and J. Kristian (Chicago: Univ. of Chicago Press), p. 1.

Thonnat, M. 1985, *Rapport de recherche INRIA*, No. **387**, March.

Thonnat, M. and Clement, V. 1988, "OCAPI: an artificial tool for the automatic selection and control of image processing procedures," in *Third International Workshop on Data Analysis in Astronomy*, in press.

Thonnat, M., Granger, C., and Berthod, M. 1985, "Design of an expert system for object classification through an application to the classification of galaxies," in *Proc. CVPR-85.*

Shortliffe, E. H. 1976, in *Computer–Based Medical Consultation: MYCIN* (New York: American–Elsevier).

de Vaucouleurs, G. 1959, in *Handbuch der Physik* **53**, ed. S. Flügge (Berlin: Springer Verlag), p. 275.

de Vaucouleurs, G. 1976, *Le monde des galaxies* (Besançon: Obs. de Besançon et Lab. d'Astron. Faculté Sci.).

Watanabe, M., Kodaira, K., and Okamura, S. 1982, *Astrophys. J. Suppl.* **50**, 1.

Discussion

J.–C. Pecker: You are applying beautiful methods to data which may be not of a high standard! On the one hand, accurate calibration of plates is essential; if it is bad, you are poorly estimating, either near the threshold or near saturation, the intensities. Hence, something such as the arm–interarm contrast will be wrong. On the other hand, Schmidt plates, even good ones, are affected by poor quality stellar images (from *e.g.* bad seeing) or by effects such as deformation of stellar images by the optics themselves. How can you take these effects into account, or correct for them, if you cannot avoid them?

Thonnat: I completely agree with you about calibration! Fortunately, for the system in progress, Prof. de Vaucouleurs now provides us with well–calibrated images. I must say, though, that even by eye, we are able to do a good classification on our own plates, using criteria such as arm–interarm contrast. Concerning the stars: since the problem that we are attempting to solve is not photometry or the study of the stars themselves, I have not attempted to precisely model these effects of poor stellar images; the stars are only detected and removed. The shape of the galaxy's profiles doesn't seem to be much affected beyond the nuclear regions and the classification is still possible.

D. Burstein: Your images obviously suffered from low signal–to–noise problems. Why not obtain CCD images for galaxies with diameters less than 3 arcmin in size? In addition, can you get CCD images from other observers? (I would estimate there are images for more than 400 galaxies now available).

Thonnat: The first thing that we had to do was to be able to recognize with the machine galaxies as we do by eye and brain, and to put this kind of knowledge in a knowledge base. Among the visual procedures, a few (*e.g.* the image preprocessing procedure) are dependent of the conditions of observation, but most of the high level pattern recognition procedures are more general. So the methodology has to be extended to various kinds of images of galaxies, not just CCD images. So, we are not, at the moment, working with observers who have CCD images, but our methods can obviously be extended to them in the future.

R. Rampazzo: 1) I would like to know technical information about the expert system, in particular the language used, the computer, and the transportability of the system. 2) Why have you preferred to start from scratch, rather than use a development shell already on the market?

Thonnat: 1) The language is Le–Lisp, and our machine is a Sun 3 under an X–window manager. The system should be easily transportable to the MicroVax, the IBM PC, and the Macintosh. 2) We started with a classical production rules system. It didn't work because we not only needed to express our criteria with rules, but also because we needed to describe the classes by objects (frames), and the hierarchy of the classes with semantic nets. We also needed fuzzy filtering and matching. All of these features were not available together with classification reasoning in a commercial package.

H. T. MacGillivray: The emulsion noise properties of photographic plates are a strong function of photographic density. 1) How are your results affected if you use plates of different density ranges? 2) Is your technique also accordingly a function of magnitude of the object on the plate? 3) Is it really possible to apply your technique uniformly from plate to plate?

Thonnat: 1) All the images are normalized between 0 and 1 to avoid these problems. 2) The density range doesn't seem to be critical, but the need to be able to see the morphology by eye in the image *is* critical. That is, resolution is important; we prefer to have an image of at least 100 × 100 pixel size. I would say that 50 × 30 is our present limit, but we have not yet studied galaxies this small. 3) Yes, the method is autoadaptive for all thresholds. We also of course work with a normalized reference level.

J. W. Sulentic: I see two major requirements for future progress in understanding galaxy morphology: 1) development of a set of objective classification criteria and 2) selection of criteria that are independent of our current theories (and biases) about galaxies. Do we need to enter the Fourier domain in order to achieve these goals?

Thonnat: While I agree completely with your two requirements, and while I believe that Fourier analysis may sometimes be useful, I feel that it is too much affected by noise and by resolution (*i.e.* size of the galaxy, kind of telescope, distance of the galaxy, digitizer, ...). I also believe that the Fourier components of an image provide some parameters, but not a sufficient number for complete classification.

M. Capaccioli: One question and one comment. Since resolution depends on distance, all else being equal, do you plan to reduce all galaxies to a standard distance? Then I should like to stress the role of a "conventional" expert system called an "astronomer." There are features of galaxies that cannot be seen either by PDS or by a CCD coupled with powerful software.

Thonnat: We believe that we have a good classification program, but as I am not an observer, we of course need cooperation with observational astronomers for providing the images. Experts are certainly important for discovering new features in the images, or — more importantly — new physical results. But the quantity of data is becoming too large for a single expert to fully integrate. We also need an objective tool, even if an imperfect one, to homogenize the reductions and remove the subjective elements.

Quantitative Classification of Galaxies

S. Okamura, M. Watanabe, and K. Kodaira

Tokyo Astronomical Observatory, University of Tokyo, Japan

Summary

A brief review is given of quantitative classification of galaxies. Results of the major studies based on principal component analysis (PCA) are summarized. Two–dimensionality of normal disk galaxies is confirmed. PCA is performed on one of the most extensive samples of elliptical galaxies available to date. This new, though preliminary, PCA study shows the potential importance of axial ratio as a parameter in a quantitative classification scheme. The role of principal component analysis in the study of quantitative classification of galaxies is discussed.

I. Introduction

It has been a long–standing dream of extragalactic astronomers to construct a sort of "HR diagram of galaxies" with which we can interpret structure, formation, and evolution of galaxies on a physical basis. Development of quantitative classification schemes is an important first step towards this goal.

Early efforts to classify galaxies resulted in a variety of morphological classification schemes based on the appearance of galaxy images on blue sensitive photographic plates. Among them, the system developed by Hubble (1936), and revised by Sandage (1975) and de Vaucouleurs (1959, 1974), is most widely used as a basic scheme.

In his three–dimensional classification scheme, de Vaucouleurs (1959, 1974) arranged galaxies along the longest axis of a classification volume, which he called the stage, according to the Hubble type, E–S0–S–Irr, and the cross section to the axis was used to represent the variety of appearance seen within a stage. De Vaucouleurs (1974) interpreted the stage axis to represent the basic physics and the cross section to represent the dynamical details of galaxies. He introduced the morphological type index T, ranging from $T = -6$ for compact ellipticals and to $T = 11$ for extragalactic H II regions, as a quantitative measure of the stage axis. The type index may be considered as a result of initial efforts toward a quantitative classification.

Many studies showed reasonably good correlations between the type index and basic properties of galaxies such as the shape of the luminosity profile, color index, gas fraction, etc. (see *e.g.* Roberts 1975; de Vaucouleurs 1974, 1977; de Vaucouleurs *et al.* 1976: RC2; Fraser 1977). These correlations demonstrated that the type index is somehow definitely related to the basic physics of galaxies. Thus, the type index was used in a variety of practical applications. For example, de Vaucouleurs (1977) introduced the luminosity index, $\Lambda = (T + L)/10$, based on the type index T and the luminosity class L, which is a quantitative measure of the luminosity class estimate in the morphological classification scheme developed by van den Bergh (1960 a,b). De Vaucouleurs (1977) showed that the correlation between Λ and the absolute magnitude M_{T}^{0} can be approximated by

$$M_{\mathrm{T}}^{0} = -19.5 + 1.40(\Lambda^2 - 1), \tag{1}$$

and later used a revised version of this expression to obtain the distance of galaxies, to map the local velocity field, and to derive the Hubble constant (de Vaucouleurs 1979; de Vaucouleurs and Bollinger 1979).

However, in spite of their great practical importance, the physical interpretation of the correlations between the quantitative measures derived from morphological classifications and other properties of galaxies, (*e.g.* Equation 1), is not straight forward. This is because the morphological type of a galaxy appears to be the product of many controlling factors. Quantitative classification schemes based on simple physical parameters are necessary for a clearer understanding of the basic physics of galaxies. Such parameters have become available only recently with reasonable accuracy, and for a large number of galaxies, due to recent advances in observational technology.

In this paper, we will give a brief review of the quantitative classification of galaxies. The background and motivation for quantitative classification were described in papers written by, among others, Brosche (1973) and de Vaucouleurs (1974, 1977) more than a decade ago when sufficient data were not available yet.

II. Principal Component Analysis and Quantitative Classification

A large number of observed properties of galaxies are mutually correlated. Accordingly, simple correlation studies of observed parameters without knowing which are fundamental would yield confusing, sometimes even misleading results. This is in fact the case we often see when dealing with these correlations. The study of galaxies is still in its infancy (Whitmore 1984), and we do not know yet what are the fundamental parameters. It is therefore essential to synthesize or summarize the observed correlations before a reasonable quantitative classification scheme can be established. This is exactly the role of principal component analysis (hereafter abbreviated PCA).

The method of PCA is described in some detail in the literature (e.g., Brosche 1973; Bujarrabal *et al.* 1981; Watanabe *et al.* 1985). We give here only a brief summary. Let $(x_1, x_2, ..., x_n)$ be the set of n observed parameters normalized for a sample of N galaxies so as to have a mean of zero and a variance of one, *i.e.* $< x_i >= 0$, and $\sigma_{x_i} = 1$ for $i = 1, 2, ..., n$. We obtain a new set of parameters, $(\xi_1, \xi_2, ..., \xi_n)$, by the linear transformation of $(x_1, x_2, ..., x_n)$ with constraints that (1) the ξ_i's are mutually independent, and (2) the ξ_i's are in order of the variances they carry, *i.e.* $\lambda_1 (= \sigma_{\xi_i}) > \lambda_2 > ... > \lambda_n$. The new variables ξ_i are called the principal components. The variances λ_i can be computed as the eigen values of the correlation matrix of $(x_1, x_2, ..., x_n)$. We can determine by PCA the dimensionality of the sample, *i.e.* how many parameters are necessary and sufficient to describe the sample.

However, PCA has several limitations. First, PCA assumes linear correlations. Second, the judgment of dimensionality is partially subjective although several criteria for judgment are proposed. Widely used criteria are the following: the dimensionality of the sample is $m(< n)$ (1) if the scatter of x_i around the relation that includes the first m principal components is comparable to the observational error, (2) if $\lambda_m > 1$ and $\lambda_{m+1} < 1$, and (3) if $\sum_{i=1}^{m} \lambda_m / n$ exceeds a certain threshold value, for example, 0.8. Third, parameter selection is crucial. Naturally, PCA works only on the parameters included in the analysis. If we omit important parameters, PCA may yield a misleading result. Fourth, selection effects and distance errors could significantly affect the result of PCA in astronomical applications. Finally, PCA is not an analytical method. This may be the most important limitation we should keep in mind in the application of PCA to astronomical problems. A textbook of statistics says that "PCA is a synthetic method to summarize the characteristics represented by mutually correlated multiple variables in order to gain useful insights to a given purpose," and that "results of PCA should be analyzed using proper knowledge of the field of study concerned." Suppose that PCA is applied to the quality control of electric parts produced by a factory using several properties of the product such as size, weight, hardness, and resistance as input parameters. If the factory has trouble in the production line that leads to poor quality, PCA could identify the machine causing the trouble, but it cannot tell us why the machine got into trouble. Because of these limitations, the results from PCA applied to an astronomical problem should be confirmed using (1) different samples and (2) different sets of parameters. We are now in the position where this confirmation can be studied.

Our purpose in applying PCA to a sample of galaxies is to know how many factors control the properties of galaxies. If we know this, a reasonable quantitative classification scheme based on a proper set of parameters can be constructed. Such a classification scheme will be a basis (but a basis only) for the understanding of structure, formation, and evolution of galaxies in terms of physics. As we have seen from the preceding discussion, any quantitative classification scheme based on PCA does not tell us much by itself without interpretations based on theories. In this context, parameters which can be compared directly with theoretical predictions are preferable for use in the classification. Another important aspect of such a classification scheme is that it will eventually allow us to find a good correlation between distance dependent parameters and distance independent parameters which can be used as a better distance estimating tool than we have now.

III. Disk Galaxies

1. Summary of Previous Studies

Based on extensive correlation studies among five observed parameters, the Nançay group (Balkowski 1973; Balkowski *et al.* 1974) pointed out the important role of morphological type and luminosity, and proposed a two-dimensional classification scheme for spiral galaxies. Following the pioneering study by Brosche (1973), PCA was applied to several different samples by different investigators. The results of these studies are summarized in Table 1 (see also Brosche and Lentes 1988 and Whitmore 1988). Table 1 shows the remarkable consistency in the relative contributions of the first three principal components to the total variance. The first, second, and third components carry roughly 60, 30, and 10 percent of the total variance, respectively. This is in fact remarkably good consistency if we consider the large differences in the samples and parameters used. Figure 1 shows the orientation of vectors corresponding to the original parameters projected onto the (ξ_1, ξ_2) plane adopted from Bujarrabal *et al.* (1981). Two clusterings of vectors are seen in Figure 1, which run almost perpendicular to each other. One consists of luminosity (L), size (A), neutral hydrogen mass ($M_{\rm H}$), and indicative mass ($M_{\rm i}$), and the other consists of color index (C) and morphological type index (T). This feature is also common to all the other studies, although a rigorous comparison of the results based on different samples would need an affine transformation (Brosche and Lentes 1988). Thus, all the studies based on PCA appear to have reached the same conclusion, which can be summarized as follows. There are two dominant axes in the manifold of normal disk galaxies. The first axis explains about 60% of total variance and has large correlation coefficients with the parameters related to the size or

Table 1. Summary of PCA for disk galaxies

Sample and variables*	Eigen Values % (cumulative %)		
Brosche (A.&Ap.,23,259,1973)	λ_1	λ_2	λ_3
(T, L_B, R, C_{BV}, R_m, V_m)			
N=31 (S0-Im)	3.225 54(54)	1.732 29(83)	0.560 9(92)
Bujarrabal, Guibert, Balkowski (A.&Ap.,104,1,1981)			
(T, L_B, R, C_{BV}, M_H, M_i)			
N=109 (Balkowski; S0-Im)	3.84 64(64)	1.37 23(87)	0.34 6(93)
N=100 (Shostak; Sbc-Im)	4.21 70(70)	1.05 18(88)	0.47 8(96)
N=90 (Dickel & Rood; S0/a-Im)	3.85 64(64)	1.35 23(87)	0.33 6(93)
(T, L_B, R, M_H, M_i)			
N=27 (Virgo; Sb-Sdm)	3.28 66(66)	1.01 20(86)	0.44 9(95)
Whitmore (Ap.J.,278,61,1984)			
(T, L_B, R, C_{BH}, V_m, B/T)			
N=51 (RFTB; Sa-Sc)	2.98 50(50)	2.04 34(84)	0.39 6(90)
(T, L_B, R, C_{BH}, V_m)			
N=51 (RFTB; Sa-Sc)	2.67 53(53)	1.70 34(87)	0.30 6(93)
N=103 (Aaronson; Sa-Im)	2.83 57(57)	1.35 27(84)	0.57 11(95)
Watanabe, Kodaira, Okamura (Ap.J.,292,72,1985)			
(L_V, R, SB, X1P)			
N=151 (Virgo+UMa; S0-Im)	2.441 61(61)	1.258 31(92)	0.301 8(100)

* T: morphological type L: luminosity or magnitude
C: color index R: photometric radius or diameter
M_H: hydrogen mass M_i: indicative total mass
V_m: maximum rotational velocity R_m: radius where $V=V_m$
B/T: bulge-to-total luminosity ratio
SB: mean surface brightness X1P: luminosity concentration index

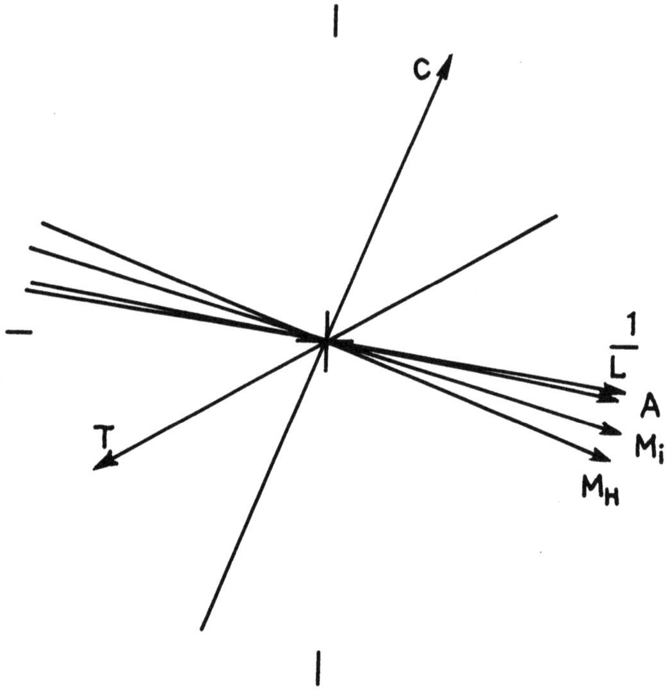

Figure 1: Vector orientations projected onto the (ξ_1, ξ_2) plane (Shostak sample in Bujarrabal *et al.* 1981).

the scale length of galaxies. The second axis explains about 30% of the total variance and has large correlation coefficients with the parameters related to the aspect or the form of galaxies. The significance of the third dimension is unclear at the moment.

2. Further Consistency Checks

In order to demonstrate further that the results of different PCA studies of disk galaxies are really consistent, we perform here a detailed intercomparison of the result by Whitmore (1984; hereafter W84) with that by Watanabe *et al.* (1985; hereafter WKO).

W84 performed PCA on a sample compiled by Rubin *et al.* (1980, 1982, 1985) using two photometric parameters (L_B, R) and one dynamical (V_m) parameter, and two other parameters related to stellar population (C_{BH}) and structure (T) (see footnote of Table 1 for explanation of the parameters). Based on the result of his PCA, W84 proposed a tentative classification scheme based on two parameters, *i.e.* scale (S) and form (F). These were defined by

$$S = m - 5\log R(\text{arcsec}) - 10\log r(\text{pc}) + 31.57, \quad (2)$$

and

$$F = 100(B - H) + 100\log B/T, \quad (3)$$

where m is the apparent magnitude, R the apparent size, D the distance, $B-H$ the blue–minus–infrared color, and B/T is the bulge–to–total light ratio. On the other hand, WKO performed PCA on a sample of galaxies in the Virgo and UMa clusters using four photometric parameters in a single color band. Based on their result, they proposed a tentative classification scheme based on the diameter and the mean surface brightness (Kodaira *et al.* 1983).

Now we classify WKO sample galaxies by the scheme proposed by W84. The resulting S–F diagram is given in Figure 2. A distance of 16.7 Mpc is assumed for the Virgo and UMa clusters. In computing F according to equation (3), we replaced $(B-H)$ with $(B-V)_T$ given in RC2 and we used B/T given in Kodaira *et al.* (1986). This S–F diagram does not look like the diameter (D) *versus* surface brightness (SB) diagram shown in Figure 3. At first sight, this is disappointing. However, we find that the diameter (D) *versus* concentration index $X1(P)$ diagram shown in Figure 4 is very similar to the S–F diagram. In fact, a careful examination of the two diagrams shows that their topological structure is almost identical. As readily seen in the vector plot from PCA by WKO, which is repeated here in Figure 5, the D–SB and the D–$X1(P)$ diagram are just two different representations of the same (ξ_1, ξ_2) plane, if we neglect the contribution of ξ_3. Accordingly, the similarity between the S–F diagram (Figure 2) and the D–$X1(P)$ diagram (Figure 4) can be considered as strong evidence that W84 and WKO reached the same fundamental plane using completely different samples and largely different sets of parameters.

The conclusions of this intercomparison can be summarized as follows. (1) Photometric parameters in a single color band may be sufficient to describe a family of normal disk galaxies. In other words, there must be strong correlations between photometric parameters and other parameters including dynamical ones. (2) Although there appears to be a unique fundamental plane in the manifold of normal disk galaxies, descriptions of the plane in terms of observed parameters are highly diverse. Any pair of parameters that are nearly orthogonal in the plane could describe the plane reasonably well. Accordingly, "theories of galaxy formation must explain all the correlations simultaneously rather than just one or two at a time" as stressed by W84. (3) Since many studies based on PCA yield the same results, further new findings will not be obtained by PCA unless new parameters are included. As far as the parameters included so far in PCA are concerned, roughly 90% of correlations have been summarized in the unique fundamental plane. This plane may already provide a strong challenge to theories of galaxy formation.

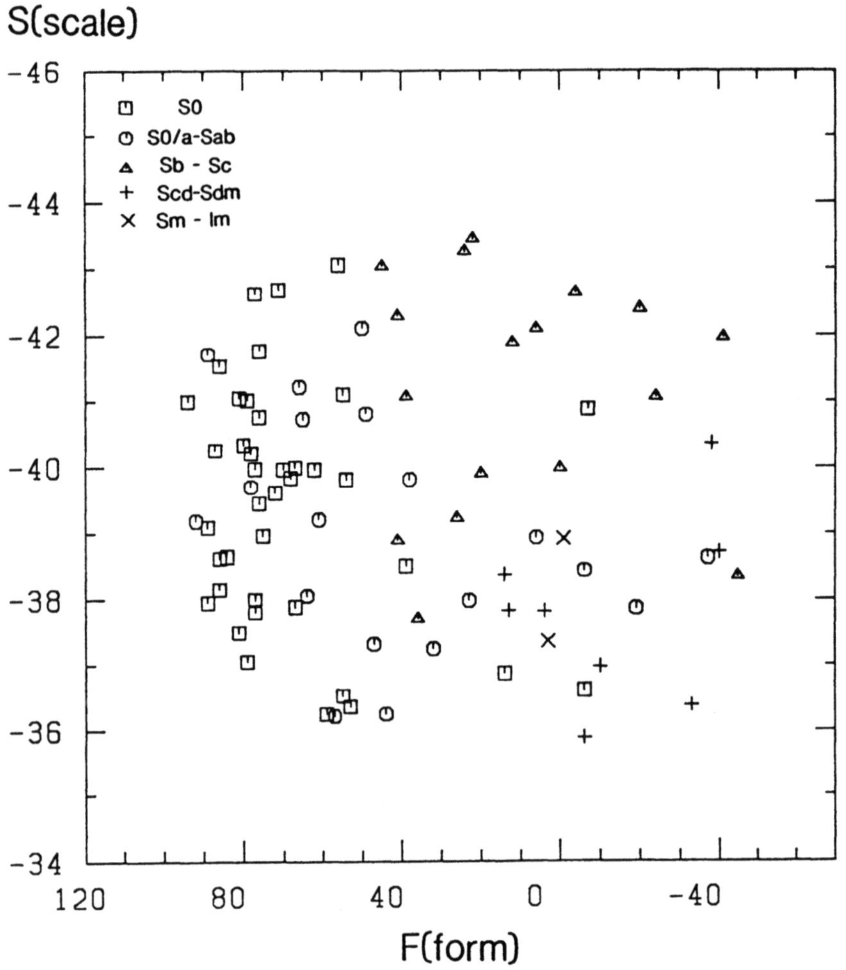

Figure 2: Scale *versus* form diagram proposed by Whitmore (1984) applied to disk galaxies in the Watanabe *et al.* (1985) sample of Virgo and Ursa Major galaxies.

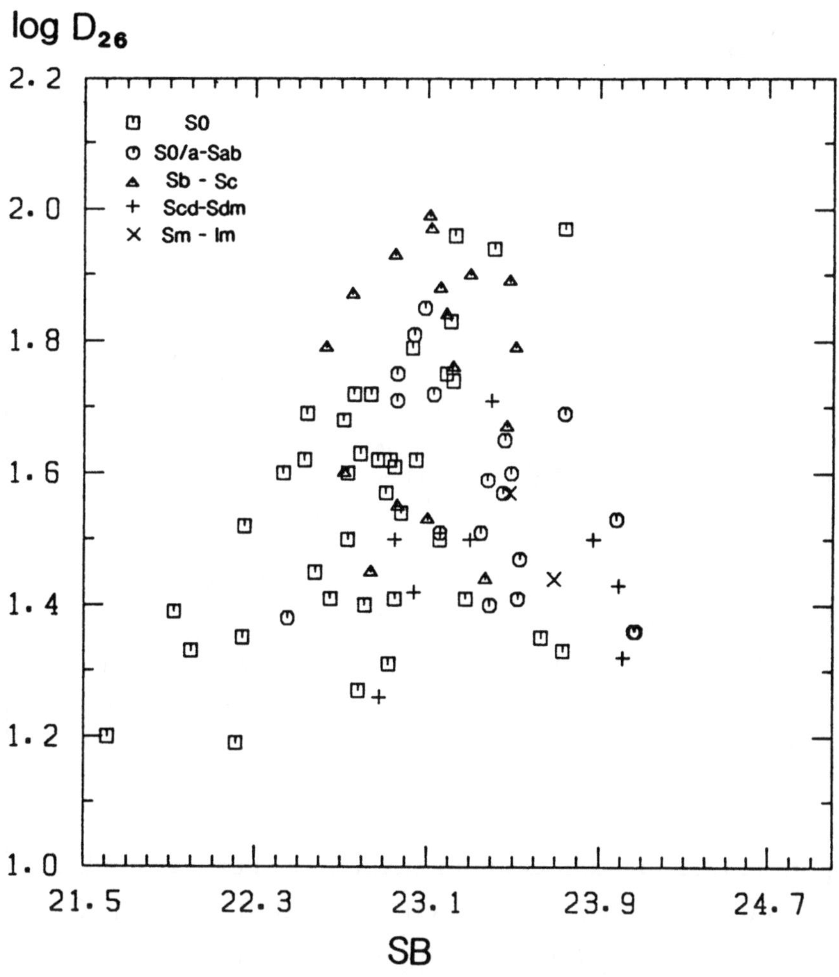

Figure 3: Diameter *versus* surface brightness diagram proposed by Kodaira *et al.* (1983) applied to the galaxies plotted in Fig.2.

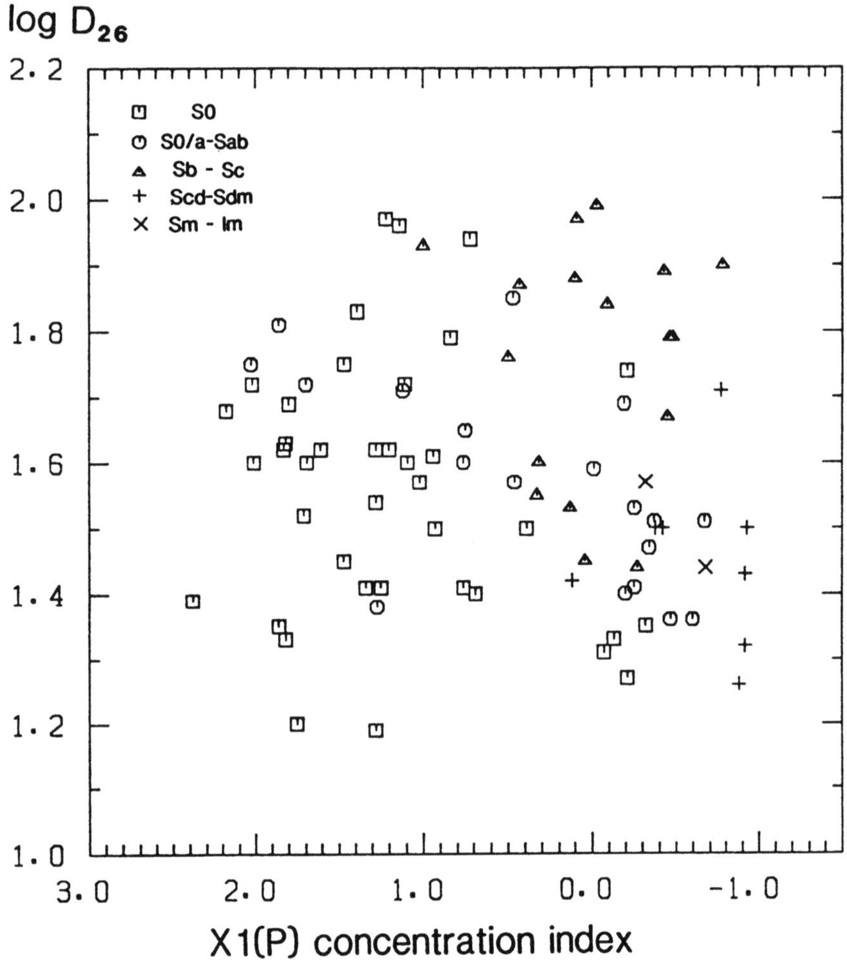

Figure 4: Diameter *versus* concentration index diagram for the disk galaxies plotted in Figure 2. Note the similarity of this diagram to Figure 2.

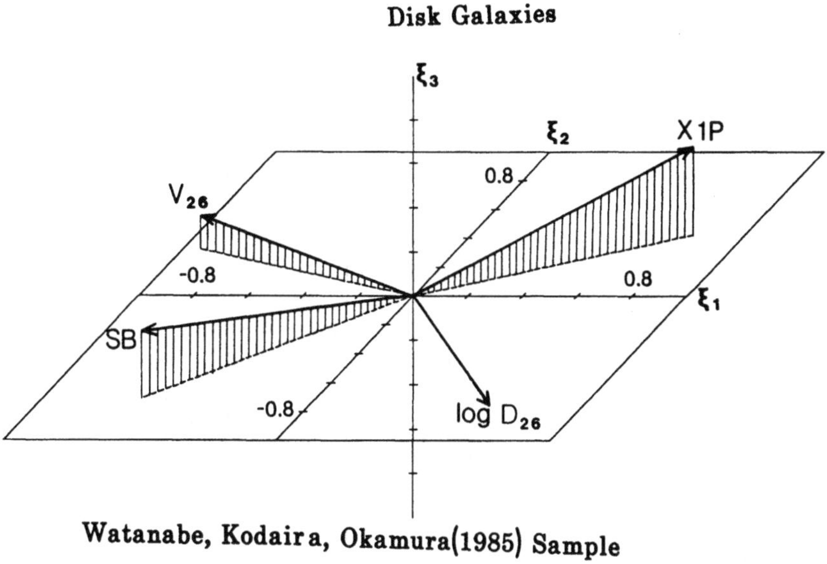

Figure 5: Configuration of unit vectors of parameters in the space of the principal components (Watanabe *et al.* 1985).

IV. Elliptical Galaxies

1. Correlation Studies

Until about 1980, elliptical galaxies were believed to form a one-parameter family because of the presence of tight correlations between the basic observed parameters such as luminosity, line strength, color, and central stellar velocity dispersion (*e.g.* Faber 1973; Faber and Jackson 1976; Visvanathan and Sandage 1977). The total luminosity, hence probably the total mass, is believed to be the most fundamental parameter of ellipticals.

However, Terlevich *et al.* (1981) made a careful analysis of the residuals in the L–σ relation and in the L–Mg_2 relation for a sample of 24 ellipticals, where L, σ, and Mg_2 are the luminosity, central velocity dispersion, and line strength, respectively. They found that at a fixed L, galaxies with larger σ tend to have stronger Mg_2. They interpreted this as due to a "second parameter" and concluded that ellipticals are at least a two-parameter family. They further suggested that intrinsic axial ratio is the second parameter because the residuals in the above relations were correlated with apparent axial ratio ε.

Tonry and Davis (1981) confirmed the two-parameter family based on an analysis of six parameters for a larger sample. However, they found no correlation between ε and any other variables. Based on the data for 97 ellipticals in six clusters Dressler *et al.* (1987) concluded that the surface brightness (μ) is the second parameter because the introduction of μ into the so called Faber–Jackson relation between L and σ improves the relation significantly. The important role of μ was also pointed out by de Vaucouleurs and Olson (1982). Djorgovski and Davis (1987) reached the same conclusion as Dressler *et al.* (1987) based on a different sample. They found that the manifold of ellipticals in the space of observed parameters forms a thin plane which is described best by σ and μ. The axial ratio ε and other parameters related to the shape showed no correlation with the plane. Accordingly, they counted σ and μ as two basic parameters and shape parameter(s) as potentially important additional parameters, reaching the conclusion that ellipticals are a "$2+N$"–parameter family.

As discussed in section II, the interpretation of the results of these correlation studies is not straightforward. In particular, the role of the "second parameter" appears to be unclear. In the following, we summarize the results of previous PCA studies and carry out PCA for an extensive sample in order to clarify the situation.

2. PCA Studies

Results of the previous studies of ellipticals based on PCA are summarized in Table 2 in the same format as in Table 1. Although the consistency among different studies appears to be poorer than in the case of disk galaxies, there is one dominant dimension carrying roughly 80% of total variance and a secondary much weaker dimension. However, the sample sizes of the previous studies are rather small.

Extensive sets of observed data have become available recently for large numbers of ellipticals (Djorgovski and Davis 1987; Burstein *et al.* 1987). We perform PCA on the Djorgovski and Davis sample. The results are given in the last part of Table 2, and the vector plots are shown in Figure 6. When we include four parameters in the PCA (R–band magnitude (M_R), effective radius (R), σ, and μ), the first principal component carries as much as 88% of the total variance. This means that all the four parameters are strongly correlated and that the manifold is almost a line rather than a plane (in the normalized sense). The line represents the Faber–Jackson relation or it may correspond to Kormendy's (1977) μ_e–r_e relation in terms of photometric parameters (cf. WKO). However, the secondary dimension is still significant because it is known that introduction of μ into Faber–Jackson relation reduces the scatter. When a shape parameter, apparent axial ratio ε, is included in the PCA in addition to the above four, shares of the first and the second principal components become 70% and 20%, respectively. An examination of Figure 6 reveals that the (ξ_1, ξ_2) plane is totally different between the two cases. Thus, ε is the parameter responsible for the new secondary dimension. It should be kept in mind, however, that ε is the apparent axial ratio subject to the projection effect. It is necessary to take the projection effect into account quantitatively for the confirmation of the signifi-

Table 2. Summary of PCA for elliptical galaxies

Sample and variables*	Eigen Values % (cumulative %)		
Efstathiou & Fall (M.N.,206,453,1984)	λ_1	λ_2	λ_3
(L_B, σ, Mg_2)			
N=37 (TDFB + DEFIS)	2.50 83(83)	0.398 13(96)	0.103 3(100)
(L_B, σ, W)			
N=53 (Tonry & Davis)	2.15 72(72)	0.605 20(92)	0.244 8(100)
Watanabe, Kodaira, Okamura (Ap.J.,292,72,1985)			
(L_V, R, SB, X1P)			
N=18 (Virgo+UMa)	3.009 75(75)	0.886 22(97)	0.105 3(100)
Vader (Ap.J.,306,390,1986)			
(L_B, σ, Mg_2)			
N=24 (Virgo+Coma)	2.604 87(87)	0.259 9(96)	0.142 4(100)
N=23 (Field)	2.328 78(78)	0.571 19(97)	0.100 3(100)
Okamura, Kodaira, Watanabe (this study)			
(L_R, R, μ, σ)			
N=106 (Djorgovski & Davis)	3.537 88(88)	0.301 8(96)	0.140 4(100)
(L_R, R, μ, σ, ε)			
N=106 (Djorgovski & Davis)	3.597 72(72)	0.948 19(91)	0.295 6(97)

* L: luminosity or magnitude
Mg_2: metallicity index
W: metallicity (equivalent width)
X1P: luminosity concentration index
ε: axial ratio
σ: central velocity dispersion
R: radius or diameter
SB,μ: surface brightness

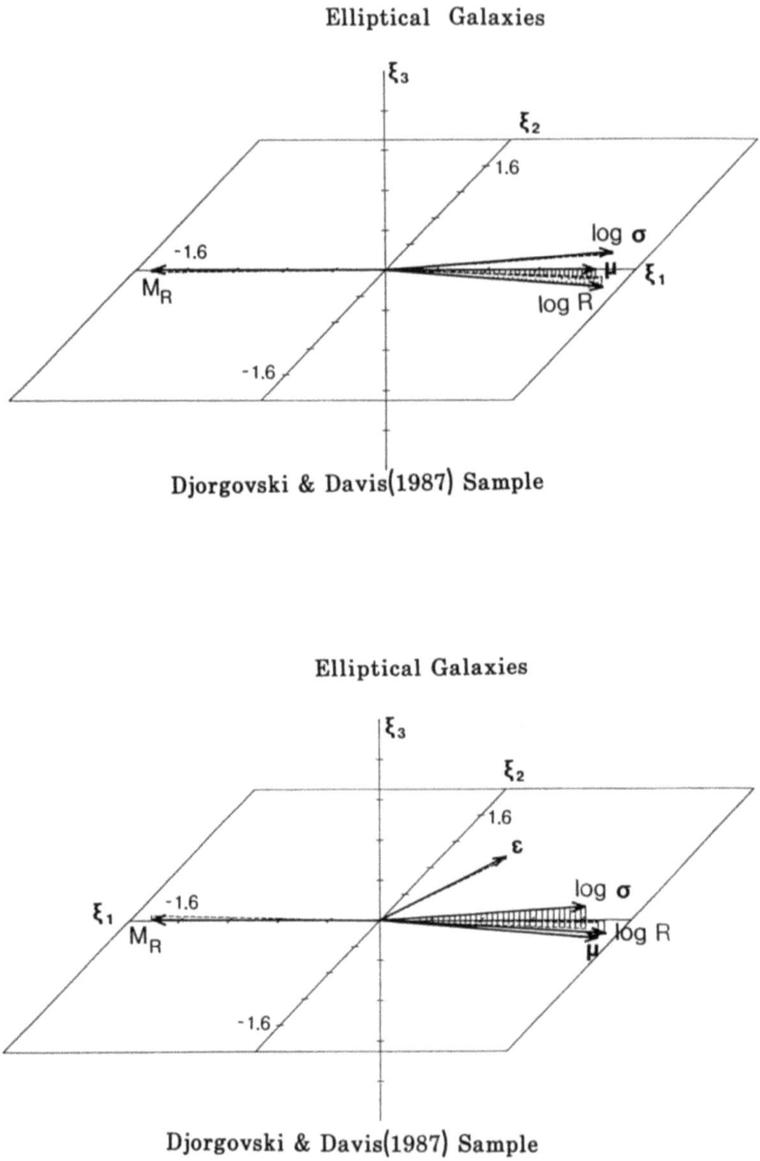

Figure 6: Configuration of unit vectors of parameters in principal component space. PCA is performed on a sample of 106 elliptical galaxies in Djorgovski and Davis (1987) based on four parameters (top) and five parameters (bottom).

cance of the new secondary dimension (cf. Brosche and Lentes 1988). This has not been done in our preliminary analysis.

The suggestions we obtain from the preliminary PCA performed on one of the most extensive samples of ellipticals available to date can be summarized as follows. There are two dominant axes and a third minor (but still significant) axis in the manifold of elliptical galaxies. They carry roughly 70%, 20%, and 5% of the total variance, respectively. The first axis is closely related to the scale length of elliptical galaxies and may represent the Faber–Jackson relation. The second axis is related to the shape of galaxies and the third axis is responsible for the scatter around the Faber–Jackson relation.

The Faber–Jackson relation between L and σ has been used as a distance indicator. The term "second parameter" is often used to mean a parameter other than σ which can improve the Faber–Jackson relation. However, an examination of Figure 6 suggests the following. If we follow the same philosophy as for disk galaxies, ε is the second parameter in terms of quantitative classification, the first parameter being any of $M_{\rm R}$, σ, μ, and R. The surface brightness μ can be used to improve the already tight correlations between any pair taken from $M_{\rm R}$, σ, and R.

These suggestions should be confirmed at least by another independent PCA study based on a different sample of similar size. Figure 7 shows the distribution of ellipticals in Djorgovski and Davis (1987) sample in $M_{\rm R}$–ε plane, which is the most reasonable classification scheme according to our preliminary PCA.

V. Problems and Future Prospects

One of the most serious concerns about the structure of galaxies is the role of dark matter. It is generally postulated that dark halos are present in individual galaxies. Compilation of structural parameters of dark halos is in progress (*e.g.* Carignan and Freeman 1985; Freeman 1987; Kent 1987a,b; Kormendy 1987). These data are still two few and often too ambiguous to be included in PCA at the moment. We note, however, that if dark matter is strongly coupled with luminous matter as suggested by, for example, Bahcall and Casertano (1985), the parameters of dark halos will not provide much new information. If this is the case, a considerable part of the various properties of galaxies has been already summarized by existing PCA studies.

It is, however, premature to say that we have included in PCA all the important parameters describing galaxies. For example, parameters related to recent star formation activity and those describing the nuclear activity have not been analyzed. Little is known about the relationship between these activities and the basic structural and dynamical properties of galaxies (*e.g.* Hunter and Gallagher 1986). PCA studies have put the primary emphasis so far on the latter. It is also expected that near infrared surface photometry may reveal yet unknown basic properties of galaxies (*cf.* Okamura 1988). PCA studies based on

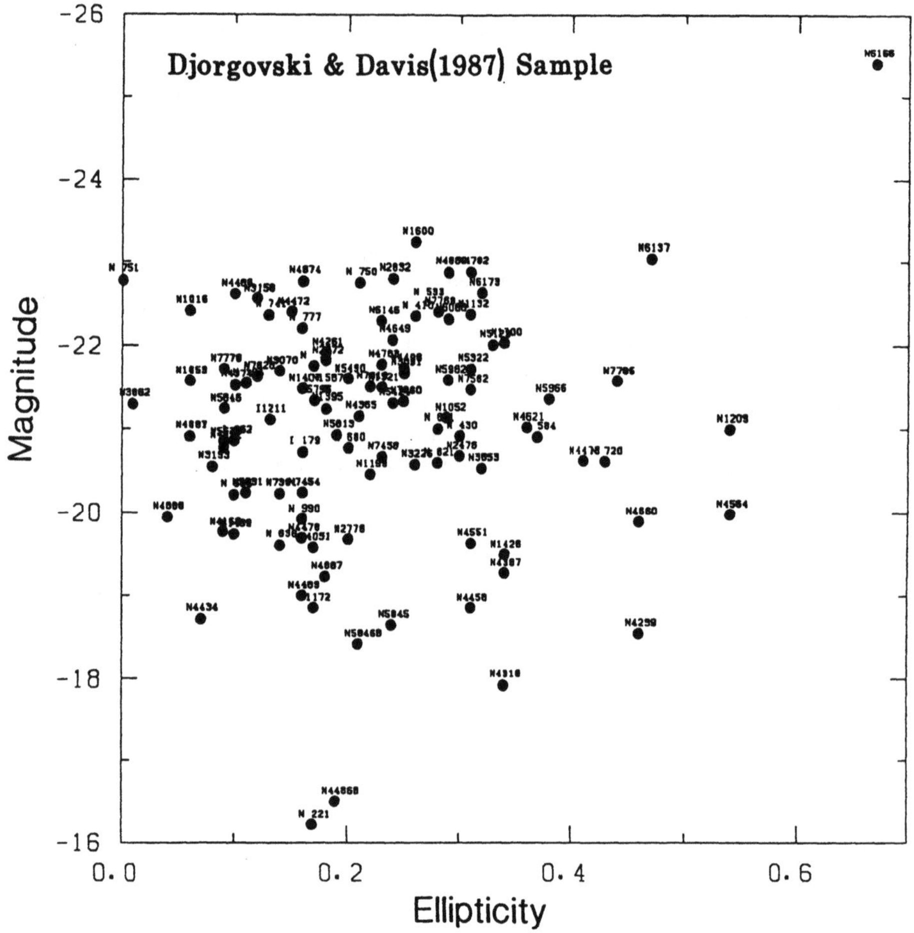

Figure 7: Magnitude (M_{R}) *versus* ellipticity (1–ε) diagram for 106 elliptical galaxies in Djorgovski and Davis (1987) sample.

structural parameters derived from near infrared surface photometry may bring new findings in the near future.

We also have the problem of how to combine data for disk galaxies with those for elliptical galaxies in terms of PCA, since the same parameters are not always available for both samples. WKO performed PCA on a composite sample consisting of 18 ellipticals and $\sim$ 180 disk galaxies using photometric parameters only. They found two-dimensionality ($\lambda_1 = 59\%$ and $\lambda_2 = 34\%$) for this sample. However, the implications of this result are not very clear. Concerning the selection effects mentioned in Section II, the importance of PCA studies for extensive samples of dwarf galaxies should be noted. To our knowledge, PCA has never been applied to dwarf galaxies. Data for them are known to show different behavior from those for giant galaxies in observed parameter space (*e.g.* Ichikawa *et al.* 1986)

Finally, we would like to stress the importance of interactions between a quantitative classification scheme and theories of galaxy formation and evolution, since as already mentioned, any quantitative classification scheme does not tell us much by itself. An interpretation of the diameter *versus* surface brightness diagram of spheroidal stellar systems by Yoshii and Arimoto (1987) is an example of such interactions.

VI. Conclusions

Conclusions of the present study can be summarized as follows.

1. PCA has been succesfully applied to samples of disk and elliptical galaxies.

2. In the case of normal disk galaxies, the existence of two dominant dimensions appears to be established. They can be interpreted as scale length (~60%) and form (~30%). The significance of a possible third dimension is unclear at the moment.

3. Based on a preliminary PCA performed on an extensive sample of elliptical galaxies, we have identified two dominant dimensions. Their tentative interpretations are scale length (72%) and shape (19%). A third dimension (6%) also seems to be significant.

4. Given that a sample of galaxies exhibits two-dimensionality, its descriptions in terms of observational parameters are highly diverse. Thus, as stressed by Whitmore (1984), theories of galaxy formation must explain all the correlations simultaneously rather than just one or two at a time. Any pair of parameters that are not significantly correlated could be used as a basis of quantitative classification. PCA is useful in searching for such a reasonable pair. However, the role of PCA, as a statistical method of synthesizing the correlations, ends at this point. Identification of the

fundamental parameters responsible for the two dimensions is the role of theories of galaxy formation and evolution.

We thank M. Hamabe for his help in preparing the manuscript in the machine readable form. One of us (S.O.) is grateful to P.Brosche for enlightening discussions about PCA, especially for making him aware of the difficulty in comparing PCA results based on different samples. He also thanks R. D. Davies, D. Burstein, and R. L. Davies for stimulating discussions, and the Organizing Committee for invitation and financial support. This work is supported by a Grant in Aid from the Ministry of Education, Science, and Culture (No.59065002).

References

Bahcall, J. N. and Casertano, S. 1985, *Astrophys. J. Lett.* **293**, L7.

Balkowski, C. 1973, *Astron. Astrophys.* **29**, 43.

Balkowski, C., Bottinelli, L., Chamaraux, P., Gonguenheim, L., and Heidmann, J. 1974, *Astron. Astrophys.* **34**, 43.

van den Bergh, S. 1960a, *Astrophys. J.* **131**, 215.

van den Bergh, S. 1960b, *Astrophys. J.* **131**, 558.

Brosche, P. 1973, *Astron. Astrophys.* **23**, 259.

Brosche, P. and Lentes, F. Th. 1989, this conference.

Bujarrabal, V., Guibert, J., and Balkowski, C. 1981, *Astron. Astrophys.* **104**, 1.

Burstein, D., Davies, R. L., Dressler, A., Faber, S. M., Stone, R. P. S., Lynden-Bell, D., Televich, R. J., and Wegner, G. 1987, *Astrophys. J. Suppl.* **64**, 601.

Carignan, C. and Freeman, K. C. 1985, *Astrophys. J.* **294**, 494.

Djorgovski, S. and Davis, M. 1987, *Astrophys. J.* **313**, 59.

Dressler, A., Faber, S. M., Burstein, D., Davies, R. L., Lynden-Bell, D., Terlevich, R. J., and Wegner, G. 1987, *Astrophys. J. Lett.* **313**, L37.

Faber, S. M. 1973, *Astrophys. J.* **179**, 731.

Faber, S. M. and Jackson, R. E. 1976, *Astrophys. J.* **204**, 668.

Fraser, C. W. 1977, *Astron. Astrophys. Suppl.* **29**, 161.

Freeman, K. C. 1987, in *I. A. U. Symp. 117, Dark Matter in the Universe*, ed. J. Kormendy and G. R. Knapp (Dordrecht: Reidel), p. 119.

Hubble, E. 1936, *The Realm of the Nebulae* (New Haven: Yale Univ. Press).

Hunter, D. A. and Gallagher, J. S. 1986, *Publ. Astron. Soc. Pac.* **98**, 5.

Ichikawa, S., Wakamatsu, K., and Okamura, S. 1986, *Astrophys. J. Suppl.* **60**, 475.

Kent, S. M. 1987a, *Astron. J.* **93**, 816.

Kent, S. M. 1987b, *Astron. J.* **94**, 306.

Kodaira, K., Okamura, S., and Watanabe, M. 1983, *Astrophys. J. Lett.* **274**, L49.

Kodaira, K., Watanabe, M., and Okamura, S. 1986, *Astrophys. J. Suppl.* **62**, 703.

Kormendy, J. 1977, *Astrophys. J.* **218**, 333.

Okamura, S. 1988, *Publ. Astron. Soc. Pac.* **100**, 524.

Roberts, M. S. 1975, in *Galaxies and the Universe*, ed. A. Sandage, M. Sandage, and J. Kristian (Chicago: Univ. Chicago Press), p. 309.

Rubin, V. C., Ford, W. K., and Thonnard, N. 1980, *Astrophys. J.* **238**, 471.

Rubin, V. C., Ford, W. K., and Thonnard, N. 1982, *Astrophys. J.* **261**, 439.

Rubin, V. C., Burstein, D., Ford, W. K., and Thonnard, N. 1985, *Astrophys. J.* **289**, 81.

Sandage, A. 1975, in *Galaxies and the Universe*, ed. A. Sandage, M. Sandage, and J. Kristian (Chicago: Univ. Chicago Press), p. 1.

Terlevich, R., Davies, R. L., Faber, S. M., and Burstein, D. 1981, *Mon. Not. Roy. Astron. Soc.* **196**, 381.

Tonry, J. L. and Davis, M. 1981, *Astrophys. J.* **246**, 680.

de Vaucouleurs, G. 1959, *Handbüch der Physik*, **53**, 311.

de Vaucouleurs, G. 1974, in *I. A. U. Symp. 58, The Formation and Dynamics of Galaxies*, ed. J. R. Shakeshaft (Dordrecht: Reidel), p. 1.

de Vaucouleurs, G. 1977, in *the Evolution of Galaxies and Stellar Populations*, ed. B. M. Tinsley and R. B. Larson (New Haven: Yale Univ. Obs). p. 43.

de Vaucouleurs, G. 1979, *Astrophys. J.* **227**, 380.

de Vaucouleurs, G. and Bollinger, G. 1979, *Astrophys. J.* **233**, 433.

de Vaucouleurs, G., de Vaucouleurs, A., and Corwin, H.G. 1976, *Second Reference Catalogue of Bright Galaxies*, (Austin: Univ. Texas Press) (RC2).

de Vaucouleurs, G. and Olson, D. W. 1982, *Astrophys. J.* **256**, 346.

Visvanathan, N. and Sandage, A. 1977, *Astrophys. J.* **216**, 214.

Watanabe, M., Kodaira, K., and Okamura, S. 1985, *Astrophys. J.* **292**, 72 (WKO).

Whitmore, B. C. 1984, *Astrophys. J.* **278**, 61 (W84).

Whitmore, B. C. 1989, this volume.

Yoshii, Y. and Arimoto, N. 1987, *Astron. Astrophys.* **188**, 13.

Discussion

R. D. Davies: Lister Staveley–Smith and I (*Mon. Not. Roy. Astron. Soc.* **231**, 833, 1988) have addressed the question of the number of independent dimensions in a classification system of a given morphological type — in our case Sb, Sbc, and Sc galaxies. The dominant dimension is *size*, measured by mass, luminosity, diameter, etc. while the second is what we call *quiescent star formation* measured by colour, $(B-V)$ and $(B-H)$. Because of the clarity of the correlation in a single morphological data set, we are able to identify a third dimension which we call *embedded activity*, which is a combination of far IR 60 to 100 μm colour and the strength of the bar. This third dimension represents recent star formation in normal galaxies which is embedded in dust, in contrast to the ongoing star formation represented by the optical colours.

S. Okamura: I can imagine that. In most PCA studies carried out so far, the primary emphasis is put on the parameters that describe basic structure of galaxies. To my knowledge, your work is the first that included in PCA the parameters that come from IRAS survey and are related to current star formation embedded in dust. Inclusion of such parameters would reveal another dimension since no clear relationship has yet been found between star forming activity and the basic structure of galaxies.

P. Brosche: The distance problem can be dealt with by using distance independent quantities only (the result is also then always two dimensions). If linearity becomes a problem, the second Hermitian polynomial of a quantity could be used as a new parameter. The term "fundamental parameter" should be reserved for a theory.

S. Okamura: Thank you for your comments. Yes, some of the limitations of PCA that I mentioned may not be very crucial. I agree with you in that PCA studies cannot identify "fundamental parameters" by themselves and that they can only give information about which observable parameters can "describe" a family of galaxies very well.

R. L. Davies: Can you identify the difference between the analysis that you have carried out and that of Djorgovski and Davis, that causes you to find a dependence on ε whereas they did not?

S. Okamura: My talk might have been slightly misleading. I was not saying that ε correlates with other variables. The lack of any significant correlation makes ε the good second parameter for use in a two dimensional quantitative classification scheme, the first parameter being any of $M_{\rm R}$, σ, μ, and R.

An Objective Classification System for Spiral Galaxies and Its Relationship to the [N II]/[S II] Ratio

Bradley C. Whitmore and Duncan A. Forbes

Space Telescope Science Institute, Baltimore, Maryland, U.S.A.

1. An Objective Classification System for Spiral Galaxies

In a poorly understood physical system such as a spiral galaxy, it is not generally obvious what the most important observational parameters are. As observational techniques improve, and the number of different types of measurements increase, the apparent complexity of the problem also grows. However, it is possible that the n different observational properties are all controlled by only m fundamental parameters, with $m \ll n$. The parameter space would then be m-dimensional, so that only the m fundamental properties need to be known to completely describe the system.

Following the pioneering work of Brosche (1973), and the study of Bujarrabel, Guibert, and Balkowski (1981), we compiled a database which contained as many reliable observational properties as available for the Rubin *et al.* (1985) sample of 60 spiral galaxies with extended Hα emission-line rotation curves. A principal component analysis was performed on this database to determine its dimensionality. Our results are published in Whitmore (1984), and are briefly described below.

Figure 1 shows the "correlation vector" diagram for six of the observational variables in our sample that show the strongest correlations (see the figure caption for the definitions of these variables). The Hubble type is also included

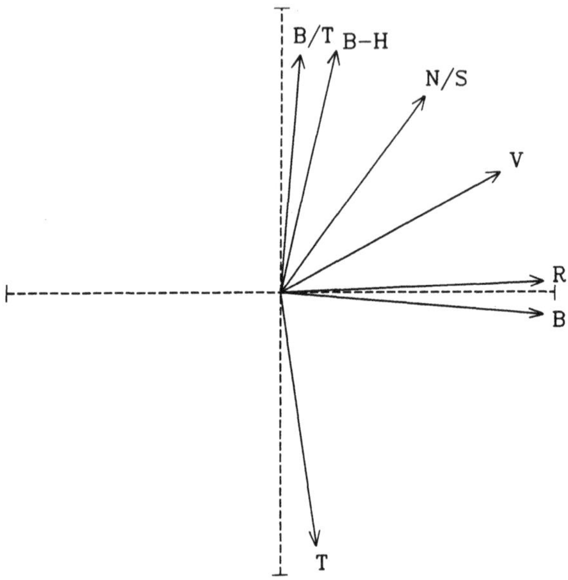

Figure 1: Correlation vector diagram (rotated) including the log of the blue luminosity (B, $\sigma = 0.451$), log of the radius at the 25th B mag arcsec^{-2} isophote (R, $\sigma = 0.300$), log of the maximum rotational velocity (V, $\sigma = 0.132$), qualitative estimate of the emission–line ratio [N II]/[S II] (N/S, $\sigma = 1.23$), the $B - H$ color ($B - H$, $\sigma = 0.614$), the log of the bulge–to–total light ratio (B/T, $\sigma = 0.378$), and the Hubble type (T, $\sigma = 1.58$). Vectors which are nearly aligned have high correlations; vectors which are nearly perpendicular have low correlations. The dotted lines show the two dominant eigenvectors; the ticks show unit variance.

to show its position relative to the main observational variables. If the parameter space is primarily two dimensional, as it is for this set of variables, then a correlation vector diagram provides an easy way to visualize all the correlations between the variables simultaneously. *The length of the vector shows how much of the variance in that parameter is explained by the first two dimensions. The angle between any two vectors indicates how well the two parameters are correlated; nearly parallel vectors are highly correlated, nearly perpendicular vectors are uncorrelated.*

We find that 81% of the variance in all seven variables can be explained by two dimensions. Since the remaining variance is consistent with our observational uncertainties, we conclude that the dimensionality of the parameter space is two. The fact that the B and R vectors are nearly perpendicular to the B/T and $B - H$ vectors provides us with a natural choice for a two–dimensional classification system. We named these dimensions the SCALE (a combination

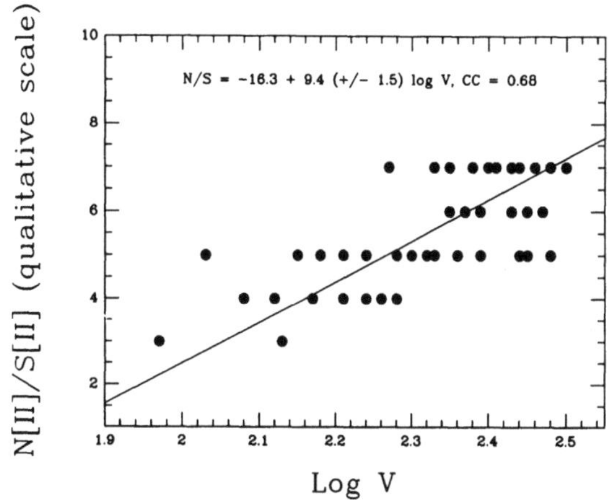

Figure 2: Plot of emission–line intensity ratio [N II]/[S II] *versus* maximum rotational velocity. The correlation coefficient is different than stated in the text since 49 galaxies had measurements of N/S and V in common (only 40 are indicated since nine galaxies have the same values), while 42 galaxies had the seven parameters included in Figure 1 in common.

of B and R) and the FORM (a combination of $B - H$ and B/T).

2. The Position of the [N II]/[S II] Vector in the Correlation Vector Diagram

About a dozen new parameters have been added to our database as a continuation of this project. These include a qualitative estimate of the emission line intensity ratio [N II](6548 Å + 6583 Å) / [S II](6717 Å + 6731 Å) (*i.e.* N/S in Figure 1) from the long–slit spectra used by Rubin *et al.* (1985). Rubin, Ford, and Whitmore (1984) originally discovered that the [N II]/[S II] ratio increased with increasing luminosity of a spiral galaxy. Since the N^+/S^+ is a good approximation to the N to S abundance ratio, this correlation indicates that brighter, more massive galaxies have more thoroughly processed their material into heavier elements.

Figure 2 indicates that while the N/S ratio does show a good cross–correlation (CC) with the two SCALE parameters (CC = 0.39 *versus* B; CC = 0.47 *versus* R), it shows a stronger correlation with the FORM parameters (CC = 0.63 *versus* $B - H$; CC = 0.59 *versus* B/T). However, the strongest correlation is between N/S and the maximum rotational velocity (CC = 0.66 *versus* V). Figure 2 shows the correlation between N/S and V.

In very luminous elliptical galaxies, the high metallicity is attributed to the deeper gravitational potential which permits the galaxy to retain more of the enriched gas which is shed by evolving stars and supernovae. In elliptical galaxies of lower mass, the enriched gas is able to leave the galaxy as a hot galactic wind because of the lower escape velocity. Perhaps the good correlation between [N II]/[S II] and the maximum rotational velocity (which is a good indication of the gravitational potential) indicates that a similar process is at work in spiral galaxies.

References

Brosche, P. 1973, *Astron. Astrophys.* **23**, 259.

Bujarrabel, V., Guibert, J., and Balkowski, C. 1981, *Astron. Astrophys.* **104**, 1.

Rubin, V. C., Ford, W. K. Jr., and Whitmore B. C. 1984, *Astrophys. J. Lett.* **281**, 21. Rubin, V. C., Burstein, D., Ford, W. K., and Thonnard, N. 1985, *Astrophys. J.* **289**, 81.

Whitmore, B. C. 1984, *Astrophys. J.* **278**, 61.

The Magellanic Clouds as Late–Type Barred Spirals

K. C. Freeman

Mount Stromlo and Siding Spring Observatories, The Australian National University, Canberra, A. C. T., Australia

Introduction

Between 1954 and 1972, Gérard de Vaucouleurs wrote a series of papers on the Magellanic Clouds which changed the perception of these systems. I will try to describe some of the steps in this process, and then mention some more recent results.

The Clouds Before 1954

Before 1954, the Magellanic Clouds were regarded as examples of a class of irregular galaxies which lack the rotational symmetry of the more familiar spirals. An early claim of spiral structure in the LMC came from H. C. Russell (1890) who photographed the clouds from Sydney using a 6–inch aperture portrait lens with a $14° \times 10°$ field. He wrote:

> "In fact, the whole of this great cloud is a complex spiral nebula ... So far as I am aware, the spiral structure of the Nubecula Major is shown for the first time in these photographs."

We will see later that John Herschel had actually beaten him to it (in drawing, if not in words) by about 40 years; Russell was aware of Herschel's chart, but commented that Herschel's drawing was very different from his (Russell's) photograph. It is difficult to know, from Russell's report, exactly what it was that he had seen as spiral structure.

Shapley (1950) recognised that there were similarities in luminosity and size between these Magellanic irregulars and late–type spirals such as M 33, and he suggested that the LMC was an irregular and open barred spiral of class SBc(pec). Baade (1950), who was *the* expert on stellar populations at the time, was convinced that the LMC was made up of pure population I, because no cluster–type variables had been found, and the stellar content of the brightest "globular cluster" in the LMC, NGC 1866, was distinctly unlike that of the galactic population II clusters. (We know now that NGC 1866 is a relatively young object, despite its brightness and globular appearance, and that the LMC has globular clusters of all ages, including very old ones.) The prevailing view, pre–1950, appears to have been that the Clouds were roughly spherical systems, in accordance with their irregular classification.

De Vaucouleurs at Mt. Stromlo

This view changed in 1953–1955. At Mt. Stromlo, de Vaucouleurs set up his twin Aero–Ektar camera to investigate the structure of the Magellanic Clouds. The objectives had a 7–inch focal length, stopped down to f/4, and a field of over 20°. In the Clouds, these objectives were comparable in power to a typical 36–inch reflector pointed at M 31 or M 33, in the sense that their linear resolution and absolute limiting stellar magnitude were similar (about 5 pc and $M_{pg} = -4.5$, respectively). He used this camera to make a mosaic of long–exposure photographs covering about 40° of sky, including both the LMC and the SMC. This mosaic is reproduced in Plate II of de Vaucouleurs (1955a) and clearly shows the LMC's major spiral arm, which emerges from the NW end of the bar and then loops around the bar.

This structure had been missed by most workers because the Clouds are so near. Photographs made with ordinary astrographs showed only a small part of each system, which gave an impression of disorder and irregularity. The particular kind of spiral structure was an important discovery, because there are

> "similar systems which share exactly the particular type of 'irregularity' of the Large Cloud, namely the great unbalanced development of one arm producing the main spiral structure, and the small embryonic arms emerging at both extremities of the axial bar."

(quote from de Vaucouleurs 1954). In other words, the characteristic "irregularity" of these Magellanic barred spirals is a strong asymmetry of the spiral pattern about the bar axis, in which the major arm dominates the spiral pattern. The LMC is not just an irregular with a random morphology, but a member of a well–defined class of "one-armed" barred spirals. NGC 4027, NGC 4618, and NGC 55 are just a few others of this class: photographs are shown in de Vaucouleurs and Freeman (1973, hereafter dVF). (We see a somewhat comparable asymmetry in some SA spirals like M 101, in which the spiral pattern is strongly asymmetrical about the nucleus). The similarity of the Magellanic Clouds and other Magellanic systems goes further: NGC 4618 has a companion, NGC 4625, and the pair are a striking replica of the LMC/SMC system. NGC 4027 and its companion NGC 4027A are also very similar, as are NGC 4631 and its companion. The dynamics of these pairs of Magellanic systems is not yet understood, but

> "it would seem to be a fair inference that the Magellanic type of barred spiral is the result of some sort of interaction with a neighbouring system in a close encounter."

(quote from de Vaucouleurs 1954). It is in this direction that the dynamics of the Magellanic Clouds is currently being investigated, by several groups.

Many people in the community were somewhat sceptical about the common barred spiral structure of the LMC and other Magellanic systems. About 1966, as one prominent sceptic became particularly vocal, Gérard de Vaucouleurs produced John Herschel's (1847) drawing of the LMC, as seen with the naked eye. This drawing, by an observer who surely had no preconceived views on the issue of spiral structure in the LMC, shows the bar *and* the major arm to the North *and* the embryonic arms to the South (see dVF for a reproduction). It made a great Christmas card.

The detection of spiral structure suggested a flat, rotating system. Rotation of the LMC was soon confirmed by Kerr and de Vaucouleurs's (1955) analysis of the Sydney 21–cm observations. Their rotation curve showed another characteristic property of Magellanic spirals: the center of symmetry of the rotation curve is significantly displaced from the center of the bar. This displacement has been amply confirmed by later rotation data (*e.g.* Rohlfs *et al.* 1984, although the position angle of the kinematic major axis is somewhat different, and the velocity field at higher spatial resolution is complex, with regions of double peaked profiles and noncircular motions). The displacement is seen in other Magellanic systems also, like NGC 55 (Robinson and van Damme 1966).

The centroid of the smoothed stellar distribution of the LMC, derived from the Aero–Ektar plates, is also displaced from the center of the bar, in the same sense. This *double* asymmetry is now recognised as one of the characteristic structural features of Magellanic barred spirals: the bar itself is the first asymmetry and the displacement of the center of the bar from the center of the disk is the second asymmetry. The displacement is typically 0.5 to 1 kpc, and the two displacements (of the center of symmetry of the rotation curve and of the photometric centroid from the bar) are well–correlated (Feitzinger 1980). The large–scale equilibrium of such doubly asymmetric systems is not yet well understood: however, we note that they are fairly common (see dVF).

Final confirmation that the LMC is not only rotating but also flat, came from de Vaucouleurs's (1955b) study of the velocity dispersion of 17 emission nebulosities in the LMC: their intrinsic vertical velocity dispersion was only about 5 km sec^{-1}. At this stage, it was clear that the LMC is a disk galaxy, approximately in centrifugal equilibrium, like most other disk galaxies. We now know that the surface brightness distributions of Magellanic systems also show the well-defined exponential disk (see dVF) typical of almost all disk galaxies. Although the origin of the exponential light distribution in disk galaxies is not well understood, it seems clear that the large scale underlying structure of the disks of the Magellanic and non-Magellanic disk galaxies is fairly similar. It is interesting to see how the flattening of disk galaxies goes with morphological type: the flattest systems are at type Sd, where the asymmetry begins to appear (*e.g.* Heidmann *et al.* 1972). In normal disk galaxies, the disk thickness is set by heating processes (encounters of stars with giant molecular clouds and spiral structure) which are not yet fully understood. However, the double asymmetry in the SBm systems provides another possible source of heating (the resonant excitation of motion perpendicular to the plane, as discussed by Binney 1981), which may explain why they are less flat.

Another interesting property of the Magellanic barred spirals is the structure of the bar itself. There is no bright central nucleus; in fact, there is a minimum of light at the center of the bar. In the LMC, one can see that this minimum is produced by a dust lane that crosses the bar from North to South. This differs from the dust lane geometry in the earlier type (Sa, Sb) barred spirals, which often show straight dust lanes that run parallel to the bar and twist near the center to cross the nucleus, *e.g.* NGC 1365 (see dVF). The structure of the dustlanes in the SBm systems is probably related to the gas dynamics in these asymmetrical systems, which is not properly understood yet.

The next issue was the question of the stellar population of the LMC, which Baade had suggested to be pure population I. De Vaucouleurs (1955a) examined the distribution of stars brighter than $m_{pg} = 14$. He noted that the radial distribution of these bright stars has a pronounced *minimum* near the center of the bar, where the surface brightness takes a *maximum*. This suggested the presence of a sizable fraction of a population older than the typical population I, *i.e.* with a luminosity function shifted towards the lower luminosities. This proved to be correct, and we now know that the stellar content of the bar is dominated by a population that is a few billion years old. At about the same time, Gascoigne and Kron (1952) showed that some of the globular clusters of the LMC were indeed red and therefore old (unlike the young cluster NGC 1866 which had misled Baade), and Thackeray and Wesselink (1953) discovered RR Lyrae stars in the LMC. It was now certain that the LMC contained older stars, and maybe some that are as old as the oldest stars in the Galaxy. The notion of a pure population I system was clearly incorrect. This was another important step in establishing the LMC as a fairly normal galaxy with an extended history of star formation. We now know that there are some significant differences between the history of star formation in the LMC and in the Galaxy. One of the most obvious is the presence in the LMC of globular clusters of all ages, from about 10^7 y to clusters that are apparently as old as the old clusters in the Galactic halo. The reason for this difference is not understood yet. However, the wide range of ages among the LMC clusters has certainly been very useful for studying the evolution of stars of different masses and the evolution of the clusters themselves.

The end product of this intense activity in the 1950's was a transformation of our view of the LMC and Magellanic systems in general. It had become clear (although not yet accepted by everyone) that the LMC is a member of a well–defined class of flat, rotating, barred disk galaxies, distinguished by their doubly asymmetric structure. Since that time, the data on the kinematics and stellar content of the Magellanic Clouds has greatly improved. There are excellent H I surveys, and much is known about the motions of stars, planetary nebulae, H II regions and clusters. High quality color–magnitude diagrams are now available for clusters and many regions of the field in the Clouds, and far more is known about stellar evolution, and the chemical properties and dynamical evolution of clusters, H II regions, and supernova remnants in the Clouds, to name just a few areas of significant progress. There was also the discovery in the 1970's of the H I Magellanic Stream [a term coined by de Vaucouleurs (1954) in a somewhat different context], which is probably a symptom of the interaction of the LMC and SMC with the Galaxy. However, it is fair to say that, by the time that Gérard de Vaucouleurs had completed his burst of activity on the Magellanic Clouds, the essential features of the Clouds, and of the LMC in particular, were known.

Some Recent Issues

1. Dynamics of Magellanic Systems

There has not been much dynamical work on these doubly asymmetric systems. The origin of the double asymmetry itself is not understood. Some possibilities include asymmetries in the protogalaxy, later accretion of material, and $m = 1$ modes if the mass ratio of the disk to the dark halo is high enough. Many (but not all) Magellanic systems occur in pairs: it is not clear if this is relevant to their doubly asymmetric structure. Several groups have used the (disk + displaced bar) mass model (see dVF) for Magellanic systems, without attempting to consider its origin or equilibrium. It has an interesting potential: for realistic parameters (the disk providing about 85% of the mass, the potential as seen in the rotating frame of the bar shows only one neutral point. There is some interesting recent work by Colin and Athanassoula (preprint) on the response of gas to such a potential. Their work is aimed at exploring effects rather than making realistic models, and uses a rather strong bar; the effect of the displaced bar on morphology of gas response is clearly demonstrated.

2. The Interaction of the LMC – SMC Galaxy System

The LMC and SMC are an interacting binary system, which is also interacting with the Galaxy. Observations of the pan–Magellanic system in H I show a well–defined flow field over the entire LMC–SMC field, disturbed by the LMC rotation only within a few degrees of the LMC, and hardly disturbed at all by the SMC (Mathewson *et al.* 1979). Therefore, the motion of H I and other objects of the young population in the outer parts of the LMC will be much affected by this pan–Magellanic flow. Feitzinger (1980) has assembled the available kinematic data for the LMC: the resulting velocity field is complex and highly distorted. This makes it very difficult to determine whether the LMC has a massive corona like some other isolated SBms (*e.g.* NGC 3109; Carignan and Freeman 1985). Also, this means that the LMC is not an ideal galaxy in which to test the dynamical theory of Magellanic systems. The Magellanic Stream is probably associated with this interaction, but the dynamics of the Magellanic Stream are not understood yet, not even at the conceptual level of tidal origin *versus* ram pressure. We note that the LMC–SMC system is not the only example of interacting SBm's: NGC 4631/4656 is another interacting Magellanic pair (Weliachew *et al.* 1978).

3. Does the LMC have a Stellar Halo?

The LMC contains RR Lyrae stars and old globular clusters, like those of the galactic halo. Because the LMC is so close, it provides a unique opportunity to see whether late–type galaxies have had a similar star forming history to the somewhat earlier type galaxies like the Milky Way; *e.g.* did a metal–weak population form in a spheroidal subsystem in the late–type galaxies during the early collapse? There is no evidence for a kinematically defined halo from the kinematics of the globular clusters (Freeman *et al.* 1983), or from the kinematics of the oldest long period variables; from their velocity dispersions, they define a disk of 300 pc scale height (Bessell *et al.* 1986).

Rotation solutions for globular clusters of different ages show that the younger clusters have similar motions to the H I, with a line of nodes near position angle (PA) $= 0°$, while the older clusters have a line of nodes near PA $= 40°$ (Freeman *et al.* 1983). Why are the PA's for the young and old clusters different ? Some theories of the Magellanic Stream and some observations predict a close approach of the LMC/SMC roughly 2×10^8 y ago (*e.g.* Fujimoto and Murai 1984). We suggest that the old clusters delineate the true old disk, while the kinematics of the young ones and the H I are affected by this recent interaction. This view is supported by analysis of the H I observations of Rohlfs *et al.* (1984), which gives a line of nodes at PA $= 28°$ in the inner parts of the LMC. These inner regions are least affected by the interaction. This PA is fairly close to the PA of the line of nodes for the older clusters. We note also that, from the red isophotes of the LMC, PA $= 40°$ looks plausible as the geometrical line of nodes for the inner parts of the LMC (Freeman *et al.* 1983).

Recent results by Meatheringham *et al.* (1988) on the kinematics of 94 planetary nebulae (PNe) in the LMC are in general agreement with this picture. Comparison of the velocity residuals of the H I and the PNe from the H I rotation curve (Rohlfs *et al.* 1984) show a similar pattern, and rotation solutions give a similar kinematic major axis for the PNe to that of the inner H I (*i.e.* close to the PA for the old clusters, which we think delineate the true old disk). However, the velocity dispersion for the PNe is significantly higher. Line–of–sight dispersion for the PNe is about 20 km sec^{-1}, similar to that of the older clusters. This makes sense because their typical ages are about 3×10^9 y.

4. The Small Magellanic Cloud

The SMC shows the characteristic Magellanic features (*e.g.* bar, beginnings of major and minor arms, asymmetry). There is evidence for rotation (*e.g.* Kerr and de Vaucouleurs 1955), although the H I motions are now known to be very complex (*e.g.* Mathewson *et al.* 1986). The interaction of the LMC and SMC appears to have had a major effect on the structure and motions in the SMC. Theoretical work by Murai and Fujimoto (1980) suggests that the SMC was significantly disrupted by the most recent encounter, about 2×10^8 y ago, with debris strung out more that 20 kpc along line of sight. In support of this recent interaction, Irwin *et al.* (1988) have reported a stellar link between the LMC and SMC: they find blue stars with ages of about 1 to 2×10^8 y and velocities consistent with Cloud membership. Several authors have found evidence for such a large line of sight extent, from studies of cepheids and other bright stars in the SMC. For example, a recent study by Mathewson *et al.* (1986) of 161 cepheids in the SMC showed that the SMC has a line of sight extent of at least 20 kpc. (The NE end of bar is closer.) This work is still somewhat contentious, but it does stress the great importance of the interaction, particularly for the structure of the SMC.

References

Baade, W. 1950, *Publ. Obs. Univ. Michigan* **10**, 7.

Bessell, M. S., Freeman, K. C., and Wood, P. R. 1986, *Astrophys. J.* **310**, 710.

Binney, J. 1981, *Mon. Not. Roy. Astron. Soc.* **196**, 455.

Carignan, C. and Freeman, K. C. 1985, *Astrophys. J.* **294**, 494.

Feitzinger, J. V. 1980, *Space. Sci. Rev.* **27**, 35.

Freeman, K.C., Illingworth, G.D., and Oemler, A. 1983, *Astrophys. J.* **272**, 488.

Fujimoto, M. and Murai, T. 1984, in *I. A. U. Symp. No. 108, Structure and Evolution of the Magellanic Clouds*, ed. S. van den Bergh and K. de Boer (Dordrecht: Reidel), p. 115.

Gascoigne, S. and Kron, G. 1952, *Publ. Astron. Soc. Pac.* **64**, 196.

Heidmann, J., Heidmann. N., and de Vaucouleurs, G. 1972, *Mem. Roy. Astron. Soc.* **75**, 85.

Herschel, J. 1847, *Results of Astronomical Observations made during the years 1834–8 at the Cape of Good Hope* (London: Smith and Co).

Holmberg, E. 1952, *Medd. Lund. Astron. Obs.*, Ser **I**, No. 176.
Irwin, M., Demers, S., and Junkel, W. 1988, *U. K. Measuring Machine Newsletter*, No. **10**, p. 23.
Kerr, F. and de Vaucouleurs, G. 1955, *Austr. J. Phys.* **8**, 508.
Mathewson, D. S., Ford, V. L., Schwarz, M. P., and Murray, J. D. 1979, in *I. A. U. Symp. No. 84, The Large-Scale Characteristics of the Galaxy*, ed. W. B. Burton (Dordrecht: Reidel), p. 547.
Mathewson, D. S., Ford, V. L., and Visvanathan, N. 1986, *Astrophys. J.* **301**, 664.
Meatheringham, S., Dopita, M., Hord, H., and Webster, B. L. 1988, *Astrophys. J.* **327**, 651.
Murai, T. and Fujimoto, M. 1980, *Publ. Astron. Soc. Japan* **32**, 581.
Robinson, B. J. and van Damme, K. J. 1966, *Austr. J. Phys.* **19**, 111.
Rohlfs, K., Kreitschmann, J., Siegmann, B. C., and Feitzinger, J. V. 1984, *Astron. Astrophys.* **137**, 343.
Russell, H. C. 1890, *Mon. Not. Roy. Astron. Soc.* **51**, 41 and 96.
Shapley, H. 1950, *Publ. Obs. Univ. Michigan* **10**, 79.
Thackeray, D. and Wesselink, A. 1953, *Nature* **171**, 693.
de Vaucouleurs, G. 1954, *Observatory* **74**, 23.
de Vaucouleurs, G. 1955a, *Astron. J.* **60**, 126.
de Vaucouleurs, G. 1955b, *Publ. Astron. Soc. Pac.* **67**, 397.
de Vaucouleurs, G. 1957, *Astron. J.* **62**, 69.
de Vaucouleurs, G. and Freeman, K. 1973, *Vistas in Astron.* **14**, 163.
Weliachew, L., Sancisi, R., and Guelin, M. 1978, *Astron. Astrophys.* **110**, 61.

Discussion

J. Binney: As one proceeds along the revised Hubble sequence post Sc, there is a rapid decrease in mean luminosity and in the characteristic circular speed. On the other hand, cooling processes set a characteristic velocity of order $10\,\mathrm{km\,sec^{-1}}$ for interstellar gas. So, in Magellanic irregulars, the random velocities of gas clouds is by no means dynamically unimportant, and since then random velocities are likely more or less isotopic, then galaxies cannot be thin. This preexisting thickness facilitates resonant coupling of vertical and planar motion as you have mentioned, and this further thickens the systems.

Freeman: This is true for most Magellanic Systems. A few, however, (*e.g.* LMC, NGC 55) are larger galaxies with rotational velocities in excess of 80 $\mathrm{km\,sec^{-1}}$; it would be interesting to know how flat they are.

D. Burstein: If the current mass fraction of H I in the Large Cloud is $\sim$ 10–20%, and if the bar came from a comparable mass, then the pre–bar gas content would have been $\sim$ 40% by mass. If this gas was originally asymmetrically distributed (as is common in spirals and irregulars), might not the periodic motion of a companion (*e.g.* , the SMC) stimulate star formation in this asymmetry, which then collapses to form a bar?

Freeman: This is possible; the dominant stellar population in the bar appears to have an intermediate age. However, not all Magellanic systems have companions at present, so you would need to argue that the companions of these systems have merged or been otherwise disrupted.

B. Whitmore: Is it possible that the apparent difference between the intrinsic flattening of Sc and Sd galaxies, and Im galaxies, is caused by the different number of galaxies in the two samples (Heidmann, Heidmann, and de Vaucouleurs 1971)? While there are probably a few hundred Sc and Sd galaxies, there are probably only a few dozen Im galaxies. If only the 2 or 3 flattest systems are plotted we might not have a large enough sample of Im galaxies to find any exactly edge–on.

We have redone this study by using the same number of galaxies for each type of galaxy. While the slope of the increase of flattening between Sa–Sd is not as steep as it was, there is still a fall–off for the irregular systems, which still appear to be intrinsically thicker than Sc and Sd galaxies.

G. de Vaucouleurs: It is true that size–of–sample effects have affected some of the statistics, but revised values from equal samples of 100 in each type bin still show a type dependence. There is no doubt that galaxies are flattest at stage Sd. The Sm and Im types are intrinsically less flat as a consequence of their slower rotation.

J. M. van der Hulst: You pointed out the lopsidedness of M 101 in relation to the asymmetric light distribution in the LMC. I would like to remark that new Westerbork H I observations by Sancisi and myself (van der Hulst and Sancisi, *Astron. J.* **95**, 1354, 1988) show that M 101's lopsidedness disappears when one examines faint enough column density levels. Single–dish observations by Huchtmeier and Witzel (*Astron. Astrophys.* **74**, 138, 1979) have already indicated that M 101 is symmetric again at low column density levels.

New Results on the Geometrical Structure of the Small Magellanic Cloud

D. Hatzidimitriou[1] and M. R. S. Hawkins[2]

[1]Department of Astronomy, University of Edinburgh, Scotland, U. K.
[2]Royal Observatory, Edinburgh, Scotland, U. K.

The large scale morphological and structural properties of the Small Magellanic Cloud (SMC) have been the subjects of numerous investigations; however, a great deal of controversy still exists. There is observational evidence suggesting a large line–of–sight extension of the SMC, or even indicating that the SMC is in the process of "irreversible disintegration" (Mathewson *et al.* 1986). It is generally accepted that the gravitational effects of the LMC and our Galaxy have played a decisive role in the SMC dynamical history, forming the Wing, Bridge, and the Magellanic Stream (Murai and Fujimoto 1984), and influencing the kinematics, as well as probably the star formation history of the SMC. However, most relevant studies to date were based on very young and young populations (blue and red supergiants, Cepheid variables) essentially confined in space to the Bar, Arms, and Wing (Brück 1982). In this project, we cover a much larger area of the SMC and a much broader range in ages, with the purpose of performing a thorough study of the structure and population synthesis in the outer regions of the SMC.

A grid of R vs. $B-R$ colour–magnitude diagrams was constructed over two areas covering a total of 48.5 square degrees in the NE and SW outer regions of the SMC (ESO/SERC survey fields 52 and 28), at projected distances of more than $\sim$ 2 kpc from the optical centre of the SMC. The grid element was a square of 0.9 deg. The data–base includes the B and R magnitudes of 550,000 stellar images, obtained from twelve UKST photographic plates (B_j and R), which were measured with the COSMOS automatic microdensitometer (at the Royal Observatory, Edinburgh), and calibrated by CCD photometric sequences. The accuracy of the photographic photometry was better than 0.12 mag, with a limiting magnitude of 21.5 and completeness greater than 95% (Hatzidimitriou *et al.* 1988). Two typical examples of colour–magnitude diagrams are shown in Figure 1.

The most outstanding feature of these CMDs is the presence of a well–populated clump of stars at the base of the red giant branch (the population I equivalent of the horizontal branch), characteristic of intermediate age populations (Cannon 1970). The luminosity of the clump is known to be constant for ages larger than 5×10^8 yr (Olszewski *et al.* 1987). In the regions considered here, there is no significant contribution from such younger populations, as can be seen clearly in Figure 1; therefore, the luminosity of the giant clump can be used as a distance indicator for the corresponding stellar populations.

The results we obtained from a preliminary study of the distribution of the magnitudes of clump stars can be summarized as follows:

(i) There is a systematic decrease in the distance modulus of the SMC in the SW–NE direction, as derived from the mean luminosity of the clump giants. This effect ammounts to approximately 0.3 mag over a projected distance of $\sim$ 8 kpc. The effect is more significant in field 52. Such a "tilt" is known to exist in the "main body" of the SMC, and it apparently persists in the outer disk and halo regions. However, a definition of a mean angle of inclination has litle meaning, since — as will be seen below — the planar model is not a good approximation of the SMC morphology (see also Caldwell and Coulson 1986).

(ii) The scatter in magnitude of the clump stars is of particular interest. Photometric errors can account for ± 0.12 mag of the scatter, while the intrinsic size of the clump contributes another $\sim \pm 0.3$ mag (Faulkner and Cannon 1973). The rest of the observed scatter can be attributed to the differences in the line–of–sight depth of the stars comprising the clump. The observed size (1σ around the mean magnitude) of the clump is $\sim \pm 0.35$ in field 28, while it rises to ± 0.48 in field 52. In F 28, the additional scatter that can be attributed to geometrical effects is low, corresponding to a maximum of 7 kpc depth along the line of sight. In F 52, however, the depth is much larger (see also fig.1), between 14–20 kpc.

We can interpret this result as indicating either a wing–like feature towards the NE of the SMC central regions linked with the SMC and directed towards us; or a localised structure, separated by $\sim$10 kpc from the SMC along the line of sight and centred north–eastwards of the SMC. This last possibility agrees at least qualitatively with the suggestion by Mathewson *et al.* (1986) of the existence of a "Mini Magellanic Cloud," which according to the present study, seems to have a dominant intermediate age population. A detailed presentation of the results and their interpretation will be given elsewhere (Hatzidimitriou and Hawkins, in preparation).

We thank the U.K. Schmidt Unit for the loan of the plates, and the COSMOS group. D.H. was supported by a University of Edinburgh Research Studentship during this work.

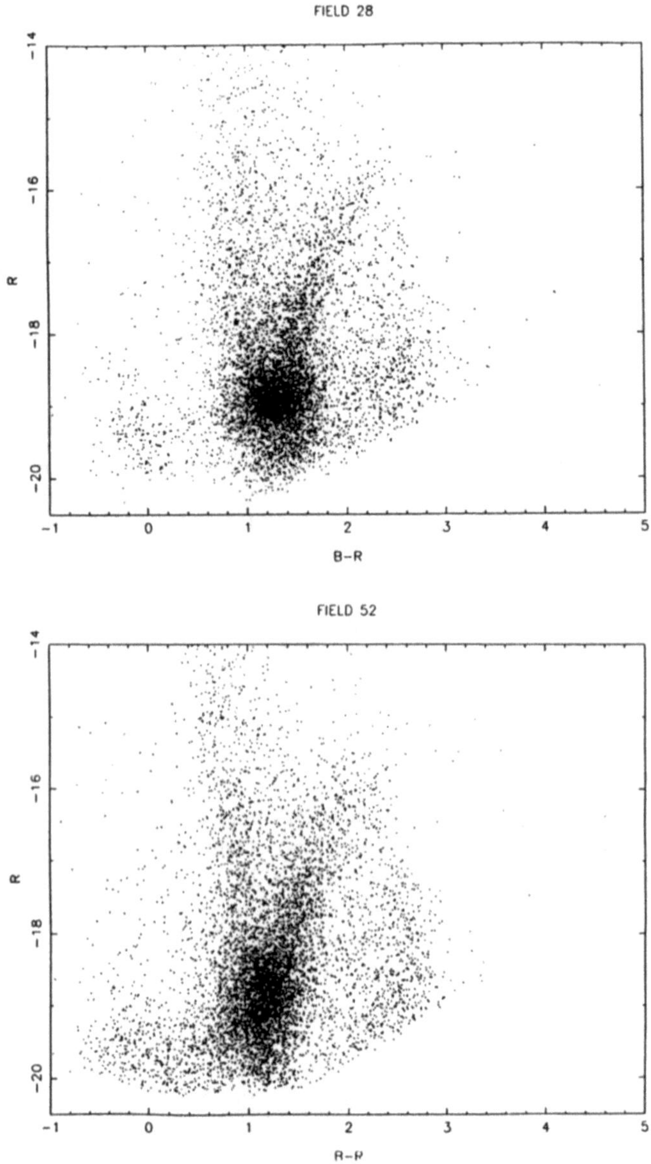

Figure 1: Colour–magnitude diagrams from two representative regions (from an area of 0.9 square degrees each) in field 28 (top) and field 52 (bottom). The projected distances of the centres of the two regions from the SMC optical centre are 2.7 deg and 3.2 deg respectively. The foreground galactic contribution has not been removed from the diagrams.

References

Brück, M. T. 1982, in *Compendium in Astronomy*, ed. E. G. Mariopoulos, P. S. Theocaris, and L. N. Mavridis (Dordrecht: Reidel), p. 297.

Caldwell, J. A. R. and Coulson, I. M. 1986, *Mon. Not. Roy. Astron. Soc.* **218**, 223.

Cannon, R. D. 1970, *Mon. Not. Roy. Astron. Soc.* **150**, 150.

Faulkner, D. J. and Cannon, R. D. 1973, *Astrophys. J.* **180**, 435.

Fujimoto, M. and Murai, T. 1984, in I.A.U. Symp. No. **108**, ed. S. van den Bergh and K. S. de Boer (Dordrecht: Reidel), p. 115.

Hatzidimitriou, D., Hawkins, M. R. S., and Gyldenkerne, K. 1988, *Mon. Not. Roy. Astron. Soc.*, submitted.

Hatzidimitriou, D. and Hawkins, M. R. S. 1988, in preparation.

Mathewson, D. S., Ford, V. L., and Visvanathan, N. 1986, *Astrophys. J.* **301**, 664.

Olszewski, E. W., Schommer, R. A., and Aaronson, M. 1987, *Astron. J.* **93**, 565.

The Magellanic Irregular Galaxy DDO 50

C. Moss and M. J. Irwin
Institute of Astronomy, Cambridge, U.K.

Introduction

The Magellanic irregular galaxy, DDO 50 (= Ho II = A 813+70 = VII Zw 223 = Arp 268), is usually assumed to be a member of the M 81 group. De Vaucouleurs (1979) has derived the distance modulus of the galaxy, $\mu_o = 27.79$, from magnitudes of the brightest blue supergiants and brightest red variable given by Sandage and Tammann (1974). The estimated absolute magnitude of the galaxy, $M_T = -17.04$, is close to that of the SMC. We have used B and V plates to obtain a colour–magnitude diagram for this galaxy. We have also obtained further estimates of the distance modulus: first, from measurements of the sizes of H II rings in the galaxy from a deep Hα plate, and second, from a comparison of the luminosity function of the brightest blue stars with the corresponding luminosity function for the SMC.

Observations

Plates of the galaxy were obtained in B (90 min) and V (210 min) by H. D. Ables using the 1–m and 1.5–m telescopes, respectively, of the U.S. Naval Observatory, Flagstaff. The plates were digitised using the Automatic Plate Measuring (APM) facility in Cambridge. Each image was processed to detect discrete objects; objects were only accepted if they had images on both plates. Calibrated B and V magnitudes were obtained for these objects using a photoelectric sequence given by Sandage and Tammann (1974). There was good agreement between APM magnitudes and the photoelectric values for these standards. A two–dimensional median filter was used to separate the diffuse unresolved stellar component of the galaxy from resolved supergiants, stellar associations, and foreground stars in the Galaxy.

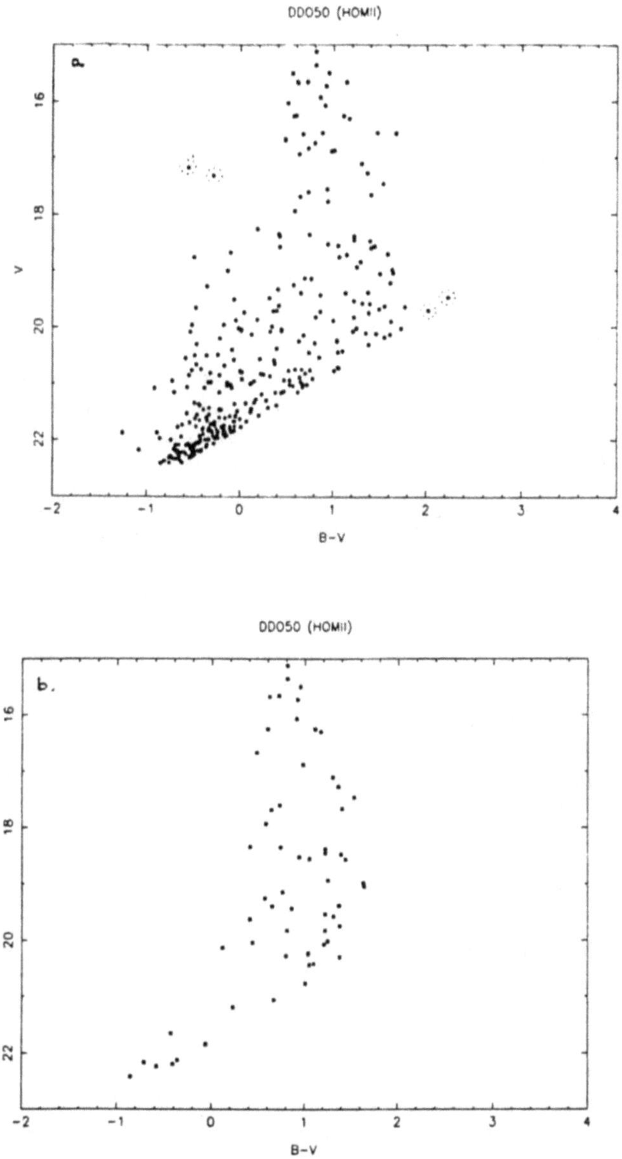

Figure 1: (top) Colour–magnitude diagram for DDO 50. (bottom) Colour–magnitude diagram for an area of the field surrounding DDO 50. Equal areas of sky were surveyed to produce these colour-magnitude diagrams.

Colour–Magnitude Diagram

The colour–magnitude diagram for the supergiants in the galaxy is shown in Figure 1a. For comparison, the colour–magnitude diagram for an equal area surrounding the galaxy is shown in Figure 1b. It is seen that for $B-V > 0.4$, the colour–magnitude diagram for the galaxy is heavily contaminated by field stars. However, for $B-V < 0.4$, there is relatively little contamination, and the upper limit of the main sequence is visible. Indeed, the few very blue stars in the colour-magnitude diagram for the field may in fact be outlying supergiants of the galaxy. Objects circled in Figure 1a are non-stellar, and are thought to be H II regions or stellar associations rather than a single supergiant. The distribution of blue supergiants closely follows the surface brightness distribution of the galaxy.

The magnitude of the brightest blue supergiant, and the mean magnitude of the three brightest such objects, are 18.25 and 18.42 respectively. Following the procedure given by de Vaucouleurs (1978a), values are obtained for the distance modulus of the galaxy, $\mu_o = 26.37$ and 26.17, respectively. A disadvantage of this method is the possible confusion of OB associations in the galaxy with the brightest supergiants which generally will lead to an underestimate of the distance of the galaxy. An alternative method which avoids this problem is to compare the luminosity functions of the brightest blue stars for different galaxies. This method also has the advantage that it uses more information from the colour–magnitude diagram. A comparison was made of the luminosity functions for the brightest blue stars ($B-V < 0.2$) for DDO 50 and the SMC. A maximum likelihood fit of the two luminosity functions gives a relative distance modulus of the two galaxies of $\Delta\mu = 8.75$, which gives a distance modulus for DDO 50 of $\mu_o = 27.32$, using the distance modulus of the SMC of $\mu_o = 18.62$ given by de Vaucouleurs (1978b).

H II rings

Using the Steward Observatory 2.1–m telescope equipped with a single–stage ITT image intensifier coupled to IIaD photographic emulsion, a 180 min Hα exposure of the galaxy was obtained. Several H II rings are clearly visible. Gum and de Vaucouleurs (1953) first demonstrated that the sizes of such rings can be used as reliable distance indicators. Using the procedure given by de Vaucouleurs (1978a), and measurements of the diameters of the two largest rings we obtain a distance modulus for the galaxy of $\mu_o = 27.28$.

Summary: Distance Modulus

Various estimates of the distance modulus of the galaxy are summarised as follows:

Luminosity function of brightest blue supergiants	$\mu_o = 27.32$
Size of largest H II rings	$\mu_o = 27.28$
Brightest red variable (de Vaucouleurs 1979)	$\mu_o = 27.52$

It is seen that these are in good agreement and a mean distance modulus may be derived of $\mu_o = 27.4$, corresponding to a distance to the galaxy of 3.0 Mpc.

Acknowledgments

It is a pleasure to thank H. D. Ables for the loan of the plates which has made this work possible. We are grateful to Steward Observatory, University of Arizona for observing time on the 2.1–m telescope. The APM facility in Cambridge is supported by the Science and Engineering Research Council.

References

Gum, C. S. and de Vaucouleurs, G. 1953, *Observatory* **73**, 152.
Sandage, A. and Tammann, G. A. 1974, *Astrophys. J.* **191**, 603.
de Vaucouleurs, G. 1978a, *Astrophys. J.* **224**, 14.
de Vaucouleurs, G. 1978b, *Astrophys. J.* **223**, 730.
de Vaucouleurs, G. 1979, *Astrophys. J.* **227**, 380.

Photometry and Kinematics of the Magellanic Type Galaxy NGC 4618

Stephen C. Odewahn

Astronomy Department, University of Texas, Austin, Texas, U.S.A.

NGC 4618 is an excellent example of an asymmetric, late–type Magellanic barred spiral (de Vaucouleurs and Freeman 1972). Detailed surface photometry of this galaxy has been obtained from photoelectrically calibrated V and I CCD images from Asiago Observatory, and from BVR CCD images, UBV electronographic plates, and one filtered IIa–O 2.1–m photographic plate from McDonald Observatory. This material, reduced with the McDonald Observatory Galaxy Photometry Package, is used to derive standard photometric parameters for comparison with previous studies of late type systems, such as that of NGC 4027 (Pence and de Vaucouleurs 1985). The luminosity distribution is decomposed into three distinct components: 1) an elliptical bar, 2) an exponential old disk, and 3) a young H II region/arm population; comprising 23, 50, and 27 percent of the total luminosity, respectively. A detailed mapping of the $V - I$ color index, and Hα imagery are used to study the distribution of young star forming regions with respect to the bar, which is significantly offset from the center of the exponential disk defined by the outer isophotes.

A two–dimensional velocity field of the ionized gas has been mapped using the McDonald Observatory Mark II Fabry–Perot interferometer. The large amount of diffuse Hα emission throughout the disk, as well the large number of H II regions in the galaxy, have resulted in well–defined fringes, providing good coverage across the face of NGC 4618, whose approximate inclination is 33 degrees. Over 400 velocity points, with an observational error of $\pm 10\,\mathrm{km\,sec^{-1}}$ per point, were extracted from two sky–subtracted interferograms. Assuming the gas is confined to a rotating, thin disk, these velocities are used to compute a rotation curve using standard methods. The derived dynamical parameters (center of rotation, inclination, and line of nodes) are compared with those obtained from the surface photometry. The kinematic center of rotation is found to be well–displaced from the center of the bar. The velocity field is complicated in the bar and outer arm regions, and shows clear signs of noncircular motion.

References

Pence, W. D. and de Vaucouleurs, G. 1985, *Astrophys. J.* **298**, 560.
de Vaucouleurs, G. and Freeman, K. C. 1972, *Vistas in Astron.* **14**, 163.

Chemical Abundance and Age in the Magellanic Clouds; with Special Reference to SN 1987A

M. W. Feast

South African Astronomical Observatory, Cape Town, South Africa

Summary

Evidence relating to the chemical abundance of objects of known age in the Magellanic Clouds is reviewed. For ages $< 2 \times 10^8$ years the mean [Fe/H] values are -0.2 (LMC) and -0.5 (SMC) where [Fe/H] is the difference between Cloud objects and similar objects in the solar neighbourhood. C and N are more deficient than this in Cloud H II regions. The relevance of these results for models for SN 1987A in the LMC is noted and recent work on this supernova is summarized. In the mean, the results show a roughly exponential increase in metal abundance with time over the last $\sim 10^{10}$ years in both Clouds. The blue globular cluster NGC 330 (age 7×10^6 years, $[Fe/H] = -1.4$) is significantly more metal deficient than other young SMC objects suggesting it may have formed remote from them, perhaps in the Magellanic Stream.

Introduction

If we are to understand the place of the Magellanic Clouds in the World of Galaxies, we need to obtain as clear a picture as possible of the relationship between chemical composition and age for the objects they contain. The aim of the present paper is to review our current understanding of this relationship. Special emphasis will be placed on the metal abundances of young objects in the Clouds. A knowledge of these latter abundances is crucial to such matters as the calibration of the Cepheid distance scale. It is also of great current interest in understanding SN 1987A in the LMC. This type II supernova was unusual

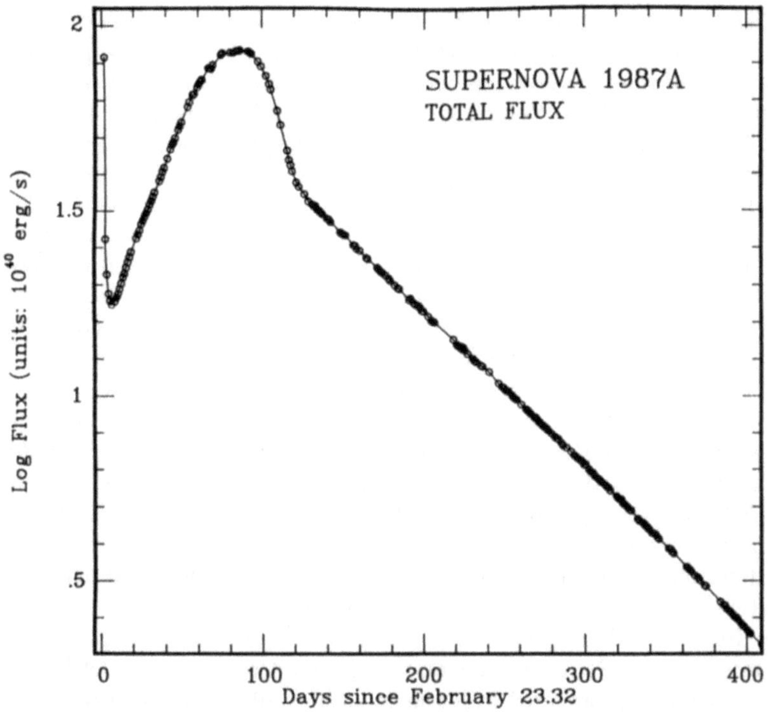

Figure 1: The bolometric light curve of SN 1987A from SAAO observations (see text for details).

in a number of ways and this has been attributed to an under-abundance of metals in the progenitor star. Our current understanding of SN 1987A will be summarized before going on to the general problem of chemical abundances.

SN 1987A in the LMC

Figure 1 shows the bolometric light curve of SN 1987A derived from SAAO *UBVRIJHKL* observations (Menzies *et al.* 1987, Catchpole *et al.* 1987, Catchpole *et al.* 1988, Whitelock *et al.* 1988). The most striking feature of this curve is the long linear decline in log flux. From day 147 after core collapse (*i.e.* neutrino detection) to day 257, the root mean square deviation from a straight line was only 0.006 mag. Within the uncertainty of a few days (cf. Catchpole *et al.* 1988) the slope of the line in this period was identical to the e-folding time for the radioactive decay of ^{56}Co (111.26 days). This provides convincing proof that ^{56}Co is the source of energy at late times in this, and presumably other SNe

II. After day 265, the bolometric flux falls off more rapidly than the ^{56}Co decay rate (Catchpole and Whitelock 1988, Whitelock *et al.* 1988). This suggests that the envelope of the supernova has become at least partially optically thin to the γ–rays emitted by ^{56}Co. Several groups (see IAU Circulars 4510, 4526, 4527, and 4535, etc.) have indeed detected γ–rays in this latter period. Nomoto and colleagues (private communication) estimate that on day 340 the flux of γ–rays and related X–rays was about 8% of the total luminosity of SN 1987A. Within the uncertainties of the high energy flux measurements, this makes up the flux deficiency shown by the bolometric light curve. Thus, the energy output from at least day 147 to day 400 is determined entirely by the radioactive decay of ^{56}Co (the daughter product of the shorter–lived ^{56}Ni). There is as yet no evidence for any significant energy input from any other sources such as a pulsar.

The earlier development of the supernova, when we expect energy input primarily from the initial shock waves, was at first more difficult to understand. The models based on the collapse of a red supergiant, which are known to fit at least approximately the relatively rough observations of some previous SNe II (*e.g.* SN 1969L Woosley and Weaver 1986), did not fit SN 1987A at all. Amongst the peculiarities of SN 1987A were: the short period (~3 hours) between neutrino detection and light outburst; the unusual shape of the initial light curve; the exceptionally high initial velocities of expansion (the H absorption minimum on the first recorded spectrum was 18,500 km sec^{-1} (Menzies *et al.* 1987) probably a record for an SNe II) and the rapid apparent deceleration ($\sim$ 780 km sec^{-1} over the first week, compared with a more usual $\sim$ 70 km sec^{-1}). It was quickly realised, however, that it was a blue (not a red) supergiant (SK −69° 202) which had collapsed, and models (*e.g.* Woosley *et al.* 1988) based on such a progenitor fit the observations much better. Even so, it appears necessary to arbitrarily amend the progenitor model by mixing helium into the hydrogen mantle (Woosley 1988) to avoid the "plateau" phase which is conspicuous in the original models but missing in SN 1987A (cf. Catchpole *et al.* 1987, Figure 6).

The question then arises: why was it a blue supergiant which was the progenitor rather than the expected red supergiant? It has been suggested that SNe II progenitors of solar composition will collapse in the red supergiant phase but that a reduction in metal abundance will allow the progenitors to evolve through this stage and back towards the blue side of the HR diagram before becoming supernovae. Models using metallicities of ~1/4 solar, ([Fe/H] ~ -0.6) have generally been adopted (*e.g.* Woosley 1988, Arnett 1987) in discussions of this effect (as usual, square brackets indicate logarithmic quantities).

Chemical Abundances

In discussing chemical abundances of different types of objects in the Magellanic Clouds, attempts have sometimes been made to express the results as a difference between the abundances in the sun and in the objects concerned. This is often a difficult matter when objects very different from the sun in physical characteristics are under consideration. It seems more satisfactory therefore, where possible, to express the results as a difference between the abundances of physically similar objects in the general solar neighbourhood and in the Magellanic Clouds (*e.g.* to compare Cepheids in the solar neighbourhood with those in the Clouds). In the case of young objects, their solar neighbourhood value is taken to define [Fe/H] = 0. By proceeding in this way, the methods of analysis are only used differentially.

A. Young Objects

(1) H II regions

A number of studies have been made of abundances in Magellanic Cloud H II regions. Perhaps the most complete is that of Pagel *et al.* (1978). The situation does not seem to have changed significantly since the review by Dufour (1984). From this we deduce the following abundances [M/H] of MC H II regions relative to those in the general solar neighbourhood.

	$[M_1/H]$	$[M_2/H]$	[He/H]
LMC H II/Local H II	−0.26	−0.58	−0.07
SMC H II/Local H II	−0.61	−1.20	−0.10

M_1 refers to the elements O, Ne, S, Cl, and Ar all of which give about the same values of [M/H], and M_1 to C and N which seem to have a distinctly greater under–abundance. Here, and in several other places, the abundance is given to two decimal places so as to avoid rounding–off errors in deriving means. It should not be inferred that the results are accurate to two figures. The H II regions offer one of the few methods of deriving helium abundances (planetary nebulae may also be used).

(2) Field Supergiants — High Dispersion Studies

In discussing high dispersion spectroscopic analyses of MC field supergiants, it is probably best to consider only the most recent studies. The earlier pioneering work dealt with very luminous objects which could not easily be compared with galactic objects. We are then limited to the SMC G (Ib) supergiants AZV 369 (Foy 1981) and AZV 121 (Thevenin and Foy 1986). Though several elements were considered in these analyses, the results may be taken to refer essentially to Fe and similar metals and are expressed conventionally as [Fe/H].

	[Fe/H]
AZV 369 (SMC)	-0.4 ± 0.3
AZV 121 (SMC)	-0.4 ± 0.2

Although these results are expressed as "with respect to the sun," they may be taken as relative to nearby Galactic G type supergiants since, for five such stars, Thevenin and Foy find $[Fe/H] = 0.0 \pm 0.2$.

(3) Cepheids

There are no high resolution spectroscopic abundance studies of Magellanic Cloud Cepheids. However, there are several estimates of chemical abundances in Cepheids. Most of these depend on comparing some colour or colours, either reddening independent or corrected for reddening, with models. These methods [(a) — (e) below] are therefore deducing abundances from overall blanketting effects and are primarily measuring [Fe/H]. The various methods are as follows.

(a) Walraven five-colour photometry (Pel *et al.* 1981, Pel 1984a, Pel 1984b). This is a differential determination of blanketting with respect to Galactic Cepheids using Kurucz models to fix the scales.

(b) Washington four-colour photometry (Harris 1981, 1983). The scale is set empirically by Galactic Cepheids and by Type II Cepheids and other galactic supergiants of known composition as well as by Kurucz and Bohm-Vitense models. Two estimates of abundance are made: one depending primarily on metal and one on metals plus CN and CH. Harris found no strong evidence for a difference in abundance between these two indices (though there are some individual outstanding cases) and he generally used mean values.

(c) Laney (1982) obtained relatively rough (low dispersion) curves of growth relative to galactic Cepheids.

(d) Wallerstein (1984) determined abundances from spectral types and colours (for an assumed reddening) compared with spectral type-colour relations for galactic Cepheids of types I and II.

To these results we can add the following.

(e) An estimate of abundances in Cepheids in the LMC cluster NGC 1866. This is based on the data given or used by Walker (1987). *UBV* photometry of the bright blue stars in the cluster gives $E_{B-V}(OB) = 0.060$, corresponding to $E_{B-V}(Cepheid) = 0.056$. Using this reddening and Walker's *BVI* measures of the seven three-day Cepheids in the cluster allows us to derive the deviation ($\Delta\ (B-V)$) of the Cepheids from the intrinsic $(B-V)_0$, $(V-I)_0$ locus adopted by Caldwell and Coulson (1985) for galactic Cepheids. Figure 2 reproduces part of Figure 1 of Caldwell and Coulson which calibrates $\Delta(B-V)$ as a function of $(V-I)_0$ using Bell-Gustafsson and Kurucz models for different metal deficiencies. The mean result for the NGC 1866 Cepheids is shown with its (internal) error. The

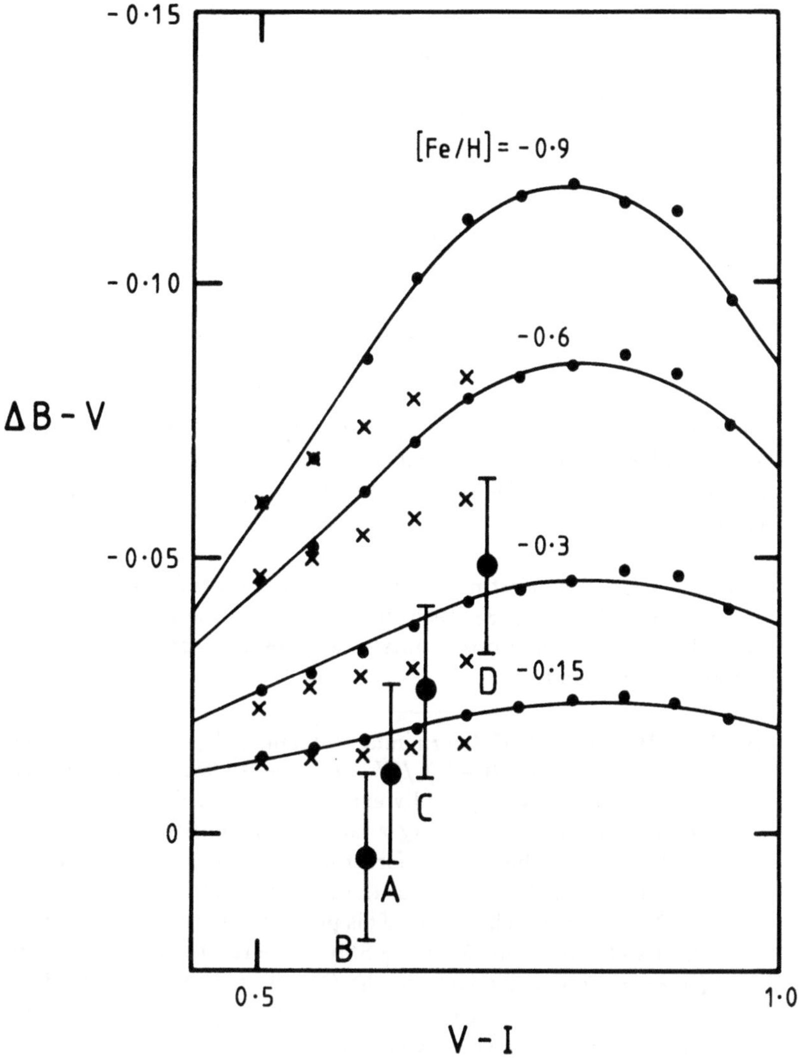

Figure 2: The effects of metallicity on $(B-V)$ as a function of $(V-I)$ for Cepheids as discussed by Caldwell and Coulson (1985), their Figure 1. Dots are Bell–Gustafsson models, crosses are Kurucz models. The positions and (internal) error bars for the NGC 1866 Cepheids are shown for (A) the reddening, $E_{B-V} = 0.056$ of other cluster members (B) mean reddening of all LMC Cepheids with *BVI*, $E_{B-V} = 0.074$ (C) $E_{B-V} = 0.030$ (D) no reddening. The [Fe/H] values are shown against the curves.

Table 1: E_{B-V} from *BVI* of Cloud Cepheids as functions of [Fe/H] (Caldwell)

[Fe/H]	E_{B-V}(LMC)	E_{B-V}(SMC)
0	0.092	0.142
−0.15	0.074	0.115
−0.30	0.059	0.091
−0.60	0.037	0.052
−0.90	0.022	0.024
−2.00	0.001	−0.029

result is also shown adopting E_{B-V}(Cepheid) = 0.074 (the mean value for LMC Cepheids generally; Caldwell and Coulson 1985) E_{B-V}(Cepheid) = 0.030 and $E_{B-V} = 0$. The results indicate [Fe/H] = −0.11 if we adopt the cluster reddening.

(f) Abundances can also be estimated from the *BVI* results (Martin *et al.* 1979, Caldwell and Coulson 1985, 1986) for Cepheids in the general field of the two Clouds. At least three estimates are possible though they are not entirely independent.

(i) The *BVI* method of determining reddening depends on metallicity since the intrinsic $(B-V)_o/(V-I)_o$ line is affected by blanketting. Dr. Caldwell has very kindly extended the calculations in Caldwell and Coulson (1985) to a larger range of metallicities and the results are shown in Table 1. The foreground reddening (E_{B-V}) has been estimated to be 0.034 mag for the LMC and 0.019 mag for the SMC (McNamara and Feltz 1980). This places limits of [Fe/H] > −0.66 for the LMC and > −0.95 for the SMC. The results of various workers for early type supergiants in the Clouds indicate mean reddenings (E_{B-V}) of ~0.08 mag (LMC) and ~0.06 mag (SMC) (cf. Caldwell and Coulson 1985) which would indicate [Fe/H] = −0.10 (LMC) and −0.53 (SMC).

(ii) The displacements of Cloud Cepheids from the Galactic Cepheid period–colour relation can be used to estimate metallicity in the way described by Caldwell and Coulson (1986). Taking into account the effects of metallicity both on blanketting and on mean temperature at a given period, their results lead to [Fe/H] = −0.12 (LMC) and −0.52 (SMC). These are mean results from $(B-V)$ and $(V-I)$ derived from their Figure 13.

(iii) Again, as is clear from Caldwell and Coulson (1986), the PL and PLC relations only give mutually consistent distance moduli for each Cloud for a particular metallicity since both these relations and the necessary reddening corrections depend on metallicity in different ways. The results of Caldwell and Coulson supplemented by new calculations of Dr. Caldwell lead to [Fe/H] = −0.13 (LMC) and −0.44 (SMC).

In method (i) the result depends on blanketting effects which can be taken to measure iron abundance rather directly. In the other two methods, both blanketting and the dependence of temperature at a given period on metallicity are important. This latter dependence is related to the effect of metallicity on the evolutionary tracks and is less directly related to the iron abundance. It may be noted that for metal deficient stars in our Galaxy, there is evidence from the study of relative abundances of various elements that the value of [M/H] appropriate to stellar evolutionary model calculations is higher than the [Fe/H] value (cf. Peterson 1981, Geisler 1984, 1986a, 1986, Nissen *et al.* 1985, Leep *et al.* 1987, Hesser *et al.* 1987). Nevertheless, the three methods give similar results for the Cloud Cepheids and we adopt mean values, [Fe/H] = −0.12 (LMC) and −0.50 (SMC).

(4) Photometric Abundances of Field Supergiants

(a) Van Genderen *et al.* (1986) have applied the Walraven five–colour method (as discussed above for Cepheids) to 51 F and G supergiants in the LMC (and a similar number in our Galaxy). They find [Fe/H] = −0.18 and a reddening of $E_{B-V} = 0.05$. If $E_{B-V} = 0$, they find [Fe/H] = −0.30 which is evidently a lower limit.

(b) H. A. Smith (1980) determined [Ca/H] for F–type supergiants using CaII K line and Balmer line strengths measured on low dispersion spectra and calibrated with Kurucz models. This is essentially a modification of the Preston ΔS method for RR Lyrae stars. It is assumed that [Ca/H] = [Fe/H].

The results from all the above determinations are given in Table 2. All these objects may be considered young ($< \sim 10^8$ years) although a precise estimate of their ages is not given here.

B. Intermediate Age and Old Objects in the Field

Abundances for old or intermediate age objects are only of use for the present purpose if some estimate of the age of the object can also be made. Apart from clusters, the only suitable objects are the following.

Table 2: Abundances in H II Regions, Supergiants and Cepheids

		[Fe/H] LMC	SMC
H II Regions (Dufour)		−0.26	−0.61
Cepheids:	Walraven Phot. (Pel *et al.*)	−0.2	−0.6
	Washington Phot. (Harris)	−0.09	−0.65
	Spectra (Laney)	−0.06	−0.50
	Spectra (Wallerstein)		−0.56
	BVI (SAAO)	−0.12	−0.50
	NGC 1866 (this paper)	−0.11	
Supergiants (high dispersion spectra)			−0.4
Supergiants (Walraven Phot. Van Genderen *et al.*)		−0.18	
Supergiants K line measures (H. A. Smith)		−0.2	−0.6

(1) The short period variables and RR Lyrae stars discussed by Butler *et al.* (1982). They determine a Preston ΔS value (from the K line) and this is used to estimate [Fe/H] through the calibration of Butler (1975) obtained from galactic RR Lyrae variables of known composition. The objects studied were RR Lyrae variables in both Clouds and, in the SMC, the "anomalous" bright short period variables. Ages (and masses) were estimated for these later objects from evolutionary considerations. The RR Lyraes are assumed to be amongst the oldest objects. We adopt an age of

Table 3: K Line Abundances of Variables (Butler *et al.* 1982)

	[Fe/H]
RR Lyraes LMC	-1.4 ± 0.1
RR Lyraes SMC	-1.8 ± 0.2
Bright short-period Variables of Age $3\text{x}10^9$ (SMC)	-1.3 ± 0.1
Bright short-period Variables of Age $2.5\text{x}10^8$ (SMC)	-0.4 ± 0.1

10^{10} years for them which is consistent with the ages adopted for some of the Cloud Clusters containing RR Lyraes. Possibly the mean age of the RR Lyrae variables is greater than this. The results are given in Table 3.

(2) The field stars near NGC 121 in the SMC for which Stryker *et al.* (1985) estimate an age of 8 — 14 Gyr and Suntzeff *et al.* (1986) estimate [M/H] = -1.6 ± 0.3 from spectral indices and the colour–magnitude diagram calibrated from globular clusters. They believe the spread of $\pm$ 0.3 to be real.

C. Clusters

Tables 4 and 5 list the data used. The ages are mainly from colour–magnitude diagrams (from the references indicated). In a few cases, the calibration of SWB classes (Searle *et al.* 1980) by Cohen (1982) has been used. Where necessary, the results have been interpolated to distance moduli of 18.5 (LMC) and 18.8 (SMC) (cf. Feast and Walker 1987). The abundances come from the following sources.

(a) A high dispersion spectroscopic analysis of one late type star in NGC 330 (Spite *et al.* 1986). Although this is a classical type of abundance analysis, the temperature used is derived from a $(B - V)$ colour together with an adopted calibration for metal–poor stars.

(b) Pseudo–equivalent widths from low and medium dispersion spectra of individual stars (Cohen 1982) calibrated against similar measures of stars in Galactic globular clusters of known metallicity (for the present paper the Zinn *I* scale has been adopted; cf. Zinn and West 1984). $(V - K)$ was used as a temperature indicator.

(c) Measurements of various features in low dispersion spectra of individual stars calibrated against globular cluster stars (Cowley and Hartwick 1982). It should be noted that the features used include CN and the G band, and that it is not entirely clear how temperature is taken into account.

(d) DDO and Washington four–colour photometry of individual stars. It should be noted that the DDO work measures primarily CN abundances.

(e) *JHK* colours (which are affected by CN abundances) and infrared CO indices of individual cool stars (McGregor and Hyland 1981, 1984, McGregor 1987).

(f) The colour magnitude diagram using, especially for intermediate age and old clusters, the colour of the red giant clump.

Bica *et al.* (1986) have estimated ages and metallicities for 51 clusters in the Clouds using integrated photometry of the cluster in the *G*–band (DDO filter)

Table 4: Metallicities and Ages of Small Magellanic Cloud Clusters

Cluster	Log Age (Years)	Method* (ref)†	[Fe/H]	Method* (ref)†
NGC 121	10.08	cm(1)(2)	−1.4	DDO, Wash, Int. Sp. etc. (see (1))
NGC 152	8.90	cm(3)	−0.8	cm(4)
NGC 330	6.84	cm(5)	−1.4	Sp(6) (see also (7)(9)(10))
NGC 411	9.26	cm(11)(2)	−0.7	cm(11) (revised cf. (12))
NGC 416	9.40	cm(13)	(−1.6)‡	Int. Sp(14) [(13) gives [Fe/H] > −1.0]
K 3	9.90	cm(2)(15)	−1.2	DDO, Wash, Int. Sp. etc. (cf. (15))
L 1	10.0	cm(16)	−1.2	cm(16) Wash(17)
L 113	9.70	cm(2)(18)(3)	−1.4	cm(18)
LW 79	9.30	cm(19)	−0.3	cm(19)
H 4	9.30	cm(20)	−0.7	cm(20)

*Notes on Methods:

cm	=	colour magnitude diagram
RR	=	RR Lyrae members
SWB	=	Cohen (1982) calibration of Searle *et al.* (1980) classes
DDO	=	DDO intermediate band photometry
Wash	=	Washington intermediate band photometry
Sp	=	Spectra of individual stars
Int. Sp.	=	Integrated spectrum
IR	=	J, H, K; CO (see text)
Ceph	=	Cepheids (see text)

‡Less certain values are given in brackets

†References to Tables 4 and 5.

(1) Stryker *et al.* 1985 (2) Seidel *et al.* 1987 (3) Aaronson/Mould 1985 (4) Hodge 1982 (5) Hodge 1983 (6) Spite *et al.* 1986 (7) Carney *et al.* 1985 (8) Schommer *et al.* 1986 (9) McGregor and Hyland 1984 (10) McGregor 1987 (11) Da Costa/Mould 1986 (12) Da Costa *et al.* 1987 (13) Durand *et al.* Ap. J. 1984 (14) Zinn/West 1984 (15) Rich *et al.* 1984 (16) Olszewski *et al.* 1987 (17) Gascoigne *et al.* 1981 (18) Mould *et al.* 1984 (19) Mateo/Hodge 1987 (20) Mateo/Hodge 1986 (21) Thackeray/Wesselink 1953 (22) Cowley/Hartwick 1982 (23) Mould *et al.* 1986 (24) Cohen 1982 (25) Mateo/Hodge 1985 (26) Da Costa *et al.* 1985 (27) Flower 1984 (28) Hodge 1981 (29) Walker/Mack 1988 (30) Hodge 1984 (31) Andersen *et al.* 1987 (32) Alcaino/Liller 1987 (33) Hodge/Lee 1984 (34) Becker/Mathews 1983 (35) Flower 1983 (36) Gascoigne 1980 (37) Olszewski 1984 (38) Flower *et al.* 1983 (39) Hodge/Schommer 1984 (40) Hardy 1981 (41) Andersen *et al.* 1986 (42) Walker 1985 (43) Stryker 1983 (44) Geisel 1987 (45) Mateo *et al.* 1986 (46) Andersen *et al.* 1984.

Table 5: Metallicities and Ages of Large Magellanic Cloud Clusters

Cluster	Log Age (Years)	Method* (ref)	[Fe/H]	Method* (ref)
NGC 1466	10.0	RR(21)	−2.0	Sp(22) Wash(17)
NGC 1651	9.30	cm(2)	−0.3	cm(23) (revised cf.(12))
NGC 1652	9.84	SWB(24)	−0.9	Sp(24)
NGC 1777	8.95	cm(25)	−0.7	cm(25)(cf.(26) (12))
NGC 1783	(9.0)	SWB(24)cm(27)(28)	−0.25	Sp(24)
NGC 1786	(10.0)	RR(29)	−2.2	Sp(22)(29)
NGC 1831	8.48	cm(30)	(−1.0)	cm(30)
NGC 1835	10.1	SWB(24)	−1.7	Sp(22)
NGC 1841	10.0	cm(31)SWB(24)	−2.1	Sp(22)(24)DDO (17)
NGC 1844	7.70	cm(5)	−0.3	Sp(24)
NGC 1846	(10.1)	SWB(24)	−0.9	Sp(24)
NGC 1850	7.48	cm(5)(32)	−0.5	IR(9)(10)
NGC 1856	7.90	cm(3)(33)(32)	−0.1	cm(33)
NGC 1866	7.93	cm(34)(3)(35)	−0.1	Ceph (this paper)
NGC 1868	8.84	cm(3)(27)(4)	−1.2	cm(36)(4)
NGC 1978	9.30	cm(37)(3)	−0.4	Sp(24) cm(37) [(22)Sp gives −1.5]
NGC 1984	6.84	cm(5)	−0.15	Sp(24)
NGC 1994	6.88	cm(5)	−0.15	Sp(24)
NGC 2004	6.90	cm(5)	−0.3	Sp(24) IR(9)(10)
NGC 2100	7.0	cm(5)	−0.5	Sp(24) IR(9)(10)
NGC 2121	8.84	cm(5)(38)(27)	−0.75	Sp(24) [(38)cm gives −1.3]
NGC 2133	8.11	cm(39)	(−1.0)	cm(39)
NGC 2134	8.04	cm(39)(3)	(−1.0)	cm(39)
NGC 2155	9.78	cm(28)	−1.5	Sp(22)
NGC 2173	9.48	cm(28)	−0.55	Sp(24)
NGC 2193	9.34	cm(12)	−0.5	Sp(24) cm(12)
NGC 2209	8.84	cm(3)(5)(27)	−1.0	DDO(17) cm(4) (Wash(40))
NGC 2210	10.0	RR(41)(42)	−1.8	Sp(22)(24) RR(42)
NGC 2213	9.11	cm(26)(12)(2)(3)	−0.4	cm(26)(12) Sp(26) Wash(44)
NGC 2231	9.08	cm(4)(5)(27)	−1.3	cm(4)
NGC 2257	10.15	cm(43)(3)(5)	−1.8	Sp(22)(34) DDO(17)
ESO 121-Sco3	9.95	cm(45)	−0.5	cm(45)
Hodge 11 (SL 868)	(10.1)	cm(46) [(27)cm has 8.9]	−2.1	Sp(24)
E 2	9.18	cm(8)	(−0.23)	cm(8)

*See notes and references at end of Table 4.

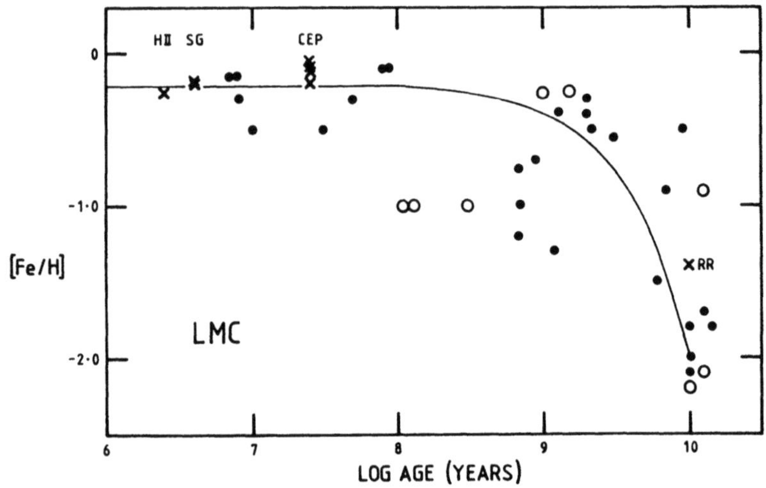

Figure 3: Relation between [Fe/H] and age in the LMC. Clusters from Tables 4 and 5 are plotted as filled circles or (for less certain data) open circles. Crosses are young objects from Table 2 plotted at arbitrary age (see text) or objects from Table 3. H II = H II Regions, sg = supergiants, Cep = Cepheids, RR = RR Lyrae variables.

and Hβ (narrow interference filter). The results are calibrated against galactic globular clusters of known abundances. Of the clusters in Tables 4 and 5, 25 are in common with Bica *et al.* There are quite large differences (of order Δ[Fe/H] ~ -1.0) between some of the values in Tables 4 and 5 and the estimates of Bica *et al.* , and we have not used their data in this paper. Nevertheless, the results of Bica *et al.* define the same general region in the age–metallicity plane (Figures 3 and 4) as the results used. The index used by Bica *et al.* depends on the carbon abundance. Taken at their face value, their results for objects in common with other estimates tentatively suggest that carbon may be under–abundant relative to other metals. However, the evidence is at present not compelling and further work is desirable.

The abundance of C and N in Cloud objects, especially young objects, is particularly interesting in view of the apparent differential under–abundance of these elements in Cloud H II regions (see above). The abundance determinations [M/H] by McGregor and Hyland (1984) and McGregor (1987) for a few young clusters are listed in Table 6. These depend on *JHK* and CO observations of cool stars in the clusters. The results give some measure of CN and CO abundances.

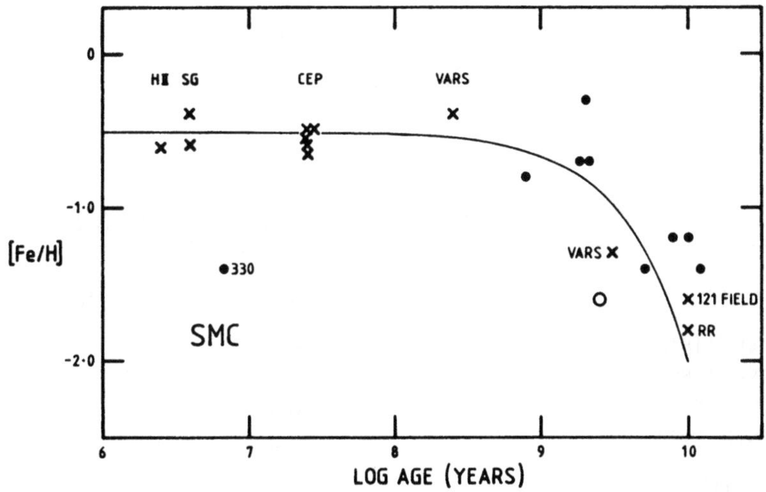

Figure 4: Relation between [Fe/H] and age in the SMC. Symbols as in Figure 2 with in addition as crosses, bright short period variables (= VARS) and the old field population near NGC 121 (= 121 field). The cluster NGC 330 is indicated.

Table 6: LMC Clusters with *JHK*, CO Abundances (McGregor and Hyland)

	[M/H]	[Fe/H]*
NGC 2100	−0.5	−0.5
NGC 2004	−0.5	−0.25
NGC 1850	−0.5	

*Cohen (1982) from spectra

Table 7: Young Clusters with [Fe/H] Estimates by Richtler and Nelles (1983)

LMC	[Fe/H]	[Fe/H]	
NGC 1818	−1.6		
NGC 1866	−1.2 ± 0.2	−0.15[a]	−0.10[b]
NGC 2157	−0.6 ± 0.3		
NGC 2214	−1.2 ± 0.2		
SMC			
NGC 330	−1.8 ± 0.2	−1.4[c]	

[a]Cepheids (this paper)
[b]Models (Becker and Mathews 1983)
[c]Spectra (Spite *et al.* 1986)

For the LMC clusters, McGregor and Hyland estimate a deficiency of a factor of 3 ([M/H] = −0.5). However, they say [*e.g.* McGregor and Hyland 1981] that the *JHK* results indicate a deficiency of "not more than a factor 3," so that, if anything, their results give a somewhat smaller deficiency than is tabulated. These very limited data give no evidence of a differential deficiency of C, N in young objects, but obviously one can draw no strong conclusion.

Richtler and Nelles (1983) have made a preliminary attempt to determine abundances for individual stars in young Cloud clusters using Stromgren four-colour photometry. Their results are shown in Table 7. In the case of NGC 1866 there is a large discrepancy with the Cepheid result discussed above. As pointed out by Carney *et al.* (1985) the Stromgren results refer primarily to C, N abundances. So, the Richtler/Nelles result may imply a large differential under-abundance of these elements in the LMC cluster stars. However, such a large under–abundance (a factor of 10 relative to other elements in the mean) is not supported by the McGregor/Hyland work (see above). In view of the preliminary nature of the Richtler/Nelles work, it seems best not to place too much emphasis on it. Evidently, further work is needed and we have not included their results in the means used here.

It may be anticipated that the situation regarding C and N abundances in the Magellanic Clouds will be complex at least in stars of low and intermediate mass. For instance, planetary nebula in both Clouds show a nitrogen over–abundance of an order of magnitude compared to Cloud H II regions (Monk *et al.* 1988). This is interpreted as indicating a nitrogen over–abundance in the atmospheres of the progenitors due to the operation of the first dredge–up phase. Similarly, an over–abundance of carbon in Cloud planetaries (Maran *et al.* 1982, Aller *et al.* 1987) is to be attributed to the operation of the third dredge–up phase in the progenitors.

Discussion

The results of Tables 2, 3, 4, and 5 are plotted in Figures 3 and 4. The results for the H II regions, supergiants, and Cepheids are plotted at arbitrary ages amongst the young clusters. These latter objects are all less than $\sim 10^8$ years old. Giving unit weight to the means for various types of object we find mean abundances of [Fe/H] $= -0.21$ (LMC) and -0.51 (SMC) for objects of age $\leq 2.5 \times 10^8$ years. In these means, we have omitted clusters marked as uncertain and NGC 330 which is discussed below.

Older than 2.5×10^8 years, there is a drop in [Fe/H] with increasing age. Qualitatively, the same result has been obtained by several earlier investigators (*e.g.* Hodge 1981, Cohen 1982). The curves plotted in Figures 3 and 4 show the expected relation if the metal abundances increase exponentially with time (*i.e.* [Fe/H] increases linearly with time) from [Fe/H] $= -2$ at a time 10^{10} years ago to the values quoted in the last paragraph for the two Clouds at the present. There is a considerable apparent scatter in abundance at a given age. Bica *et al.* (1986) find a similar scatter in their results. In view of the considerable differences in age and/or abundance derived by different investigators for the same cluster it is not yet clear how much (if any) of the scatter is real. (Cowley *et al.* (1979) have tentatively suggested that for young objects there may be a small abundance gradient across the LMC, but see Harris 1983). Within the (large) uncertainties, the mean results do not deviate significantly from a simple exponential increase in metal abundance with time.

There is one discrepant case amongst the young objects. NGC 330, the brightest cluster in the SMC, is apparently very metal deficient. Spite *et al.* (1986) give [Fe/H] $= -1.4$ from a high dispersion spectral analysis of one late type member and this result is supported by other evidence (Tables 4 and 5) though not by the colour–magnitude diagram (Carney *et al.* 1985) which suggests a near solar abundance. It would be highly desirable to obtain further evidence on the metallicity of this cluster, but the balance of the present evidence suggests a large metal deficiency. The result is of special interest because NGC 330 is one of the best examples in either Cloud of a "blue globular" cluster. The cluster is also notable for its high frequency of Be stars (Feast 1972) and its low frequency of spectroscopic binaries (Feast and Black 1980). It is likely that these two latter effects are related. If NGC 330 does indeed have an unusually low metallicity, we may need to ask whether it is separated in space from other young SMC objects. A possibility would be that it was formed out of low metallicity gas on the outskirts of the SMC or in the Magellanic Stream and is merely seen projected against the body of the SMC. Carney *et al.* (1985) have pointed out that the velocity of NGC 330 measured by Feast and Black lies between that of the two main H I concentrations in this direction, so that it is not easily associated with either.

Conclusions

For young objects (age $< \sim 2.5 \times 10^8$ years) the mean [Fe/H] is -0.2 (LMC) and -0.5 (SMC) when the results are derived from a comparison of similar objects in the Clouds and in the solar neighbourhood. These values are close to those adopted in recent discussions of Magellanic Cloud Cepheids (-0.15 (LMC) and -0.6 (SMC) (cf. Feast and Walker 1987 for a summary). If metal-poor objects in our Galaxy are any guide, the metal abundances of relevance for stellar evolutionary calculations would be equal to, or somewhat greater, than the [Fe/H] values (cf. Hesser *et al.* 1987 and the discussion above). The value derived for the LMC is significantly less metal-deficient than that adopted in some recent models of SN 1987A (*i.e.* [M/H] = -0.6). The present results imply that if the blue (rather than red) nature of the progenitor is to be attributed to metallicity effects, then either the late stages of stellar evolution must be very sensitive to small changes in metallicity or we must assume that the previous best-observed SNe II have been super-metal-rich. Our Galaxy shows a radial metallicity gradient $\Delta[M/H] \sim -0.07\,\mathrm{kpc}^{-1}$ and other spirals are similar (cf. Shaver *et al.* 1983, Pagel and Edmunds, 1981). The metallicity conditions appropriate to the occurrence of a SN 1987A-type progenitor in our Galaxy thus presumably exist only [$\sim 3\,$kpc] further than the Sun from the Galactic Centre. If the galaxies illustrated in Figure 6 of Pagel and Edmunds (1981) or Figure 27 of Shaver *et al.* (1983) are typical of galaxies in general, then metallicities appropriate to the LMC occur well within the isophotal radii of a significant fraction of spiral galaxies. Young and Branch (1988) have made the interesting comment that the light curve of SN 1909A in M 101 may possibly be like that of SN 1987A. This supernova occurred ~ 0.8 isophotal radii from the centre of M 101. At this radial distance, the metal abundance of M 101 is intermediate between that of the LMC and the SMC (Pagel and Edmunds 1981, Evans 1986). However SN 1969L, which has been considered the text–book example of an object matching a model with a red supergiant progenitor (Weaver and Woosley 1980, Woosley and Weaver 1986), is unusually far out from the centre of the parent Galaxy (NGC 1058) for a SNII (Ciatti *et al.* 1971 see also Tammann 1974). NGC 1058 is similar to M 33, both galaxies having the same morphological type (Sc(s) II–III) and similar absolute blue magnitudes (−19.27 and −19.07 respectively) according to Sandage and Tammann (1981). SN 1969L was well outside the main body of NGC 1058; a minimum distance of 2.4× the corrected isophotal radius (de Vaucouleurs *et al.* 1976) or 16 kpc at the distance of the galaxy used by Sandage and Tammann (1981). It would perhaps be reasonable to expect that at this distance from the centre of the galaxy, the metal abundance would be less than solar. This is of course not certain, and one should remember that measurements of abundance gradients in the disks of spirals have generally been carried out only to about the isophotal radius (cf. Pagel and Edmunds 1981, Shaver *et al.* 1983). Thus, the quite mild metallicity deficiency of the LMC discussed here and the large radial distance of SN 1969L from the centre of its parent galaxy do not entirely rule out metallicity as the basic cause of the observed differences between SN 1987A and SN 1969L. However, they do indicate that this hypotheses is by no means proved and that it might be wise to examine in more detail other possible causes of the differences (such as different progenitor masses, for instance) especially as the predictions of models are very sensitive to the input physics (cf. Renzini 1987).

Results from Cloud H II regions suggest that the deficiencies of C and N are greater than that of other "metals" ([C,N/H] = −0.6 (LMC) and −1.2 (SMC)). The available stellar observations show no conclusive evidence for such under–abundances, whilst Cloud planetary nebulae show relative over–abundances of C and N presumably due to nuclear processing in their progenitors.

Present evidence suggests that the blue globular cluster, NGC 330 (SMC), with an age of 7×10^6 years and [Fe/H] = −1.4, is significantly more metal deficient than other young SMC objects. This raises the possibility that it was formed outside the main groupings of young SMC objects, perhaps in the Magellanic Stream.

Within the considerable scatter, the abundance–age correlation in both Clouds is consistent with an exponential increase in metal abundance with time.

I am grateful to Dr J. A. R. Caldwell for carrying out the additional calculations mentioned in the text, to Dr P. A. Whitelock for a number of helpful suggestions, and to the visitors and staff involved in the SAAO SN 1987A collaboration for access to results in advance of publication.

References

Aaronson, M. and Mould, J. 1985, *Astrophys. J.* **288**, 551.

Alcaino, G. and Liller, W. 1987, *Astron. J.* **94**, 372.
Aller, L. H., Keyes, C. D., Maran, S. P., Gull, T. R., Michalitsianos, A. G., and Stecher, T. P. 1987, *Astrophys. J.* **320**, 159.
Andersen, J., Blecha, A., and Walker, M. F. 1984, *Mon. Not. Roy. Astron. Soc.* **211**, 695.
Andersen, J., Blecha, A., and Walker, M. F. 1986, *Astron. Astrophys. Suppl.* **64**, 189.
Andersen, J., Blecha, A., and Walker, M. F. 1987, *Mon. Not. Roy. Astron. Soc.* **229**, 1.
Arnett, W. D. 1987, *Astrophys. J.* **319**, 136.
Becker, S. A. and Mathews, G. T. 1983, *Astrophys. J.* **270**, 155.
Bica, E., Dottori, H., and Pastoriza, M. 1986, *Astron. Astrophys.* **156**, 261.
Butler, D. 1975, *Astrophys. J.* **200**, 68.
Butler, D., Demarque, P., and Smith, H. A. 1982, *Astrophys. J.* **257**, 592.
Caldwell, J. A. R. and Coulson, I. M. 1985, *Mon. Not. Roy. Astron. Soc.* **212**, 879.
Caldwell, J. A. R. and Coulson, I. M. 1986, *Mon. Not. Roy. Astron. Soc.* **218**, 223.
Carney, B. W., Janes, K. A., and Flower, P. J. 1985, *Astron. J.* **90**, 1196.
Catchpole, R. M., Menzies, J. W., Monk, A. S., Wargau, W. F., Pollacco, D., Carter, B. S., Whitelock, P. A., Marang, F., Laney, C. D., Balona, L. A., Feast, M. W., Lloyd Evans, T. H. H., Sekiguchi, K., Laing, J. D., Kilkenny, D. M., Spencer Jones, J., Roberts, G., Cousins, A. W. J., van Vuuren, G., and Walker, H. 1987, *Mon. Not. Roy. Astron. Soc.* **229**, 15p.
Catchpole, R. M. and Whitelock, P. A. 1988, *I. A. U. Circ.* **4544**.
Catchpole, R. M., Whitelock, P. A., Feast, M. W., Menzies, J. W., Glass, I. S., Marang, F., Laing, J. D., Spencer Jones, J. H., Roberts, G., Balona, L. A., Carter, B. S., Laney, C. D., Lloyd Evans, T., Sekiguchgi, K., Hutchinson, M. G., Maddison, R., Albinson, J., Evans, A., Allen, D. A., Winkler, H., Fairall, A., Corbally, C., Davies, J. K., and Parker, Q. A. 1988, *Mon. Not. Roy. Astron. Soc.* **231**, 75p.
Ciatti, F., Rosino, L., and Bertola, F. 1971, *Mem. Soc. Astron. Ital.* **42**, 163.
Cohen, J. G. 1982, *Astrophys. J.* **258**, 143.
Cowley, A. P., Dawson, P., and Hartwick, F. D. A. 1979, *Publ. Astron. Soc. Pac.* **91**, 628.
Cowley, A. P. and Hartwick, F. D. A. 1982, *Astrophys. J.* **259**, 89.
Da Costa, G. S., King, C. R., and Mould, J. R. 1987, *Astrophys. J.* **321**, 735.
Da Costa, G. S. and Mould, J. R. 1986, *Astrophys. J.* **305**, 214.
Da Costa, G. S., Mould, J. R., and Crawford, M. D. 1985, *Astrophys. J.* **297**, 582.
Dufour, R. J., 1984. In *I. A. U. Symp. 108, Structure and Evolution of the Magellanic Clouds*, ed. S. van den Bergh and K. S. de Boer (Dordrecht: Reidel), p. 353.
Durand, D., Hardy, E., and Melnick, J. 1984, *Astrophys. J.* **283**, 552.

Evans, I. N. 1986, *Astrophys. J.* **309**, 544.
Feast, M. W. 1972, *Mon. Not. Roy. Astron. Soc.* **159**, 113.
Feast, M. W. and Black, C. 1980, *Mon. Not. Roy. Astron. Soc.* **191**, 285.
Feast, M. W. and Walker, A. R. 1987, *Ann. Rev. Astron. Astrophys.* **25**, 345.
Flower, P. J. 1983, *Publ. Astron. Soc. Pac.* **95**, 122.
Flower, P. J. 1984, *Astrophys. J.* **278**, 582.
Flower, P., Geisler, D., Hodge, P., Olszewski, E., and Schommer, R. 1983, *Astrophys. J.* **275**, 15.
Foy, R. 1981, *Astron. Astrophys.* **103**, 135.
Gascoigne, S. C. B. 1980, in *I. A. U. Symp. 85, Star Clusters*, ed. J. E. Hesser (Dordrecht: Reidel), p. 305.
Gascoigne, S. C. B., Bessell, M. S., and Norris, J. 1981, in *I. A. U. Coll. 68, Astrophysical Parameters for Globular Clusters*, ed. A. G. Davis Philip and D. S. Hayes (Schenectady: L. Davis Press), p. 223.
Geisler, D. 1984, *Astrophys. J.* **287**, L85.
Geisler, D. 1986a, *Publ. Astron. Soc. Pac.* **98**, 762.
Geisler, D. 1986b, *Publ. Astron. Soc. Pac.* **98**, 847.
Geisler, D. 1987, *Astron. J.* **93**, 1081.
van Genderen, A. M., van Driel, W. and Greidanus, H. 1986, *Astron. Astrophys.* **155**, 72.
Hardy, E. 1981, *Astron. J.* **86**, 217.
Harris, H. C. 1981, *Astron. J.* **86**, 1192.
Harris, H. C. 1983, *Astron. J.* **88**, 507.
Hesser, J. E., Harris, W. E., Van den Berg, D. A., Allwright, J. W. B., Shott, P., and Stetson, P. B. 1987, *Publ. Astron. Soc. Pac.* **99**, 739.
Hodge, P. 1981, in *I. A. U. Coll. 68, Astrophysical Parameters for Globular Clusters*, ed. A. G. Davis Philip and D. S. Hayes, (Schenectady: L. Davis Press), p. 205.
Hodge, P. W. 1982, *Astrophys. J.* **256**, 447.
Hodge, P. W. 1983, *Astrophys. J.* **264**, 470.
Hodge, P. W. 1984, *Publ. Astron. Soc. Pac.* **96**, 947.
Hodge, P. W. and Lee, S.-O. 1984, *Astrophys. J.* **276**, 509.
Hodge, P. W. and Schommer, R. A. 1984, *Publ. Astron. Soc. Pac.* **96**, 28.
Laney, C. D. 1982, Ph. D. Thesis, Brigham Young University (Ann Arbor: University Microfilms Int.)
Leep, E. M., Oke, J. B., and Wallerstein, G. 1987, *Astron. J.* **93**, 338.
Maran, S. P., Aller, L. H., Gull, T. R., and Stecher, T. P. 1982, *Astrophys. J.* **253**, L43.
Martin, W. L., Warren, P. R., and Feast, M. W. 1979, *Mon. Not. Roy. Astron. Soc.* **188**, 139.
Mateo, M. and Hodge, P. 1985, *Publ. Astron. Soc. Pac.* **97**, 753.
Mateo, M. and Hodge, P. 1986, *Astrophys. J. Suppl.* **60**, 893.
Mateo, M. and Hodge, P. 1987, *Astrophys. J.* **320**, 626.
Mateo, M., Hodge, P., and Schommer, R. A. 1986, *Astrophys. J.* **311**, 113.

McGregor, P. J. 1987, *Astrophys. J.* **312**, 195.
McGregor, P. J. and Hyland, A. R. 1981, *Astrophys. J.* **250**, 116.
McGregor, P. J. and Hyland, A. R. 1984, *Astrophys. J.* **277**, 149.
McNamara, D. M. and Feltz, K. A. 1980, *Publ. Astron. Soc. Pac.* **92**, 587.
Menzies, J. W., Catchpole, R. M., van Vuuren, G., Winkler, H., Laney, C. D., Whitelock, P. A., Cousins, A. W. J., Carter, B. S., Marang, F., Lloyd Evans, T. H. H., Roberts, G., Kilkenny, D., Spencer Jones, J., Sekiguchi, K., Fairall, A. P., and Wolstencroft, R. 1987, *Mon. Not. Roy. Astron. Soc.* **227**, 39P.
Monk, D. J., Barlow, M. J., and Clegg, R. E. S. 1988, *Mon. Not. Roy. Astron. Soc.*, in press.
Mould, J. R., Da Costa, G. S., and Crawford, M. D. 1984, *Astrophys. J.* **280**, 595.
Mould, J. R., Da Costa, G. S., and Crawford, M. D. 1986, *Astrophys. J.* **304**, 265.
Nissen, P. E., Edvardsson, B., and Gustafsson, B. 1985, in *Production and Distribution of C, N, O Elements*, ed. I. J. Danziger, F. Mattencci, and K. Kjar (Garching: ESO), p. 131.
Olszewski, E. W. 1984, *Astrophys. J.* **284**, 108.
Olszewski, E. W., Schommer, R. A., and Aaronson, M. 1987, *Astron. J.* **93**, 565.
Pagel, B. E. J. and Edmunds, M. G. 1981, *Ann. Rev. Astron. Astrophys.* **19**, 77.
Pagel, B. E. J., Edmunds, M. G., Fosbury, R. A. E., and Webster, B. L. 1978, *Mon. Not. Roy. Astron. Soc.* **184**, 569.
Pel, J. W. 1984a, in *Proc. Second Asian-Pacific Regional Meeting on Astron.*, ed. B. Hidayat and M. W. Feast, (Jakarta: Tira Pustaka), p. 411.
Pel, J. W. 1984b, in *I. A. U. Symp. 108, Structure and Evolution of the Magellanic Clouds*, ed. S. van den Bergh and K. S. de Boer, (Dordrecht: Reidel), p. 170.
Pel, J. W., van Genderen, A. M., and Lub, J. 1981, *Astron. Astrophys.* **99**, L1.
Peterson, R. C. 1981, in *I. A. U. Coll. 68, Astrophysical Parameters for Globular Clusters*, ed. A. G. Davis Philip and D. S. Hayes, (Schenectady: L. Davis Press), p. 121.
Renzini, A. 1987, *ESO Workshop on SN1987A*, ed. I. J. Danziger (Garching: ESO), p. 295.
Rich, R. M., Da Costa, G. S., and Mould, J. R. 1984, *Astrophys. J.* **286**, 517.
Richtler, T. and Nelles, B. 1983, *Astron. Astrophys.* **119**, 75.
Sandage, A. and Tammann, G. A. 1981, *Revised Shapley-Ames Catalog of Bright Galaxies*, (Washington: Carnegie Institution).
Schommer, R. A., Olszewski, E. W., and Aaronson, M. 1986, *Astron. J.* **92**, 1334.
Searle, L., Wilkinson, A., and Bagnuolo, W. G. 1980, *Astrophys. J.* **239**, 803.
Seidel, E., Da Costa, G. S., and Demarque, P. 1987, *Astrophys. J.* **313**, 192.

Shaver, P. A., McGee, R. X., Newton, L. M., Danks, A. C., and Pottasch, S. R. 1983, *Mon. Not. Roy. Astron. Soc.* **204**, 53.

Smith, H. A. 1980, *Astron. J.* **85**, 848.

Spite, M., Cayrel, R., Francois, P., Richtler, T., and Spite, F. 1986, *Astron. Astrophys.* **168**, 197.

Stryker, L. L. 1983, *Astrophys. J.* **266**, 82.

Stryker, L. L., Da Costa, G. S., and Mould, J. R. 1985, *Astrophys. J.* **298**, 544.

Suntzeff, N. B., Friel, E., Klemola, A., Kraft, R. P., and Graham, J. A. 1986, *Astron. J.* **91**, 275.

Tammann, G. A. 1974, in *Supernovae and Supernova Remnants*, ed. G. R. Cosmovici, (Reidel: Dordrecht), p. 215.

Thackeray, A. D. and Wesselink, A. J. 1953, *Nature* **171**, 693.

Thevenin, F. and Foy, R. 1986, *Astron. Astrophys.* **155**, 145.

de Vaucouleurs, G., de Vaucouleurs, A., and Corwin, H. G. 1976, *Second Reference Catalogue of Bright Galaxies*, (Austin: Univ. of Texas Press).

Walker, A. R. 1985, *Mon. Not. Roy. Astron. Soc.* **212**, 343.

Walker, A. R. 1987, *Mon. Not. Roy. Astron. Soc.* **225**, 627.

Walker, A. R. and Mack, P. 1988, *SAAO Preprint* **579**.

Wallerstein, G. 1984, *Astron. J.* **89**, 1705.

Weaver, T. A. and Woosley, S. E. 1980, in *Ninth Texas Symposium on Relativistic Astrophysics.* (*Ann. N. Y. Acad. Sci.* **336**, 335).

Whitelock, P. A., Catchpole, R. M., Menzies, J. W., Feast, M. W., Winkler, H., Marang, F., Glass, I. S., Balona, L. A., Egan, J., Carter, B. S., Roberts, G., Sekiguchi, K., Laney, C. D., Lloyd Evans, T., Laing, J. D., Spencer Jones, J., Fernley, J., James P., Fairall, A. P., Monk, A. S., and van Wyk, F. 1988, *SAAO Preprint* **583**.

Woosley, S. E. 1988, Preprint.

Woosley, S. E., Pinto, P. A., and Ensman, L. 1988, *Astrophys. J.*, in press.

Woosley, S. E. and Weaver, T. A. 1986, *Ann. Rev. Astron. Astrophys.* **24**, 205.

Young, T. R. and Branch, D. 1988, *Nature*, submitted.

Zinn, R. and West, M. J. 1984, *Astrophys. J. Suppl.* **55**, 45.

Discussion

J.–C. Pecker: (1) Although the qualitative abundances have a meaning, I would say that even such data as (Fe/H) imply that departures from LTE are the same in the sun and in the stars under study. I would strongly hesitate to accept the second digits in the (Fe/H) data for the LMC and SMC.

(2) It may be relevant to say that in his thesis (to be defended very soon), Jacques Breysacher (ESO) has conclusively shown that the number of WR stars is much smaller in LMC than either in our galaxy or in SMC. The two questions (metallicity, and WR star counts) may be related.

Feast: I agree that comparing the sun with very different objects may introduce uncertainties. That is why I have relied primarily on a comparison of similar objects in the Clouds and in the solar neighborhood. Especially in the case of the LMC where the under–abundance of metals is very mild, the errors introduced by the models will then be quite small.

In the tables, the second decimal figure is not significant. It is given merely to avoid rounding off errors when taking overall means.

Ages of Star Clusters in the Bok Region of the Large Magellanic Cloud

Gonzalo Alcaino and William Liller

Instituto Isaac Newton, Santiago, Chile

The importance of the Magellanic cloud clusters cannot be overemphasized. At one distance lie a great variety of clusters with ages ranging from the youngest to the oldest found so far in our own Galaxy. At the same time, the relatively large distance to these nearest galaxies creates problems not so often found in investigations in our own Galaxy: this distance makes it difficult to reach to the main sequence of the older clusters, and crowding is severe, especially in globular clusters.

The surest way to proceed is to study, at one time and with a large telescope, a number of clusters that lie close to one another in the sky. Thus, intercomparisons become a relative matter; uncertainties tend to cancel out. The comparison of the observed color–magnitude diagrams (CMDs) with theoretical isochrones can provide more reliable results.

Consequently, we have carried out a simultaneous investigation of the CMDs of 14 clusters located within a sky area of about 1 square degree. We chose the so–called Bok region (Bok and Bok 1969) in the northwestern part of the LMC bar, an area rich in clusters and where numerous photoelectric standards exist (Alcaino and Liller 1982). Only four of the selected clusters have had previous *BV* photometry; none have been observed in *RI* passbands.

Our observations were made with three different telescopes: the 3.6–m reflector at La Silla (prime focus); the 2.5–m du Pont telescope at Las Campanas (RC focus); and the 1.5–m reflector at CTIO (Cassegrain focus). To extend the photometry beyond $V = 15.3$, the limit of the photoelectric sequence, we used judiciously a Pickering–Racine wedge.

In the resulting CMDs, we see clearly the main sequences plus numerous blue and red giants and supergiants that have evolved off the upper part of the main sequence. After making corrections for reddening, we have derived ages by using

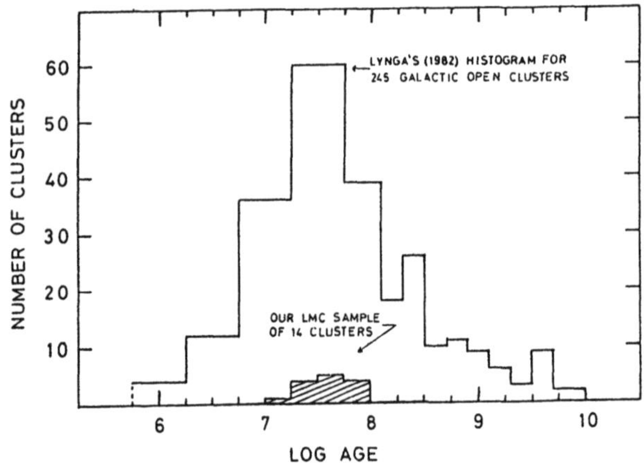

Figure 1: Lyngå's (1982) histogram of galactic cluster ages together with the 14 clusters studied here.

the methods described by Hodge (1983), and by comparing the CMDs with the isochrones of Maeder and Mermilliod (1981), adopting $X = 0.70, Z = 0.03$ with $@_c = 0$. The results are presented in Table 1 together with further explanation of our procedure.

The ages in our sample of clusters range from $17 \pm 6 \times 10^6$y for NGC 1858 to $90 \pm 30 \times 10^6$y for NGC 1860. An age histogram (see Figure 1) peaks at $\log t \sim 7.5$, similar to the result found by Lyngå (1982). Figure 2, a photograph of the region identifying the clusters, includes, under the name, the age in millions of years.

References

Alcaino, G. and Liller, W. 1982, *Astron. Astrophys.* **114**, 213.
Bok, J. B. and Bok, P. F. 1969, *Astron. J.* **74**, 1125.
Hodge, P. W. 1983, *Astrophys. J.* **264**, 470.
Lyngå, G. 1982, *Astron. Astrophys.* **109**, 213.
Maeder, A. and Mermilliod, J. C. 1981, *Astron. Astrophys.* **93**, 136.
Sandage, A. R. and Tammann, G. 1974, *Astrophys. J.* **190**, 525.

Table 1.

Reddening and Ages for the Clusters Studied

Cluster (NGC, SL) (1)	E_{B-V} (2)	$(m\text{-}M)_v$ (3)	Mean Age MST + BBS 10^6y (4)	Age Isochrone 10^6y (5)	Age Adopted 10^6y (6)
1834	0.10:	18.89	52 ± 20:	45 ± 15	48 ± 15
1836	0.20:	19.19	31 ± 10	50 ± 15	38 ± 10
1839	0.27	19.40	22 ± 8	45 ± 10	33 ± 8
1847	0.25	19.34	18 ± 8	35 ± 20	24 ± 10
1850	0.18	19.13	19 ± 10	24 ± 6	21 ± 5
1854	0.20	19.19	20 ± 5	35 ± 15	25 ± 6
1856	0.26	19.37	66 ± 18	90 ± 50	73 ± 20
1858	0.15	19.04	10 ± 15	32 ± 15	17 ± 6
1860	0.18:	19.13	115 ± 56	70 ± 30	90 ± 30
1863	0.20	19.19	57 ± 20	60 ± 20	58 ± 17
1870	0.14	19.01	55 ± 42	85 ± 30	72 ± 30
SL 234	0.15:	19.04	57 ± 22	40 ± 20	48 ± 20
SL 237	0.17:	19.10	20 ± 10	40 ± 20	27 ± 9
SL 304	0.20	19.19	36 ± 16	50 ± 20	42 ± 15

Column 1 identifies the cluster with NGC or SL number. Column 2 lists the reddening obtained by shifting the cluster main sequence to the unreddened ZAMS shown in the V *vs.* $B-V$ diagrams. Colons signify estimated errors larger than ±0.03. Column 3 gives the apparent distance modulus $(m\text{-}M)_V$, deduced by adopting the corrected distance modulus to the LMC of $(m\text{-}M)_o = 18.59$ (Sandage and Tammann 1974) and $A_V = 3E_{B-V}$. In column 4 are the mean ages in millions of years, as obtained from the main sequence turnoff (MST) and the brightest blue star (BBS), following the method of Hodge (1983); standard deviations are from the mean of the two values and colons indicate values from only one of the two methods. Column 5 gives our best age estimate in millions of years from the superimposed isochrones in the V *vs.* $B-V$ diagram, and again the estimated uncertainty. Column 6 gives our final adopted age in millions of years, plus estimated errors.

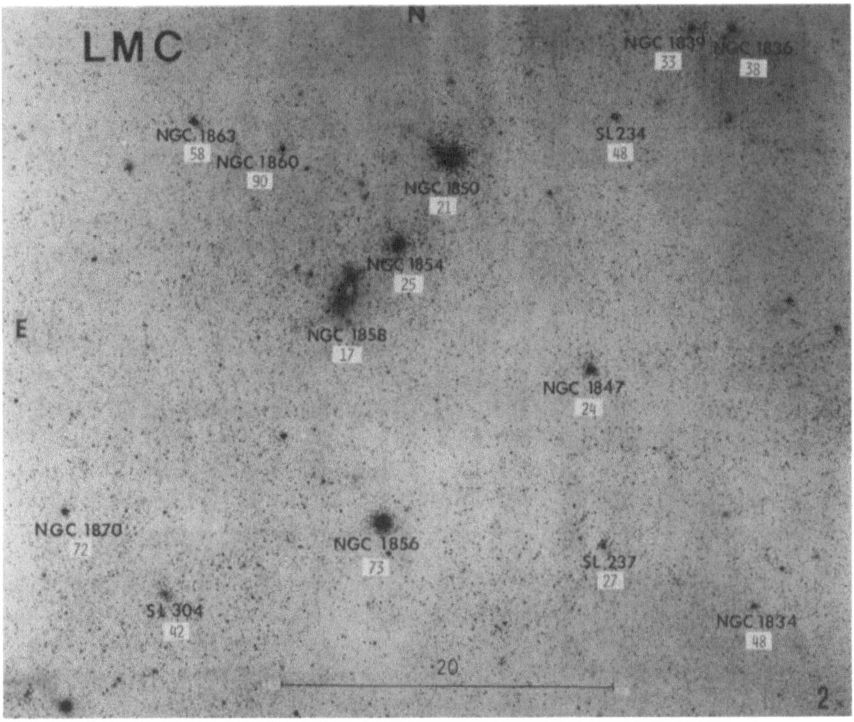

Figure 2: Identification of LMC clusters in the Bok Region. The age in millions of years is given below the cluster name.

Dynamical and Spectral Properties of Some Double Clusters in the LMC

M. Kontizas[1], M. Chrysovergis[1], E. Kontizas[2], and D. Hatzidimitriou[3]

[1]Department of Astrophysics, University of Athens, Greece
[2]National Observatory, Athens, Greece
[3]Department of Astronomy, University of Edinburgh, Scotland, U. K.

Introduction

The existence of binary star clusters in the Large Magellanic Cloud has been recently suggested (Hatzidimitriou and Bhatia 1988, Bhatia and Hatzidimitriou 1988 (BH)). A statistical study of close pairs of clusters (centre–to–centre separation of less than 18 pc) in the LMC yielded a significant number of such pairs that cannot be accounted for by projection effects. Photometric observations of some of these clusters showed that clusters belonging to the same pair often have very similar ages (Bhatia *et al.* 1988 (BCH)). However, the physical association of the pair members still remains to be proved.

In an attempt to investigate the "binarity" of these cluster pairs, we have undertaken a dynamical and stellar population study of some of the candidate pairs catalogued by BH. Density profiles by means of star counts were produced initially for six of the pairs (NGC 2011a and b, NGC 2006 and SL 538, NGC 1775a and b, SL 23 and HS 24, SL 111a and b and LW 43a and b), in order to search for evidence of tidal interaction between the members of each pair. Additionally, their stellar content was examined, using the distribution of the spectral types of their stars.

We present here some preliminary results of this study, which reinforce the suggestion of physical association of at least some of the double clusters.

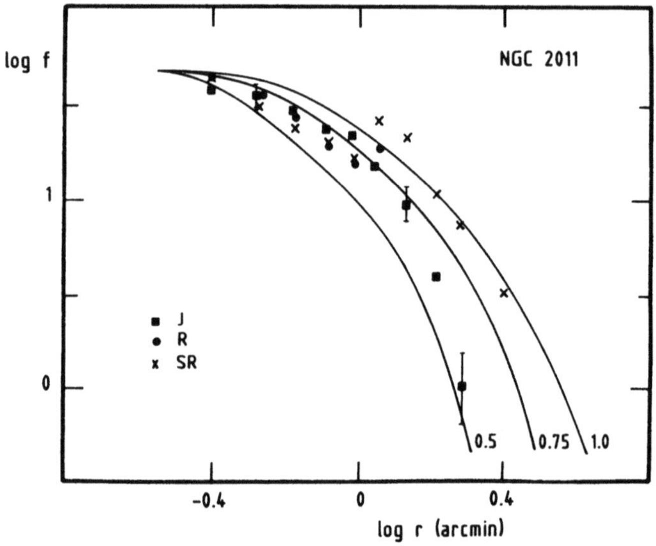

Figure 1: Surface density profiles for the pair NGC2011b, derived from star counts (squares for the J plate, crosses for the SR plate and dots for the I plate). The nearest King models are also illustrated (solid lines).

Dynamical structure

The dynamical structure of a cluster is defined by its own gravity, the influence of the surrounding field and tidal effects from the parent galaxy. Star counts have been proved to be a very effective method for the determination of the dynamical properties of star clusters. We therefore applied this method to investigate the dynamics of the binary cluster candidates in the LMC. The counts were carried out on GG395/IIIa–J (J) and RG715/IV–N (I) photographic plates obtained with the 1.2–m U. K. Schmidt Telescope. The limiting magnitude of the counts reached $B \sim 22.5$. A short exposure RG630/IIIa–F (SR) plate was also used ($B_{lim} \sim 20.0$) in order to examine the distribution of the brightest stars in the clusters. A circular reseau was centred by eye on each cluster and the counts were conducted inside an approximately semicircular area opposite the assumed gravitational centre of the pair and clear of other clusters in the immediate neighbourhood. The method of deriving the structural papameters from the density profiles is described in detail in a previous paper (Kontizas and Kontizas 1983). A more complete presentation of the results will be presented elsewhere (Kontizas, Chrysovergis, Kontizas, and Hatzidimitriou, in preparation). Briefly, *most of the density profiles show evidence of significant disturbances, mostly in the outer regions, and do not seem to fit all the way to any one of the single-mass King models.* Figure 1 shows a typical example of such a density profile. The observed distortions of the density profiles cannot possibly be explained by

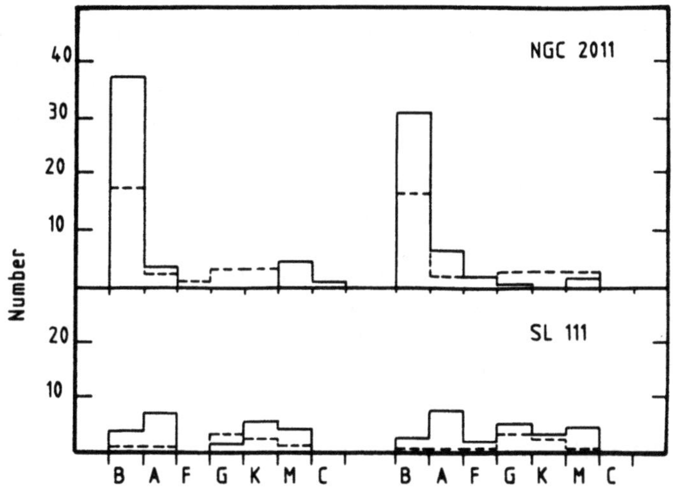

Figure 2: The distribution of spectral types for the stars of the cluster pairs NGC 2011a and b, and SL 111a and b (solid lines). The distributions for the corresponding fields (normalised to the same area) are also shown (dashed lines). A completeness factor was also taken into consideration.

projection considerations; therefore, there must be some kind of gravitational interaction between the cluster pair members. A quantitative study of this interaction would be desirable and for this reason we are performing similar star counts for all catalogued pairs. It should also be emphasized that some of the clusters are imbedded in a highly non-uniform background; therefore, the studied sample must be as large as possible to minimize any random effects from the surrounding fields.

Stellar content and Ages

The stellar content and ages of some of the pairs were examined by means of spectral classification of their brighter stars ($14.5 < B < 18.5$). High quality film copies of medium and low dispersion objective prism plates taken with the UK Schmidt Telescope were used for this purpose (see Dapergolas, Kontizas, and Kontizas 1986). For each pair member, stars were classified in the same area where the star counts were performed and within a radius of 3 arc-min. The spectra of the stars in neighbouring fields were also classified, so that foreground and background objects could be removed from the cluster spectral-type distributions. As an example, Figure 2 presents the distribution of spectral types for the cluster pairs NGC 2011a and b, and SL 111a and b. A very important result of this study is that most of the pairs measured displayed *very similar spectral-type distributions for their members, which implies comparable ages and metallicities.* This justifies the suggestion made by BH and BCH that the "binary" clusters should be more or less coeval and have formed from the same molecular cloud. A rough estimate of the age of a cluster can be derived from the distribution of the spectral types of its stars (Kontizas *et al.* 1987). The clusters NGC 2011a and b were thus found to have ages around 1–2 $\times 10^7$ years, whereas SL 111a and b are older (3–8 $\times 10^7$ yr). However, some of the pairs seem to have ages of several 10^8 yr (Kontizas *et al.* 1988).

It would be very interesting to examine the evolution in time of the dynamical interaction between the cluster pair members, by combining the results on the dynamical structure from the star counts and their ages.

Conclusions

The dynamical properties and stellar content of some cluster pairs in the LMC were studied, using UKST plates. Positive evidence was found for significant distortions in the density profiles of the clusters, particularly in the outer regions. These disturbances were intrerpreted as indications of dynamical interaction between the pair members. Some of the pairs were found to have very similar stellar content, and ages ranging from a few 10^7 yr to several 10^8 yr.

We thank the U. K. Schmidt Unit for the loan of the plates. D. H. was supported by a University of Edinburgh Research Studentship during this work. E.K. acknowledges the financial support of the Greek General Secretariat of research and technology.

References

Bhatia, R. K. and Hatzidimitriou, D. 1988, *Mon. Not. Roy. Astron. Soc.* **230**, 215.

Bhatia, R. K., Cannon, R. D., and Hatzidimitriou, D. 1988, *Stellar Evolution and Dynamics in the Outer Halo of the Galaxy*, ESO Conference and Workshop Proceedings No. **27**, ed. M. Azzopardi and F. Matteucci (Garching bei München: ESO), p. 489.

Dapergolas, A., Kontizas, E., and Kontizas, M. 1986, *Astron. Astrophys. Suppl.* **65**, 283.

Hatzidimitriou, D. and Bhatia, R. K. 1988, I.A.U. Symp. No. **126**, in press.

Kontizas, E. and Kontizas, M. 1983, *Astron. Astrophys. Suppl.* **52**, 143.

Kontizas, E., Kontizas, M., and Xiradaki, E. 1987, *Astron. Astrophys. Suppl.* **71**, 575.

Kontizas, E., Xiradaki, E., and Kontizas, M. 1988, *Astron. Astrophys. Suppl.*, submitted.

Clusters of the Small Magellanic Cloud: L 113 and NGC 411

Antonella Vallenari

Departement of Astronomy, Padova, Italy

Introduction

The purpose of this paper is to determine the age of two Small Magellanic Cloud (SMC) clusters, L 113 and NGC 411, not only by fitting the main sequence turnoff, but by using at the same time all of the characteristic features of their colour–magnitude diagrams (C–M), and comparing the experimental and the theoretical luminosity functions. Our analysis is based on the photometrical data $(B-R)$ and R already published by Da Costa and Mould (1986) for NGC 411 and by Mould *et al.* (1984) for L 113. Previous dating of L 113 (Da Costa *et al.* 1984) and NGC 411 (Da Costa and Mould 1986) indicates that they are both intermediate age clusters.

These determinations are based on the properties of the main sequence turnoff using classical stellar evolution models. Among the various sources of stellar models existing in literature, we adopt those incorporating a non–local view of the convective overshoot in stellar interiors following the method developed by Bressan *et al.* (1981). Starting from the stellar tracks calculated by Bertelli *et al.* (1986), who fixed to 1 the ratio λ of the mean free path of convective elements to the scale height of the pressure, a numerical code has been developed which generates theoretical isochrones and synthetic HR diagrams including all evolutionary stages up to the AGB phase. Theoretical luminosity and temperatures are translated into R magnitude and $(B-R)$ colours making use of the conversion tables published by Green *et al.* (1987). Owing to their location in SMC, it seems reasonable to suppose that these clusters are both affected only by the interstellar reddening due to the Galaxy. Making use of the determination of Azzopardi and Vigneau (1977), we derive $E(B-R)=0.1$.

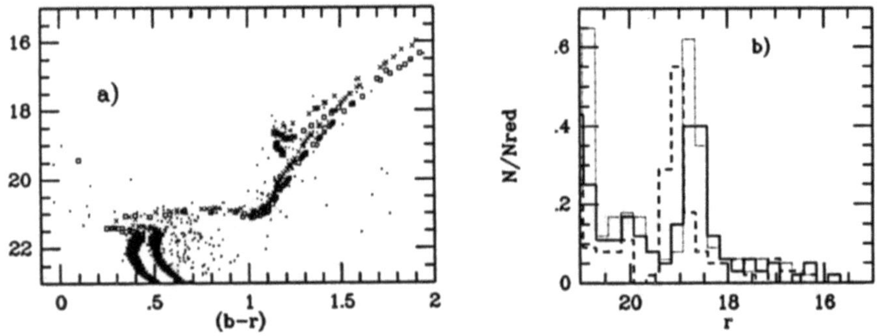

Figure 1: (a) C–M diagram of L 113 fitted to a model of 4 Gyr, $Z = 0.001$, $(m-M)_0 = 18.8$ (crosses; model a); and fitted to a model of 3 Gyr, $Z = 0.001$ and a distance modulus $(m-M)_0 = 19.3$ (stars; model b). (b) Luminosity functions of L 113. Solid line represents the experimental luminosity function, dotted line indicates the theoretical luminosity function for the model a, whereas dashed line indicates the theoretical luminosity function for the model b.

L 113

If we adopt a metal content $Z = 0.001$, following the determination of Mould *et al.* (1984), the C–M diagram of L 113 (Figure 1a) is matched by isochrones with ages of 3 to 4 Gyr depending on the assumed distance modulus of the SMC. The good agreement with the observational and the theoretical luminosity functions for the red stars relative to the ages in question is shown in Figure 1b. The luminosity functions are all normalized to the number of red stars in the clump. On the basis of the photometric data, no correction for contamination by field stars is necessary.

NGC 411

This object is affected by a severe contamination by field stars. We subtract the field stars from the C–M diagram of NGC 411 following the procedure described in Chiosi *et al.* (1988). Unfortunately, the published photometric data do not allow us to properly take crowding into account. This subsequently changes the completeness with respect to magnitude and radius, which should be considered in this type of study (Chiosi *et al.* 1988). We partially obviate this by fitting the luminosity profile of NGC 411, calculated by summing up the contribution of individual stars, with a King law, and by retaining only the region of the cluster where the agreement with the King law is satisfactory. The photometry is reasonably complete outside of 60 pixels (38 arcseconds) for magnitudes ranging from 20 to 24. Using $Z = 0.001$, in good agreement with the determination of

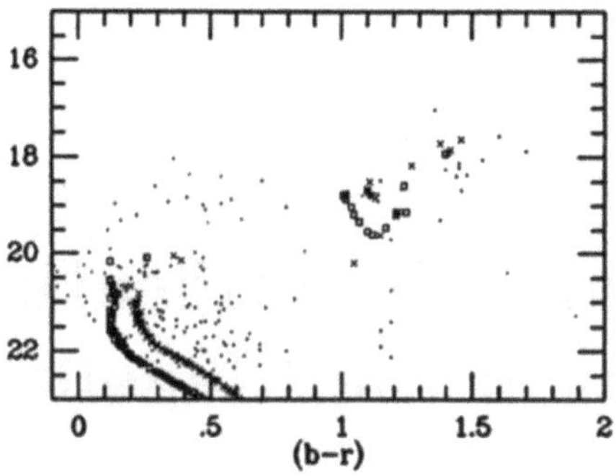

Figure 2: C–M diagram of NGC 411 fit to a synthetic HR diagram calculated for various ages. Squares show the model calculated for an age of 1.4 Gyr, $(m-M)_0 = 19.3$ and $Z = 0.001$; crosses indicate the model of age 1.9 Gyr, $(m-M)_0 = 18.8$, $Z = 0.001$.

Bica *et al.* (1986), we estimate for NGC 411 an age of 1.9 — 1.4 Gyr depending on the distance modulus of the SMC (Figure 2). Notice that for both the clusters, the fit worsens if the larger distance modulus is adopted.

The comparison of the theoretical and experimental luminosity functions is complicated by the presence of a large number of objects bluer than HB stars, whose tip luminosity is located at $R = 18.8$, with a $(B-R)$ ranging from 0 to 1. We point out that no field stars are found in the same region of the C–M diagram. So, it is difficult to avoid the conclusion that these anomalous objects are likely not field stars. However, further observations will be necessary to draw any other conclusion. Work is in progress using a new set of photometric data.

Our determinations of the ages of NGC 411 and L 113 are in good agreement with the previous ones obtained by the authors quoted here.

References

Azzopardi, M. and Vigneau, J. 1977, *Astron. Astrophys.* **56**, 151.
Bertelli, G., Bressan, A., Chiosi, C., and Angerer, K. 1986, *Astron. Astrophys. Suppl.* **66**, 191.
Bressan, A., Bertelli, G., and Chiosi, C. 1981, *Astron. Astrophys.* **102**, 25.
Bica, E., Dottori, H., and Pastoriza, M. 1986, *Astron. Astrophys.* **156**, 261.
Chiosi, C., Bertelli, G., and Bressan, A. 1988, *Astron. Astrophys.*, submitted.
Da Costa, G. S., Mould, J. R. 1986, *Astrophys. J.* **305**, 214.
Green, M. E., Demarque, P., and King, C. R., 1987, *The Revised Yale Isochrones and Luminosity Functions* (New Haven: Yale Univ. Observatory).
Mould, J. R., Da Costa, G. S., Crawford, M. D. 1984, *Astrophys. J.* **280**, 595.

Automated Identification of Star Clusters in the Magellanic Clouds

H. T. MacGillivray[1] and R. K. Bhatia[2]

[1]Royal Observatory, Edinburgh, Scotland, U.K.
[2]Dept. of Astronomy, University of Osmania, Hyderabad, India

Summary

We are carrying out a programme to identify (from completely objective, machine–produced data) the star clusters of the Magellanic Cloud system. Our technique has been tested in a 6 square degree field near the north–eastern end of the bar of the Large Cloud. We have analysed the properties of the clusters detected and present some of our preliminary findings in this paper. A more detailed analysis is presented elsewhere.

1. Introduction

The relative proximity of the Large and Small Magellanic Clouds affords an excellent opportunity for detailed study of their star cluster systems, and for a comparison of the properties of their star clusters with those of our own Galaxy. In this way, examination can be made of formation and destruction rates of clusters in the different galaxies, as well as an investigation of stellar evolutionary patterns, and of the nature of these systems from a global viewpoint.

From a study of clusters from 4–m telescope plates of the LMC region, Hodge (1980) has estimated that the total detectable cluster population of the LMC is ∼5100; a sizeable quantity should exist also for the SMC. Clearly, it is important to identify these clusters and to obtain information about them (such as ages, sizes, shapes, orientations, etc.) so that a meaningful analysis can be carried out on their properties.

Identification of all of the clusters in the Magellanic Cloud system from manual means is a difficult task, as anyone familiar with the LMC topography will testify. Such searches must be conducted over a fairly large area of sky, will contain subjective biases and varying selection effects, and run into serious problems in the regions of crowded stellar background in the inner parts of the Magellanic Clouds. Moreover, once clusters have been identified visually, considerable further effort has to be put into getting the basic data. Even fairly straightforward parameters such as accurate celestial coordinates, ellipticities, etc., are known only for a few hundred of their number. For these reasons, it is ultimately desirable to have an automated technique for identifying the clusters and for obtaining accurate and reliable information which is subsequently suitable for statistical investigation.

We have initiated a major programme to survey the clusters in the Magellanic Clouds from the use of purely objective, machine–produced data. Furthermore, the technique for cluster detection we have developed is applied automatically, systematically, and homogeneously to the data for the whole of the Magellanic Cloud region, providing (as a matter of course) accurate and reliable parameters (coordinates, sizes, shapes, orientations, etc.), thus enabling an analysis of the properties of the clusters to be carried out by completely objective means.

2. The Observations

We use for our investigation automatic scans of glass copy plates of the SERC/ESO Southern Sky IIIa–J survey made with the COSMOS machine (MacGillivray and Stobie 1984) at the Royal Observatory, Edinburgh (ROE). The "crowded–field" software package developed at the ROE is used for the data processing. The resulting output for plates of the LMC region represents a catalogue of several million images per plate (stars and galaxies) down to a limiting magnitude of $B \sim 23$.

The technique used for cluster detection is described fully elsewhere (Bhatia and MacGillivray 1988). We estimate that the method is ~99% complete down to a size limit of 5 pc, and can be extended to smaller sizes with little loss in completeness. We regard 4pc (~0.3 arcmin), however, as a practical lower limit.

3. Results

We have applied the cluster–detection technique to a 6 square degree field (of rectangular shape) to the north of the eastern end of the LMC bar. The field centre has approximate coordinates of R. A. (1950) = $06^h\ 06^m$, and Declination (1950) = $-68°$. 282 clusters were detected in this region down to the limit of 4pc (with approximate limiting magnitude of $B \sim 18$).

We have carried out a preliminary analysis of the properties of these clusters detected. Our main results can be summarised as follows.

a) The LMC clusters are generally larger than their Galactic counterparts; the mode of sizes is 8pc for the LMC clusters while the modal size for Galactic open clusters is 4pc (after Becker and Fenkart 1971). The distributions are similar, with an apparent scale shift being required to produce equality.

b) The ellipticities of the LMC clusters are also larger than their Galactic counterparts, as has been found by other authors. We find no difference in ellipticity between large and small clusters.

c) The position angles of the major axes of the LMC clusters show an anti-correlation with the direction of the bar of the LMC (which we adopt as 120° from de Vaucouleurs and Freeman 1972). The larger clusters and the more spherical clusters show a more random distribution of position angles (see Figure 1).

d) Nearest-neighbouring clusters show a tendency to be aligned with their major axes parallel. Again, the effect is weaker for the larger and more spherical clusters. However, since we find that nearest-neighbours are not preferentially pointing towards each other, we interpret this observation as indicating the result of fragmentation from common initial gas clouds rather than the result of tidal interaction.

References

Bhatia, R. K. and MacGillivray, H. T. 1988, *Astron. Astrophys.*, submitted.
Becker, W. and Fenkart, R. 1971, *Astron. Astrophys. Suppl.* **4**, 241.
Hodge, P. W. 1980, *Astron. J.* **85**, 423.
MacGillivray, H. T. and Stobie, R. S. 1984, *Vistas in Astron.* **27**, 433.
de Vaucouleurs, G. and Freeman, K. C. 1972, *Vistas in Astron.* **14**, 163.

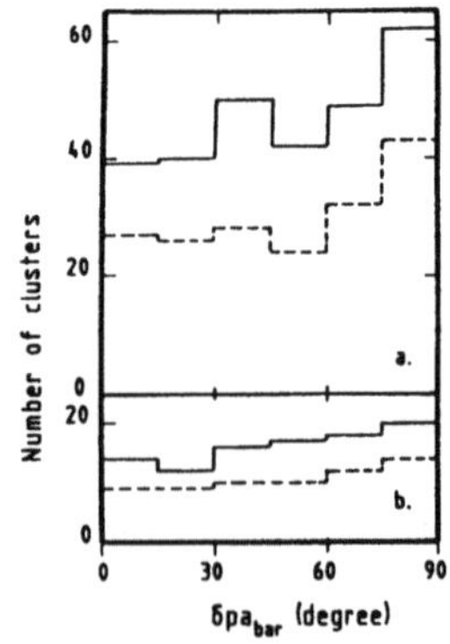

Figure 1: The position angles for the clusters with respect to the position angle of the bar of the LMC. Figure 1a shows the case for all clusters, while figure 1b shows the case for clusters with diameter larger than 10pc. The broken-line histogram shows the distribution when only clusters with ellipticity > 0.3 are examined.

Star Cluster Population of M 33

C. A. Christian

CFHT Corporation and University of Hawaii, U. S. A.

Abstract

The study of M 33 star clusters is pursued in an effort to understand the cluster formation processes in that disk galaxy. The clusters can be used as tracers of the kinematics, chemical evolution, and formation history of M 33, one of the few disk galaxies where such a complete picture can be obtained.

Age Criteria

The determination of cluster ages is accomplished by convolving optical, infrared, and model information to obtain a generalized picture of the cluster population. The data and theory include: (1) Optical Photometry (*UBVRI* Hiltner 1960, Kron and Mayall 1960, Christian and Schommer 1982, 1988) (2) IR Photometry (*JHK* Cohen *et al.* 1984, Christian *et al.* 1988) (3) Spectroscopy (Cohen *et al.* 1984, Christian and Schommer 1983a) and (4) Models (Christian and Schommer 1988, VandenBerg 1985, Chiosi *et al.* 1987).

Calibrations: Optical

The predictions from the theoretical models are compared against published open cluster (Kron and Mayall 1960, Janes and Adler 1982) and globular cluster (Hanes and Brodie 1986) data for the Galaxy, and data for clusters in the Magellanic Clouds (Elson and Fall 1985, Madore 1987). The first set of models (Christian and Schommer 1988) was constructed by calculating integrated colors from theoretical models (VandenBergh 1985) using a power law mass function (Salpeter 1955) supplemented by the empirical giant branch for M 67 (Janes and Smith 1984) with metallicity adjustments from model expectations (Ciardullo and Demarque 1977). The second set is based on a set of convective–overshoot models (Chiosi *et al.* 1987). Ages derived can be checked for consistency by examining integrated spectroscopy obtained for a subset of the clusters by Christian and Schommer (1983a, 1988). The M 33 clusters are subsequently assigned ages by their positions in the *BVI* two-color plane (Christian and Schommer 1988). The tests of the age calibrations are:

Cluster data	Model output	Comment
V, $(B-V)$	fading lines	Normalization (N_{TOT}) arbitrary (models)
$(B-V)$, $\log age$	color–age relation	Stochaistic variations important: for smaller ($N_{TOT} \leq 10^3$) clusters (models, open cl.); also aperture effects, field contamination (LMC, SMC)
$(V-I)$, $\log age$	color–age relation	Stochaistic variations important
$(U-B)$, $(B-V)$	color–color relation	"Turnback" at large ages and varying metallicities ⇒ ambiguous (Christian and Schommer 1988, Elson and Fall 1985, Searle *et al.* 1980)
$(B-V)$, $(V-I)$	color–color relation	"Turnback" at large ages and varying metallicities ⇒ ambiguous

Calibrations: IR

IR photometry is used for dating cluster populations only in that the integrated $(V-K)$ color is roughly a function of age. *JHK* colors are more useful for determining carbon star content in a mixed population (Persson *et al.* 1983) which is derived primarily through consideration of LMC clusters and globulars in M 31 and the Galaxy. Clusters with suspected IR enhancements may contain carbon stars and *most likely* are intermediate aged clusters. The clusters that *do* exhibit such enhancements indeed have intermediate ages as assigned from the optical data.

Metallicity Calibration

Metallicity estimates are made using spectrophotmetry from Cohen *et al.* (1984) and Christian and Schommer (1983a), empirically calibrated against data for galactic clusters (abundances from individual stars) and LMC clusters (primarily from integrated spectra). Abundance estimates are measured using a variety of metallicity indices measuring the strength of features such as Hβ, Mg b, CN, CH, and Fe.

Velocities

Velocities for M 33 clusters have been obtained using low resolution spectroscopy at KPNO and Palomar by Christian and Schommer (1983b) and Cohen *et al.* (1984). These data have been supplemented by observations acquired at the MMT by Schommer *et al.* (1988). The velocity structure of the M 33 clusters is such that the clusters classified previously as "globulars" have a higher velocity dispersion relative to the H I disk than the other clusters with ages $\leq 10^{10}$ yrs (see Schommer *et al.* 1988). The old clusters do not appear to exhibit a disk–type kinematic structure.

Age–Metallicity Relation

With the assigned ages and metallicities, the age–metallicity relation of the M 33 clusters can be examined. Based on an admitttedly small sample, the clusters exhibit an abundance gradient with age. That is, clusters *at a given age* are more metal rich in the inner parts of M 33 than in the outskirts (Christian and Schommer 1988).

Results

(1) An approximate age calibration was derived from *BVI* data for galactic and LMC clusters compared to model predictions and applied to 85 M 33 clusters.
(2) Averaged over the entire M 33 disk, the clusters have formed more or less continuously, with an *apparent* formation rate more smooth than that of the LMC clusters (*cf.* Christian and Schommer 1988 for details).
(3) IR photometry suggests that some of the M 33 clusters with ages $\sim 10^9$ yrs have carbon stars.
(4) The outer cluster system exhibits a slower chemical enrichment history than that exhibited by the inner clusters. Put another way, *at each age*, an abundance gradient exists in the M 33 cluster system.
(5) The cluster system is divided into two distinct kinematic groups: a halo population with a higher velocity dispersion — these clusters have red colors and ages $\geq 10^{10}$, and the younger group exhibiting a disk structure.

Problems and Future Work

Models do not adequately exhibit "turnback" in two–color diagram. *For the M 33 clusters, a set of models at several metallicities and a grid of ages from* 10^7 *to* 10^{10} *would be useful to study the effects of the turnback as a function of metallicity (cf. Figure 1).* Heretofore, the models have been aimed at producing a coherent picture of the age–metallicity relations of the Galaxy (*i.e.*

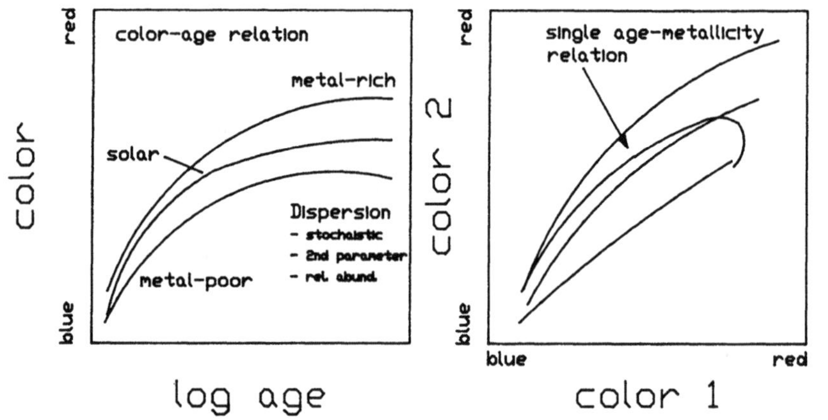

Figure 1: Schematic representation of age–color relations contributing to "turn-back" in integrated two–color diagrams of cluster systems.

solar neighborhood and globular clusters) and the LMC clusters. M 33 has an abundant population of clusters which *appears to contain objects with a wide range of metallicities at each age.* The sample should be increased to investigate both the cluster luminosity fucntion evolution as a function of age, as well as the metallicity gradients as a function of age.

References

Chiosi, C., Bertelli, G. and Bressan, A. 1987, *Astron. Astrophys.*, submitted.

Christian, C. A. and Schommer, R. A. 1982, *Astrophys. J. Suppl.* **49**, 405.

Christian, C. A. and Schommer, R. A. 1983a, *Astrophys. J.* **275**, 92.

Christian, C. A. and Schommer, R. A. 1983b, in *Internal Kinematics and Dynamics of Galaxies*, I.A.U. Symp. No. **100**, ed E. Athanassoula (Dordrecht: Reidel), p. 365.

Christian, C. A. and Schommer, R. A. 1988 *Astron. J.* **95**, 704.

Christian, C. A., Zinnecker, H. and Cannon, R. (1988), in preparation.

Ciardullo, R. and Demarque, P. 1977, *Trans. Yale Univ. Obs.* **33**, 1.

Cohen, J. G., Persson, S. E., and Searle, L. 1984, *Astrophys. J.* **281**, 141.

Elson, R. A. W. and Fall, S. M. 1985, *Astrophys. J.* **299**, 211.

Hiltner, W. 1960, *Astrophys. J.* **131**, 163.

Hanes, D. and Brodie, J. 1986, *Mon. Not. Roy. Astron. Soc.* **214**, 491.

Janes, K. and Adler, D. 1982, *Astrophys. J. Suppl.* **49**, 425.

Janes, K. and Smith, G. 1984, *Astron. J.* **89**, 487.

Kron, G. E., and Mayall, N. U. 1960, *Astron. J.* **65**, 581.

Madore, B. 1987, *preprint.*

Persson, S. E., Aaronson, M., Cohen, J. G., Frogel, J. A., and Matthews, K. 1983, *Astrophys. J.* **266**, 105.
Salpeter, E. E. 1955, *Astrophys. J.* **121**, 161.
Schommer, R. A., Christian, C. A., Caldwell, N., Huchra, J., and Bothun, G. 1988, in preparation.
Searle, L., Wilkinson, A., and Bagnuolo, W. G. 1980, *Astrophys. J.* **239**, 603.
VandenBerg, D. 1985, *Astrophys. J. Suppl.* **58**, 711.

Neutral Hydrogen in the Magellanic System

F. J. Kerr

Astronomy Program, University of Maryland, College Park, Maryland, U.S.A.

The term Magellanic System is being used because in H I observations, the two Clouds are clearly joined together and in fact, recent work has shown that the overall system is much bigger still, covering a quite large area of the sky.

History

Kerr, Hindman, and Robinson (1954) first discovered hydrogen in the Clouds in 1952. This in fact was the first discovery of H I in an external system, the Small Cloud having the distinction of being first. The work was done on a 36-foot dish, mounted as a transit instrument (the biggest radio telescope in the world for a short while), with a receiver which was not very sensitive by modern standards. The first map of the H I distribution is shown in Figure 1. As the amount of dust in the Clouds was known to be low, it was expected that the radio signal might be fairly small. We were surprised to see how strong it was, indicating a higher gas/dust ratio than in the Galaxy. We saw, too, that the Clouds were somewhat bigger in the sky in hydrogen than in the optical image.

The early work showed, too, that both Clouds are rotating, and here is where Gérard comes into the story (Kerr and de Vaucouleurs 1955). We had been working together already, as he was at Mount Stromlo Observatory in Canberra, and I was at the Radiophysics Laboratory in Sydney. He had just been looking at rotational patterns in other contexts, and was able to see immediately that, especially in the LMC, the distribution of the H I radial velocities showed a clear rotational pattern.

Prior to this work, the popular view was that the radial velocities observed in the Clouds could best be interpreted as indicating a common translational motion of the two objects through space, nearly at right angles to the line of sight. This suggestion dates back to Hertzsprung (1920), and it was refined

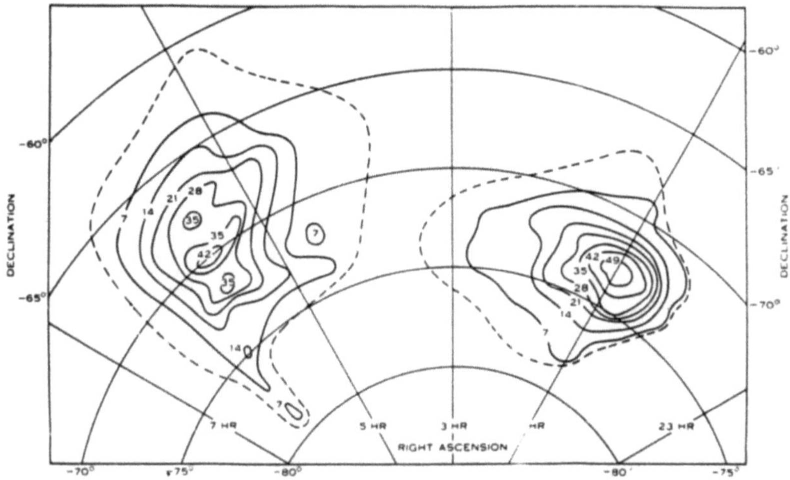

Figure 1: The first map of the distribution of integrated hydrogen over the Magellanic Clouds (Kerr *et al.* 1954; reprinted with permission). The contour unit is $10^{-16}\,\mathrm{W\,m^{-2}\,sterad^{-1}}$. The dashed line approximately encloses the areas within which radiation was detected.

later by Luyten (1928) and Wilson (1944). This interpretation requires a translational velocity of $550\,\mathrm{km\,s^{-1}}$, which seems much too high for any motion in the Local Group, so the demonstration of rotation of each Cloud gave a much more reasonable picture. On the new basis, the supposed translational motion was mostly a reflection of the Galactic rotation, and the variation in its radial component across the sky. Following the rotation study, Gérard and I went on to discuss the total mass of each Cloud (Kerr and de Vaucouleurs 1956).

A second survey in 1963, discussed by Hindman, Kerr, and McGee (1963), showed clearly the existence of a bridge between the two Clouds, joining the two into a single system. Studying this map showed also that the brightest parts of the H I distribution correlated well with the positions of the main H II regions in each Cloud. This map was done with somewhat better receiver sensitivity than was the case for the first survey, but still with a small telescope. In the next stage, surveys of the two Clouds were carried out with the Parkes 64-meter telescope, for the LMC by McGee and Milton (1966a), and for the SMC by Hindman and Balnaves (1967). With the higher angular resolution and better receiver sensitivity, considerably more structural and kinematic detail was seen.

Since 1966 there have been a number of surveys of each Cloud and the various surrounding regions by McGee and Newton (1981, 1986), Mathewson and Ford (1984), Rohlfs *et al.* (1984), all at Parkes; and by Erkes and Turner (1973) in

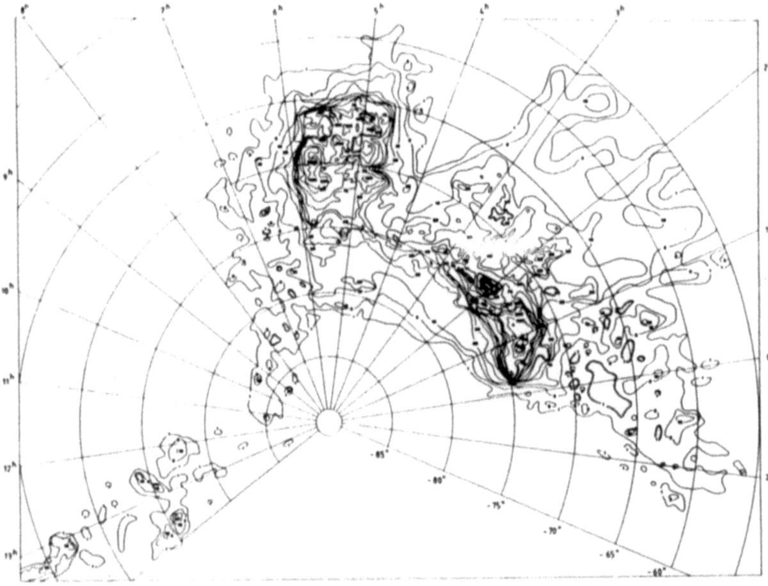

Figure 2: A diagram compiled by Mathewson and Ford (1984) of the integrated H I surface density contours of the Magellanic System. The contour unit is 10^{19} atom cm^{-2}. The data have been taken from a number of surveys by various groups with the Parkes 64–m telescope. Reprinted with permission.

Argentina. Mathewson and Ford (1984) have presented a valuable compilation of all of these in a diagram which is presented as Figure 2. This covers the LMC, the SMC, the Bridge, and a new feature of great interest which has been studied in recent years, the Magellanic Stream. Overall, the gas which is believed to be associated with the Magellanic System occupies about 10,000 square degrees (a quarter of the sky), with a hydrogen mass of approximately $1.8 \times 10^9 M_\odot$, split amongst the four components as 5, 5, 5, and $3 \times 10^8 M_\odot$, respectively. Now we will examine these components in turn.

Large Magellanic Cloud

H I is seen all over the surface of the LMC, at strengths which are often comparable with those seen in the Milky Way. The strongest points have peak brightness temperatures about a third to a half of those at the brightest points in the Galaxy. In the distribution, the bar is not very important, but the Shapley constellations show up well, and there are clear correlations of H I maxima with H II regions, both in distribution and kinematics. There is a striking structure which starts around 30 Doradus and extends a long way southwards, with no optical counterpart. This will be discussed again later.

The overall structure is complex in depth, as shown by the presence of multiple peaks in the line profiles. McGee and Milton (1966b) described the Cloud in terms of two sets of rudimentary spiral arms. More recently, attention has been turned to features described by Meaburn *et al.* (1987) as H I sheets, which are at different depths in the Clouds and are moving away from us at different velocities. They seem to be correlated with the very large shells of Hα emission discovered by Meaburn and others in recent years. Meaburn *et al.* suggest that the sheets have been blown away from the main plane of the LMC by large supernova explosions and stellar winds.

The LMC rotation curve has not changed much from the earlier determinations. An important point is that a turnover is seen in the outer parts, rather than the flat section found in so many larger galaxies, indicating perhaps that there is not much dark matter surrounding tha LMC.

Small Magellanic Cloud

The SMC also has the complexity of structure implied by double or multiple profiles over most of the area. The surface distribution is smoother than that of the LMC, the main feature of interest being the wing prominence. After the early Parkes work, Hindman (1967) talked of 3 large shells, but later work has shown that the situation is more complicated than that.

In 1972, Schmidt questioned whether there might be a difference in the coordinates of H I and Hα maps because there seemed to be a displacement between the most intense regions of H I and H II. In 1981, McGee and Newton (1981, 1982) re–examined the H I in the SMC, and found excellent agreement with Hindman's earlier picture. Their conclusion was that the H I and Hα positions agreed well, provided features with the same radial velocities were compared.

Several authors have spoken of the SMC having great extent in depth, working from the range of the distance moduli of blue supergiant stars and Cepheids. Extending this approach, Mathewson and Ford (1984) introduced the idea of the SMC having two major parts, one behind the other. Then in 1986 Mathewson *et al.* carried out a more systematic study of Cepheids and demonstrated the great extent in depth, especially in one particular section. Details of this picture are apparently rather controversial.

As with the LMC, the SMC rotation curve seems to show a turnover in the outer parts, implying that there is not much dark matter, but in this case the result is less clear, because of the complicated and very asymmetrical structure of the SMC.

Bridge

The Bridge is one of the most striking features of the Magellanic System, and progressive improvements in sensitivity have revealed more and more material there. A recent picture of the detailed structure in this region has been given by McGee and Newton (1986). The most striking structural element is what might be called a dark halo associated with the SMC. On the kinematic side, the survey shows that the median velocity moves clearly across from the SMC value to the LMC value as one proceeds along the Bridge. However, several major velocity components can be identified, showing that the Bridge also is complex in this way.

The total H I mass in the Bridge is $\sim 5 \times 10^8 M_\odot$, as mentioned earlier, an amount which is almost equal to the total in either the LMC or the SMC. It is noteworthy that no stars have yet been found in the Bridge region as far as I know, except very close to the SMC; this indicates that the gas in the bridge is "young."

Magellanic Stream

High velocity clouds of neutral hydrogen have been studied for many years. As most of the early ones had negative radial velocities, the idea arose that these were material falling into the Galaxy. However, Kerr and Sullivan (1969) pointed out that the observed radial velocities of the HVC's contained strong components from galactic rotation, and first proposed that most of these objects are at Magellanic distances. At about the same time, Verschuur (1969) suggested that they might be gas clouds or protogalaxies in the Local Group of galaxies.

Mathewson, Cleary, and Murray (1974) first showed the rather continuous pattern that became known as the Magellanic Stream, and then Davies (1975) and others extended the pattern. Broadly, the Stream is composed of a series of individual concentrations (six principal ones). Arguments were made for the end of the feature being at about the same distance as the Clouds.

Various theories have been proposed for the origin of the Stream, either as a tidal phenomenon or as a primeval structure, strewn along the orbit of the Magellanic Clouds. The latest theory of Mathewson *et al.* (1987) is that the Stream results from first a collision between the SMC and LMC 4×10^8 years ago, followed by a collision between the inter–Cloud gas and HVC's in the galactic halo. The general population of HVC's in this view could be the result of a collision between another galaxy and our own 6×10^9 years ago. Certainly the great mass of gas that can be seen to be moving around out there in regions not so far away is most likely the result of one or more collisions, but the details are sure to be controversial for some time to come.

CO Observations

Although H I is the main subject of this paper, CO is strongly related to it as another method for studying large-scale structure. There have been two surveys, the first by Israel *et al.* (1985) on the 4-meter telescope at La Silla. They saw CO in 11 positions in the LMC near bright H II regions and in six positions in the SMC.

A more recent and more elaborate survey has been carried out by Thaddeus and his colleagues (1988) in the Columbia University survey with a 1.2-meter telescope at Cerro Tololo. This study covered the whole area of the LMC and most of the main body of the SMC. Note that CO observations require longer integration times than for H I, so this was a really monumental survey. In the LMC, the group lists 40 main molecular clouds, some of which are closely related to the Shapley constellations. The most striking object is the very pronounced feature extending southwards from 30 Doradus, closely resembling an important feature in any H I map. This object is presumably due to a recent large burst of star formation; it is quite unlike the appearance of our own Galactic nucleus. In the SMC, an early report by Rubio *et al.* (1984) described CO detections in a small number of positions near known dark clouds and H II regions.

A striking result for both Clouds is that the CO emission is rather weak compared with that from our Galaxy. The brightness temperatures are only about 20% of what one would expect if a typical giant molecular cloud were moved from our Galaxy to the Magellanic distance. This difference is apparently connected with the lower metallicity and lower dust/gas ratio in the Clouds.

Conclusion

The outstanding thing from this review of a long succession of surveys of Magellanic gas is the very large volume of space which contains hydrogen associated with the Magellanic System. Looking to the future, we can expect that more sensitive observations will presumably reveal even more gas which is related to the System. We also see that both Clouds are complex in structure, and can be said to have a number of different pieces, probably as a result of the rather disturbed history of the various parts of the System.

Because it is so close, we can clearly learn a lot from the Magellanic System about the interactions in one type of three-body system, and the progressive evolution of such a system.

References

Davies, R. D. 1975, *Mon. Not. Roy. Astron. Soc.* **170**, 45P.
Erkes, J. W. and Turner, K. C. 1973, *Bull. Am. Astron. Soc. 5*, 430.
Hertzsprung, E. 1920, *Mon. Not. Roy. Astron. Soc.* **80**, 782.

Hindman, J. V. 1967, *Aust. J. Phys.* **20**, 147.

Hindman, J. V. and Balnaves, K. M. 1967, *Aust. J. Phys. Astrophys. Suppl.* No. **4**.

Hindman, J. V., Kerr, F. J., and McGee, R. X. 1963, *Aust. J. Phys.*

Israel, F. P., de Graauw, Th., van de Stadt, H., and de Vries, C. P. 1985, in *I. A. U. Symp. No. 106, The Milky Way Galaxy*, ed. H. van Woerden, R. J. Allen, and W. B. Burton (Dordrecht: Reidel), p. 333.

Kerr, F. J. and Sullivan, W. T. 1969, *Astrophys. J.* **158**, 115.

Kerr, F. J. and de Vaucouleurs, G. 1955, *Aust. J. Phys.* **8**, 508.

Kerr, F. J. and de Vaucouleurs, G. 1956, *Aust. J. Phys.* **9**, 90.

Kerr, F. J., Hindman, J. V., and Robinson, B. J. 1954, *Aust. J. Phys.* **7**, 297.

Luyten, W. J. 1928, *Proc. Nat. Acad. Sci.* **14**, 241.

Mathewson, D. S. and Ford, V. L. 1984, in *I. A. U. Symp. No. 108, Structure and Evolution of the Magellanic Clouds*, ed. S. van den Bergh and K. S. de Boer (Dordrecht: Reidel), p. 125.

Mathewson, D. S., Cleary, M. N., and Murray, J. D. 1974, *Astrophys. J. Lett.* **190**, 291.

Mathewson, D. S., Ford, V. L., and Visvanathan, N. 1986, *Astrophys. J.* **301**, 664.

Mathewson, D. S., Wayte, S. R., Ford, V. L., and Ruan, K. 1987, *Proc. Astron. Soc. Aust.* **7**, 19.

McGee, R. X. and Milton, J. A. 1966a, *Aust. J. Phys. Astrophys. Suppl.* No. **2**.

McGee, R. X. and Milton, J. A. 1966b, *Aust. J. Phys.* **19**, 343.

McGee, R. X. and Newton, L. M. 1981, *Proc. Astron. Soc. Aust.* **4**, 189.

McGee, R. X. and Newton, L. M. 1982, *Proc. Astron. Soc. Aust.* **4**, 308.

McGee, R. X. and Newton, L. M. 1986, *Proc. Astron. Soc. Aust.* **6**, 471.

Meaburn, J., Marston, A. P., McGee, R. X., and Newton, L. M. 1987, *Mon. Not. Roy. Astron. Soc.* **225**, 591.

Rohlfs, K., Kreitschmann, J., and Feitzinger, J. V. 1984, in *I. A. U. Symp. No. 108, Structure and Evolution of the Magellanic Clouds*, ed. S. van den Bergh and K. S. de Boer (Dordrecht: Reidel), p. 395.

Rubio, M., Cohen, R., and Montani, J. 1984, in *I. A. U. Symp. No. 108, Structure and Evolution of the Magellanic Clouds*, ed. S. van den Bergh and K. S. de Boer (Dordrecht: Reidel), p. 399.

Schmidt, Th. 1972, *Astron. Astrophys.* **16**, 95.

Thaddeus, P. 1988, in *The Outer Galaxy*, ed. L. Blitz and F. J. Lockman (New York: Springer Verlag), in press.

Verschuur, G. L. 1969, *Astrophys. J.* **156**, 771.

Wilson, R. E. 1944, *Publ. Astr. Soc. Pac.* **56**, 102. **16**, 570.

Discussion

J.–C. Pecker: You noted in the LMC an interesting and intense structure, observable in H I and in CO, though the bar is much less bright in these wavelengths. Do you have any information on the results which come from IRAS scans of these two different areas of the LMC, and which could be markers for dust, but not for gas, as CO and H I?

Kerr: This has not really been studied yet, but it certainly should be.

F. Bash: Has anyone prepared a map of the LMC and SMC showing the ratios of the H I and CO line strengths, at similar angular resolution, over the face of the Clouds?

Kerr: Not as far as I know.

J. Binney: Is there not evidence that in the solar neighborhood there is a lot of H_2 not associated with CO in clouds dense enough for H_2 molecules to form, but not sufficiently thick to shield CO from destruction by UV photons? Is it not likely that this effect contributes to the small amount of CO emission detected in the Clouds?

Kerr: Yes, this probably occurs in the Clouds as well.

The Origin of the Magellanic Stream

Simon Wayte

Mt. Stromlo and Siding Spring Observatories, Canberra, A. C. T., Australia

Summary

New observations of the Magellanic Stream are used in a recent discrete ram–pressure theory which proposes that the signs of interaction in the Magellanic System point to a collision event some 2–4 $\times 10^8$ years ago, and that the collisions between the inter–Cloud gas and a discrete halo produce the Magellanic Stream.

Introduction

The Magellanic Stream is a band of H I about 8° wide extending some 100° across the sky from the Magellanic Clouds. The Stream is essentially continuous, but there are six main concentrations, MS I to MS VI. There are various theories for the origin of the Stream and the discrete ram–pressure theory is presented here. The discrete ram–pressure theory states that the density enhancements in the halo (possibly visible as high velocity clouds (HVCs)) collide with the gas between the Magellanic Clouds and, mixed with this gas, they form the Magellanic Stream.

The Stream is well–endowed with observational features which severely constrain theories for it's origin. New obseravtions described here (to be published elsewhere) add to these observational characteristics of the Stream. Of particular note is the bifurication of the Stream throughout its entire length (*i.e.* like a pair of tram lines).

Results

The tip of the Stream, mapped using the Parkes 64–m telescope, was found to consist of three main velocity components. These components are centred at $-380\,\mathrm{km\,s^{-1}}$, $-330\,\mathrm{km\,s^{-1}}$, and $-280\,\mathrm{km\,s^{-1}}$. The two components at high negative velocities are complementary in position and the $-330\,\mathrm{km\,s^{-1}}$ component clearly has a bifuricated structure. This indicates that the $-330\,\mathrm{km\,s^{-1}}$ component is truly Stream material and not just a superimposed extreme velocity cloud (EVC).

The core of the $-380\,\mathrm{km\,s^{-1}}$ component was mapped at full spatial resolution. In this core, a number of positions were found that had two–component profiles. These two–component profiles were only found in a small region of high intensity in MS V. A high velocity resolution two–component profile obtained is shown to consist of a broad component of half–width $20\,\mathrm{km\,s^{-1}}$ with a narrow component of $6\,\mathrm{km\,s^{-1}}$ asymmetrically placed towards more negative velocity with respect to the broad peak. The profiles in this region are remarkably similar to the extreme velocity clouds (Wright 1979a,b).

As with HVCs in general, the narrow component occurs at the core of an intense region of H I. A narrow half–width component was found by Morras (1985) in a higher velocity cloud alongside the main Magellanic Stream. The profile shown by Morras has a shape that is typical of profiles in MS V at the position of the core. Thus, it is my opinion that the cloud found by Morras beside the traditional Stream is in fact Stream material similar to the core I found in MS V. The overall pattern of components is that the intensity increases with increasing negative velocity, and only the most negative peak has a narrow component; this is a possible indication of differential motion toward the Galaxy.

Cross–cuts (*i.e.* contour maps of right ascension *versus* velocity) through the Magellanic Stream at various declinations have been obtained. The cross–cuts through the inter–Cloud region show its characteristic high velocity together with lower velocity components superposed.

The R.A.–velocity cross–cuts through the top of the inter–Cloud region and MS I show evidence of continuing interaction. For example, in the Dec. $= -48°$ cross–cut at R.A. $= 0^{\mathrm{h}}\,6^{\mathrm{m}}\,40^{\mathrm{s}}$, interaction is in progress. Gas at velocities different from the Stream is also visible. Further down the Stream, this gas at different velocities becomes less intense. For most cross–cuts, a bifurication of the Stream is seen. This is most beautifully illustrated by the cross–cut at Dec. $= -36°$.

References

Morras, R. 1985, *Astron. J.* **90**, 1801.
Wright, M. C. H. 1979, *Astrophys. J.* **233**, 35.
Wright, M. C. H. 1979, *Astrophys. J.* **234**, 27.

H II Regions in the Magellanic Clouds

G. Courtès

Laboratoire d'Astronomie Spatiale du CNRS and Observatoire de Marseille, Marseille, France

Summary

The H II regions are, in the LMC and SMC, a dominant phenomenon rich in various structures. The relative proximity of the Clouds offers an exceptional opportunity for detailed morphological and spectrographic observations. New results on kinematics and physics of the ionized interstellar medium are described. New directions for future observations are suggested.

I. H II Region Observations

Most of the observations reported here have been obtained from the Hα and [N II] lines. The first inspection shows three broad types:

Type 1) The condensed classical "Strömgren spheres", very similar to the ones of the Milky Way, M 33, M 31, etc. (Courtès 1977), some of them grouped in obvious pseudo–spiral structures.

Type 2) The bubbles and super bubbles, especially detected beginning with the work of Meaburn (1978, 1980), as well as the high contrast Fabry–Perot pictures and the first velocity fields of 43 shells in the LMC (Georgelin *et al.* 1983).

Type 3) The diffuse and filamentary large–scale extended emission regions, similar to the ones recently detected in M 33 (Courtès *et al.* 1987).

The 200–pc diameter bubbles of the Magellanic Clouds have very thin shells in comparison of the ones of M 33. The smaller dust density is certainly in accord with this difference. Absorbing clouds and "bright rim effects" around bubbles of this size are more common in M 33.

A. Comments on the Three Types of H II Regions

Type 1: In spite of the traditional MC classification as Irregular galaxies, there is among the Type 1 H II regions strong evidence for spiral or pseudo–spiral arcs (de Vaucouleurs and Freeman 1972, Schmidt–Kaler 1977) in both clouds. These peculiar structures are apparently related to the formation of a relatively large fraction of the LMC–SMC early stars and their H II regions. A significant example is given in the B1, B2, and B3 arcs noted by de Vaucouleurs and Freeman (1972). Hα pictures (Cruvellier 1967, Courtès and Sivan 1972) exhibit arcs and spiral arrangements.

One of the best ways to test if these spiral arcs have the same origin as the ones observed in spiral galaxies is to verify that the radial velocity gradient is related to the effects of the density waves. These have been observed in Hα emission of spiral galaxies between the front of the arms and the disk emission as we did for the first time in M 33 (Carranza *et al.* 1968, Courtès and Dubout–Crillon 1971, Dubout–Crillon 1976).

Unfortunately, the slow rotational velocities, the ill–determined "pitch angle" (Courtès 1977) of the Magellanic Clouds, and the very common strong expansion phenomenon that we shall describe for each H II region in and out of the arms, seriously compromise the hope of separating the fossil arms and of measuring the elapsed evolution time. The dispersion of the velocities in the H II region LMC rotation curve of Cheriguene and Monnet (1972) is certainly due to the individual expansion effects.

Type 2: There are many intermediate morphologies between Types 1, 2, and 3, often due to simple evolution. The difference between 2 and 3 is often very vague in spite of a relatively clear difference between 200–pc diameter bubbles, and 750 to 1144–pc super bubbles (Goudis and Meaburn 1972, Meaburn 1978). The stellar content is also very important.

Type 3: The diffuse and filamentary large–scale structures are the superposition of many very extended bubbles at various stages of evolution and deterioration. From the first observations in the Galaxy (Courtès 1960), it was obvious that these filaments were, most of the time, edge–on, bidimensional shells seen tangent to the line of sight.

1) Interpretation

All these morphologies are mainly dependent on expansion phenomena of various origins:

Table 1*:Distinctions between evolved supernova remnants and radiatively ionized, wind–driven shells.

Old supernova remnants	Radiatively ionized H II/H I MOL shells
Non-thermal radio shell-source $\alpha \leq -0.5$	Thermal radio shell-source $\alpha \geq -0.2$.
Perhaps central O and B stars.	Always central O and B stars
Fine filamentary nebulosity associated with shocks (up to 250 pc diameter).	Low excitation filaments – bright rims – associated with ionization fronts (up to 250 pc diameter).
Collisionally ionized material – $T_e \geq 20,000$ K, perhaps [S II]/Hα high, etc.	Radiatively ionized material $T_e \leq 10,000$ K
Sometimes pulsar	—
Soft X-ray source.	Stellar wind bubbles may produce something similar, but less energetic.
Expanding neutral and ionized shells $\leq$100 km s^{-1}.	Expanding (or contracting) neutral and ionized shells $\leq$100 km s^{-1}.
—	On edge of huge, neutral clouds

Larger line profiles revealed by Fabry–Perot methods (this paper)

*From Meaburn (1978); reprinted with permission.

a) Classical symmetrical Strömgren sphere expansion, already found for Type 1 (Deharveng 1973) (see Figure 1), and for Type 2 (Georgelin *et al.* 1983);

b) Interaction with stellar winds from more or less condensed star clusters;

c) Bow shocks, convolved with the velocity of the star, generating parabolic profile layers [*e.g.* bubble "p" (N185) in the LMC (Georgelin *et al.* 1983)] as in M 33 (see NN 261–249 and Z100 in Courtès *et al.* 1987); see also Van Buren and McCray (1988);

d) Supernovae blast waves;

e) "Champagne bottle effect" (Tenorio–Taglé 1976) as in NN 88, 218, 632 and ZZ 112, 115, 197 of M 33 (Courtès *et al.* 1987).

A classification is needed; one can, first, imagine a morphology's coefficient of symmetry, starting from the almost perfect spherical (or ring) shape to some

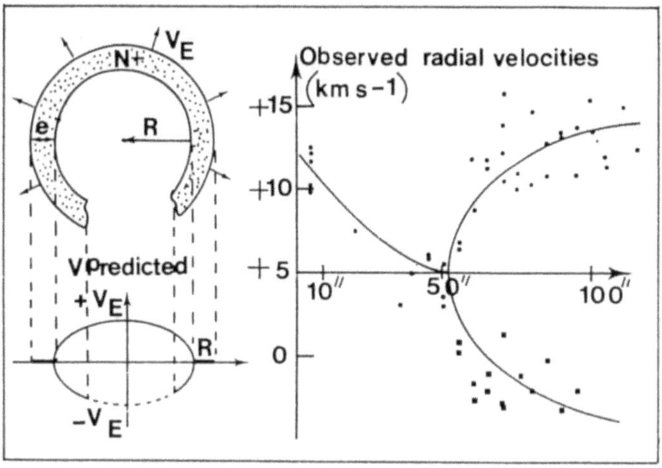

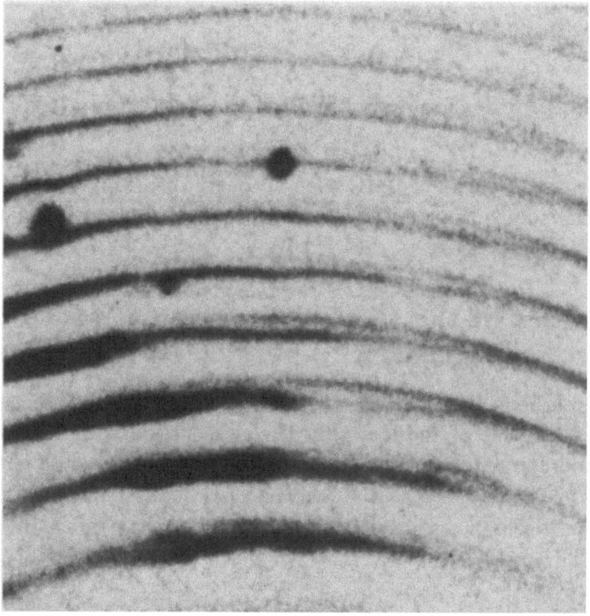

Figure 1: Top: Detail of interferogram I 733 in the region of θ^2 Orionis. Bottom: Radial velocities predicted from an expanding shell model of the Orion Nebula; radial velocities observed in region A; the abscissa is the distance from the outline of the Nebula in a direction perpendicular to this outline. From Deharveng (1973).

degree of segmentaion, and final complete dissipation. The relation between the diameter of the bubble in parsecs and, if possible, the nature of expansion energy would be considered. The best hope is in the most quantitative and accurate methods. We give the following examples:

2) Interference Radial Velocites

The very deep and extended radial velocity mapping due to wide field Fabry–Perot methods (Georgelin *et al.* 1983, Caulet *et al.* 1982, Laval *et al.* 1987) permit us to derive the initial energy from the mass motion measures (Meaburn 1978).

3) Interference Monochromatic Photometry

Dust absorption mixed with the optical emitting regions of the gas is obtained from $H\beta/H\alpha$ intensity ratios by Fabry–Perot interegrated line profiles independent of any stray continuum light (Caplan and Deharveng 1986). This method could be systematically applied to the differences between MC and M 33 H II and H I shell stratification.

4) Spectrographic Radial Velocities of the Absorbing [Ca II] and Na I Lines

Spectra of distant stars in the MC seen through absorbing intersellar material (Maurice 1988) could provide information about the rest of the shells after the optical emission has ceased.

B. Kinematics of the Optical Emission Lines — Need for Resolution and Accuracy

Interference methods have, for the first time, given us the precise profiles of emission lines because of the fact that their resolution is independent of their brightness, and can reach beyond the expected width of the line without loss of sensitivity. This is not the case in a conventional nebular spectrograph in which the line profile, for high resolution, is much broader than the geometrical width of the slit (Courtès 1972).

C. The Quality of the $H\alpha$ Emission

The width of the $H\alpha$ line is affected by:

a) Fine structure of H atoms ($\Delta\lambda = 0.14\AA$),

b) Temperature: $\Delta\lambda = \lambda 0.78 \times 10^{-6}\sqrt{T/M}$,

c) Turbulence. This is often complicated by superposition along the line of sight of emission completely independent of the observed structure. In this case, mean radial velocities, profiles, and other lines (*e.g.* [O III], [S II]) can be used to help sort out the different effects.

d) General internal motions at the scale of the space resolution of the instruments.

Effects b) and c) can be sorted out by comparing profiles of Hα with heavier atoms like N or S. For example, since temperature broadening is a function of the square root of the atomic weight, the [N II] 6584 Å line is about 3.7 times narrower than Hα (Courtès *et al.* 1968b) (see Figure 2). When [N II] is bright enough, this is the best way to detect and measure with great accuracy expansion splitting of the bubbles (Deharveng 1973) (Figure 1). One notes, in the case of Orion Nebula, the remarkable continuity of the expansion.

In Hα, this method applied by the Marseilles Observatory leads to the discrimination of several components with expansion of $30\,\mathrm{km\,s^{-1}}$ for the LMC shell situated at $\alpha_{1975} = 5^{\mathrm{h}}41^{\mathrm{m}}5^{\mathrm{s}}, \delta = -69°25'$. The radius of the sphere is ρ = 475–pc (Caulet *et al.* 1982). The famous complex shell of 30 Doradus shows some areas of empty shells (Cox and Deharveng, 1983) (Figure 3), or at least evidence of a larger density at the inner surface of the spheres as was noted from the beginning of the Fabry–Perot observations (Courtès 1960, p. 95; Courtès *et al.* 1968a). Seen "edge on," these shells give the bright rims of the absorbing clouds temperatures higher than the classical Type 1 H II regions. The interpretation depends on each individual case. On average, the temperature of the H II regions was found to be below the 10×10^3 K predicted by theory. For example, $T_e = 5.5 \times 10^3$ in IC 405 (Courtès *et al.* 1968a). These lower temperatures, obtained from pure Hα profiles, have been confirmed by 109α radio measures.

D. New Methods of Observations

An important recent improvement has been the Field Scanning Interferometer with a photon–counting detector, known as CIGALE (Boulesteix *et al.* 1983; for very wide fields – 15° – also see Caplan *et al.* 1985). A stellar wind bubble N62B was observed with this new instrument, providing imagery at any radial velocity (Figure 4) (Laval *et al.* 1987). The authors found a semi–spherical expanding cavity (V_e = 35 km s^1) around an O8 I star, open towards the observer. The mean diameter is 50–pc. The main advantage of CIGALE, as with all Fabry–Perot designs (Courtès 1977), is to provide a very selective filter (only a few tenths of an Å) resulting in a perfect elimination of the stellar continuum (Figure 4 without star) (Boulesteix *et al.* 1987).

CIGALE could detect, in its survey mode, all the various emission features owing to their different radial velocities and line broadenings; this gives us another way of determining the differing physical natures of the H II regions.

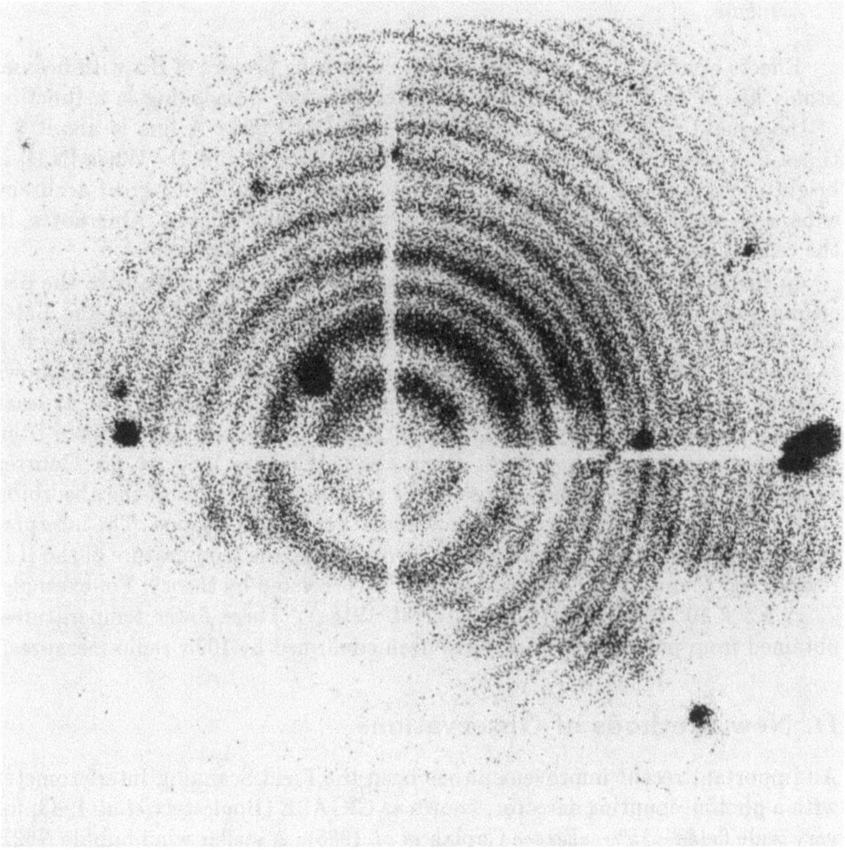

Figure 2: Hα line broadening in comparison with [N II] 6584 Å, in NGC 7000 — bright rim — $T_e = 14 \times 10^3$ K. From Courtès *et al.* (1968b); reprinted with permission.

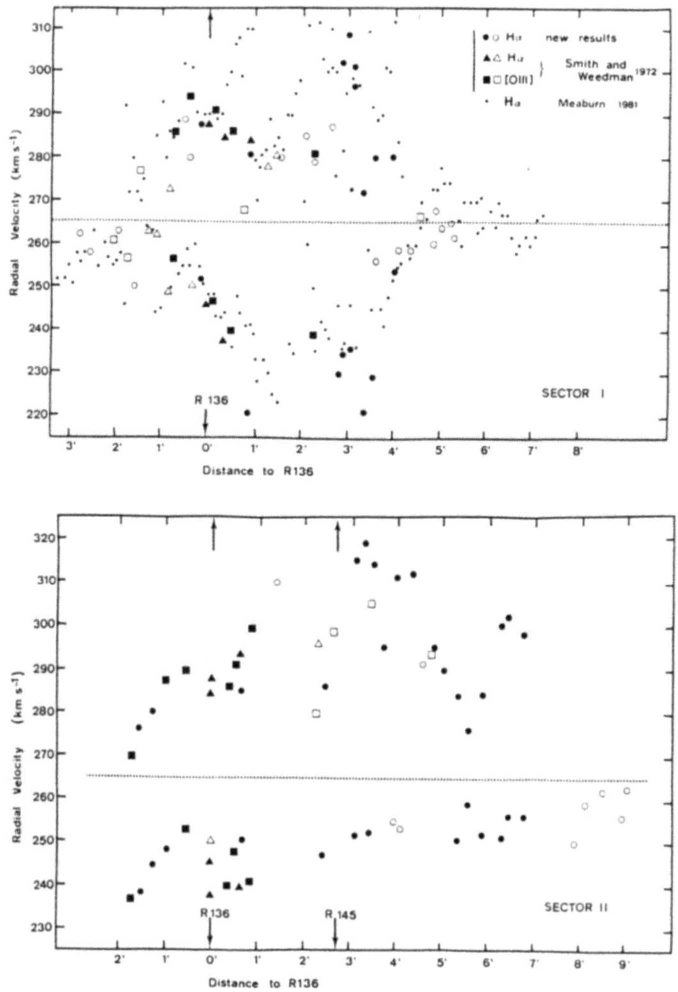

Figure 3: Variation of radial velocity *versus* distance to R136 for sectors 1 (top) and 2 (bottom). The filled symbols represent the components of a split line. From Cox and Deharveng (1983).

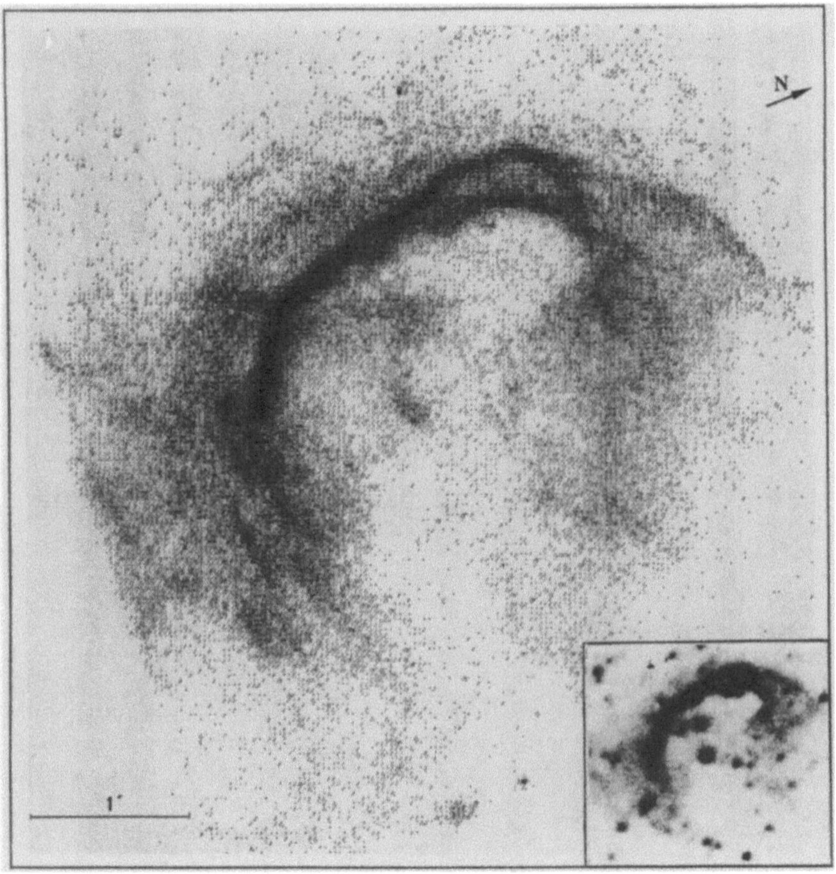

Figure 4: Monochromatic Hα map of N 62B restored using the CIGALE data processing. The first level of the grey scale encoding corresponds to a mean value of 2 counts/pixel (1 pixel = 2.6″× 2.6″), and the step between two levels is 8 counts/pixel. One can see the contours of the square interference filter used. The inset shows the best Hα picture of N 62B (from Lasker 1979, 3h exposure through a 100 Å filter) available until the present study. From Laval *et al.* (1987).

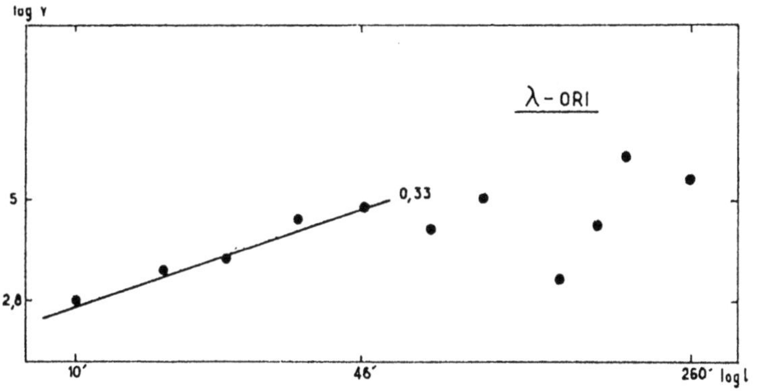

Figure 5: The Kolmogoroff law in the λ Ori H II region. The limit of the relation at 46′ can be interpreted as the limit of the turbulent element. From Courtès (1960); reprinted with permission.

E. The Real Geometry of the Expanding Shells

The high spectral resolution of the Fabry–Perot method shows that the thickness of the H II layer is very small in comparison to the mean radius of the spherical or pseudo–spherical shells. This is seen by the clear separation of the emission lines in two radial velocities components. Another method, used long ago, was derived from the measure of velocity fields and their statistics, in terms of verification of the Kolmogoroff turbulence law (Von Hoerner 1955; Courtès 1953; Courtès *et al.* 1968b; Louise and Monnet 1970): "in an homogeneous and isotropic medium, the mean velocity differences for a distance is given by $u(\Lambda) = k\Lambda^n$ (n being $= 1/3$)." The Kolmogoroff law was verified for several H II regions (Table 2 and Figure 5) showing that the direct morphology of monochromatic imagery, bright rims, and concave evidence of thin emission layers was giving a qualitative reason for the bidimensional repartition of the gas in addition to the statistical verification of the Kolmogoroff law. The primary bidimensional nature of the H II region was then established and has been confirmed in recent years as we have seen above. These kind of turbulence studies, very laborious in 1955, should be now easier owing to the Field Scanning Fabry–Perot Interferometer (CIGALE) and modern computers; it could provide more information on the various effects of disruption of the layers, compressibility phenomenon, expanding condensations, rings, spheres or cylinders, etc. — all important for the physical understanding of the H II regions. There are often secant bubbles corresponding to several expansion events happening at successive times (for example, bubble "g N70" in Georgelin *et al.* 1983; *idem* "l N135").

Table 2.*

Nebula	n	$< V^2 > - < V'^2 >$	$u(\Lambda$ min) (km sec^{-1})	Λ min (arcmin)
NGC 281	—	428.7		
NGC 2174–5	—	143.3		
IC 434†	0.33	33.2	2	2′
Rosette	—	246.4		
NGC 1976†	0.4	141.2	2.3	2′
NGC 7000	—	47.3		
M 8	0.32	30	2	2′
M 17	—	111		
IC 405	—	223		
NGC 1499		128.6		
Ori†	0.33	—	3	2′

*From Courtès (1960) and Louise and Monnet (1970).
†Courtès (1960).

F. Relations with Star Repartition, UV Space Experiments

There are also new observational methods for the interpretation of the expansion velocities and the identification of stars at the origin of the stellar winds (see Table 3, from Meaburn 1978; see also Georgelin *et al.* 1983). Measurments of the integrated UV stellar flux from space experiments (Carruthers and Page 1977, Apollo 16 on the Moon; Vuillemin 1988, Skylab S–183 Experiment of the Laboratoire d'Astronomie Spatiale) will improve our knowledge of the general diffuse emission as well as showing us the cases of expansion around individual OB associations.

The detection, by the Very Wide Field Camera of the Laboratoire d'Astronomie Spatiale aboard Spacelab–I, of a large star forming area (2.5 x 3.0 Kpc) in the wing of the SMC was obtained at 1650 and 1930Å($\Delta\lambda = 400$Å) (Courtès *et al.* 1984) (Figure 6). This stellar cloud, rich in hot stars $< 30 M_\odot$ (Pierre *et al.* 1986) in a gradient of H I intensity in the LMC–SMC bridge, was certainly radiating its UV flux with ionisation of the neutral hydrogen. The verification was easy owing to CIGALE (Marcelin *et al.* 1985) after we had suggested trying to detect Hα and measuring its radial velocities, in order to make the optical identification of the corresponding H I maxima.

The UV maximum flux of the stellar cloud at $\alpha = 1^h50^m$; $\delta = -74°17'$ is well–situated in the LMC–SMC bridge with an Hα radial velocity of $160 \pm 5\,\mathrm{km\,s^{-1}}$, in very good agreement with 21–cm (McGee *et al.* 1985) and stellar (Carrozzi *et al.* 1971) radial velocity data.

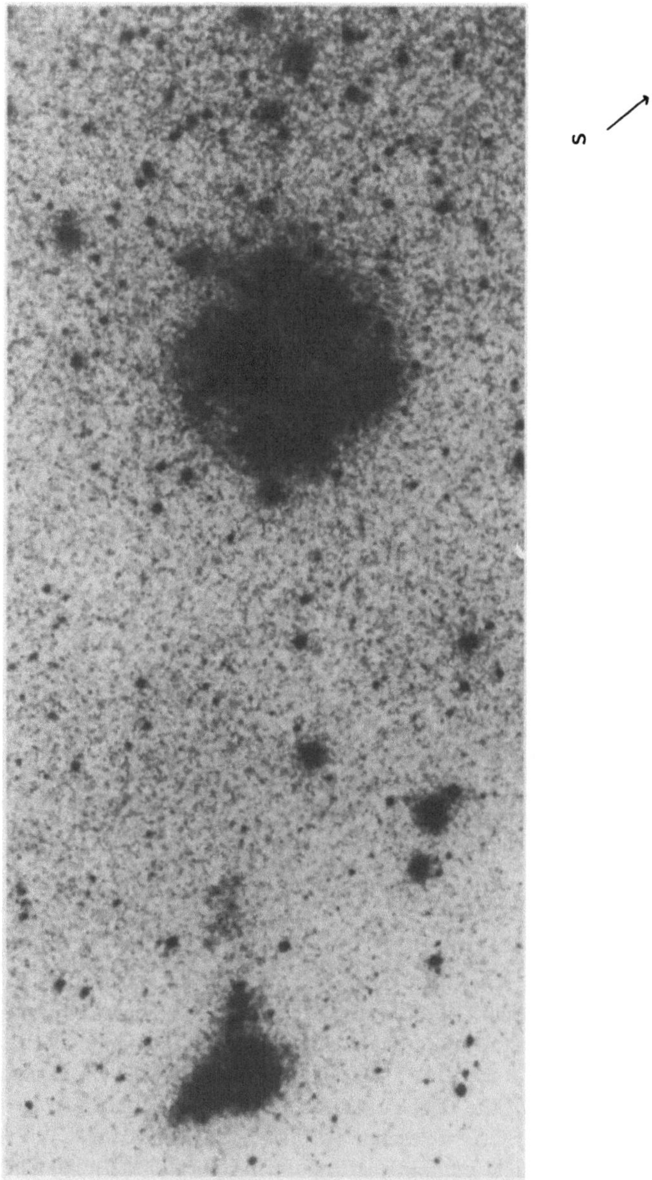

Figure 6: The extension of the UV stellar radiation of the LMC and the SMC. Note the large arc south of the LMC and the star forming cloud in the LMC–SMC bridge detected by the Laboratoire d'Astronomie Spatiale Very Wide Field Camera on Spacelab–1. From Courtès *et al.* 1984, and Pierre *et al.* 1986.

Table 3.*

	1 star	10 stars	100 stars	500 stars
$n_0 = 100$	21–pc	34	54	74
	(40)	(66)	(104)	(142)
	6 km s^{-1}	10	16	22
	(4)	(7)	(10)	(14)
$n_0 = 10$	33–pc	54	85	117
	(64)	(104)	(164)	(226)
	10 km s^{-1}	16	25	35
	(6)	(10)	(16)	(23)
$n_0 = 1$	54–pc	84	134	185
	(104)	(164)	(260)	(358)
	16 km s^{-1}	24	40	55
	(10)	(16)	(26)	(36)
$n_0 = 0.1$	84–pc	133	212	296
	(164)	(258)	(413)	(571)
	25 km s^{-1}	39	64	88
	(16)	(26)	(41)	(57)
$n_0 = 0.01$	133–pc	211	340	470
	(258)	(412)	(656)	(907)
	40 km s^{-1}	64	101	140
	(26)	(41)	(65)	(90)

Values of shell diameters ($2R$) and expansion velocities ($\dot{R}$) after 10^6 yr and (in parentheses) 3×10^6 yr which is the expected lifetime of an average early–type star. These are shown for various numbers of central stars each with $L_w = 10^{36}$ erg s^{-1} in their stellar winds and for various densities (n_0) of the original material of hydrogen atoms from 100 to 0.01 atom cm^{-3}.
***From Meaburn (1978); reprinted with permission.**

II. Extensions — The Ionized LMC–SMC Complex

First, observations of the Clouds with wide field cameras equipped with interference filters are showing optical atomic emission, not only over the whole surface of the clouds, but also down to the limits of the optical UV continuum radiation.

In the case of the LMC, one sees, in the images obtained by the UV balloon 2000 Å peaked SCAP telescope (Milliard 1984) (passband 120 Å), as well as in the very wide field camera pictures from Spacelab (λ 1680 and 1950 Å), that the stellar continuum gives the LMC an almost circular shape with relatively sharp limits (Courtès *et al.* 1984; Pierre *et al.* 1977). Wide field Hα photographs (Figure 7) show approximately the same shape; one concludes that

a) the ionisation is mainly due to the UV radiation of the hot stars (the only

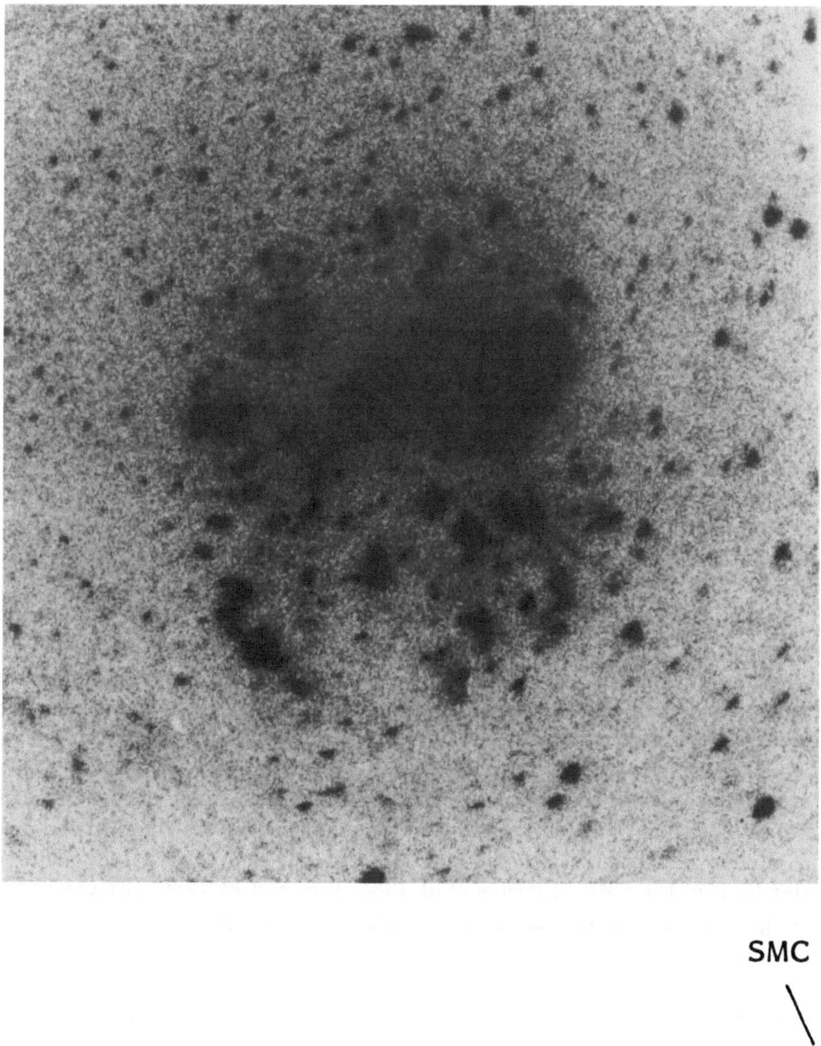

Figure 7: Wide field narrow $H\alpha$ interference filter of the LMC. Compare with the UV picture of Figure 6. The $H\alpha$ emission has the same extension as the UV stellar radiation. From Sivan (1974).

ones $\leq$ 2000 Å) detected by Spacelab and balloon flights,

b) the gas, at least, exhibits a relatively homogeneous density all around the LMC. This is not the case in M 31 where stellar associations like NGC 206 have lost their gas.

Similar remarks can be made about the SMC, but with a large departure from a circular shape. Special assymmetry is due to the "Shapley Wing" and its extensions; the very faint stellar cloud at $\alpha = 1^h57^m$; $\delta = -74°30'$ was detected from the ground in the visible on wide field photographs by de Vaucouleurs; the mean surface brightness is $V = 26.5 \pm 0.5$ mag arcsec^{-2} (de Vaucouleurs and Freeman 1972). Our detection from Spacelab was about $m_{1950\text{Å}} = 24$ mag arcsec^{-2} (Courtès *et al.* 1984; Viton *et al.* 1985) and the ratio of the faint Hα emission to the UV radiation of the stellar clouds (Marcelin *et al.* 1985) is close to the expected value. At our suggestion, observations in the IR (in the 60– and 100–μm bands by IRAS) have been made by Ungerer (1986) but did not lead to any detections. These examples of observations from the IR, Hα (visible), and the near and far UV show that, for these types of hot stellar clouds with stars $\leq 30M_{\odot}$ and their dusty environments, the UV, as well as detection in the visible, is not at a disadvantage. In the same 60–μm band, Van Buren and McCray (1988) detected the IR counterpart of the large, bubble–like, Galactic H II regions. This detection could apply to the LMC Bubble, which would be much brighter than the diffuse H II regions of the Shapley wing. The full extension of the H II complex of the Clouds is not yet well–defined; to date, the deepest Hα imagery survey is that by Johnson *et al.* (1982). The best selection between the various components (some of which could be in the foreground from the Galaxy) of the diffuse Hα emission will be obtained from a detailed Fabry–Perot radial velocity mapping now being prepared by Georgelin using CIGALE. Special attention to possible expansion phenomena will be given to the large UV stellar arc structure (about 60° centered on the LMC) at the southern outer limit of the LMC (Courtès *et al.* 1984; Pierre *et al.* 1986) (Figure 5).

This last suggestion leads to the problem of a possible corona or, at least, a "cosmic bubble bath" extending outside of the main disk of the LMC. We are preparing at LAS a Fabry–Perot experiment ("SCAP 1909") for detecting the C III] doublet, emitted around 10^5 K (Dopita 1977; Deharveng *et al.* 1986). This monochromatic emission is the only one in the UV balloon range (Dopita 1977). The C III] intensity line ratios will be measureable and will permit us to obtain the density of the medium deficient in the inside of the bubble but increasing toward the border shells. At the same time, this C III] observation will detect extensions of the gas emission of the SNR in the LMC. Ground based spectrographs, like the Integral Field Spectrograph on small telescopes (Courtès *et al.* 1989) looking at the Hβ, [O III] and [S II] emission lines will complete this study.

III. Conclusions

The H II regions of the LMC–SMC complex, as well as the ones in M 33, exhibit a detailed sample of the various stages of H II region formation and evolution. The responses to the often contradictory interpretations of these "cosmic bubble baths" with radii from 100 to 1000–pc, recently analysed (with an appreciated touch of humour) by Tenorio–Taglé and Bodenheimer (1988, p.37), is obviously in a field where quantitative and accurate measurements are especially significant (Fabry–Perot gas radial velocities, turbulence, and temperature). The dust medium, more or less rich, changes the appearance of the shells (thick in M 33, thin in LMC). These can be evaluated from H β/H α intensity ratios and stellar reddening (Caplan and Deharveng 1986). Criteria to distinguish between SNR and H II fragments of various origins (Georgelin *et al.* 1983) need an extended spectrographic survey, and comparisons with radio and X–ray data. They are also leading us to discover optical morphology at the scale of the filamentary structures (Laval *et al.* 1987). Absolute spectrophotometry of stars, and evaluation of the expansion energy (up to 10^{54} ergs) are two keys to a better understanding of the observed phenomena, which are sufficiently general in galaxies to obviously represent a major astrophysical interest.

Since the main mechanisms are being at least described if not really understood, observing time at the large telescopes will be in competition with other important studies. In spite of the remarkable efficiency (about 10^2 times conventional methods) of the new instruments (Fabry–Perot, multi–long slits, integral field spectrographs, etc. Courtès *et al.* 1989), there persists a risk, because of the fantastically large number of necessary data, their acquisition, and their reduction times, to stop before reaching a final conclusion. Fortunately, the large scale of the Magellanic clouds leaves us with the possibility of using very efficiently small telescopes (ground based and balloons). Combined with the high quantitative quality of the kinematics, spectrophotometry, and geometry of the new instruments, this fundamental program will have its chance to succeed.

References

Boulesteix, J., Georgelin, Y. P., Marcelin, M., and Monnet, G. 1983, *S. P. I. E. Conference Instr. Astron. V* **445**, 37.

Boulesteix, J., Georgelin, Y. P., Lecoarer, E., Marcelin, M., and Monnet, G. 1987, *Astron. Astrophys.* **178**, 91.

Caplan, J. and Deharveng, L. 1986, *Astron. Astrophys.* **155**, 297.

Caplan, J., Perrin, J. M., and Sivan, J. P. 1985, *Astron. Astrophys.* **145**, 221.

Carranza, G., Courtès, G., Georgelin, Y. P., Monnet, G., and Pourcelot, A. 1968, *Ann. d'Astrophys.* **31**, 63 (1968).

Carrozzi, N., Peyrin, Y., and Robin, A. 1971, *Astron. Astrophys. Suppl.* **4**, 231.

Carruthers, G. R. and Page, T. 1977, *Astrophys. J.* **211**, 728.
Caulet, A., Deharveng, L., Georgelin, Y. M., and Georgelin, Y. P. 1982, *Astron. Astrophys.* **110**, 185.
Cheriguène, M. F. and Monnet, G. 1972, *Astron. Astrophys.* **16**, 28.
Courtès, G. 1953, *Comptes-Rendus Acad. Sci.* **1237**, 378.
Courtès, G. 1960, *Ann. d'Astrophys.* **23**, 115.
Courtès, G. 1972, *Vistas in Astron.* **14**, 81.
Courtès, G. 1977, in *Topics in Interstellar Matter*, ed. H. van Woerden (Dordrecht: Reidel), p. 209.
Courtès, G. and Dubout–Crillon, R. 1971, *Astron. Astrophys.* **11**, 468.
Courtès, G. and Sivan, J. P. 1972, *Astrophys. Letters* **11**, 159.
Courtès, G., Louise, R., and Monnet, G. 1968a, *Ann. d'Astrophys.* **31**, 1.
Courtès, G., Louise, R., and Monnet, G. 1968b, *Ann. d'Astrophys.* **32**, 493.
Courtès, G., Viton, M., Sivan, J. P., Decher, R., and Gary, A. 1984, *Science* **225**, 179.
Courtès, G., Petit, O., Sivan, J. P., Dodonov, S., and Petit, M. 1987, *Astron. Astrophys.* **174**, 28.
Courtès, G., Georgelin, Y., Bacon, R., Boulesteix, J., and Monnet, G. 1989, *I. A. U. Symp. 134, Active Galactic Nuclei* (Dordrecht: Reidel), in press.
Cox, P. and Deharveng, L. 1983, *Astron. Astrophys.* **110**, 185.
Cruvellier, P. 1967, *Ann. d'Astrophys.* **30**, 1059.
Deharveng, L. 1973, *Astron. Astrophys.* **29**, 341.
Deharveng, J. M., Bixler, J., Joubert, M., Bowyer, S., and Malina, R. 1986, *Astron. Astrophys.* **154**, 119.
Dopita, M. A. 1977, *Astrophys. J. Suppl.* **33**, 437.
Dopita, M. A., Mathewson, D. S., and Ford, V. L. 1985, *Astrophys. J.* **297**, 599.
Dubout–Crillon, R. 1976, *Astron. Astrophys. Suppl.* **25**, 25.
Georgelin, Y. M., Georgelin, Y. P., Laval, A., Monnet, G., and Rosado, M. 1983, *Astron. Astrophys. Suppl.* **54**, 459.
Goudis, C. and Meaburn, J. 1978, *Astron. Astrophys.* **68**, 189.
von Hoerner, S. 1955, *I. A. U. Symp. 2, Gas Dynamics of Cosmic Clouds*, ed. J. M. Burgers and H. C. van de Hulst (Amsterdam: North Holland), p. 172.
Johnson, P. G., Meaburn, J., and Osman, A. M. I. 1982, *Mon. Not. Roy. Astron. Soc.* **198**, 985.
Laval, A., Boulesteix, J., Georgelin, Y. P., Georgelin, Y. M., and Marcelin, M. 1987, *Astron. Astrophys.* **178**, 199.
Louise, R. and Monnet, G. 1970, *Astron. Astrophys.* **8**, 486.
Marcelin, M., Boulesteix, J., and Georgelin, Y. 1985, *Nature* **316**, 705.
Maurice, E. 1988, private communication.
McGee, R. X., Newton, L. M., and Mathewson, D. S. 1985, private communication.
Meaburn, J. 1978, *Astron. Space Sci.* **59**, 193.

Meaburn, J. 1984, *Mon. Not. Roy. Astron. Soc.* **211**, 521.
Milliard, B. 1984, Ph. D. thesis.
Pierre, M., Viton, M., Sivan, J. P., and Courtès, G. 1986, *Astron. Astrophys.* **154**, 249.
Schmidt–Kaler, Th. 1977, *Astron. Astrophys.* **54**, 771.
Sivan, J. P. 1974, *Astron. Astrophys. Suppl.* **16**, 163.
Tenorio–Taglé, G. 1979, *Astron. Astrophys.* **71**, 59.
Tenorio–Taglé, G. and Bodenheimer, P. 1988, *Ann. Rev. Astron. Astrophys.* **26**, 145.
Ungerer, V. 1986, private communication.
Van Buren, D. and McCray, R. 1988, *Astrophys. J. Lett.* **329**, L93.
de Vaucouleurs, G. and Freeman, K. C. 1972, *Vistas in Astron.* **14**, 163.
Viton, M., Sivan, J. P., Courtès, G., Gary, A., and Decher, R. 1985, *Adv. Space Res.* **5**, 207.
Vuillemin, A. 1988, *Astron. Astrophys. Suppl.* **72**, 249.

De Vaucouleurs's Galaxy

John N. Bahcall

Institute for Advanced Study, Princeton, New Jersey, U. S. A.

Summary

Gérard de Vaucouleurs proposed the standard geometerical descriptions of both the disk and spheroid populations of spiral galaxies, and applied this description in a pioneering model of the Galaxy. This talk describes the results of computer modeling of the Galaxy using the de Vaucouleurs distributions for the disk and spheroid.

1. Introduction

Star counts is a subject whose time has past and come again. The reasons for the resusitation are: (1) the use of automatic measuring devices; (2) a change of theoretical tactics; and (3) a detailed model which can be used with modern computers. All three of these factors were developed at approximately the same time and have led to a renewal of interest in Galactic star counts.

Important collections of star counts and colors are contained in the older work of Seares and in the catalogs of the Basel survey. The observational situation has changed dramatically in recent years. Modern and (in most cases) more accurate data on star counts are being published with increasing frequency in highly readable papers. One could more than double the available published high quality data for star count analysis, acquired over the past 60 years of research, by devoting one month of continuous scanning with one of the modern devices for automatically measuring Schmidt plates (like the Cosmos machine at the ROE or the APM in Cambridge), or by making deep dedicated surveys with charge coupled devices. As a by-product, the guide star selection system for the Hubble Space Telescope and the surveys of the cameras aboard the HST observatory will result in a great increase in our knowledge of well-calibrated star distributions.

Now that lots of good data is available and much more is on the horizon, what can one do with it? There are well-known problems in using star counts for learning about the Galaxy. The pervasive and patchy nature of obscuration (which was Kapteyn's undoing) and the instability of the inversion (with imperfect data) of the integral equation for the projected number of stars on the sky are major obstacles to progress along traditional paths.

The problems with the previous methods of analyzing star counts can be avoided by *assuming* that the geometrical distribution of stars in the Galaxy is similar to that observed in other galaxies of the same Hubble type, and by only comparing the model with observations in regions in which the obscuration is known to be small. This change of tactics uses star counts to determine parameters in functions whose overall shape is assumed known, and avoids obscuration rather than trying to determine it.

Ray Soneira I have championed this approach to the theoretical analysis of star counts. We do *not* invert the equation of stellar statistics. Instead, we assume a model, calculate the expected results in terms of the observational parameters (numbers per unit magnitude per color bin), and iterate the model with the aid of a computer until the calculated results agree with the observations to within the estimated errors. This procedure does not lead to a unique model, since small components with special properties can always be added without destroying significantly the agreement between calculation and observation. But, it does determine relatively well some of the basic characteristics of the Galaxy.

In summary, with the tactics described above, we can determine only relatively few things, but the things we do determine are relatively well established.

I shall spend most of this review discussing what can be learned about the Galaxy and its stellar content from the simplest version of a "copy-cat" Galaxy model (one whose shape is chosen by analogy with other galaxies) with a Population I disk and a Population II spheroid. Only at the end of this review shall I discuss extensions and improvements of this model. This simple (and conventional) model gives a surprisingly accurate description of the observed distributions in brightness and color of hundreds of thousands of stars in different directions, in different magnitude ranges, and in different broad-band colors.

2. De Vaucouleurs's Galaxy

The basic idea of a "copy-cat" Galaxy model was described by de Vaucouleurs and Pence (1978). This paper was the inspiration for the development of the detailed Bahcall and Soneira Galaxy model. The models in their simplest form contain an exponential disk and a spheroidal halo of older stars. Everyone knows that a de Vaucouleurs $r^{1/4}$ law is an excellent approximation to the distribution of spheroid stars in other galaxies and, not surprisingly, in our own Galaxy. Not everyone knows that the appropriateness of the distribution was first recognized by Gérard in 1948 [see de Vaucouleurs (1948) and de Vaucouleurs (1959)]. What may be surprising to some is that the representation of the disk as an exponential in the distance from the galaxy's center was first proposed in de Vaucouleurs (1958).

Since the structure of the two major ingredients of the Galaxy were first recognized by Gérard, it seems to me appropriate to refer to the Galaxy as "de Vaucouleurs's Galaxy."

3. Comparison with Observations

For the past several years, many observatories have used a version of the Galaxy model that is known as the Bahcall and Soneira (B & S) Export Code. This is a user-friendly computer code that predicts the number of galactic stars in different directions, wave length bands, and brightnesses, using the geometry of the stars that has been established on the large scale by de Vaucouleurs, and with luminosity functions and scale heights determined by many workers from observations of nearby stars and of globular clusters.

Figures 2–11 of Bahcall (1986) show a number of different directions, magnitude ranges, and colors in which the observed and predicted star distributions are compared. Both the *shapes* of the distributions and the *absolute numbers* of the stars are important. The predicted star counts and the observed distributions both change drastically with the color range, the magnitude range, and the direction considered, but they are always in a reasonable agreement. I think anyone who studies these Figures [and the many other comparisons discussed in other papers reviewed in Bahcall (1986)] will be convinced that the two-component model is a reasonable first approximation to the observed stellar content of the Galaxy.

Now that we have seen that there is lots of data and that a simple model gives an acceptable description of the observations, we need to ask: what is this model good for?

4. Applications

4.1 The Spheriod Axis Ratio

The axis ratio of the spheroidal star distribution can be determined by comparing the number of spheroid stars that are observed in different fields that lie in the plane perpendicular to the vector directed from the Sun to the galactic center (*i.e.* in the $l = 90°, 270°$ plane). The number of spheroid stars with a given magnitude or color will be independent of galactic latitude in this plane if the spheroid is perfectly round (neglecting obscuration, which is appropriate for the case that is considered below).

The key to the analysis is that spheroid stars can be separated from disk stars in a frequency–color histogram for stars in certain magnitudes and directions. A double–peaked star distribution is predicted by the two–component Galaxy model for a limited magnitude and color range which just happens to include the range studied carefully by Richard Kron and David Koo. In the faint magnitude range they worked, nearly all the model stars bluer than $J - F = 1.35$ mag are predicted to be spheroid stars and nearly all the model stars redder than $J - F = 1.6$ mag are disk stars. The best fit to the Koo–Kron data corresponds to an axis ratio of 0.80, with a rather large uncertainty:

$$\frac{b}{a} = 0.80^{+0.20}_{-0.05}. \tag{1}$$

This result could be checked and improved by making observations in other fields which are sensitive to the spheroid axis ratio.

The result given above is in good agreement with the determination by Oort and Plaut (1975), who found $0.8 \leq b/a \leq 1.0$ for RR Lyrae variables in fields near the galactic center. We conclude that there is no evidence for a large gradient in the axial ratio of the spheroid stars between 2 and 10 kpc. Frenk and White (1982) have reached similar conclusions for the globular cluster system, for which they find that $b/a \simeq 0.85 \pm 0.13$.

4.2 The Spheroid Normalization

The separation of spheroid and disk stars in the frequency–color diagram permits one to determine the normalization of the spheroid density function by fitting the model results to the absolute number of observed stars. This determination of the spheroid density is of special interest because, unlike most other estimates of the spheroid normalization, the value obtained from star counts is *independent of kinematic assumptions* about the spheroid field population.

The number of spheroid stars between $M_V = 4$ and $M_V = 8$ at the solar position can be calculated from the best fit to the Koo–Kron data in either SA 57 or SA 68. One finds, using a de Vaucouleurs form for the spheroid mass distribution, that

$$n(4 \leq M_V \leq 8) = 2.65 \times 10^{-5}\mathrm{pc}^{-3}. \tag{2}$$

The uncertainty in this normalization is at least as large as 25% with the available data. The total number of spheroid stars at the solar position can be estimated by integrating the assumed luminosity function using the density normalization given above:

$$n(M_V \leq 16.5) \simeq 3 \times 10^{-4}\mathrm{pc}^{-3}. \tag{3}$$

Stars outside the observationally accessible range could contribute a larger number density. The total mass and luminosity of the spheroid have been estimated using various assumptions about the spheroid luminosity functions outside the observed region. The results are: $M_{\text{spheroid}}(\leq 16.5 M_V) \approx (0.9 \text{ to } 3.2) \times 10^9 M_\odot$, and $-16.9 \geq M_{V,\text{spheroid}} \geq -19.0$.

The spheroid normalization given above is about 1/500 times the number density of disk stars locally.

4.3 The Slope of the Spheroid Luminosity Function

The slope of the spheroid luminosity function can be determined in a region in which the spheroid dominates the differential star counts by comparing the observed number of stars per unit of magnitude with that predicted by the model. For certain magnitude ranges and directions that are indicated by the model, the slope of the spheroid luminosity function can be read directly from the observed differential star counts.

This procedure has only yielded one number so far, the slope of the spheroid luminosity function in the absolute visual magnitude range between +4 and +8. Suppose that the spheroid luminosity function has the form $\Phi(M_V) \propto 10^{\gamma M_V}$, then $\gamma_{\text{eff}} = 0.145 \pm 0.035(1\sigma)$.

4.4 Absence of the Blue Tip of the Horizontal Branch

A surprising result of the comparison of the Galaxy model with observations has been the conclusion that the blue tip of the horizontal branch is greatly depleted for spheroid field stars. The expected number of blue stars was calculated by assuming that the spheroid field population is represented by the luminosity function and color–magnitude diagram of each of the classical globular clusters M 3, M 13, and M 92. There are an order of magnitude fewer spheroid field stars with $B - V < 0.2$ than are present proportionately in the three classical clusters listed above. The observations are consistent with the spheroid field stars having a stubby horizontal branch similar to what is seen in 47 Tuc.

4.5 The Wielen Dip in the Disk Luminosity Function

The Wielen dip in the disk luminosity function near $M_V = 7$ mag is a characteristic feature of the stars in the local neighborhood. It is of great interest to know if this feature is present in the luminosity function of disk stars at other positions in the Galaxy.

This question has been answered for the field SA 51, which is almost exactly at the anticenter, and is at a relatively low galactic latitude ($b = 21°$). The theoretical and observational results are in excellent agreement when the Wielen luminosity function is used. The agreement is less satisfactory, especially at fainter magnitudes, when the Luyten function is used. This result is significant since it shows that the Wielen feature is present in the luminosity function of disk stars that are at a characteristic distance of 0.6 kpc (according to the B & S Galaxy model).

4.6 Scale Heights, Density Fluctutations, and Scale Lengths

We can separate disk and spheroid stars by using the bimodal color distribution that appears at intermediate apparent magnitudes. Once the disk stars are isolated, we can put observational constraints on scale heights, volume fluctuations, and the scale length. Three fields (SA 57, SA 68, and SA 141) that have been extensively studied show a clear separation in color of disk and spheroid stars. In addition, SA 51 (which is in the anti-center direction) has a relatively unimportant spheroid contamination in the magnitude range for which observations are available. By comparing model predictions obtained with different values of the assumed disk scale heights with observations in each of the fields mentioned above, one obtains the following 1-σ limit on the scale heights of main sequence disk stars: $\mathrm{H_{disk}} = 350 \pm 50pc, 5 \leq M_V \leq 13.5$. A similar argument shows that the fluctuations in the volume density of disk stars are not large in the four directions considered, namely: $\mathrm{n_{disk}/n_{solar\ vicinity}} = 1.0 \pm 0.15, 5 \leq M_V \leq 13.5$. The disk giants are expected to be prominent only at apparent visual magnitudes brighter than 10. McLaughlin has assembled data in this magnitude range. An analysis of his results yields: $\mathrm{H_{disk}} = 250 \pm 100pc, -1 \leq M_V \leq 3$. One can set a lower limit on the value of the exponential scale length, h, of the observed star density in the plane of the disk. This scale length cannot be so small that, contrary to observations, the disk densities that are inferred by comparison with observations in the direction of the galactic center and the anti-center are very different. The quantitative constraint implied by existing observations is $2.5\mathrm{kpc} \leq \mathrm{h}$.

4.7 A Thick Disk?

Does the Galaxy have a thick disk?

Yes, of course. In some improved approximation, the Galaxy must have many components (both for the disk and the spheroid). Some of these will stick up well above the younger disk components. (In fact, the thin disk has an exponential tail that extends far above the main disk.) We have already embodied this knowledge to a limited extent in the assumed dependence of scale height upon absolute magnitude that was built into the Export Code of the B & S Galaxy model. In order for the question to have quantitative meaning, one must test whether or not a specific model for a thick disk is necessary for describing the data and whether or not it offers a better improvement than other possible generalizations.

Gilmore and Reid (1983) proposed a specific model: a thick disk with a characteristic scale height that is about 1.5 kpc and which contains approximately 2% of the stars in the solar neighborhood. Gilmore and Reid concluded that the stellar population associated with their proposed thick disk is not the same as the Population II which was defined, for example, by Schmidt (1975), because the local number density of thick disk stars is almost an order of magnitude larger than was estimated for Population II stars by other authors.

The original argument of Gilmore and Reid is incorrect. It was based on the assumption that all of the stars they were looking at were dwarfs, when in fact, many were giants. An examination of the argument shows how important it is to compare models and observations in the space of the observational parameters, without the introduction of additional assumptions. Gilmore and Reid (1983) used photometric parallaxes, assuming that all of the stars were on the main sequence. They argued that the B & S Galaxy model could not fit their observations interpreted using main sequence photometric parallaxes. But they made the comparison of their observations with the B & S model in the theoretical plane of ***absolute magnitudes*** (see Figure 8 of Gilmore and Reid (1983)). The assumption that all of the stars were on the main sequence was criticized by Ray Soneira and myself; we pointed out that most of the spheroid stars in the magnitude range studied by Gilmore and Reid were expected to be giants — not dwarfs —according to the model. The Gilmore and Reid observations are well-fit by the standard B & S model [see Figure 9 of Bahcall(1986)].

Gilmore and Reid (1983) derived the scale height of their thick disk by fitting with two exponentials the observed distribution of the number of stars versus apparent magnitude (or height above the plane). This argument is also misleading since the predictions of the two component (thin disk plus spheroid) B & S Galaxy model can be fit equally well by two exponentials. In other words, you can infer a thick disk from exponential fits to the star counts of a model that contains only a thin disk and a spheroid.

More recently, Gilmore has proposed a Galaxy model that is similar to the Bahcall and Soneira (1980) model except for the addition of a thick disk with a spheroid-like luminosity function. This suggested thick disk has no unique implications for star counts that would allow it to be isolated easily from the prominent thin disk and spheroid.

If there is a thick disk with the revised parameters, then it must be discovered by careful analyses of data that includes kinematic information.

4.8 The Faint End of the Disk Luminosity Function

Dynamics of stars prependicular to the galactic disk shows that about 0.1 $M_{\odot} pc^{-3}$ of material has not yet been observed. It is possible that this unseen mass could be in the form of stars of mass $\leq 0.1 M_{\odot}$ which would not be detected without special programs in the red or near-infrared. In order to account for the missing mass, the luminosity function would have to increase as $\Phi \propto 10^{\gamma M}$, where for absolute visual magnitudes $\gamma \simeq 0.01$ to 0.05, depending somewhat upon the mass-luminosity relation for these faint stars. The expected number of faint red disk stars can be calculated, for the luminosity function given above, using the Galaxy model.

Several observers have suggested that the luminosity function turns over below about $M_V = 13$. I am cautious about accepting these conclusions as final because of the difficulty in calibrating the photographic (or CCD) colors in terms of absolute magnitude, and because of the sensitivity of the conclusions to the kinematic selection effects in proper-motion samples. Both these effects will be investigated more extensively over the next few years, during which time the answer will presumably become apparent.

Of course, the unseen mass could all be contained in stars too faint to burn hydrogen even if the luminosity function turns over in a range accessible to optical observers. A two-hump distribution, with a second peak somewhere in the range $0.01 \leq M \leq 0.08 M_\odot$, might best be studied by infrared surveys such as were obtained by IRAS.

4.9 Sample Definition and Field Contamination

The Galaxy model can be used to help plan observations. For many systematic programs, it is useful to have an *a priori* estimate of the number of stars of a particular type (luminosity class, color, or population) that are present in specified directions and in given apparent magnitude ranges. This information can easily be obtained from the Export Version of the B & S Galaxy model. For example, suppose you wanted to identify a sample of distant disk K dwarfs (in order to determine K_z). You could run the Export Version of the Galaxy model to calculate the expected number of candidate stars in the field of interest as a function of color and of apparent magnitude. With this information, you could choose the color range of the sample to maximize the ratio of K dwarfs to other stars, and could select the size of the field to be studied so that it would give you approximately the desired number of sample stars.

The model can also be used to estimate the contamination of field galactic stars in a variety of observations. For example, if one is interested in the galaxy or quasar content of a specific field, in the globular cluster luminosity function around a distant galaxy, or the color-magnitude diagram of galactic globular clusters, one would like to have an estimate for the likely contamination by galactic stars. Very deep, wide-field exposures represent another class of applications. It is useful to have a standard set of expected number densities (and color distributions) with which to compare future deep exposures that will be obtained with ground-based or space telescopes.

References

Bahcall, J. N. 1986, *Ann. Rev. Astron. Astrophys.* **24**, 577.
Bahcall, J. N. and Soneira, R. M. 1980, *Astrophys. J. Suppl.* **44**, 73.
de Vaucouleurs, G. 1948, *Ann. d'Astrophys.* **11**, 247.

de Vaucouleurs, G. 1958, *Astrophys. J.* **128**, 465.
de Vaucouleurs, G. 1959, in *Handbuch der Physik*, **53**, ed. S. Flügge (Springer–Verlag: Berlin), p. 275.
de Vaucouleurs, G. and Pence, W. D. 1978, *Astron. J.* **83**, 1163.
Frenk, C. S. and White, S. D. M. 1982, *Mon. Not. Roy. Astron. Soc.* **198**, 173.
Gilmore, G. and Reid, N. 1983, *Mon. Not. Roy. Astron. Soc.* **202**, 1025.
Oort, J. H. and Plaut, L. 1975, *Astron. Astrophys.* **41**, 71.
Schmidt, M. 1975, *Astrophys. J.* **202**, 22.

Discussion

C. Christian: One important question is: if the thick disk exists, how is it related to the "disk globulars" (or is it)? In an effort to study this question, Norris and Green looked at a large number of pole stars photometrically (using DDO photometry). They find that those stars are actually analagous to clump stars in open clusters, rather than to globular giants as Rose has suggested. They further suggest that there is no evidence for a thick disk in their sample, and that their work supports the previous models of Larson where the spheroid forms first and the disk grows outward much later. There is, therefore, a gradient in age and metallicity outward in the thin disk. No thick disk is required.

I also note that kinematically, the group of clusters in M 31 which are "disk globulars," are distinct from the spheroid and the thin disk, as they are in the Galaxy and perhaps in M 33. Proponents of the thick disk (*e.g.* Gilmore and Reid) need to explain how the four components of these galaxies are related and formed (the four components are spheroid, disk globulars, thick disk, and thin disk). Perhaps there are not many field stars related to the "disk globulars."

P. van der Kruit: Both Kylafis and you and also myself have shown that the photometry of NGC 891 can be fitted as well with or without a thick disk. Yet many people in the literature quote our results as evidence for the existence of a thick disk in spite of the fact that we believed the result is inconclusive. On the other hand, I believe that new kinematic data gives us a body of evidence that a component with intermediate metallicity and kinematics exists in our Galaxy.

Bahcall: Thank you. I agree.

M. Capaccioli: I confirm that to within state–of–the–art accuracy, there is no evidence for a thick disk in NGC 3115 and in five other edge–on lenticulars that I have studied.

Bahcall: Very interesting.

Burstein: In my thesis, I looked at five edge–on S0's — NGC 4111, NGC 4350, NGC 4474, NGC 4570, and NGC 4762 — in terms of standard $r^{1/4}$ law and exponential disks. When I compared the two–dimensional luminosity distributions of these galaxies to these components, I found that they could not account for all of the light. A further component, which I named a "thick disk," was needed. (Of these five galaxies, NGC 4570 shows evidence of this component the least, NGC 4474 and NGC 4762 the most.) After 10 years, I now think (but cannot prove) that what I interpreted as a "thick disk" in edge-on S0's is seen as an "outer lens" in face-on S0's (John Kormendy's terminology). It is therefore significant that these outer lenses are primarily seen in S0's, and not in spirals. I have objected from the beginning to the analogy of the "thick disk" in S0's relative to the hypothesized "thick disk" in our galaxy.

Dark Halos In Virialized Two–Component Systems of Galactic Mass

P. Brosche[1], R. Caimmi[2], and L. Secco[2]

[1]Observatorium Hoher–List der Universitäts–Sternwarte Bonn, F.R.G.
[2]Dipartimento di Astronomia, Università di Padova, Italy

Introduction

We propose an application of the tensor virial theorem to two–component galactic systems, which could suggest a possible range for the mass and flatness of the dark halo in which the bright component is likely embedded.

The Equations

Following the formulation of Brosche *et al.* (1983, Paper I) and Caimmi *et al.* (1984, Paper II), a galaxy is modelled by two spheroidal homogeneous components, dark and bright, one completely inside the other, in virial equilibrium; the two kinetic energy tensor equations are (G is the gravitational constant):

$$\begin{aligned}(T_B)_{xx} &= 0.3GM_B^2(K_{BD})_x/a_B; \; (T_D)_{xx} = 0.3GM_D^2(K_{DB})_x/a_D \\ (T_B)_{zz} &= 0.3GM_B^2(K_{BD})_z/a_B; \; (T_D)_{zz} = 0.3GM_D^2(K_{DB})_z/a_D \end{aligned} \quad (1)$$

D and B mean dark (outer) and bright (inner) components, respectively; $M_{B,D}$ and $a_{B,D}$ are the corresponding masses and semi–major axes. $(K_{ij})_{x,z}$ ($i = B, D; j = D, B$) are functions of the axis ratios ϵ_i, and the ratios $m = M_D/M_B$, $y = a_D/a_B$. Altogether, we have to deal with ten variables: a pair of masses, axis ratios, semi–major axes, and four tensorial total velocities: v_{ix}, v_{iz}. We choose as four unknowns: M_i, v_{Dx}, v_{Dz}; as observables: a_B, ϵ_B, v_{Bx}, v_{Bz}; and the remaining a_D, ϵ_D as parameters. We will look for the values of the parameters in order to produce tensorial dark velocities consistent, within the errors, with the experimental velocity dispersion tensor of globular clusters considered as probe particles in the dark–plus–bright matter field.

By definition, and using the first and second equations (B) of system (1), we find:

$$m = -y^3((\epsilon_B v_{Bx})^2\Gamma_B - v_{Bz}^2\alpha_B)\epsilon_D/(((\epsilon_B v_{Bx})^2\Gamma_D - v_{Bz}^2\alpha_D)\epsilon_B)$$

($\Gamma_i = A_{3i}$ and $\alpha_i = A_{1i}$ of Paper II.) Now, the $(K_{ij})_{x,z}$ are available and also the unknowns

$$M_B = 5a_B v_{Bz}^2/(3G(K_{BD})_z);\ M_D = mM_B \tag{2}$$

The third and fourth equations (D) of system (1) yield the other solutions:

$$v_{Dx} = (3GM_D(K_{DB})_x/(5a_D))^{1/2},$$

$$v_{Dz} = (3GM_D(K_{DB})_z/(5a_D))^{1/2}.$$

An Application

We begin, as an exercise, by applying the method to the Galaxy considering it, in a rough approximation, as a two component system: a bright "mean disk" and a dark halo. The possible values for the input are: $a_B = 15$ Kpc; $\epsilon_B = 0.09$ (a compromise between the "flat disk" and the bulge); $v_{Bx} = 220/\sqrt{2}\,\mathrm{km\,sec^{-1}}$; and $v_{Bz} = 51.55\,\mathrm{km\,sec^{-1}}$ (Norris 1986, Freeman 1986, and van der Kruit 1986).

For the range of the parameters ϵ_D, a_D, we have some experimental constraints due to the physical meaning of the parameters themselves and the velocity output we have to obtain. If we refer to Webbink's (1986) results, the dispersion velocity tensor for the globular clusters system, once the systematic effects due to the low rotation have been removed, is, within the errors, consistent with an isotropic cluster velocity ellipsoid. Assuming a mean common value for the three components:

$$\sigma \approx 110 \pm 66\,\mathrm{km\,sec^{-1}},$$

we have to look for the values of a_D and ϵ_D in such a way that:

$$v_{Dx}(= v_{Dy}) \approx v_{Dz} \approx \sigma.$$

The distances in the Webbink's sample, containing 90 objects including four spheroidals, are very wide but, because we do not refer to the nearby halo stars which could be dominated by a flattened component (Freeman 1986, Hartwick 1986), we fix:

$$\epsilon_D \geq 0.5;\ b_D, a_D \approx 50 \div 100\ Kpc\ (b = \epsilon a)$$

Table 1.

Input			Output			
a_D (Kpc)	b_D (Kpc)	ϵ_D	M_D $10^{11} m_{\odot}$	v_{Dx} (km sec^{-1})	v_{Dz} (km sec^{-1})	M_{tot} $10^{11} m_{\odot}$
52.5	46.5	0.88	2.8	82	79	4.6
60	46.5	0.77	4.0	86	78	
	54	0.89	4.3	83	81	6.1
75	46.5	0.62	7.2	95	78	
	54	0.72	7.6	93	82	
	61.5	0.81	8.0	92	85	
	69	0.92	8.5	90	88	10.3
90	46.5	0.51	12.0	107	81	
	54	0.60	12.4	105	85	
	61.5	0.68	12.9	104	89	
	69	0.76	13.5	102	92	
	76.5	0.85	14.1	101	95	
	84	0.93	14.7	100	98	16.5
105	54	0.51	19.0	119	90	
	61.5	0.58	19.6	117	94	
	69	0.65	20.3	116	97	
	76.5	0.72	21.1	114	100	
	84	0.80	21.9	113	103	
	91.5	0.87	22.7	112	106	
	99	0.94	23.6	111	109	25.4

$M_B = 1.9x10^{11} m_{\odot}$ in all cases.

(the velocity dispersion seems to remain sensibly constant with radius; Hartwick *et al.* 1978).

With these constraints, Table 1 shows, for the different a_D and ϵ_D, the corresponding solutions of the system's equations.

Conclusion

For every a_D, the cases with $\epsilon_D \approx 0.9$ are the nearest to isotropy, all of them with velocity values within the very wide range of σ. M_B is actually an invariant due to both the trend of the function Γ/ϵ diverging when ϵ goes to 0, and the values considered for the ratios m/y^3. The total mass corresponding to different values of a_D (put in Table 1 only for the different isotropic models) is not in disagreement with that which Faber and Gallagher (1979) proposed from obsevational considerations. In addition, the total mass inside 60 Kpc obtained by Hartwick and Sargent (1978), in the case of isotropy, is well–fit by our model.

References

Brosche, P., Caimmi, R., and Secco, L. 1983, *Astron. Astrophys.* **125**, 338 (Paper I).

Caimmi, R., Secco, L., and Brosche, P. 1984, *Astron. Astrophys.* **139**, 411 (Paper II).

Faber, S. and Gallagher, J. 1979, *Ann. Rev. Astron. Astrophys.* **17**, 135.

Freeman, K. C. 1986, in *The Galaxy*, ed. G. Gilmore and R. Carswell (Dordrecht: Reidel), p. 291.

Hartwick, F. 1986, in *The Galaxy*, ed. G. Gilmore and B. Carswell (Dordrecht: Reidel), p. 281.

Norris, J. 1986, in *The Galaxy*, ed. G. Gilmore and B. Carswell (Dordrecht: Reidel), p. 297.

van der Kruit, P. C. 1986, *The Galaxy*, ed. G. Gilmore and B. Carswell (Dordrecht: Reidel), p. 27.

Hartwick, F. and Sargent, W. 1978, Astrophys. J. **221**, 512.

Webbink, R. 1986, in *The Harlow Shapley Symposium on Globular Cluster Systems in Galaxies*, I.A.U. Symp. No. **126** (Dordrecht: Kluwer), p. 49.

Kinematics and Chemical Properties of the Old Disk of the Galaxy

James R. Lewis[1,2] and K.C. Freeman[1]

[1]Mount Stromlo and Siding Spring Observatories, Australia
[2]Institute of Astronomy, Cambridge, U.K.

Late type spiral galaxies such as the Milky Way have two major luminous components, the disk and the bulge. In this paper, we discuss the kinematics and chemical properties of the disk component of the Milky Way. The local stability of a uniform cold disk was discussed by Toomre (1964). He found that, unless a certain amount of random motions was present, the disk would be unstable to local axisymmetric disturbances. Disks are also prone to barlike instabilities. Obviously, not all galaxies are stabilized against these modes; at least 30% of bright spirals have bars in their disks. But at least another 30% have no hint of such a bar.

So what stabilizes those disks which do not have bars? It has been suggested that a cold disk would be stabilized by placing it in a massive dark corona. As the disk of the Milky Way is locally cold ($\sigma_R = 35\,\mathrm{km\,sec}^{-1}$ for old stars), and some sort of unseen corona is needed to account for flat rotation curves, this appeared to be an attractive way to stabilize the disk.

Recently, Athanassoula and Sellwood (1986) made N–body experiments on Kuz'min/Toomre disks, which were aimed at evaluating the contribution of random motions to the stability of bi–symmetric modes. They showed that the growth rate of these modes was a simple function of the corona mass and the random velocities. They found that if σ_R is large enough in the inner parts of the disk, the disk is stabilized against these modes. (This dispersion corresponds to a value of Q of about 2.0 to 2.5). This is interesting because recent work by Carignan and Freeman (1985) and van Albada *et al.* (1985) supports the view that halos do exist, but are only dynamically important in the outer regions.

In the Milky Way, it is possible to measure the radial distribution of velocity dispersion directly, and this is the subject of this paper. The procedure is

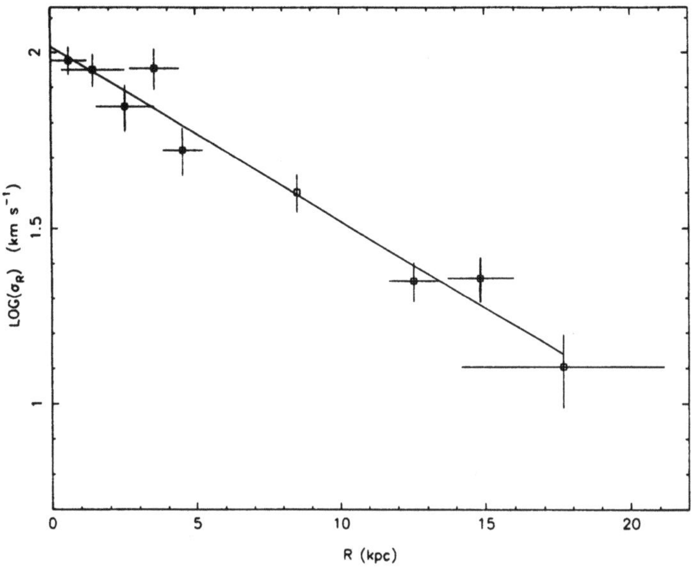

Figure 1: Logσ_R *versus* R for our sample of ~600 red giants in six Galactic windows.

to measure radial velocities of a large sample of individual stars in various directions in low absorption windows close to the galactic plane. In this way, we can build up a picture of the kinematic state of the old disk of this spiral galaxy over a wide interval of galactocentric distance (from near the center to about 17 kpc).

We should also consider the evolution of chemical abundances and chemical gradients in the Galactic disk. Many authors have investigated the metallicity distribution of young and intermediate age objects and have found a metallicity gradient in the disk which is generally about -0.05 dex kpc^{-1} in [Fe/H]. It would be interesting to know the run of abundance in the old disk as well.

To address these questions we acquired medium resolution spectra of about 600 red giant stars in six Galactic windows. From the spectra, we measured line–of–sight velocities and equivalent widths for abundance estimates.

Figure 1 shows the run of $\log \sigma_R$ *versus* R for our stars. The exponential decrease of σ_R is interesting in the context of the models of van der Kruit and Searle (1981). If we assume that the anisotropy σ_R/σ_z is constant throughout the disk, then we can determine the disk's exponential scale–length. This kinematically determined value is $h_R = 4370 \pm 320$ pc whose consistency with other photometric determinations supports the assumption of uniform anisotropy.

Figure 2 shows the run of Toomre's Q parameter with radius. This rises to a value of about 2.5 at half a disk scale–length from the galactic center. This

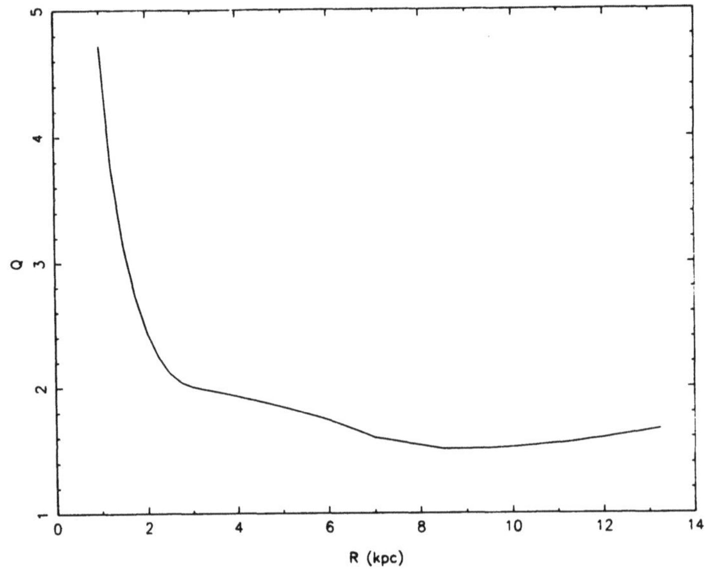

Figure 2: Toomre's Q parameter *versus* Galactic radius.

suggests that the disk is hot enough in its inner parts to be stable to global bi-symmetric modes.

Figure 3 shows the run of $\log \sigma_\varphi$ *versus* R for our stars. The scale-length implied by the fit is $h_\varphi = 3360 \pm 620$ pc. The fact that h_R and h_φ are different suggests that the galactic rotation curve is not precisely flat over the region bounded by our data.

Figure 4 shows the metallicity distribution for our stars. The radial gradient which has been observed for both young and intermediate age objects is not present in this figure. This may be indicative of a star formation history in which star formation procedes radially from the center as a function of time, with the star formation rate and metallicity depending on local gas fraction.

References

van Albada, T. S., Bahcall, J. N., Begeman, K., and Sanscisi, R. 1985, *Astrophys. J.* **295**, 305.

Athanassoula, E. and Sellwood, J. A. 1986, *Mon. Not. Roy. Astron. Soc.* **221**, 2.

Carignan, C. and Freeman, K. C. 1985, *Astrophys. J.* **294**, 494.

van der Kruit, P. C. and Searle, L. 1981, *Astron. Astrophys.* **95**, 105.

Toomre, A. 1964, *Astrophys. J.* **139**, 1217.

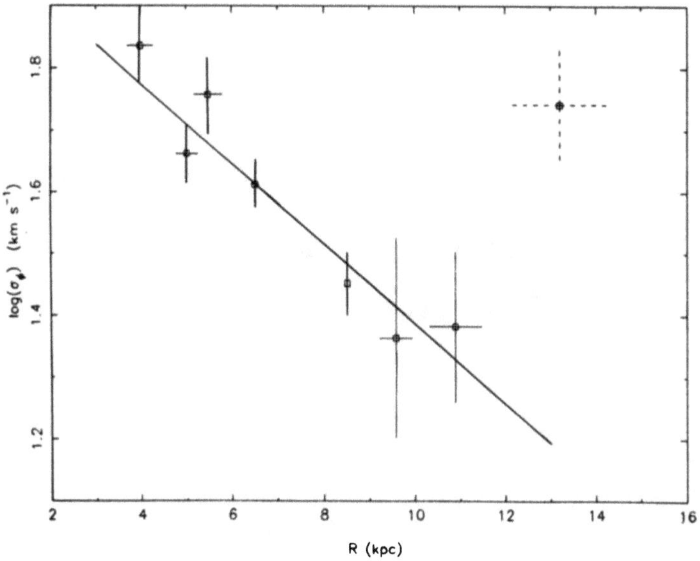

Figure 3: Logσ_φ *versus* R for our sample of ~600 red giants.

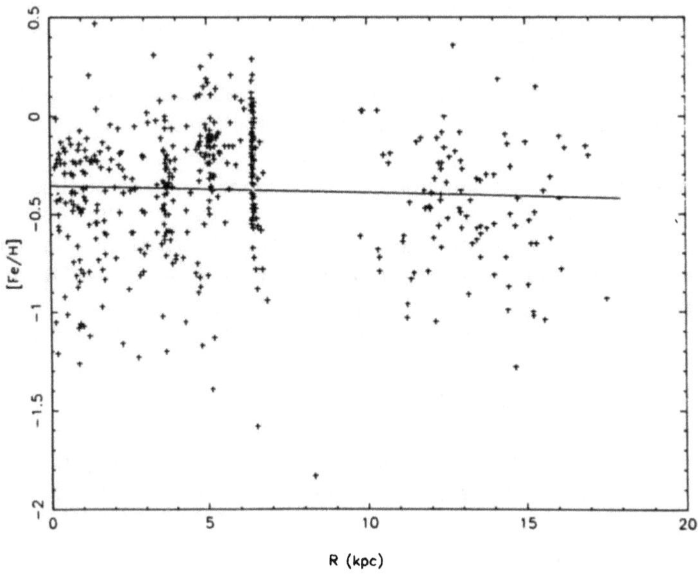

Figure 4: Metallicity *versus* Galactic radius for our ~600 red giants.

Complex Instability of Simple Periodic Orbits in a Realistic Galactic Potential

P.A. Patsis and L. Zachilas

Department of Astronomy, University of Athens, Greece

The realistic, two component, triaxial, rotating potential

$$\begin{aligned} V(x,y,z) = V_* + v_0^2/2\ln[1+x^2+(y/q_a)^2+(z/q_b)^2] \\ -GM_{DISC}/\{x^2+(y/q_a)^2+[a+(z^2+b^2)^{1/2}]^2\}^{1/2} \end{aligned} \quad (1)$$

is used to study numerically simple periodic orbits and their stability. It consists of a Miyamoto disc (Miyamoto and Nagai 1975) surrounded by a triaxial halo that changes its geometrical shape via the parameters q_a and q_b.

Following a method introduced by Broucke (1969), an orbit can be characterized as stable (S), simple unstable (U), doubly unstable (D), or complex unstable (Δ).

In our study, we found families that correspond to these of Heisler *et al.* (1982) and other important families of periodic orbits. For the z-axis family, for increasing values of energy H and for constant Ω_p (pattern rotation), the sequence of stability/instability regions changes from S $\longrightarrow$ U $\longrightarrow$ DU to S $\longrightarrow$ U $\longrightarrow$ DU $\longrightarrow$ Δ $\longrightarrow$ DU $\longrightarrow$ Δ $\longrightarrow$ DU $\longrightarrow$ Δ, and later to S $\longrightarrow$ U $\longrightarrow$ S $\longrightarrow$ Δ $\longrightarrow$ DU $\longrightarrow$ Δ $\longrightarrow$ DU $\longrightarrow$ Δ, as we see in Figure 1a. We notice that in the cases we examined (for various q_a and q_b), we did not find a realistic value of Ω_p for which there occurs a direct transition from S $\longrightarrow$ Δ $\longrightarrow$ ...

We also found the families of elliptical orbits on the equatorial plane and the "anomalous" orbits which are related to the stable polar rings in polar ring galaxies. In our potential, the orbits of this latter family do not lie entirely in an orbital plane.

Two other important families have been studied. The first of them is analogous with the 1a family of the potential of three coupled unharmonic oscillators

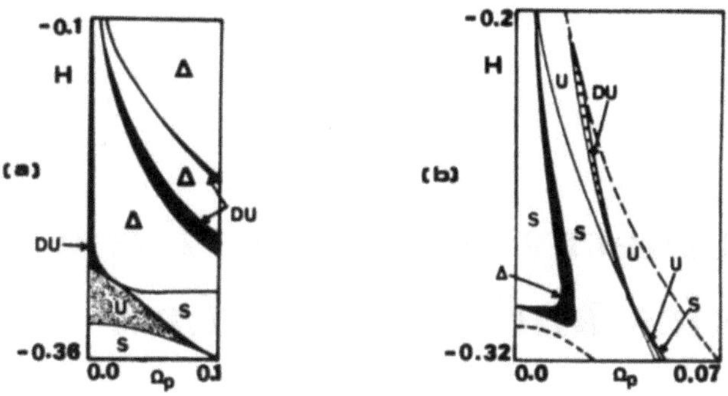

Figure 1: Existence diagrams for (a) the z-axis family for $q_a = 0.7$, $q_b = 0.6$, and for (b) 1y′ with $q_a = 1.05$, $q_b = 1.2$.

of Contopoulos and Magnenat (1985), and remains stable as we vary our parameters over a wide range. The second occurs as a bifurcation of the y–axis family and in its "existence diagram" (Figure 1b), we see that it has a large complex region for realistic values of Ω_p. We called it 1y′. Extending our study to extreme values of Ω_p, we found a collision of bifurcations, a phenomenon of great interest in the study of Hamiltonian systems of three degrees of freedom.

References

Broucke, R. 1969, *Am. Inst. Aero. Astr. J.* **7**, 1003.
Contopoulos, G. and Magnenat, P. 1985, *Cel. Mech.* **37**, 387.
Heissler, J., Merrit, D., and Schwarzschild, M. 1982, *Astrophys. J.* **258**, 490.
Miyamoto, M. and Nagai, R. 1975, *Publ. Astron. Soc. Japan* **27**, 533.

Photometry of Early–Type Galaxies and the $r^{1/4}$ Law

Massimo Capaccioli

Osservatorio Astronomico di Padova, Padova, Italy

Introduction

Surface photometry was applied to the study of galaxies even before the true nature of these objects was clearly established. In 1914, Reynolds succeeded in measuring the luminosity profile of the central part of the Andromeda nebula, and proposed a fitting formula which eventually became very popular through the systematic applications and related theoretical speculations made by Edwin Hubble (1930; see also de Vaucouleurs 1979, 1987, for an historical review and a comprehensive list of references). In dimensionless units, the generalized Reynolds–Hubble formula is written as

$$J(\xi) = 4\,(1+\xi)^{-2}, \qquad (1)$$

where $\xi = r/r_\circ$, and $J = I(r)/I(r_\circ)$. The "characteristic" length, $r_\circ$, and the corresponding surface brightness $I_\circ = I(r_\circ)$ (1.505 mag fainter than the peak brightness), cannot be related to a certain fraction of the total luminosity since the area integral of Equation 1 diverges with ξ (a fact that, otherwise, has no physical consequence since galaxies have finite sizes; this statement is lucidly stressed by Baum, 1955). Let us consider, for instance, the *standard* model for an elliptical galaxy, whose isophotes are similar, homocentric, and co–axial ellipses with constant flattening $\epsilon = 1 - b/a$; we obtain

$$\begin{aligned} \mathcal{L}(a = a_\circ\xi) &= 8\pi\, I_\circ\, a_\circ^2\,(1-\epsilon) \int_0^\xi \frac{x\,dx}{(1+x)^2} = \\ &= 8\pi\, I_\circ\, a_\circ^2\,(1-\epsilon)\left[\ln(1+\xi) - \frac{\xi}{1+\xi}\right], \end{aligned} \qquad (2)$$

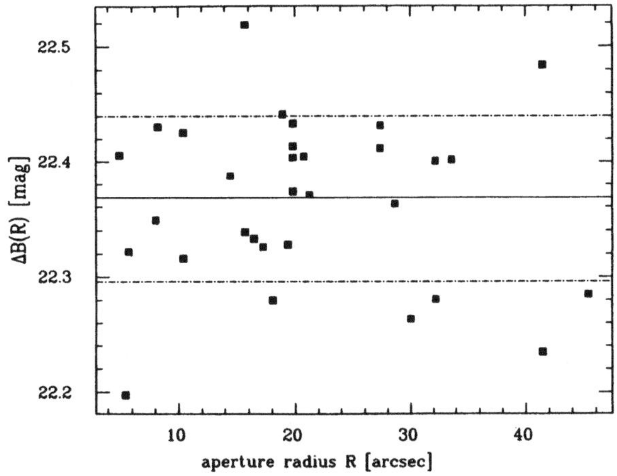

Figure 1: Accuracy of modern surface photometry: differences between the photoelectric magnitudes of NGC 3379 integrated over centered circular apertures of radius R (from Longo and de Vaucouleurs 1983) and the corresponding magnitudes (in units of the night sky surface brightness, μ_s), extracted from a 10^{min} B-band CCD exposure of the galaxy taken with the ESO–MPI 2.2–m telescope at La Silla (the CCD background subtraction is based upon comparison with a deep Schmidt photometry). The dispersion about the mean μ_s (0.072 mag) is of the same order of the scatter of the photoelectric data (see, *e.g.*, the cluster of values at $R \sim 20''$). Note the absence of any significant gradient with the aperture radius. This result (and the excellent agreement with the CCD photometry of Davis *et al.* 1985, reported by Capaccioli *et al.* 1987) reinforces the perplexity concerning the strong gradient shown by a similar plot for the standard S0 galaxy NGC 3115 and based on the same type of material (Capaccioli *et al.* 1987).

where a_0 indicates the scale length of the major axis light profile.

For several years the Reynolds–Hubble formula had no challengers. In 1948, Gérard de Vaucouleurs published a classical study entitled *Recherches sur les Nébuleuses Extragalactiques: I. Sur la technique de l'Analyse Microphotométrique des Nébuleuses Brillantes*. This 40 page–long paper is divided into two main sections. The first section deals with the technical aspects of galaxy photometry. Following a previous extensive study by Redman and Shirley (1936, 1938), all known sources of accidental and systematic errors are thoroughly analyzed and discussed — a subject that de Vaucouleurs will tackle again on several subsequent occasions (*e.g.*, de Vaucouleurs 1983; Capaccioli and de Vaucouleurs 1983; see also Jones *et al.* 1967).

The second section of the paper reports on a new photometry of the spiral NGC 4594, and of three early–types galaxies, namely NGC 3115, 3379, and 4649 (then classified as E7, E0, and E2 respectively), based on photographic material collected at the Cassegrain focus of the 80–cm Haute Provence telescope. Previous studies of the same objects were carried out by Hubble (1930), Redman and Shirley (1938), and Oort (1940, 1946). At the end of this part, while discussing the trend of the light profiles of his three early–type objects, de Vaucouleurs presents a new fitting formula, now called the de Vaucouleursor $r^{1/4}$ law. Introduced as a better alternative to the Reynolds–Hubble formula, the $r^{1/4}$ law is proposed for what it is, an *empirical* relation to be judged against observations "*sans qu'on préjuge de sa signification physique*".

Since its birth, the $r^{1/4}$ law has gained an increasing reputation for its proved ability to represent fairly accurately the luminosity profiles of early–type galaxies (*cf.* de Vaucouleurs 1958, 1959, 1961; Fish 1964; de Vaucouleurs and Capaccioli 1979; Schweizer 1979; Kormendy 1977, 1982, 1987; Michard 1984; Capaccioli 1987; Nieto 1988) and of bulges of S0s and spirals (de Vaucouleurs 1958, 1959; Simien, *this volume*) over an ample interval of surface brightness. This simple formula has been extensively — while not always properly — used as a standard ruler by observers, and either as a reference (van Albada 1982; Bertin and Stiavelli 1988) or as an input constraint (Binney 1982; Newton and Binney 1984; Merritt 1985) by theoreticians. In other words, it has shared, with ups and downs, the dramatic growth of galaxy photometry, both in quantity and quality (see Figure 1); a process fostered by the need of large data sets as well as by a series of well-known improvements in the technology of observations and data reduction. As an example of significant astrophysical findings bound to the $r^{1/4}$ law, we shall mention the discovery of the two principal photometric components of spiral galaxies, the exponential disk and the $r^{1/4}$–like bulge (de Vaucouleurs 1958, 1959; van Houten 1961).

Taking advantage of the present celebration of Gérard and Antoinette de Vaucouleurs's superb career in science, which coincides in time with the fortieth birthday of the $r^{1/4}$ formula, in this paper I shall revisit the subject of photometric fitting laws for early–type galaxies. I shall resist the temptation of reviewing the latest astrophysical results related to galaxy photometry, since this subject has been amply treated in numerous recent papers (Kormendy 1987, Capaccioli 1987; Jedrzejewski 1987; Nieto 1988). I will also touch lightly upon technical aspects of modern surface photometry, given the excellent compendium by Okamura (1988; see also Capaccioli 1988).

The $r^{1/4}$ Law as a Tool

Using the same dimensionless variables as in Equations 1 and 2, the de Vaucouleurs law is written as

$$\log J(\xi) = -\beta \left(\xi^{1/4} - 1\right), \text{with } \beta > 0. \tag{3}$$

Contrary to the case of Equation 1, the area integral the de Vaucouleurs formula does not diverge. For the standard model, it becomes

$$\begin{aligned} \mathcal{L}_T &= 2\pi I_o a_o^2 (1-\epsilon) \int_0^{+\infty} \exp\Big[-\gamma(\xi^{1/4}-1)\Big]\, d\xi = \\ &= 8!\,\pi I_o a_o^2 (1-\epsilon)\frac{e^a}{a^8} \qquad \text{with } \gamma = \beta \ln 10. \end{aligned} \tag{4}$$

This property allows us to define the fractional luminosity, $k(\xi) = \mathcal{L}(\xi)/\mathcal{L}_T$, whose analytical expression is (*cf.* de Vaucouleurs 1961, Young 1976)

$$0 \le k(\xi) = 1 - \exp(-\gamma\xi^{1/4}) \sum_{n=0}^{7} \frac{\gamma^n \xi^{n/4}}{n!} \le 1. \tag{5}$$

It is apparent that the constant $\beta = \gamma/\ln 10$ appearing in Equation 3 may be fixed by assigning an arbitrary value to $k(\xi = 1)$ within the limits where k is defined, *i.e.* by choosing which fraction of the total luminosity[1] has to be encircled by the isophote with semi-major axis a_o. For $k(1) = 1/2$ (de Vaucouleurs 1948), Equation 5 gives $\beta = 3.33071$, with which the $r^{1/4}$ law is rewritten as

$$\log J(r) = \log I(r)/I_e = -3.33071 \left[\left(r/r_e\right)^{1/4} - 1\right]. \tag{6}$$

The scale length r_e and the surface brightness of the corresponding isophote are called *effective* parameters. With these conventions, the total luminosity is

$$\mathcal{L}_T = 7.21457\,\pi I_e a_e^2 (1-\epsilon). \tag{7}$$

The $r^{1/4}$ formula is free from shape parameters, r_e and I_e being *scale factors*. In logarithmic form, it is linear against $\xi^{1/4}$:

$$-2.5 \log J(\xi) = \mu(r/r_e) - \mu_e = 8.3268 \left[(r/r_e)^{1/4} - 1\right]. \tag{8}$$

Therefore, one can easily verify whether an observed light profile follows the de Vaucouleurs law by checking if the plot of the measured surface brightness $\mu(r)$ *vs.* $r^{1/4}$ is a straight line: $\mu(r) = p + qr^{1/4}$. If so, then r_e and μ_e result in $r_e = (8.3268/q)^4$, and $\mu_e = p + 8.3268$. In the practice, however, this test, and the corresponding easy estimate of the effective parameters, are not so straightforward and deserve some comments. In the following discussion, unless differently stated, we will assume that galaxies consist of one photometric component only; we will thus explicitly ignore disk components, lenses, dust rings, *etc.*

First of all, we notice that the $r^{1/4}$ scale stretches a light profile in its inner part and compresses it in the outer region. This coordinate transformation has

[1] Obviously one might choose a value for β and determine the corresponding value of $k(\beta)$; for instance, if $\beta = 1$, then $k(\beta) \simeq 10^{-5}$, a figure which is impractical.

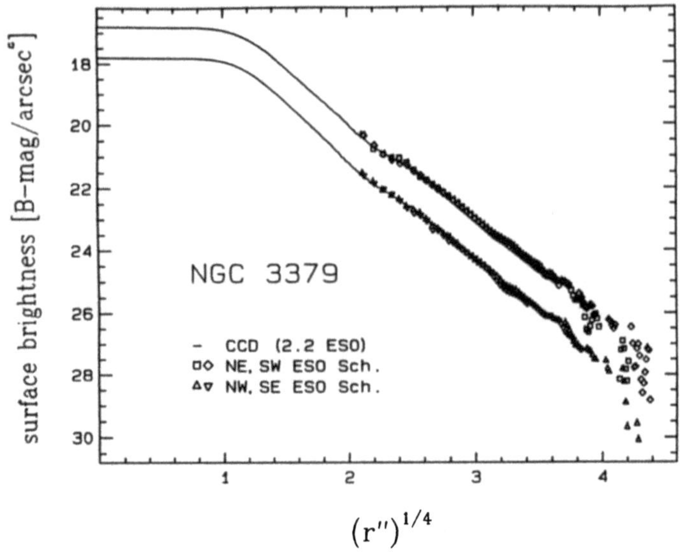

Figure 2: B–band luminosity profiles along the main axes of the E1 galaxy NGC 3379, obtained by combining good resolution ($\sigma_* = 0''.65$) CCD exposures with deep Schmidt material (Capaccioli *et al.* 1989*b*). The minor axis profile is shifted by +1 mag to avoid overlapping with the major axis data. The $r^{1/4}$ behaviour of the outer profiles breaks down at $r \lesssim 15''$; this feature is particularly evident along the minor axis. The inner "core" has been modelled by de Vaucouleurs and Capaccioli (1979) by adding a gaussian component (dispersion=$3''.7$) to the main $r^{1/4}$ component best fitting the outer E–W profile.

the following consequences: imagine that an observed light profile is rebinned at equal $r^{1/4}$ steps; such a procedure is useful to obtain a better signal–to–noise ratio (S/N) in the faint outer parts (S/N will improve by a factor $2\xi^{3/8}$), and is needed to perform an unbiased linear fit of the data. In the best possible case and for nearby large galaxies, these photometric measurements will span the interval from $0.3\,r_e^{1/4}$ to $1.7\,r_e^{1/4}$ — the lower limit being set by seeing convolution, the upper (corresponding to $\mu = 27.8$, if $\mu_e = 22.0$) by the cumulative effects of several sources of noise, which prevent measurements at fainter levels (*cf.* Capaccioli and de Vaucouleurs 1983). Therefore, the bins at all $r < r_e$ will contribute to establishing the linearity of the light profile (and to estimate the constants p and q) with just the same weight as the whole of the bins at $r > r_e$. This consideration implies that no meaningful $r^{1/4}$ fit can be made on light profiles obtained with narrow field images such as those provided by present–technology CCDs at conventional focal plane scales. In turn, it also points out the danger of large photometric scale errors at the bright end of light profiles, *i.e.* in that photometric range where, for instance, photographic emulsions suffer by saturation (in long exposures) and/or adjacency effects (due to the sharp luminosity gradients).

In conclusion, accurate tests of the de Vaucouleurs law can at present be made only with light profiles built by combining high resolution CCD and large–field photographic observations (*cf.* Capaccioli *et al.* 1987, 1988*a*, 1989*a*). It is clear that these observations must be corrected for all the effects which contribute to modifying the *shape* of the composite light profile: mainly the color term in the CCD photometry (which may introduce an artificial change of slope; *cf.* Capaccioli *et al.* 1987), and the scattered light, which produces a spurious *corona* whose significance depends on the values of $\mu(0)$ and r_e (Capaccioli and de Vaucouleurs 1983; see Capaccioli 1987, 1988, for a simple recipe to evaluate this effect).

Innermost light profiles, which are heavily modified by atmospheric and instrumental convolution, deserve special attention. First, the consequence of the energy redistribution is detectable over a radial range which is at least $\sim 15\%$ of the entire $r^{1/4}$ axis coverable by observations. Second, an inner light profile may actually depart from the extrapolation of the de Vaucouleurs law best fitting the outer profile whether *(a)* the galaxy nucleus is a distinct component (de Vaucouleurs and Capaccioli 1979; Schweizer 1979; Figure 2) or *(b)* its profile is shaped according to an isothermal density distribution (Kormendy 1985, 1987). The latter case is particularly shifty because isothermal profiles mimic seeing–convolved $r^{1/4}$ profiles, which in turn are not dissimilar from other empirical laws (such as Reynolds–Hubble's or King's; see Figure 3).

The reconstruction of the "true" profile, $I_i(r)$, from the observed one, $I_{obs}(r)$, can be made (to the extent allowed by the Sampling Theorem) by applying one of the several restoration algorithms available on the market to the observed image of the object. However, such algorithms are generally unstable and/or have uncertain convergence criteria (*cf.* Capaccioli 1988; but see also Lauer 1985). If the problem is to test the validity of the inward extrapolation of the de Vaucouleurs law, there is an alternative approach which is simpler and safer than direct deconvolution (a statement which applies as well to any other photometric model). Using the photometric and geometrical data coming from the region of the galaxy unaffected by seeing, one may construct an $r^{1/4}$ two–dimensional model, convolve it with the appropriate PSF, and compare the extracted profile with the observed one. This is the method used by de Vaucouleurs and Capaccioli (1979) to establish their model of NGC 3379, the photometric standard for ellipticals.

Unlike deconvolution algorithms, direct convolution of an analytical model, $I_m(r)$, presents no practical difficulty provided that the PSF and the inner isophotal structure are both known. However, one should be aware that the residual differences in magnitudes, $\Delta m_c = -2.5 \log[I_{obs}/I_m^c]$, between the observed and the convolved $r^{1/4}$ model profiles, have a rather complex dependence on the corresponding differences, Δm_i between the *intrinsic* profile, I_i (deconvolved I_{obs}), and the model profile. Formally, let $I_i(r) = I_m(r)*(1 + f(r)) =$

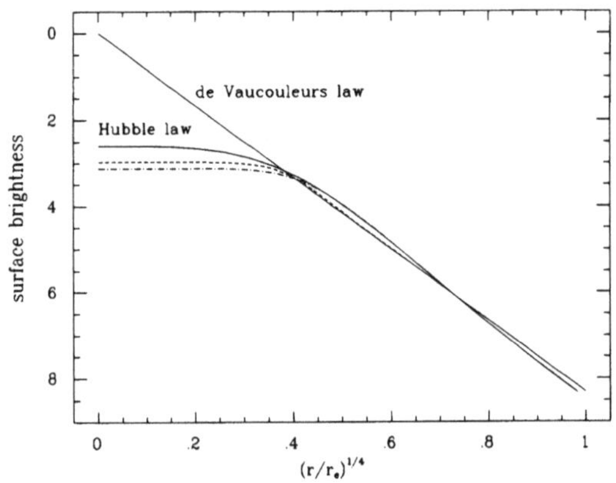

Figure 3: A round galaxy model with light profile following the de Vaucouleurs law [*upper solid line*] has been convolved with single gaussian PSF with $\sigma_* = (1/60)\, r_e$; the abscissa is in units of the effective radius. The profile of the convolved model [*dashed line*] is compared to the similar curve [*dot–dashed line*] obtained by convolving, with the same PSF, a round model following a Hubble law [*lower solid line*] with $r_o = 0.0717\, r_e$. Note that the agreement between the convolved profiles is far better than that between the original formulae.

$I_m(r) + W(r)$; then

$$I_{obs}(r) = I_i^c(r) = I_m^c(r) * \left\{1 + \frac{W^c(r)}{I_m^c(r)}\right\} = I_m^c(r) * \{1 + f(r')\}, \qquad (9)$$

where we have applied the Lagrange theorem to extract $f(r')$ from the convolution integral: $W^c(r) = \Big[f(r) * I_m(r)\Big]^c = f(r') * I_m^c$. In general, $r' \neq r$, and thus $\Delta m_i(r) \neq \Delta m_c(r)$. In plain words, the agreement between I_{obs} and I_m^c does not assure that I_m is a good representation of I_i, nor does a disagreement rule out the model completely.

As an analytical example of this statement, let us consider, for the sake of simplicity, $I_i = \exp(-r^2/2\sigma_i^2)$, $I_m = \exp(-r^2/2\sigma_m^2)$, with $\sigma_i \simeq \sigma_m$, and also a gaussian PSF, $G = \exp(-r^2/2\sigma_*^2)$. We obtain

$$\Delta m_i(r) \propto \frac{r^2}{2}\left[\frac{1}{\sigma_i^2} - \frac{1}{\sigma_m^2}\right], \qquad \Delta m_c(r) \propto \frac{r^2}{2}\left[\frac{1}{(\sigma_i^2 + \sigma_*^2)} - \frac{1}{(\sigma_m^2 + \sigma_*^2)}\right].$$

If $\sigma_* = k\,\sigma_i \simeq k\,\sigma_m$, with $k =$ const., then $\Delta m_c(r) = (1+k)^{-2} \Delta m_i(r)$. In this example, Δm_c will be always smaller then Δm_i. Two other numerical examples are shown in Figures 3 and 4.

Let us now examine the accuracy of the determination of r_e. The effective radius appears at the 1/4–*th* power in the linear relation of Equation 6; thus

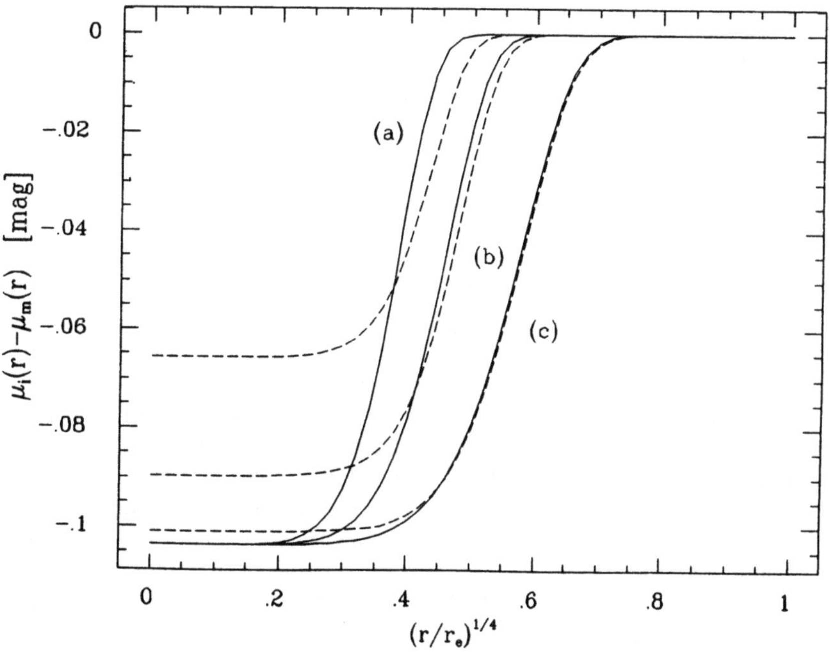

Figure 4: Comparison between $\Delta m_i(r) = -2.5 \log \{I_i(r)/I_m(r)\}$ [*solid lines*], and $\Delta m_c(r) = -2.5 \log \{I_{obs}(r)/I_m^c(r)\}$ [*dashed lines*], for the case in which the model profile, I_m, is an $r^{1/4}$ law, and the "intrinsic" profile is $I_i = I_m\,(1+f)$, with $f = 0.1 \exp\left(-r^2/2\sigma_G^2\right)$. The three sets of curves differ for the value of the dispersion of the inner gaussian component: *(a)* $\sigma_G = (1/60)\,r_e$, *(b)* $\sigma_G = (2/60)\,r_e$, and *(c)* $\sigma_G = (5/60)\,r_e$. In all cases the galaxy model has been assumed to be circularly symmetric, and the PSF to be a single gaussian with $\sigma_* = (1/60)\,r_e$. Note that the agreement between the composite ("intrinsic') profile and the $r^{1/4}$ model improves after convolution. However, the differences between the two cases are marginal until σ_G becomes comparable to σ_*. Analytically, it can be shown that the convolution of the function $W(r) = I_m(r;r_e)\,f(r;\sigma_G)$ with the single gaussian PSF (dispersion= σ_*) is the product of the convolved gaussian function, $f^c \propto \exp\left[-r^2/2(\sigma_g^2+\sigma_*^2)\right]$, with the result of the convolution of another $r^{1/4}$ model, $I_m\left(r; r_e' = r_e\,(1+\sigma_*^2/\sigma_G^2)\right)$, made by changing the dispersion of the PSF to $\sigma_*' = \sigma_*\sqrt{1+\sigma_*^2/\sigma_G^2}$.

Table 1: Reduced radius *vs.* relative integrated luminosity for the $r^{1/4}$ law

$k(r/r_e)$	0.25	0.50	0.75	0.90	0.95	0.99	0.999
$(r/r_e)^{1/4}$	0.7766	1.0000	1.2626	1.5345	1.7142	2.0871	2.5500

$\delta r_e/r_e = 4\,\delta q/q$, where $\delta q/q$ is the relative error of the slope of the line best fitting the observed surface brightness data of a given light profile (*vs.* $r^{1/4}$). Since $\delta q/q$ can be easily larger than 5% even in the best studies (this figure corresponds to an overall error of 0.25 mag over a 10 mag interval), the effective radius determined directly from a light profile has a typical uncertainty which is greater than 20%. It should be understood that this fact is intrinsic to the trend of the light profiles of early–type galaxies/bulges, and it is simply reflected by an empirical law which represents them well.

An alternative (and better) way to measure r_e — which is applicable to any galaxy irrespective of the trend of its light profiles — is through the relative integrated luminosity $k(r) = \mathcal{L}(r)/\mathcal{L}_T$, since $k(r_e) = 1/2$ by definition. There is no difficulty computing $\mathcal{L}(r)$ from two–dimensional photometric maps[2]. In the case of deep photometry ($\mu_B(lim) \simeq 28$) of an object with light profiles following reasonably well the $r^{1/4}$ law, we see from Table 1 that the extrapolation required to estimate $\mathcal{L}_T$ is of the order of 5%. Thus, the formal error on r_e is $\sim 3\%$ (if we ignore photometric scale errors).

In conclusion, this second method of computing r_e gives more accurate results than a plain linear fit of a light profile (*vs.* $r^{1/4}$) even for good $r^{1/4}$ galaxies, since the isophotal integration smooths out all minor deviations of the azimuthal profiles from the model. To a lesser extent, such an isophotal smoothing is achieved by the so–called *equivalent profile*. This is the light profile, $\mu(r_e^*)$, of a round galaxy model whose isophotes encircle the same area as the corresponding isophotes in the real galaxy. As a tool for calculations, the equivalent profile (or any "mean" profile such as in King's, 1978) is quite effective; but it should never be used to test an empirical law, since it may wash out systematic differences in the trends of the light profiles along individual axes (*cf.* Capaccioli 1987). The common practice of producing equivalent profiles is co–responsible for the late discovery of faint disks in E galaxies (Carter 1987; Bender and Möllenhoff 1987; Capaccioli 1987; Capaccioli and Vietri 1988; Capaccioli *et al.* 1988*b*), which has postponed until recently the formulation of a basic question: do E galaxies form a homogeneous class?

The spatial emissivity, $\rho(s)$, corresponding to the $r^{1/4}$ law in projection, has been investigated by Poveda *et al.* (1960) and Young (1976). Both have used

[2]Since seeing convolution preserves the total energy, it does not affect $\mathcal{L}(r)$ appreciably for $r \gtrsim 3\sigma_*$; note that the single gaussian convolution correction for the $r^{1/4}$ law is smaller than 0.05 mag at all $r > 11.4\sigma_*^{4/3}$.

Table 2: Photometric laws

Baum: 1955	$I(r)/I_o = 2\,r_o^2\big[r\,(r+r_o)\big]^{-1}$
Oemler: 1976	$I(r)/I_o = 4\,r_o^2(r+r_o)^{-2}\exp\big[-(r^2-r_o^2)/\beta^2\big]$
King: 1966	$(I(r)/k) = \Big\{\big[1+(r/r_c)^2\big]^{-1/2} - \big[1+(r_t/r_c)^2\big]^{-1/2}\Big\}^2$
Sersic: 1968	$I(r)/I(0) = \mathrm{dex}\{A\,r^{2/n}\}$

the standard formula

$$\rho(s) = -\frac{1}{\pi}\int_s^{+\infty}\frac{\left(dI(t)/dt\right)_{t=r}}{\sqrt{r^2-s^2}}\,dr, \tag{10}$$

which is the inverted Von Zeipel integral applying to the principal plane of absorption–free spheroids (obvious projection corrections are needed for other cross sections of spheroids and for triaxial ellipsoids; see Sandage *et al.* 1970). The solution of Equation 10 has a rather simple asymptotic expansion (Young 1976)

$$\rho(s) \propto s^{-7/8}\exp(-3.33071\,\ln 10\,s^{1/4}), \tag{11}$$

which is sufficiently accurate for radial distances $r > 0.1r_e$.

Other Fitting Formulae

Several luminosity laws other than Reynolds–Hubble and de Vaucouleurshave been proposed to represent the manifold of light profiles of early–type galaxies, a manifold which is partly intrinsic to the nature of these objects and partly spurious (due to systematic errors and/or lack of resolution). Some of these alternative formulae, listed in Table 2, have already been compared by de Vaucouleurs and Capaccioli (1979) with their deconvolved East–West profile of NGC 3379.

Baum's (1955) formula coincides with Reynolds–Hubble for $r \gg r_o$, and is steeper at small galactocentric distances, diverging at $r = 0$ (which is irrelevant from the physical point of view). In addition, its area integral (total luminosity) diverges with r. Oemler's (1976) formula is instead almost coincident with Reynolds–Hubble in the core region, and is progressively steeper at increasing distances (to account for the fact that the observed surface brightness of E galaxies falls off more rapidly than an r^{-2} power law). The tapering (gaussian) factor brings with it a *free* parameter β which is viewed by the author as a convenient tool for modelling the variety of shapes of the outer light profiles of cluster ellipticals[3].

[3] *"The weakness of de Vaucouleurs's law is that it has only two free parameters, for length*

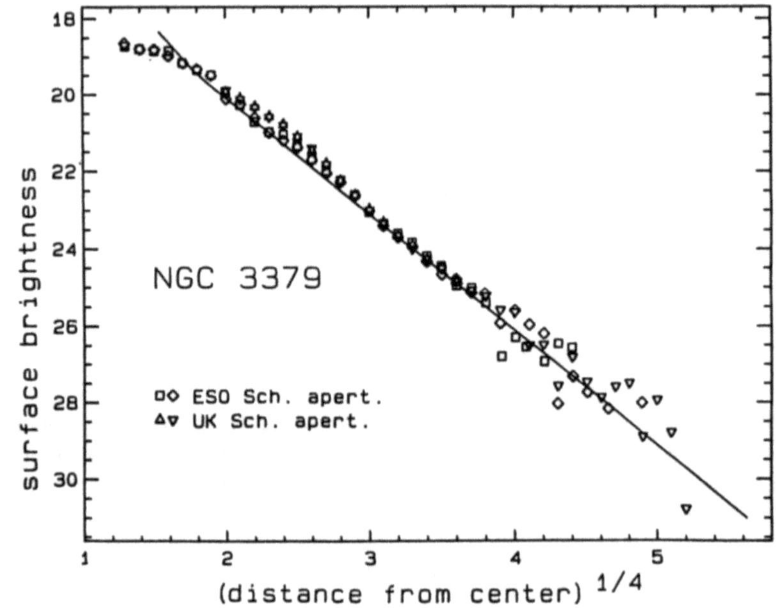

Figure 5: Deep photometry of the standard E1 galaxy NGC 3379 based on ESO and UK Schmidt material (Capaccioli *et al.* 1989*b*)). The major axis light profile [*open symbols*] has been sampled with a software–simulated aperture whose size is allowed to increase with distance from the galaxy center. It is apparent that the $r^{1/4}$ model of de Vaucouleurs and Capaccioli (1979; *solid line*) fits quite well the outer profile (the systematic large $(O-C)$ residuals at all $a \lesssim 80''$ are mainly due to the convolution operated by the aperture integration) with no sign of any tidal truncation.

Table 3: Parameter β in Equation 12

$1/\alpha$	0.20	0.22	0.24	0.27	0.30	0.33
β	4.19907	3.80435	3.45543	3.07345	2.75189	2.48882

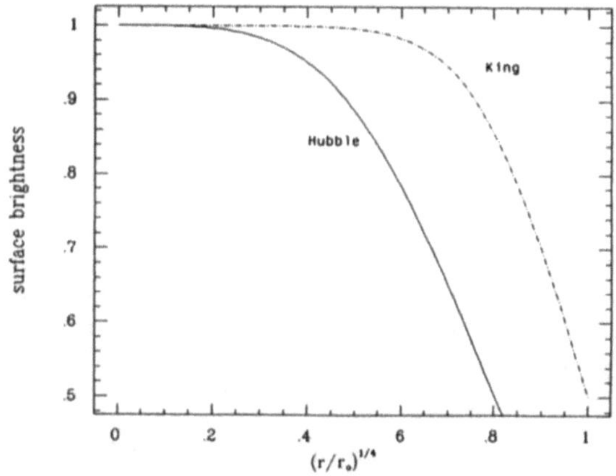

Figure 6: Reynolds–Hubble and King formulae for the same value of the "characteristic" length r_o (or r_c).

King's (1962) formula was developed to model the projected density distribution of star clusters, and it was later applied to describe the luminosity profiles of elliptical galaxies (King 1966); it is semi–empirical in the sense that it is related to quasi–isothermal isotropic dynamical models. This formula has two characteristic lengths, one of which is a free parameter: the *core radius*, r_c, is where the surface brightness is $2.5 \log 2 = 0.753$ mag fainter than the peak brightness $I(0)$, and the *tidal radius*, r_t, is where $I = 0$. The tidal radius defines the physical boundary of a galaxy. Though dwarf spheroidals are indeed tidally truncated (but see the case of M32–like galaxies reviewed by Nieto, 1988), normal ellipticals do not show, in general, any such effect within the range of present photometric measurements (Figure 5; see also the "evolution" of r_t summarized by de Vaucouleurs, 1987).

For $r \ll r_t$ (and $r_t \gg r_c$) King's formula simplifies into $I(r) \propto 1/(1+r^2)$, which is similar but not identical to the Reynolds–Hubble law. A way to appreciate this difference is to plot both formulae stretching the radial coordinate (Figure 6) (note that, in general, both of them are slightly modified by seeing convolution; see Figure 3 in Kormendy, 1987). It is apparent that King's law produces a more prominent "flat core'.

and surface brightness scale; the shape is fixed." (Oemler 1976, pg. 701). de Vaucouleurs, however, regards this as one of its strengths!

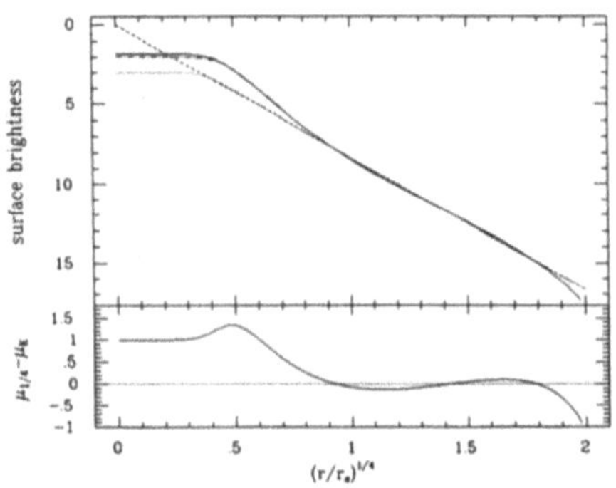

Figure 7: Comparison between the $r^{1/4}$ law [*dashed line*] and King's formula [*solid line*] having $r_c = (1/20)\, r_e$ and $r_t = 20\, r_e$. Convoluted profiles with a single gaussian PSF ($\sigma_* = (1/60)\, r_e$) are also shown together with their magnitude differences [*lower panel*].

Well outside the core region and down to the current limit of surface photometry measurements, King's formula may be adjusted to match closely the the $r^{1/4}$ law by an appropriate selection of the tidal radius (Figure 7; *upper panel*); the role of r_t is to taper the r^{-2} trend of the first term in the formula. In the same radial range, the magnitude differences between the two empirical laws, $\Delta m = \mu_{r^{1/4}} - \mu_K$ (Figure 7; *lower panel*), have a systematic trend, which is actually the trend of magnitude residuals of light profiles of E galaxies with respect to the $r^{1/4}$ law (de Vaucouleurs 1953, Capaccioli 1985).

Sérsic's (1968) formula is the generalization of the $r^{1/4}$ law ($n = 8$); it contemplates also the case of exponential light-profiles ($n = 2$) holding for face–on disks (Freeman 1970; van der Kruit 1987, and references therein). Actually, since there is no theoretical reason for the exponent of the de Vaucouleurs law to be exactly 1/4, it is of interest to investigate the more general form

$$\log J(r) = -\beta\,\left(\xi^{2/\alpha} - 1\right), \text{with}\;\; \beta > 0, \tag{12}$$

where α is a fixed number close to 8. This exercise, which can be based upon the modern large set of good photometric profiles, is justified by the presumption that Gérard de Vaucouleurs had little chance to explore all possibilities for the exponent of his photometric law, given the lack of computing facilities in 1948!

If $\xi = r/r_e$ as usual, the parameter β is given by the condition $k(\xi = 1) = \mathcal{L}(1)/\mathcal{L}_T = 1/2$ of the fractional luminosity. Results of the numerical integration on sampled values of α are given in Table 3. An interpolation formula, accurate to better than 10^{-4} in the range $0.15 < 2/\alpha < 0.5$, is

$$\beta = -0.142058 + 434121\,\alpha. \tag{13}$$

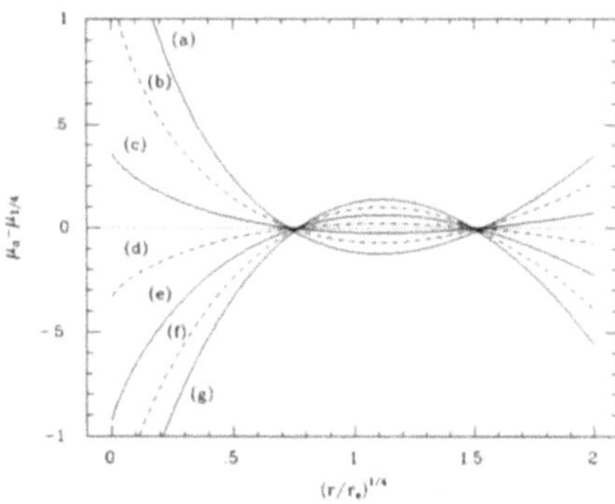

Figure 8: Magnitude differences between the photometric law expressed by Equation 12 and the $r^{1/4}$ formula, for a set of values of the exponent $2/\alpha = 0.20$ to 0.32, step 0.2; curves from *(a)* to *(g)*. The effective radii of the $r^{2/\alpha}$ law have been chosen to give the same total magnitude as the $r^{1/4}$ law, in units of I_e. The abscissa is in units of the effective radius proper to the $r^{1/4}$ law.

Finally, the total luminosity is

$$\mathcal{L}_T(\alpha) = \pi\, I_e\, r_e^2\, \frac{\alpha \exp\left(\beta \ln 10\right)}{\left(\beta \ln 10\right)^{\alpha}}\, \Gamma(\alpha). \tag{14}$$

Examples of the magnitude differences between Equation 12 and a standard $r^{1/4}$ law with the same total magnitude (Figure 8) show that a possible improvement of the classical law is achieved for $\alpha/2 > 4$; For instance, an $r^{2/9}$ law (Figure 9) reproduces the typical "concavity" of the residuals of observed light profiles with respect to the best fitting $r^{1/4}$ law (see above) and creates a little "core', which is also often noted in real galaxies.

We conclude this review of the photometric law for early–type galaxies with a formula recently proposed by Jaffe (1983). Rather than fitting the observed profiles with some empirical formula, Jaffe has produced an expression for the spatial emissivity,

$$\rho(\zeta) = \frac{1}{4\pi}\zeta^{-2}\left(1+\zeta\right)^{-2}; \tag{15}$$

here ζ is the radius normalized to 1 at the isophotal *surface* enclosing half of the total light; numerical integration shows that this radius is $r_e/0.763$, if r_e is the effective radius of the projected light distribution. The total luminosity (volume integral of Equation 15) is $\mathcal{L}_T = \zeta/(1+\zeta)$; the surface brightness at the projected distance $\eta = 0.763\,(r/r_e)$, computed through the inverse of Equation

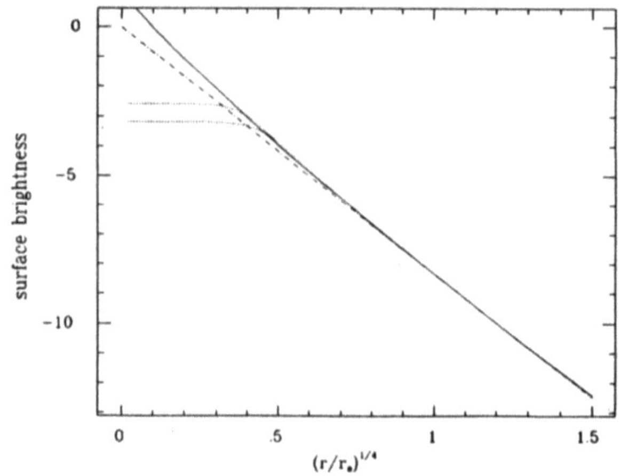

Figure 9: de Vaucouleurs law [*dot–dashed line*] and Equation 12 [*solid line*] with $2/\alpha = 2/9$ and effective radius giving the same total magnitude as the $r^{1/4}$ law for the same value of I_e. The dotted lines represent the convolution of the $r^{2/9}$ formula with a single gaussian PSF having $\sigma_* = (1/60)\,r_e$ and $(1/30)\,r_e$.

10, follows:

$$I(\eta) = \begin{cases} \dfrac{1}{4\eta} + \left[\dfrac{1}{1-\eta^2} - \dfrac{2-\eta^2}{(1-\eta^2)^{-3/2}}\,\mathrm{arcosh}\left(\dfrac{1}{\eta}\right)\right]\dfrac{1}{2\pi} & \eta < 1, \\[2ex] \dfrac{1}{4\eta} - \left[\dfrac{1}{\eta^2-1} + \dfrac{\eta^2-2}{(\eta^2-1)^{-3/2}}\,\mathrm{arcos}\left(\dfrac{1}{\eta}\right)\right]\dfrac{1}{2\pi} & \eta > 1. \end{cases} \tag{16}$$

The double expression avoids problems with complex values of *arcosh*.

A comparison between Jaffe's and de Vaucouleurs's laws given in Figure 10 for the same value of the effective radius r_e. Figure 11 shows the ability of Jaffe's law to model the inner change of slope in the light profiles of ellipticals *vs.* $r^{1/4}$, noted by Capaccioli *et al.* (1988*a*).

Concluding Remarks

Which is the "best" fitting formula for early–type galaxy profiles? At the end of the above *excursus* in the land of empirical photometric laws we are still unable to answer this question, given the large variety of behaviours displayed by observations of Es and S0/S bulges. Possibly, a better question is: how many free parameters are needed to describe, with just one formula, the manifold of light profile shapes? We know that, at least for normal (non–dwarf) ellipticals (Binggelli *et al.* 1984) that the (linear) "core radii" and "effective radii" correlate with the absolute luminosity M_T (Kormendy 1982, 1985, 1987;

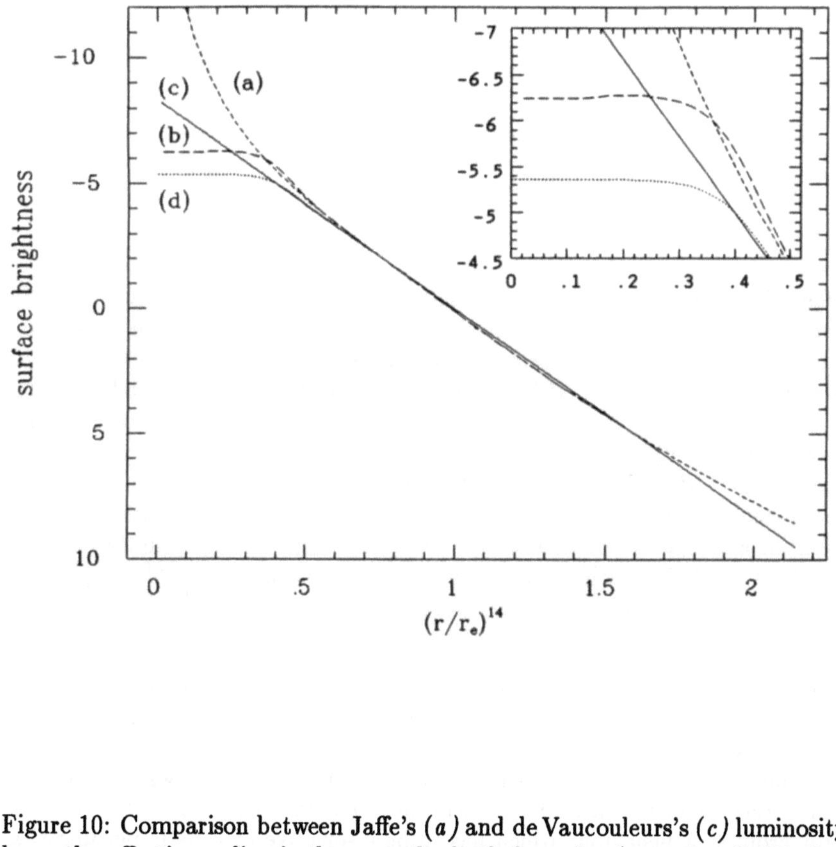

Figure 10: Comparison between Jaffe's *(a)* and de Vaucouleurs's *(c)* luminosity laws; the effective radius is the same for both formulae (note that Jaffe's law is expressed in units of the spatial "effective radius', which is 0.763 larger than the corresponding projected quantity). The two laws are very similar from $r \simeq 0.13\, r_e$ to $r \simeq 8.3\, r_e$, *i.e.* over the total radial range covered by observations of large galaxies except the region affected by seeing. In this range the residual difference indicates a small but clear concavity in Jaffe's law plotted *vs.* $r^{1/4}$. Both formulae have been convolved with the same single gaussian PSF having a dispersion $\sigma_* = (1/60)\, r_e$. It is apparent that, while the convolved $r^{1/4}$ law *(d)* becomes flat at small radii, Jaffe's law *(b)* maintains an innermost steeper gradient (see also the blow-up in the inset).

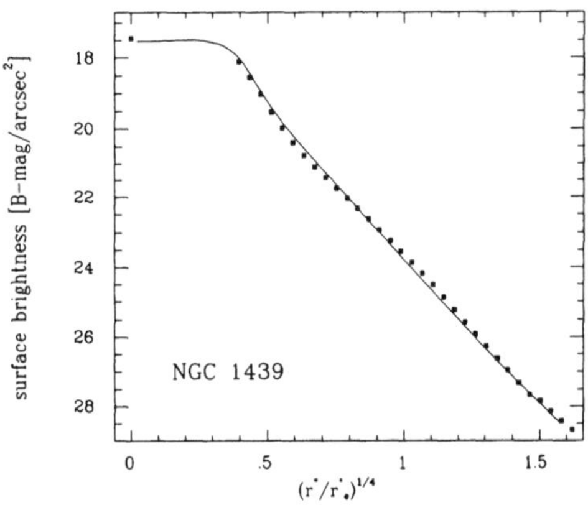

Figure 11: Equivalent luminosity profile [*full dots*] of the E0 galaxy NGC 1439, obtained by combining high resolution CCD and deep Schmidt material (from Capaccioli *et al.* 1988*a*; reprinted with permission). A Jaffe (1983) law with effective radius $r_e(J) = 41''$ (same as the equivalent effective radius of the $r^{1/4}$ law best fitting the outer mean profile galaxy) has been convolved with a single gaussian PSF with $\sigma_* = 0''.683$, 10% larger than the value reported by Capaccioli *et al.* (1988*a*; their Table 9) for the dispersion of the first gaussian component of their measured PSF. Circular symmetry has been assumed. The resulting curve [*solid line*] matches the change of slope occurring in the observed profile at $r \simeq 3''.5$, which is not accounted for by a convolved $r^{1/4}$ law (see also the inset in Figure 10).

Schombert 1986). Therefore, M_T should be the only parameter needed by a quite general formula; but it is *not* a free parameter, being determined entirely by observations.

Thus, as a first approximation, shape–free photometric laws seem too be quite acceptable, at least to represent the "general" behaviour of early–type light profiles: a behaviour which has to be insensitive to genetic peculiarities, to modifications induced by dynamical evolution and interactions with the environment (*e.g.*, the tidal classes of Kormendy, 1977), and even to the presence of faint components (such as disks). Evaluated in this sense, shape–free photometric laws should *not* be used as morphological descriptors.

Among shape–free laws, the $r^{1/4}$ (or possibly the $r^{2/9}$) formula is quite interesting for:

1. it is simple (but Jaffe's law has a much easier representation for the spatial emissivity);

2. it fits reasonably well the light profiles of normal Es and of bulges over an

ample interval of surface brightness;

3. in any case it provides the key to a coordinate transformation which turns most part of a light profile into a straight line.

These are the main reasons of its success and of its remarkable longevity. Happy birthday, $r^{1/4}$!

References

Baum, W. A. 1955, *Pub. Astron. Soc. Pac.* **83**, 199.

Bender, R. and Möllenhoff, C. 1987, *Astron. Astrophys.* **177**, 71.

Bertin, G. and Stiavelli, M. 1988, *Mon. Not. Roy. Astron. Soc.*, in press.

Binggeli, B., Sandage, A., and Tarenghi, M. 1984, *Astron. J.* **89**, 64.

Binney, J. 1982, *Ann. Rev. Astron. Astrophys.* **20**, 399.

Capaccioli, M. 1985, in *New Aspects of Galaxy Photometry*, Lectures on Phys. Ser. No. **232**, ed. J.-L. Nieto, (Berlin: Springer Verlag), p. 53.

Capaccioli, M. 1987, in *I. A. U. Symp. No. 127, Structure and Dynamics of Elliptical Galaxies*, ed. T. P. de Zeeuw (Dordrecht: Reidel), p. 47.

Capaccioli, M. 1988, *Proceedings of the Second Summer School on Extragalactic Astronomy, (Cordoba)*, in press.

Capaccioli, M. and de Vaucouleurs, G. 1983, *Astrophys. J. Suppl.* **52**, 465.

Capaccioli, M. and Vietri, M. 1988, *Proc. of the Yellow Mountain Summer School, (Yellow Mountain: China)*, in press.

Capaccioli, M., Held, E. V., and Nieto, J.-L. 1987, *Astron. J.* **94**, 1519.

Capaccioli, M., Piotto, G., and Rampazzo, R. 1988a, *Astron. J.*, in press.

Capaccioli, M., Vietri, M., and Held, E.V. 1988b, *Mon. Not. Roy. Astron. Soc.*, in press.

Capaccioli, M., Caon, N., and Rampazzo, R. 1989a, in preparation.

Capaccioli, M., Held, E. V., and Lorenz, H. 1989b, in preparation.

Carter, D. 1987, *Astrophys. J.* **312**, 514.

Davis, L. E., Cawson, M., Davies, R. L., and Illingworth, G. 1985, *Astron. J.* **90**, 169.

Fish, R. A. 1964, *Astrophys. J.* **139**, 284.

Freeman, K. C. 1970, *Astrophys. J.* **160**, 811.

van Houten, C. J. 1961, *Bull. Astron. Netherlands* **16**, 1.

Hubble, E. 1930, *Astrophys. J.* **71**, 231.

Jaffe, W. 1983, *Mon. Not. Roy. Astron. Soc.* **202**, 995

Jedrzejewski, R. I. 1987, *I. A. U. Symposium No. 127, Structure and Dynamics of Elliptical Galaxies*, ed. T. P. de Zeew (Dordrecht: Reidel), p. 37.

Jones, W. B., Obbits, D. L., Gallet, R. M., and de Vaucouleurs, G. 1967, *Publ. Dept. Astron. Univ. Texas, Austin*, Ser. 2, **1**, No. 8.

King, I. R. 1962, *Astron. J.* **67**, 471.

King, I. R. 1966, *Astron. J.* **71**, 276.

King, I. R. 1978, *Astrophys. J.* **222**, 1.
Kormendy, J. 1977, *Astrophys. J.* **218**, 333.
Kormendy, J. 1982, *Morphology and Dynamics of Galaxies*, XII Advanced Course of Swiss Soc. Astron. Astrophys., ed. L. Martinet and M. Mayor (Sauverny: Geneva Obs.), p. 115.
Kormendy, J. 1985, *Astrophys. J. Lett.* **292**, L9.
Kormendy, J. 1987, *I. A. U. Symp. No. 127, Structure and Dynamics of Elliptical Galaxies*, ed. T. P. de Zeeuw (Dordrecht: Reidel), p. 27.
van der Kruit, P. C. 1987, *Astron. Astrophys.* **173**, 59.
Lauer, T. R. 1985, *Astrophys. J. Suppl.* **57**, 473.
Longo, G. and de Vaucouleurs, A. 1983, *A General Catalogue of Photoelectric Magnitudes and Colors in the U, B, V System*, Univ. of Texas Monographs in Astron. No. **3**.
Michard, R. 1984, *Astron. Astrophys.* **140**, L39.
Merritt, D. 1985, *Astron. J.* **100**, 1027.
Newton, A. J. and Binney, J. 1984, *Mon. Not. Roy. Astron. Soc.* **210**, 711.
Nieto, J.–L. 1988, *Proceedings of the Second Summer School on Extragalactic Astronomy, (Cordoba)*, in press.
Nieto, J.–L., Capaccioli, M., and Held, E. V. 1988, *Astron. Astrophys.* **195**, L1.
Oemler, A. 1976, *Astrophys. J.* **209**, 693.
Okamura, S. 1988, *Publ. Astron. Soc. Pac.* **100**, 524.
Oort, J. H. 1940, *Astrophys. J.* **91**, 273.
Oort, J. H. 1946, *Mon. Not. Roy. Astron. Soc.* **106**, 171.
Poveda, A., Iturriaga, R., and Orozco, I. 1960, *Bull. Obs. Tonantzintla* No. **20**, p. 3.
Reynolds, J. H. 1914, *Mon. Not. Roy. Astron. Soc.* **74**, 132.
Redman, R. O. and Shirley, E. G. 1936, *Mon. Not. R. Astron. Soc.* **96**, 588.
Redman, R. O. and Shirley, E. G. 1938, *Mon. Not. R. Astron. Soc.* **98**, 613.
Sandage, A., Freeman, K. C., and Stokes, N. R. 1970, *Astrophys. J.* **160**, 831.
Schombert, J. M. 1986, *Astrophys. J. Suppl.* **60**, 603.
Schweizer, F. 1979, *Astrophys. J.* **233**, 23.
Sersic, J.–L. 1968, *Atlas de Galaxias Australes* (Cordoba: Observatorio Astronomico).
Van Albada, T. 1982, *Mon. Not. R. Astron. Soc.* **201**, 939.
de Vaucouleurs, G. 1948, *Ann. d'Astrophys.* **11**, 247.
de Vaucouleurs, G. 1953, *Mon. Not. Roy. Astron. Soc.* **113**, 134.
de Vaucouleurs, G. 1958, *Astrophys. J.* **128**, 65.
de Vaucouleurs, G. 1959, *Handbuch der Phys.* **53**, 331.
de Vaucouleurs, G. 1961, *I. A. U. Symposium No. 15, Problems of Extragalactic Research*, ed. G. C. McVittie (New York: Macmillan), p. 3.
de Vaucouleurs, G. 1979, *Photometry, Kinematics and Dynamics of Galaxies*, ed. D. S. Evans, (Austin: Univ. Texas Press), p. 1.
de Vaucouleurs, G. 1983, *I. A. U. Coll. No. 78, Astronomy with Schmidt-type Telescopes*, ed. M. Capaccioli (Dordrecht: Reidel), p. 367.

de Vaucouleurs, G. 1987, *I. A. U. Symp. No. 127, Structure and Dynamics of Elliptical Galaxies*, ed. T. P. de Zeeuw (Dordrecht: Reidel), p. 3.
de Vaucouleurs, G. and Capaccioli, M. 1979, *Astrophys. J. Suppl.* **40**, 699.
Young, P. J. 1976, *Astron. J.* **81**, 807.

Discussion

F. Bertola: What is the photometric evidence of the presence of faint disks in elliptical galaxies?

Capaccioli: It comes first from the analysis of the deviations of the isophotal contours from pure ellipses, and then from the fact that the residuals from purely elliptical models are "images" whose major axis light profiles are well–fitted by an exponential law. Moreover, these "disks" are found mostly in flat E's, as expected for flat components embedded in almost oblate bulges.

F. Simien: Do you think that faint disks in elliptical are a distinct stellar population?

Capaccioli: I have no observational fact that I can use to answer your question. Disks in ellipticals are very faint and so far they can only just be detected. However, if there is continuity in disk–to–bulge ratios from S0's to E's, one may temptatively extrapolate to faint–disk ellipticals what we learn about lenticulars.

Simien: What about the systematic residuals of the profiles with respect to the $r^{1/4}$ law?

Capaccioli: They occur at any surface brightness level and are usually correlated with the absolute luminosity. Shallow cores are present in superluminous E's, which also possess extended coronae, while tidal truncations may be found in dwarfs. In any case, even the light profiles of almost isolated giant ellipticals and bulges present systematic deviations from a pure $r^{1/4}$ law in the sense that the gradient $d\mu/dr^{1/4}$ is not constant, but is rather decreasing slowly (in absolute value).

R. A. E. Fosbury: One of the most significant results from X–ray astronomy is the discovery of hot gas in at least the most luminous of the nearby early–type galaxies. The mass of this hot gas is comparable to that of cooler gas in large spirals. We don't know the angular momentum of the hot gas. If it is comparable to that in spirals, then this would raise some very fundamental questions about their formation properties.

Capaccioli: This is certainly another important element to keep in mind while one tries to solve the "mistery" of ellipticals.

D. Burstein: Out of the $\sim$ 400 galaxies for which we have determined magnitudes and diameters, I would estimate that 10–15% might have disks that contribute $< 10\%$ of the light. This does not seem to affect the overall dynamics of the galaxies. A more personal observation: I think that it is possible that every E could have a disk that contributes 0.1–10% of the total light. We will have to accurately map every elliptical to find out!

Capaccioli: There is obviously no claim that a faint and light disk may influence the overall dynamics. The question is another one: do disky E's constitute a family separate from the rest of ellipticals, maybe a family closely related to that of the S0's? In other words, are these faint and dynamically insignificant disks pointing out something which is much more fundamental than their simple existence. If so, I would tend to disagree with your personal observation. In any case, we need more information than we have at present, and not necessarily kinematical, to establish the significance of faint disks.

Box– and Peanut–Shaped Bulges of Disk Galaxies

Ralf–Jürgen Dettmar

Radioastronomisches Institut der Universität Bonn

Introduction

Box– and peanut–shaped bulges have been described in the literature since the late 1950's, but it was G. de Vaucouleurs (1974) who has drawn attention to the fact that this shape might be a general property of the spheroidal components of disk galaxies. This supposition might be still valid as the box–shaped struture recently found in ellipticals is much less pronounced in most cases. In addition to the classical one dozen examples of box– or peanut–shaped bulges, recent surveys have found another 100 galaxies with such spheroids, but the derived statistics are discrepant by more than one order of magnitude. Jarvis (1986) concludes that 1.2% of all disk galaxies have box– and peanut–shaped bulges while Shaw (1987) and de Souza and dos Anjos (1987) find 20% and 23%, respectively. These are lower limits due to selection effects.

These statistics were obtained from distributed survey material [Palomar Observatory Sky Survey (POSS) and the ESO/SRC Southern Sky Survey]. Here we present data from CCD surface photometry with higher dynamic range, better S/N, and larger scale.

Observations and Reduction

We have obtained CCD images of 73 edge–on galaxies with the 2.2–m telescopes at ESO/La Silla and Calar Alto. The objects were selected by visual inspection of the SRC J film copies and POSS prints and their major axis diameters were between 2 and 4 arc–minutes. Morphological types are taken from Lauberts's catalogue (1982).

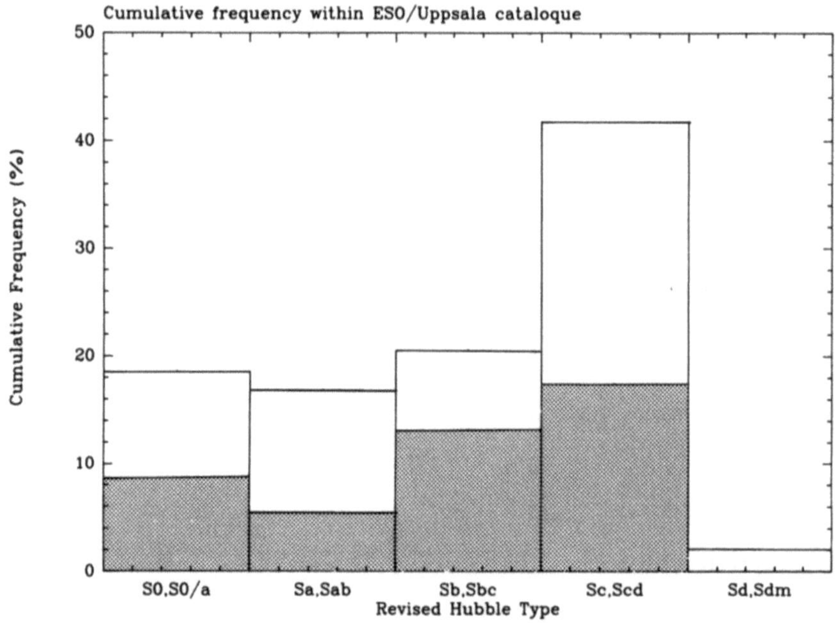

Figure 1: Cumulative frequency of morphological types (upper histogram) and box– and peanut–shaped bulges within each bin (hatched histogram) scaled to the distribution within ESO/Uppsala catalogue.

Results

35 or 48% of the galaxies observed show box– and peanut–shaped bulges. If we scale our sample to the distribution of morphological types in the ES0/Uppsala catalogue (Figure 1, upper histogram) this results in a total of 45±8% of disk galaxies having a box– or peanut–shaped bulge (Figure 1, hatched histogram). Compared to the earlier studies, the most surprising result is the rather high percentage of late–type spirals ($T = 5 - 6$) that show box– or peanut–shaped bulges. Even with the large statistical errors ($\sqrt{N}$) of the small sample, this seems to be significant, and the distribution for earlier types given by Shaw is reproduced. The larger dynamic range of the CCD photometry easily explains the higher detection rate in later type spirals.

The unusual bulge of IC 4745

The box–shaped bulge of the S0/a galaxy IC 4745 is unusual in several respects. It is asymmetric in the inner part and the outskirts show signs of interaction (see Dettmar 1986 for a contour plot). The asymmetry is also visible in the rotation curve, indicating that the object is heavily disturbed. This supports the idea that at least some of the box- and peanut-shaped bulges are due to accretion of material from a companion galaxy as is discussed for IC 4767 by Whitmore and Bell (1988).

Acknowledgements

The author wishes to thank A. Barteldrees for help with part of the observations. Travel support was granted by the DFG under Me 745/5-1 and De 385/1-1.

References

Dettmar, R.-J. 1986, *Mitt. Astron. Gesellschaft* **67**, 380.
Jarvis, B. J. 1986, *Astrophys. J.* **91**, 65.
Lauberts, A. 1982, *The ESO/Uppsala Survey of the ESO(B) Atlas* (Garching bei München: ESO).
Shaw, M. A. 1987, *Mon. Not. Roy. Astr. Soc.* **229**, 691.
de Souza, R. E. and dos Anjos, S. 1987, *Astron. Astrophys. Suppl.* **70**, 465.
de Vaucouleurs, G. 1974, in *The Formation and Dynamics of Galaxies*, I.A.U. Symp. No. **58**, ed. J. R. Shakeshaft (Dordrecht: Reidel), p. 335.
Whitmore, B. C. and Bell, M. 1988, *Astrophys. J.* **324**, 741.

The Luminosity Law of Ellipticals; A Test of a Family of Anisotropic Models on Eight Galaxies

G. Bertin, R. P. Saglia, and M. Stiavelli

Scuola Normale Superiore, Pisa, Italy

An important clue to the structure and dynamics of elliptical galaxies is provided by the empirical $r^{1/4}$ luminosity law proposed by de Vaucouleurs (1948). The existence of such a law is indicative of a common underlying mass distribution in these galaxies. The fact that this law is universal suggests that essentially a single physical mechanism characterizes the formation of ellipticals. Here, we report on a recent study where we have analyzed published photometric and kinematical data for a set of bright elliptical galaxies (NGC 3379, NGC 4374, NGC 4472, NGC 4486, NGC 4636, NGC 7562, NGC 7619, and NGC 7626) in terms of self-consistent anisotropic models (f_∞), under the assumption of constant mass-to-light ratio (Bertin, Saglia, and Stiavelli 1988).

The adopted models are based on a simple distribution function

$$f_\infty = \begin{cases} A(-E)^{3/2} exp[-aE - cJ^2/2], & \text{if } E \leq 0 \\ 0, & \text{otherwise} \end{cases}$$

which were argued to incorporate the essential features of collisionless collapse. These models define an equilibrium sequence parameterized by the value of the dimensionless central potential $\Psi = -a\Phi(0)$. Concentrated models ($\Psi \geq 7$) are found to possess realistic density profiles. In fact, the photometric fit based on the f_∞-models proves to be very good (in the majority of cases, the differences between the projected density of the model and the data are less than 0.15 mag) and better than fits based on King models or the $r^{1/4}$ law. In the case of NGC 3379, a complete fit over 11 magnitudes is obtained, with residuals mostly confined to 0.05 mag (see Figures). The fit to the kinematical data allows us to determine masses and mass-to-light ratios.

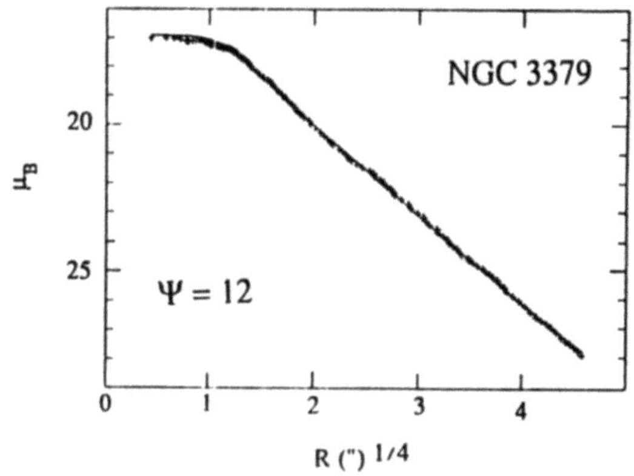

Figure 1: Magnitudes μ_B (plus signs) from de Vaucouleurs and Capaccioli (1979), and magnitudes μ_∞ of the seeing-convolved f_∞-model (solid line) selected by the photometric fit.

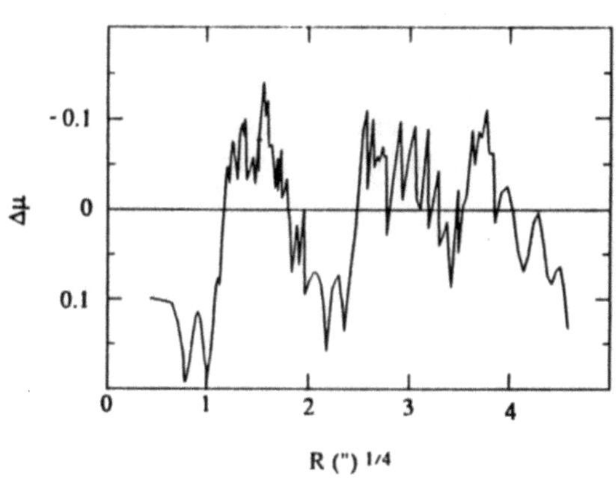

Figure 2: The residuals $\Delta\mu = \mu - \mu_\infty$ between the data and the model.

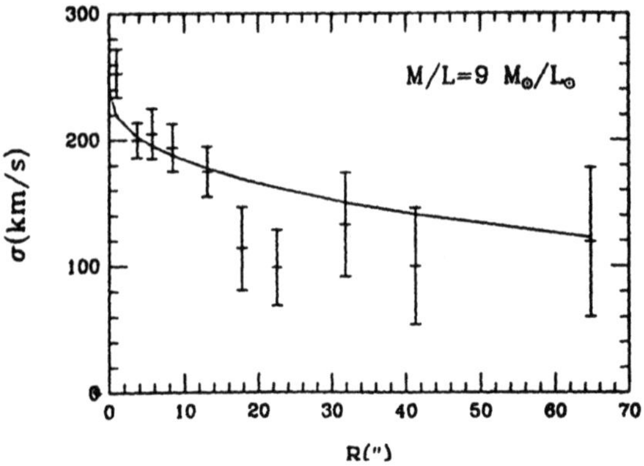

Figure 3: The velocity dispersion along the line of sight (with error bars) from Davies (1981) and the projected velocity dispersion of the f_∞-model (solid line).

References

Bertin, G., Saglia, R. P., and Stiavelli, M. 1988, *Astrophys. J.* **330**, in press.
Davies, R. L. 1981, *Mon. Not. Roy. Astro. Soc.* **194**, 879.
de Vaucouleurs, G. 1948, *Ann. d'Astrophys.* **11**, 147.
de Vaucouleurs, G. and Capaccioli, M. 1979, *Astrophys. J. Suppl.* **40**, 699.

"Box/Peanut"– Shaped Galactic Bulges

Martin Shaw

Department of Astronomy, University of Manchester, England, U.K.

Overall goal

By a visual inspection of photographic plate material for all 504 of the largest, normal spiral and lenticular galaxies contained within the *Second Reference Catalogue* (de Vaucouleurs *et al.* 1976; RC2) and supplementary lists, we have constructed a subset of those 117 systems which are most edge– on to the line–of–sight. This has been used as a basic database for our investigation of the nature of the galaxian non–thin disc population in general, but it has also proved an ideal means by which to study general characteristics of those galaxies which display "box" or "peanut"–shaped central isophotes.

Sample completeness

- All–sky insofar as the catalogues used here allow;
- Major axis dimensions in excess of $\log_{10} D_{25} = 1.55$, or ~ 5 arcmin;
- $m_B \sim 13.5$ mag (an approximate limit imposed by use of the Second Reference Catalogue), and $M_B \sim -20$;
- Inclination angles $\geq 70°$. This defining characteristic is, however, a strong function of morphological type: 85% of the disc dominated systems have inclination angles in excess of 80°, compared to 75% of the bulge dominated ones and 65% of the box/peanuts.

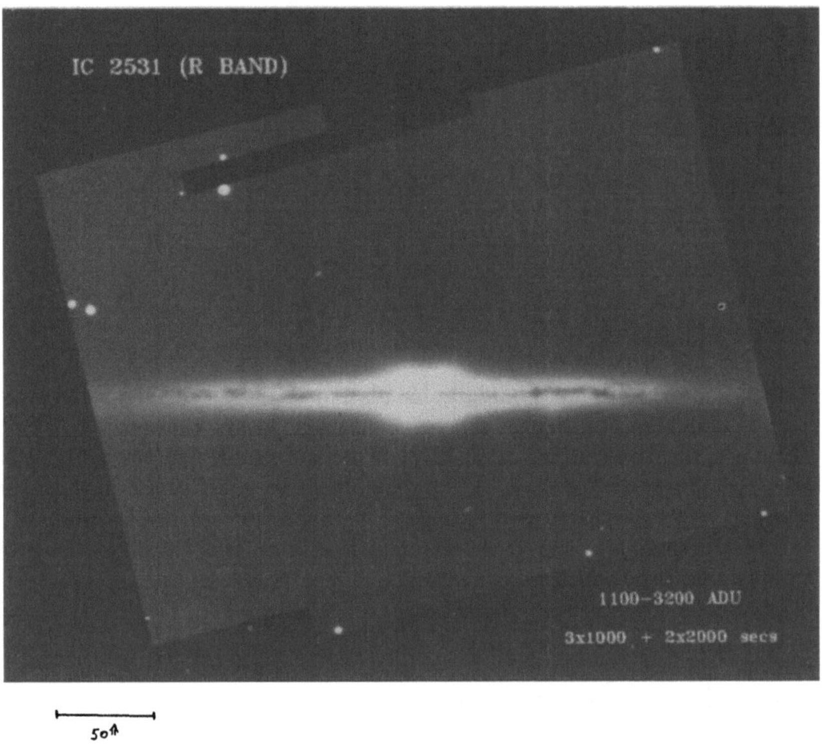

Figure 1: Prime focus AAT CCD frame mosaic in the Cousins R passband of the Sc galaxy IC 2531 (after calibration and removal of field stars) showing compact, high surface brightness box/peanut morphology. The box/peanut here comprises an additional ~35% light in excess of a "normal" (*e.g.* $r^{1/4}$ law) bulge of such a morphological type.

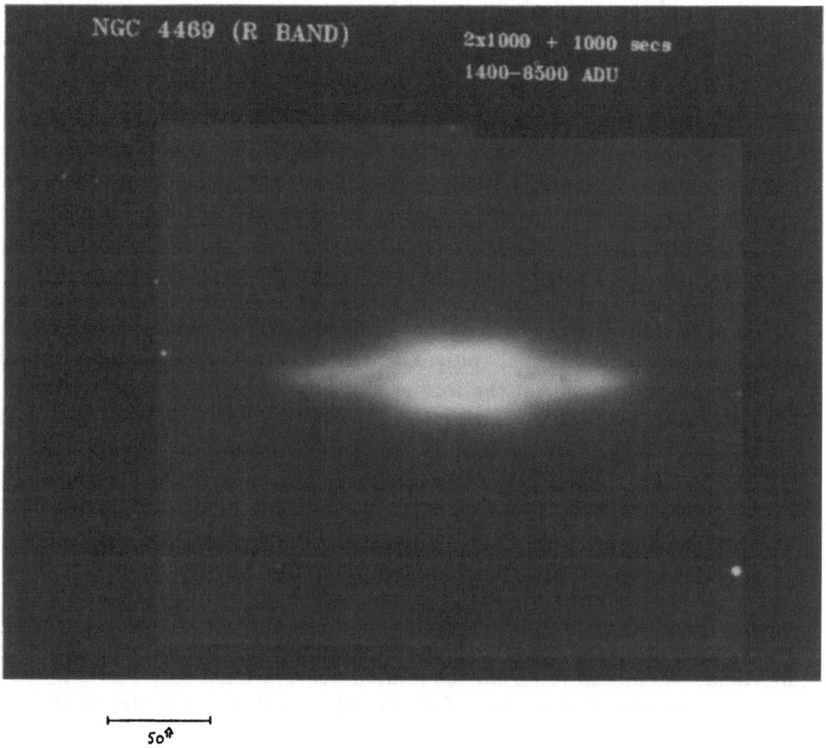

Figure 2: As Figure 1, but for the edge-on S0/a galaxy NGC 4469, showing the other extreme of box/peanut bulge morphology. The diffuse, low surface brightness distortions here contribute only ~10% excess luminosity over an $r^{1/4}$ law bulge. Such low surface brightness distortions are not necessarily restricted to the earliest types. Assuming occupancy of the Virgo Cluster means the scale bar of 50" corresponds to ~2.6 kpc.

Results

1. We find that 20 ($\pm$4) % of the sample show box/peanut shapes to the limiting surface brightness of the plate material used (*i.e.* $\sim$ 25.0 mag arcsec^{-2}), implying that as many as 750 normal spirals/lenticulars in the RC2 may show such morphology. This must be an absolute lower limit to the frequency of these objects in that ultra–high contrast prints from the plate material used show many more candidates at much fainter surface brightness levels (*e.g.* NGC 1055).

2. The distribution of morphological types in these box/peanuts shows a strong peak within the range of late–type lenticulars to intermediate (Sbc) type spirals, with ~70% between Sa – Sbc. Restricting our analysis to the very large systems ensures minimal uncertainties of bulge identification, particularly for the early–type systems, but also, we feel, for the systems later than Sbc (despite their inherently smaller bulge components).

3. No differences are evident in the derived radio or optical properties between boxy and "normal" bulges, in marked contrast to kinematic studies of a (small) number of the former systems which reveal the existence of cylindrical rotation. However, the sample sizes for both box/peanut and "normal" bulges are sufficiently small as to make even the most general conclusions concerning the similarities or otherwise of their respective velocity fields as very uncertain. Work is currently in progress to investigate both the kinematic and radio properties for a representative subset of our edge–on galaxy sample in more detail.

4. Box/peanuts are not preferentially found in clusters.

5. As far as one can tell from the limited observational data currently available, the non–disc regions of these galaxies show no evidence of colour gradients in $B - R$.

6. There is no preference for such bulges to be surrounded by more faint satellites (~6 mag fainter than the primary) than are seen in the non–box/peanuts. This implies that the accretion of a relatively large number of small satellites along the appropriate orbits is an unlikely means of forming such systems, although the possible merger/accretion of an SMC–sized neighbour is a more distinct possibility.

7. Based on the distribution of morphological types in the current sample, and also the theoretical modelling of Barnes and White (1984), discs seem an unlikely source of sufficient torques to yield the bulge shapes identified here. This is particularly the case in such objects as NGC 1055, where the box/peanut morphology is most evident at a radial distance of ~7 kpc and a Z–height above the disc plane of 6 kpc.

8. If one calculates the excess integrated luminosity contained within the box/peanut (at distances well–removed from the dust lane) over and above that seen within a "normal" bulge population of equivalent dimension, concentration index, and absolute magnitude, one finds the box/peanut to contribute on average $\sim$ 20% additional light, and as much as 35% in the more extreme cases such as IC 2531 (Figure 1). In the markedly more diffuse distortions – such as those seen in NGC 4469 (Figure 2) and NGC 1055 – the box/peanut only accounts for an additional $\sim$15% of the non–disc light.

A more complete presentation of this work will be found in Shaw (1987).

References

Barnes, J. and White, S. D. M. 1984, *Mon. Not. Roy. Astron. Soc.* **211**, 753.
Shaw, M. 1987, *Mon. Not. Roy. Astron. Soc.* **229**, 691.
de Vaucouleurs, G., de Vaucouleurs, A., and Corwin, H. G. 1976, *Second Reference Catalogue of Bright Galaxies* (Austin: Univ. of Texas Press).

Surface Photometry of NGC 3379 with a Tektronix 2048 × 2048 CCD Camera — Comparison with the Luminosity Profile of de Vaucouleurs and Capaccioli

Harold D. Ables, Hugh C. Harris, and David G. Monet
U.S. Naval Observatory, Flagstaff Station, Flagstaff, Arizona, U.S.A.

I. Introduction

The development of large–format CCDs offers a new opportunity to study the luminosity distribution in galaxies. In this paper we present the first astronomical photometry with a Tektronix 2048 × 2048 CCD. The U. S. Naval Observatory large format CCD camera with an engineering setup grade, front illuminated Tektronix 2048 × 2048 CCD was used on the 40–inch Ritchey–Chretien telescope at the Flagstaff Station to measure the luminosity profile in NGC 3379. We compare the result with the E–W luminosity profile published by de Vaucouleurs and Capaccioli (1979, hereafter dV+C), and we describe some of the features of the CCD.

II. Observations

The observations consist of four 20–minute V exposures of NGC 3379, each shifted by roughly 1 arc–minute from the others; three 20–minute V dark sky exposures, each of a different region of the sky some 30 arc–minutes away from the galaxy center; and one dome flat field exposure. The CCD pixels are 27μm square and the overall dimensions of the CCD are 55.3 × 55.3 mm. In the focal plane of the 40–inch telescope, these dimensions correspond to 0.76 arcsec and 26 × 26 arcmin. A high–contrast print (Figure 1) of an unprocessed 1 hour exposure of NGC 3379 shows both the large field of the CCD, which is ideal for studying a nearby galaxy like NGC 3379, and some of the defects in our engineering setup grade device. We describe its characteristics in the following paragraph.

The engineering setup grade CCD exhibits four types of problems. (1) Charge pockets in the serial (horizontal) register require that a large charge (fat zero) be added to each pixel during serial readout to achieve acceptable charge transfer. (2) A few columns are blocked and provide no useful data. (3) Many columns respond nonlinearly and so do not flatten properly. (4) Many pixels respond nonlinearly, often apparently gaining or losing charge along columns for a few pixels, and so do not flatten properly. However, the CCD response is quite uniform over large scales, and most pixels behave linearly and flatten very well. With suitable analysis procedures, we can take advantage of the wide field for surface photometry.

III. Results

A field flattening frame composed of the low signal–to–noise (S/N) dark sky flats for correcting large–scale nonuniformities and the high S/N dome flats for correcting small–scale nonuniformities was used to flatten of the 4 galaxy frames. Unfortunately, most of the defective columns and pixels have a nonlinear response and this field flattening procedure did not eliminate them. The frames were registered for superposition and trimmed to a size of 1898 × 1948 pixels or 24 × 24.7 arcmin. The frames were scaled to have equivalent exposure values, averages were taken of congruent pixels, and the most deviant value in each average was discarded. The median of the remaining three exposure values at each pixel location was adopted for the final galaxy frame (Figure 2). This procedure reduced the sensitivity variations of the defective columns and pixels to about ±1% of the background level.

Ellipses having the position angle and ellipticities specified by dV+C were centered on the nucleus of NGC 3379 in the data frame of Figure 2. Elliptical rings were generated for the entire data frame, spaced in equal (0.025) steps of $\log r$, where r is the distance along the semi–major axis. The adopted sky level is the average of the mean pixel values in the outermost six complete elliptical rings ($8.'8 < r < 12.'5$).

The integrated V magnitudes measured through circular apertures centered on the nucleus of NGC 3379 from Table 3 of dV+C were compared to our CCD magnitudes measured through the same circular apertures in order to establish the zero–point of our instrumental magnitudes. We have made a least–squares fit for our zero–point using aperture diameters from 0.′340 to 4.′074, and the result is shown in Figure 3. The zero point calibration has a mean error of ±1%.

The mean V luminosity profile was determined from the average pixel value for each elliptical ring after subtraction of the adopted sky background and removal of all pixel values exceeding 4σ in each ring. For comparison with the B luminosity profile of dV+C, a $(B-V)$ color profile of NGC 3379 is required. Unfortunately, we were unable to find a suitable $(B-V)$ profile in the literature. Instead, we adopted the $(B-R_C)$ color profile from Davis *et al.* (1985), converted it to a $(B-V)$ profile through the transformation $(B-V) = 0.645(B-R_C)$, and parameterized it as $(B-V) = 1.08 - 0.08(\log r)$.

The resulting B luminosity profile is compared with the observed profile from dV+C (Table 2B) in Figure 4, where our radii have been scaled to distances along the E–W axis of the galaxy. The error bars include contributions from the dispersion of pixel values in the elliptical rings and the uncertainty in the value of the sky background. Except near the center where seeing effects dominate and in the outermost regions where errors in the sky level dominate, the profiles agree very well.

We conclude that this engineering setup grade CCD with all its problems can be used successfully for large field astronomical photometry.

References

Davis, L. E., Cawson, M., Davis, R. L., and Illingworth, G. 1985, *Astron. J.* **90**, 169.

de Vaucouleurs, G. and Capaccioli, M. 1979, *Astrophys. J. Suppl.* **40**, 699.

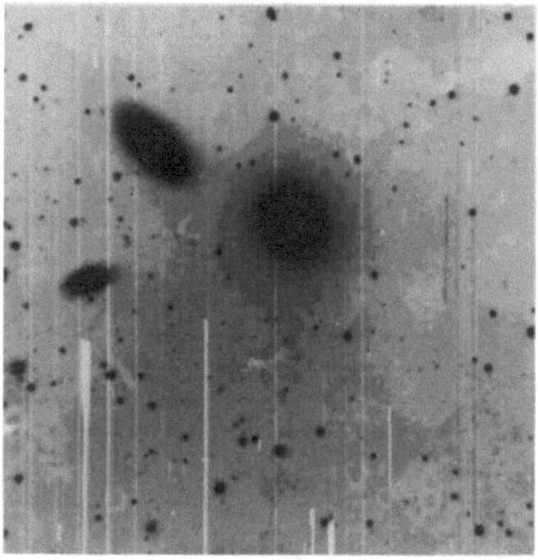

Figure 1.

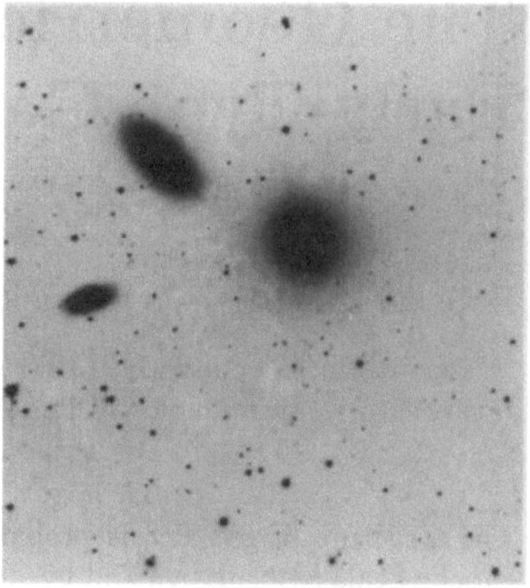

Figure 2.

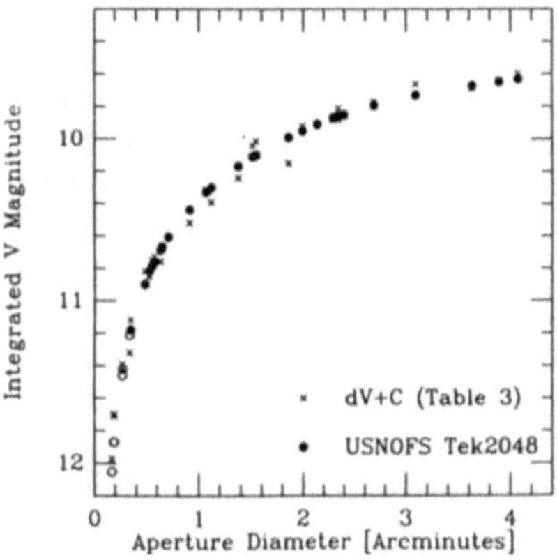

Figure 3.

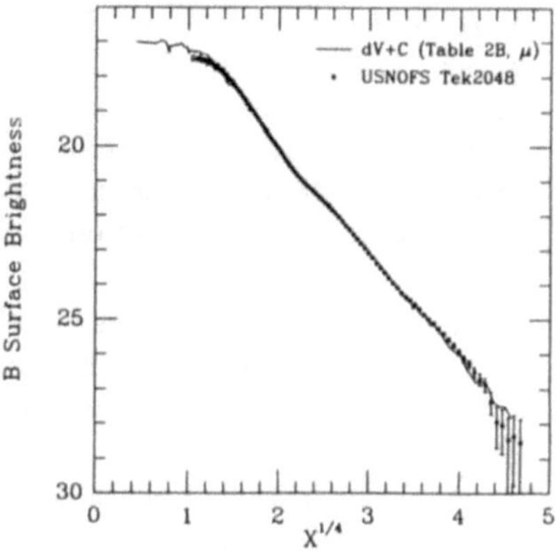

Figure 4.

The Geometrical Parameters of Early-Type Galaxies and the Local Density

Roberto Rampazzo[1] and Lucio M. Buson[2]

[1]Osservatorio Astronomico di Brera, Italy
[2]Osservatorio Astronomico di Padova, Italy

The structure of galaxies is probably affected both by conditions during the epoch of galaxy formation and by subsequent environmental processes, *e.g.* tidal influence and mergers.

Insights may come from the association of spatial density catalogues and the growing number of galaxies with detailed surface photometry. We used the catalogue of spatial densities of De Souza *et al.* (1985). The density estimator ρ (Gal/Mpc^{-3}) has been computed from a sample of galaxies including all morphological types brighter than $m_p = 14.5$ (Zwicky magnitude) from the Uppsala Catalogue (UGC; Nilson 1973). The photometry of the galaxies is taken from Barbon *et al.* (1984), Kent (1984), Michard (1985), Davies *et al.* (1985), Djorgovski (1985), Lauer (1985), Jedrzejewsky (1987) and Boroson and Thompson (1987). We excluded galaxies with a clearly–present of dust lane, and also those with signs of bars. Furthermore, we considered only galaxies with profile data at a distance from the center of the galaxy $r > 5"$, to avoid the major influence of seeing in the determination of twisting and ellipticity, and to avoid central dust features often noted.

The final sample includes 110 objects classified E, 19 classified E/S0, and 48 S0's, observed in the B and R bands. The quantities we derived are: the maximum ellipticity, ϵ_{max}; the variation of position angle, $\Delta P.A.$; and the variation of the ellipticity, $\Delta\epsilon$.

As shown in Table 1, all the geometrical parameters appear independent from the local density ρ, over a range of about four orders of magnitude, corresponding to nearly isolated to clustered objects. In effect, all the objects in the sample fall in a region below $\log\rho = 0.7$ for which the collapse time ($t_c = 1.43/(G\rho^{0.5})$)

Table 1.

$\log \rho$	$< \Delta P.A. >$	$\sigma_{\Delta P.A.}$	$< \Delta\epsilon >$	$\sigma_{\Delta\epsilon}$
-4:-3	23.7	18.1	0.257	0.132
-3:-2	32.8	40.0	0.135	0.079
-2:-1	22.5	26.2	0.156	0.091
-1:0	20.9	23.9	0.142	0.084

is greater than the Hubble time, and the morphology is thought to be independent from the local density (Postman and Geller 1984). On the other hand, the great spread observed in the geometrical parameters suggests that subsequent environmental processes like tidal interaction with companions (Kormendy 1982, Rampazzo 1987) play a major role in the twisting effect, while conditions during galaxy formation have played a minor role.

If we add to the previous statistics data from Benacchio and Galletta (1980), we populate the right part of the plot since their data refer to nine clusters, including Coma with a central density of $\log \rho = 2.7$. In these regimes of density, S0's predominate over other morphological types (Postman and Geller 1984), and the formation conditions act to produce objects with small twisting (91.7% of these objects have a twisting lower than 10°).

References

Barbon, R., Capaccioli, M., and Rampazzo, R. 1984, *Astron. Astrophys.* **115**, 388.

Boroson, T. A. and Thompson, I. B. 1987, *Astron. J.* **92**, 33.

Carter, D. 1978, *Mon. Not. Roy. Astron. Soc.* **182**, 897.

Davies, L. E., Cawson, M., Davies, R. L., and Illingworth, G. 1985, *Astron. J.* **90**, 169.

De Souza, R. E., Vettolani, G., and Chincarini, G. 1985, *Astron. Astrophys.* **143**, 143.

Djorgovski, S. B. 1985, Ph. D. Thesis, Univ. of California, Berkeley.

Jedrzejewski, R. 1987, *Mon. Not. Roy. Astron. Soc.* **226**, 474.

Kent, S. M. 1984, *Astrophys. J. Suppl.* **56**, 105.

Kormendy, J. 1982, "Observations of Galaxy Structure and Dynamics," in *Morphology and Dynamics of Galaxies*, ed. L. Martinet and M. Mayor (Sauverny: Geneva Observatory), p. 113.

Lauer, T. 1985, *Astrophys. J. Suppl.* **57**, 473.

Nilson, P. 1973, *Uppsala General Catalogue of Galaxies*, Uppsala Astron. Obs. Ann., **6**.
Michard, R. 1985, *Astron. Astrophys. Suppl.* **59**, 205.
Postman, M. and Geller M. J. 1984, *Astrophys. J.* **281**, 95.
Rampazzo, R. 1987, *Atti Ist. Ven. SS. LL. AA.* **CXLV**, 82.

On the Nature of Compact Elliptical Galaxies

Ph. Prugniel

Observatoire Midi-Pyrénées, Toulouse, France

Summary

We present observational evidence that faint elliptical galaxies in pairs evolve toward compact ellipticals. The latter are the normal continuation of the elliptical family toward low masses. This evolution is due to a dynamical heating which weakens the rotational support.

Introduction

The sparse class of compact elliptical galaxies (cE, T = −6) introduced by de Vaucouleurs (1961), contains galaxies characterized by a high surface brightness. Most of the cE's listed in the *Second Reference Catalogue* (de Vaucouleurs *et al.* 1976; RC2) are companions to massive galaxies. Examples include M 32 and NGC 4486B. This association is the starting point for the classical interpretation of the origin of cE's: Faber (1973) has argued that the high surface brightness is the result of the stripping of a massive galaxy. However, the origin of cE's is still controversial (see review in Nieto and Prugniel 1987a).

We have conducted a detailed investigation of 8 pairs of interacting elliptical galaxies with unequal masses, including some galaxies classified cE in RC2 (Table 1). Our goal was to investigate the possibility that the interaction leads to the formation of a cE. This program was carried out with CCD images obtained at the 2–m telescope of Observatoire du Pic du Midi and high–dispersion spectra obtained with the Carelec spectrograph at the 1.93–m telescope of Observatoire de Haute–Provence.

The galaxies of each pair overlap and this complicates the image analysis. We have thus developed an algorithm for fitting the blended galaxies with elliptical

Table 1

NGC	Separation		Lum.	Size
	arcsec	kpc	ratio	ratio
507-8	90	19	0.15	0.27
741-2	50	11	0.09	0.14
2672-3	33	7	0.08	0.16
2693-4	54	12	0.07	0.14
2831-2	32	10	0.09	0.13
3640-1	150	11	0.10	0.27
3862/A	58	18	0.05	0.10
7274-6	144	54	0.25	0.20

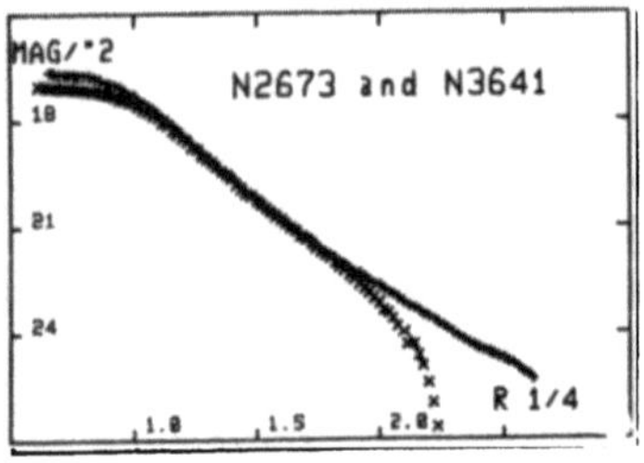

Figure 1. Typical photometric profiles.

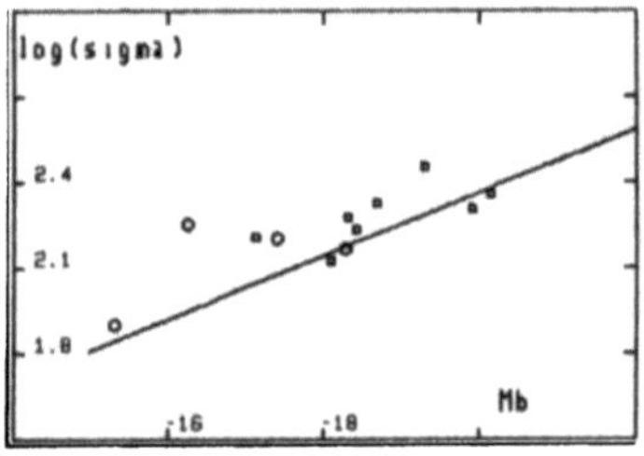

Figure 2. log(σ) *versus* absolute magnitude.

models. After they are minimized, the residuals indicate the non–elliptical shape of the galaxies, such as in the pair NGC 3640–1 (Prugniel *et al.* 1988).

Characteristics of Faint E Galaxies in Pairs

How gravitational interactions affect the photometric profiles is a matter of debate. The tidal field is expected to produce a radial truncation and indeed several profiles falling below the $r^{1/4}$ law favor this hypothesis. But other profiles indicate the presence of tidal extensions, as often found in numerical experiments. Figure 1 gives examples of both cases.

The observed galaxies do not show a trend toward effective surface brightnesses higher than for isolated ellipticals, as would be expected if Faber's hypothesis is correct.

All the small galaxies of the pairs in our sample show large isophote twists, greater than 30 degrees. Three also present asymmetric deformations (see Davoust *et al.* this volume). This is undoubtedly a sign of interaction.

The faint ellipticals of our list (squares), and the known cE's (open circles) are plotted on the central velocity dispersion (σ) — luminosity (L) diagram (Figure 2), with the standard relation given by de Vaucouleurs and Olson (1982). The departure from this relation is remarkably small, at variance with other low–luminosity E's that tend to have systematically lower σ's. We interpret this as a sign of dynamical heating induced by the interaction. It is nevertheless surprising that the observed σ's correspond accurately to those implied by the standard relation.

Conclusion: the Nature of cE's

We have found evidence for gravitational interaction within the pairs of our sample, but these galaxies do not show any trend toward a surface brightness higher than other low luminosity ellipticals, or any significant departure from the σ–L relation. These results, together with time scale arguments (Nieto and Prugniel 1987b), suggest to us that tidal stripping is not the essential cause of compactness. The interaction induces a redistribution of mass and energy that leads to galaxies whose surface brightnesses and central velocity dispersions are continuous with those of bright E's. Rotational support, which is very important in most faint E's (Davies *et al.* 1983; Bender 1988), is weakened by dynamical heating. cE's result from this process and are the natural low–luminosity extension of the family of E's. A full account of this study will appear in a forthcoming paper.

References

Bender, R. 1988, *Astron. Astrophys. Lett.* **193**, L7.

Davies, R. L., Efstathiou, G., Fall, S. M., Illingworth, G., and Schechter, P. 1983, *Astrophys. J.* **266**, 41.

Faber, S. M. 1973, *Astrophys. J.* **179**, 423.

Nieto, J.–L. and Prugniel, Ph. 1987a, in *Structure and Dynamics of Elliptical Galaxies*, I.A.U. Symposium **127** (ed T. de Zeeuw), p. 99.

Nieto, J.–L. and Prugniel, Ph. 1987b, *Astron. Astrophys.* **186**, 30.

Prugniel, Ph., Nieto, J.–L., Bender, R., and Davoust, E. 1988, *Astron. Astrophys.*, in press.

de Vaucouleurs, G. 1961, *Astrophys. J. Suppl.* **5**, 233.

de Vaucouleurs, G. and Olson, D. W. 1982, *Astrophys. J.* **256**, 346.

Two–Color Studies of Isophotal Contours in E – S0 Galaxies

A. Bijaoui, J. Marchal, and R. Michard

Observatoire de Nice, France

The occurence of "parent structures" in E and S0 galaxies has been described by several authors including one of us (Michard 1984, 1985; and Michard and Simien 1986, 1988). We are looking for subtle color variations that might be associated with obvious or incipient morphological features in both types of objects.

The observations were made in B and R at the Pic du Midi 2–m telescope, with a focal reducer and a CCD camera. The plate scale is 1.2 arcsec per pixel, which gives a field of 6.5 arc–minutes. Suitable frames have been collected for about 30 galaxies in 1987–88, but we have just started reduction of the data. This is a first report of our results.

The analysis technique includes a parametrization of the isophotes as distorted ellipses of semi–major axis a, semi–minor axis c, major axis orientation P, and distortion parameters ϵ_4, ϵ_6, ϵ_8, ..., such that

$$x = a(1 + \epsilon_4 \cos 4\omega + \epsilon_6 \cos 6\omega + \ldots) \cos \omega$$

and

$$y = c(1 + \epsilon_4 \cos 4\omega + \epsilon_6 \cos 6\omega + \ldots) \sin \omega.$$

The distributions in B and R of the surface brightness μ, and of the shape parameters c/a, P, ϵ_4, ϵ_6, ..., against the scaled radius $r = \sqrt{ac}$, are then intercompared (see Figure 1).

The results for the few objects so far reduced show very flat color distributions through a range of 5 magnitudes or more in surface brightness. Deviations occur near the centers of the galaxies (due to seeing or a nuclear point source?) and in the outskirts (due to imperfect sky corrections?).

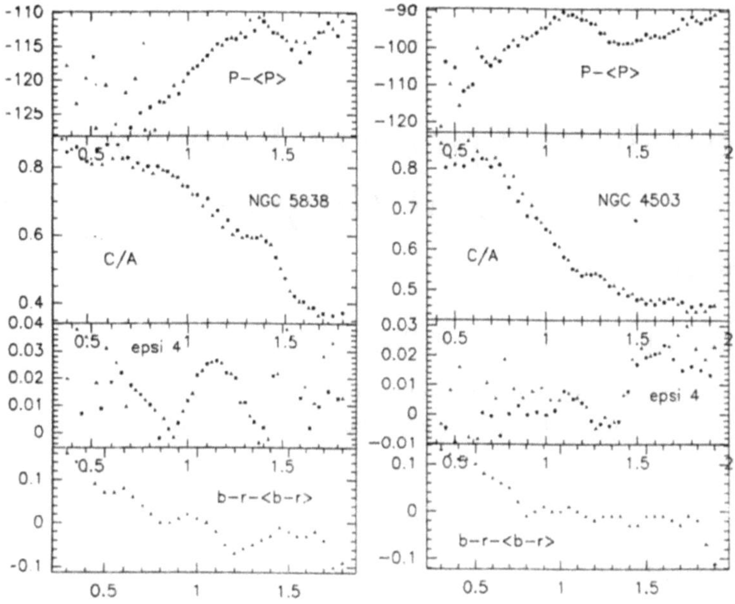

Figure 1: Left: NGC 5838 (S0). Right: NGC 4503 (S0). Three-pointed markers show B data, while six-pointed markers show R data.

References

Bender, R. and Möllendorf, C. 1987, *Astron. Astrophys.* **177**, 71.

Boroson, T. A. and Thompson, I. B. 1987, *Astron. J.* **92**, 33.

Boroson, T. A., Thompson, I. B., and Shectman, S. A. 1983, *Astron. J.* **88**, 1707.

Carter, D. 1987, *Astrophys. J.* **312**, 514.

Jedrzejewski, R. I. 1987, *Mon. Not. Roy. Astron. Soc.* **216**, 747.

Michard, R. 1984, *Astron. Astrophys.* **140**, L39.

Michard, R. 1985, *Astron. Astrophys. Suppl.* **59**, 205.

Michard, R. and Simien, F. 1986, in *Structure and Dynamics of Elliptical Galaxies*, I.A.U. Symp. No. **127**, ed. T. de Zeeuw (Dordrecht: Reidel), p. 393.

Michard, R. and Simien, F. 1988, *Astron. Astrophys. Suppl.*, in press.

Models of Spectral Energy Distribution of Elliptical Galaxies

Guido Barbaro[1] and Fabrizio M. Olivi[2]

[1]Department of Astronomy, University of Padova, Italy
[2]International School for Advanced Studies, Trieste, Italy

Summary

Spectral energy distributions of elliptical galaxies have been computed, with the method of evolutionary synthesis, in the range 1000 – 7000Å, taking into account all the evolutionary phases from the Main Sequence to the post–AGB phase. A fundamental quantity is the mass M_f of stars in the post–AGB phase of old stellar generations. Values for M_f have been derived in the frame of the adopted evolutionary scheme and are strongly dependent on the mass loss rate and on the AGB evolution. By assuming that in the chemical evolution Y and Z are related by

$$Y = 0.24 + 3Z \tag{1}$$

M_f decreases with the metallicity, ranging from 0.56 to 0.54 $M_\odot$ in the metallicity interval typical of ellipticals.

Model spectra of generations 16 Gyr old have been derived, and from them the $(1550 - V)$ colours have been computed. Figure 1 shows the behaviour of $(1550 - V)$ as a function of the composition. Full lines give M_f as a function of Z at constant Y; the dashed line refers to chemical compositions in which Y and Z are correlated by (1). The decreasing branch of the dashed line corresponds to realistic compositions for ellipticals. For the lowest values of Z and Y, the UV flux of post–AGB stars is masked by the more abundant flux from HB stars. Crosses represent the "normal" elliptical galaxies of the sample of Burstein *et al.* (1988): Mg 2 has been converted into Z according to Terlevich *et al.*

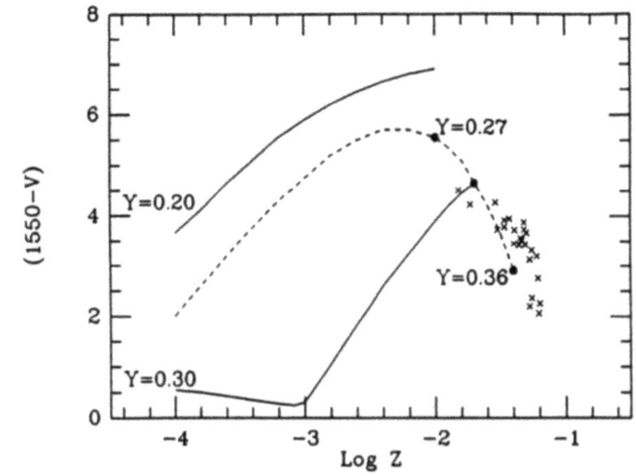

Figure 1: The $(1550 - V)$ color index plotted against Z for 16 Gyr old generations. Full lines connect points with the same Y values, the dashed line refers to compositions obeying equation (1). Crosses are normal ellipticals from the sample of Burstein *et al.* (1988).

(1981). From the figure, it can be seen that the models reproduce fairly well the behaviour of the observed colours and metallicity.

The last point of our calibration, corresponding to the highest metallicity, is uncertain both because the definition of the evolutionary background required some extrapolation and because atmospheric models with solar composition have been used. For these reasons, we cannot exclude the possibility that the far UV flux of giant ellipticals requires, besides post–AGB stars, the contribution of young massive stars.

The spectrum of the dwarf galaxy M 32 has been analysed in detail. The UV data are from Burstein *et al.* (1988), and the visible flux from Oke *et al.* (1981). In Figure 2, the observed spectrum is compared with a model galaxy characterized by the following parameters: a) Salpeter's mass distribution function; b) exponentially decreasing stellar birth–rate with an e–folding time of 1 Gyr; c) average metallicity $< Z >= 0.01$; d) age of 16 Gyr. On the whole, the model reproduces the behaviour of the observed spectrum. The far UV flux is influenced both by the post–AGB and blue HB stars. This superimposition of different sources, due to the weakness of the post–AGB flux at relatively low metallicities, is responsible for the rather peculiar UV energy distribution of M 32, whose slope is very irregular and differs from that of giant ellipticals. Burstein *et al.* (1984) and Rose (1985), by analysing high resolution visible spectra of M 32 and other elliptical galaxies, found evidence for the presence of a young population about 5–7 Gyr old, whose contribution to the light at 4000Å is equally important to the old population; Rocca–Volmerange and Guiderdoni

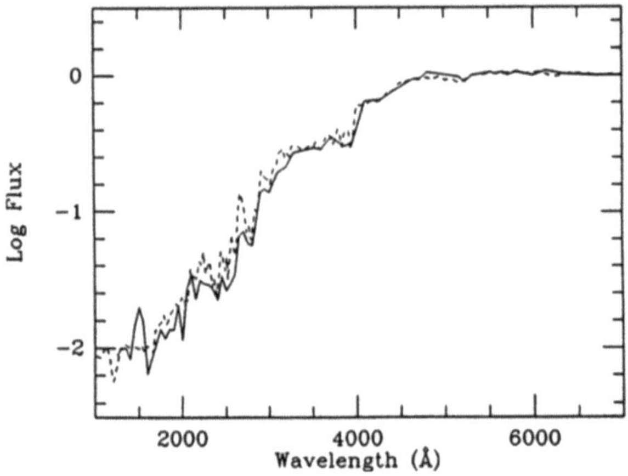

Figure 2: The spectrum of M 32 (full line), compared with a galaxy model with $< Z >= 0.01$ and an exponential birth rate (dashed line). The two curves are matched at 7000Å.

(1987) have reached the same conclusion by analysing the UV spectrum. Following this suggestion, we have computed a model with the following birth rate:

$$
\begin{array}{ll}
y(t) = \text{const} = yo, & 0 < \text{t(Gyr)} < 1 \\
y(t) = 0, & 1 < \text{t(Gyr)} < 6 \\
y(t) = 0.2yo, & 6 < \text{t(Gyr)} < 11 \\
y(t) = 0, & 11 < \text{t(Gyr)} < 16
\end{array}
$$

The spectral energy distributions of this model and of the model discussed in Figure 2, differ only in the far UV by about 0.5 magnitude, while longwards they are practically indistinguishable.

Unless one attributes a particular weight to the detailed behaviour of the UV flux, which however is the most uncertain feature of both the observed and computed spectra, we are led to the conclusion that the synthesis of the spectral energy distribution cannot define in a unique way the stellar content of elliptical galaxies: the consideration of the spectral features or more stringent constraints on the stellar birth rate can help in eliminating this ambiguity.

In this regard, for M 32, a non-monotonic birth rate could be realistic because of the interactions with the companion M 31: tidal stripping and gas infall could be responsible. In giant ellipticals, galactic winds and cooling flows, for which observational evidence is growing, are of interest in this context.

References

Burstein, D., Bertola, F., Buson, L., Faber, S. M., and Lauer, T. R. 1988, *Astrophys. J.*, in press.

Burstein, D., Faber, S. M., Gaskell, C. M., and Krumm, N. 1984, *Astrophys. J.* **287**, 586.

Oke, J. B., Bertola, F., and Capaccioli, M. 1981, *Astrophys. J.* **243**, 453.

Rocca-Volmerange, B. and Guiderdoni, B. 1987, *Astron. Astrophys.* **175**, 15.

Rose, J. A. 1985, *Astron. J.* **90**, 1927.

Terlevich, R., Davies, R. L., Faber, S. M., and Burstein, D. 1981, *Mon. Not. Roy. Astron. Soc.* **196**, 381.

Photometry of Disks in Galaxies

P. C. van der Kruit

Kapteyn Astronomical Institute, Groningen, The Netherlands

Abstract

In this review, four aspects of the light distribution in galactic disks are addressed: (1) Following the pioneering work by de Vaucouleurs in the 1950's, we know that the majority of disks have a radial surface brightness distribution that follows an exponential law. (2) Freeman's law of a small spread in the extrapolated central face–on surface brightness of disks in non–dwarf disk galaxies shows up in a statistically complete sample after correction for the known selection effects. (3) Observed radial truncations in edge–on spirals are now also seen in moderately inclined systems. These fail to show up in azimuthally averaged profiles, because their precise galactocentric radius varies somewhat with azimuthal angle in the plane. Collapse from a uniformly rotating, uniform sphere with detailed conservation of angular momentum explains the exponential nature of the radial profiles and at the same time the occurrence of radial truncations at about 4.5 exponential scalelengths. (4) The vertical scale parameter of the old disk population does not change with radius. A change from the sech^2 isothermal description to a sech function seems to take account of the expected deviations from isothermallity at small z due to younger populations with smaller velocity dispersions.

Introduction

Central to any understanding of the structure and origin of disks in galaxies is naturally a general knowledge of the three–dimensional distribution of stars in these disks. This information can be derived in principle from studies of surface photometry. The fundamental questions behind such scientific enquiries relate to the equilibrium structure of the stellar disk components and involve eventually also the kinematics in order to study the dynamical structure and stability. It is not surprising that kinematic studies of stellar components are difficult due to the necessary high wavelength resolution in the spectroscopy required to determine bulk velocities and dispersions. Now it is true that the discovery of the rotation of the stellar disks (Pease 1916, 1918) preceded that of the exponential nature of the radial surface brightness distributions (see below for references), but it is equally true that the kinematic data necessary for a thorough dynamical analysis — *i.e.* including stellar velocity dispersions — have become available only in recent times (e.g. van der Kruit and Freeman, 1986). In this review, I will concentrate on the questions related to the distribution of stars from surface photometry and refer to kinematic and dynamical matters only where appropriate.

Surface photometry has traditionally been performed mainly from photographic observations, although some early exceptions involving photoelectric observations exist. It is well known that de Vaucouleurs has been one of the prime founders of this field. It is true that light profiles in bulges of disk galaxies and in elliptical galaxies were measured in early times. This has been reviewed by de Vaucouleurs himself in his introductory lecture at the 1979 meeting in Austin, Texas (de Vaucouleurs 1979) and can be traced back to the work of Reynolds (1913) on the bulge of M 31. However, the work on disks started much later as a predictable result of the much fainter surface brightness of these components, and de Vaucouleurs has initiated and dominated this field for many years. Two developments have finally opened this to a wider community, namely the wide-spread availability of digital techniques (involving fast scanning machines and large computers) and the recent advent of CCD's and other digital detectors. But in the 1950's and 1960's, de Vaucouleurs almost entirely owned the field, with notable exceptions for example the detailed work by van Houten (1961) and Ables (1971).

Before reviewing photometry of disks in galaxies, I would like to make a general observation. In the 1970's, it was generally believed by the astronomical community that photographic surface photometry was unreliable. This was a result of the poor comparison of the published profiles at faint levels between observers and certainly was justified. However, it was also believed that it was a difficult art that had better not be attempted by the general astronomer. In this context, it is worthwhile to recall that Ivan King once said that photographic surface photometry was not difficult, but it just had to be done right! Observers now generally agree even at such faint levels as 5 magnitudes below sky, and there is excellent agreement between photographic work and CCD photometry.

In the following, I will concentrate on four aspects of light distributions in galactic disks: (1) the exponential nature of the radial light distribution, (2) the variation in face-on central surface brightness, (3) the question of the radial truncations of stellar disks, and (4) the vertical distribution of stars in disks.

The Exponential Nature of the Radial Light Distributions

I will be rather brief in this section, because the exponential nature of radial profiles of surface brightness in disks has by now been demonstrated in numerous cases and is generally accepted. As noted above, the study of profiles of disks was preceded by that in bulges by about 3 decades. A notable difference is also that in the course of time a number of "laws" have been proposed for spheroids and elliptical galaxies (the Reynolds-Hubble law, King's law of a truncated isothermal sphere and de Vaucouleurs's famous $r^{1/4}$ law); while for disks, only the most simple exponential has received attention in practice. In the following, I will designate the latter as $I(r) = I_o(r)\exp(-r/h)$, and I will use the symbol $\mu(r)$ when the surface brightness is expressed in magnitudes per square arcsecond.

The exponential law was, according to de Vaucouleurs, described first in an unpublished Harvard thesis based on observations of M 33. A short report is available in Patterson (1940) and a plot of the data appears in the extensive description of light distributions in galaxies by de Vaucouleurs (1959a, Figure 10). The first detailed work was performed by de Vaucouleurs in a series of papers in the late 1950's that involved the suitable members of the Local Group: LMC (de Vaucouleurs 1957), M 31 (de Vaucouleurs 1958) and M 33 (de Vaucouleurs 1959b). This fundamental work established once and for all that exponentials are the rule and occur in Sb, Sc, and Irregular galaxies. As a tribute to this epochal work, the exponential light profile of the disk of M 33 is reproduced from the original paper in Figure 1. Note that the data reach a level of about 27 B–magnitudes arcsec^{-2}, which is not much brighter than is attainable today in such studies. The two curves show the major and minor axes separately.

Many studies have in the intervening time demonstrated the universal applicability of the exponential as a fitting function. I will have more to say on its possible origin below, but here I want to note two things: (1) in all galaxies, for which an exponential has been fitted to the disk surface brightness profile, there are important, non–local deviations from the fit at a level of a few tenths of a magnitude; while often the fit can only be performed over a few magnitude (or about e–foldings) intervals in the first place. The reason for the last point is that often the bulge dominates the inner parts and then there is little fall–off in surface brightness left before reaching the noise level introduced by the sky background. This means that the exponential is convenient to use because of its simplicity as well as because of the limited dynamic range over which luminosity profiles can be determined. (2) The significance of the facts just mentioned is that one should be hesitant to attach too much physical meaning to the exponential nature. Any functional behavior that over a few e–foldings reasonably resembles an exponential can serve as a useful approximation to the observations.

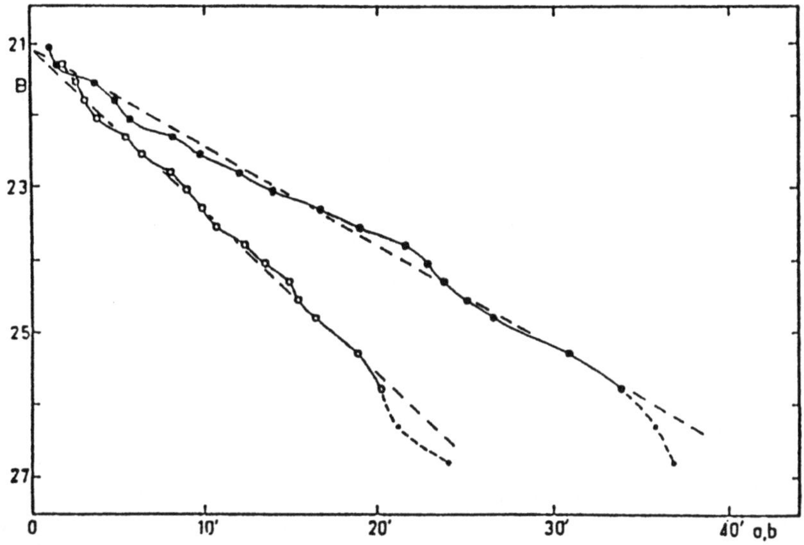

Figure 1: The surface brightness distribution of the disk of the Local Group Sc galaxy M 33 from the original publication by de Vaucouleurs (1959b). This constitutes one of the earliest published fits to the now generally accepted exponential. The two curves show the distributions measured along the minor and major axis separately. Note the good fit to an exponential over about 4 scalelengths and the faintest level of the photometry (about 26 B–mag arcsec^{-2}), which is only one or two magnitudes brighter than what is attainable today in such studies.

The Central Surface Brightness Problem

In a classical study, Freeman (1970) used published surface photometry — mainly supplied by de Vaucouleurs — to demonstrate an unexpected fact. For the majority of the systems, it turned out that the extrapolated, face-on, central surface brightness μ_o fell in the surprisingly small range of 21.67±0.30 B–mag arcsec^{-2}. Exceptions were some S0's on the bright side and dwarf galaxies on the faint side. Immediately it was suspected that this was some subtle or maybe even obvious selection effect. Indeed, Arp (1965) and de Vaucouleurs (1974, but expressed in various other ways earlier) had pointed out that the natural sky background selects against faint, extended objects and will make bright, small objects appear starlike. Disney (1976) took this up in an ingenious paper, in which he demonstrated that the actual levels of sky surface brightness involved were such to give Freeman's values. Further, the difference between the exponential law for disks and the $r^{1/4}$ law gives rise to the observed difference of about 6.5 mag between Freeman's values and the equivalent (Fish's law) in ellipticals. These striking coincidences have been of major influence in the controversy over this subject.

However, two points should be made before discussing this issue. (1) It has been known from the start that dwarfs with fainter values of μ_o occur and their existence is not an issue. (What is an issue is whether these occur in such numbers so as to make a major contribution to the cosmic luminosity density.) The ones known were Local Group members recognized as conglomerations of faint stars and therefore selected in a completely independent way. (2) Most studies, starting with Freeman's, were based on samples that were not complete with respect to well-defined selection criteria and consequently the resulting distributions could not be corrected for these effects. The extensive study of Grosbol (1985) is also not complete in a statistical sense, because the selection is done on the basis of inclusion in the *Second Reference Catalogue* (de Vaucouleurs *et al.* 1976).

An attempt to remedy this situation was the study reported in van der Kruit (1987a). Here galaxies were selected from background fields on deep IIIa-J Schmidt plates and after scanning, the selection was made quantitative as follows: all disk galaxies (with inclination and morphological type restrictions) which had a major axis diameter at the isophote of 26.5 mag arcsec^{-2} in excess of 2 arcmin were chosen. This then confirmed that for so-called non-dwarf galaxies, Freeman's law held up, namely that $\mu_o = 21.52\pm0.39$ mag arcsec^{-2} (in about the B-band). Dwarfs, which are of morphological class later than Sc, and which turn out to be small in physical size as well when redshifts are available, are fainter and have $\mu_o = 22.61\pm0.47$ mag arcsec^{-2}. Furthermore, the selection as indicated above with the values appropriate to the present sample, predicted a peak at a different surface brightness than observed, and the narrow distributions survived after correction for these effects. Also, although in a distance limited sample the dwarfs dominate in number by a large factor, it turned out that the dwarfs provide only about one quarter of the cosmic luminosity density. This analysis resulted in a first guess at the bi-variate distribution function of central surface brightness μ_o and scalelength h. This can be summarized in the following table, which contains space densities of galaxies per Mpc^3 for a Hubble constant of 75 km s^{-1} Mpc^{-1}:

h(kpc)	0.5–2.0	2.0–3.5	3.5–5.0	5.0–7.0
μ_o (B–mag arcsec^{-2})				
20.5–21.0	6.0E–3	4.8E–4	–	–
21.0–21.5	3.3E–2	3.6E–3	7.8E–4	1.1E–4
21.5–22.0	2.3E–2	1.1E–3	9.0E–4	8.6E–5
22.0–22.5	1.1E–1	2.7E–3	1.4E–3	5.6E–5
22.5–23.0	7.1E–2	1.0E-3	–	–

The conclusions from this then are: (1) Although Disney's selection must operate on chosen samples, Freeman's law is a physical property at least of non-dwarf (Sc and earlier) galaxies. (2) In a volume limited sample, dwarf galaxies indeed dominate in number, but are not the major contributors to the luminosity density. (3) Scalelengths exist up to about 7 kpc, but the density distribution has a very large drop–off by a factor 10 in about 2 kpc.

At this stage, is it of interest to see where our Galaxy fits into this whole. Values for the photometric parameters of our disk have been derived by van der Kruit (1986) from the data provided by the background starlight experiment with the Pioneer 10 spacecraft near Jupiter. These data and other information summarized and discussed in van der Kruit (1987b) leads to $\mu_o = 22.1 \pm 0.3 B$–mag arcsec^{-2} and $h = 5.0 \pm 1.0$ kpc. This puts our Galaxy among the larger spirals; in fact, this value for the scalelength and that for M 31 compared to spirals in the Virgo cluster, rule out values for the Hubble constant in excess of 75 km s^{-1} Mpc^{-1}. This does make our position in one of the largest spirals somewhat unusual, although not extremely so: about 10% of all stars should occur according to the table above in spirals with h larger than 4 kpc.

A major question, of course, is what happens at redder wavelengths, where the contribution from young populations and effects of dust absorption should be much less. One approach is to repeat the analysis just given on data derived from IIIa–F plates rather then IIIa–J. This has been done now for the sample discussed above in an unpublished study by van der Kruit and Westerhof. Note that the J–magnitudes are about B and those in F are between standard V and R such that $(J - F) = 1.25\ (B - V)$. The face–on distributions resulting then are shown in Figure 2. The means and dispersions are

	J	F
33 non–dwarfs	21.54±0.39	20.63±0.49
14 dwarfs	22.52±0.32	21.99±0.44

The ratio of the J– and F–scalelengths is 1.07±0.13, so there is no significant change in h with wavelength. The distributions given are somewhat narrower in J, but the effect is not significant. However, if young populations and absorption were important effects, we would expect significant narrowing at F compared to J, and this is certainly not observed. In agreement with the bluer colors of the dwarfs, the means separate indeed by going from J to F.

In a first attempt, Giovanardi and Hunt (1988) have performed infrared photometry of 9 Sc's in J, H, and K, and confirmed dispersions comparable to those in B (0.6 to 0.9 mag), but it should be warned that this is not a statistically well–defined sample. However, it appears that the observed constancy of central surface brightness is a property that is visible at all wavelengths, and therefore applies to the old disk population, which of course contains almost all of the stellar mass.

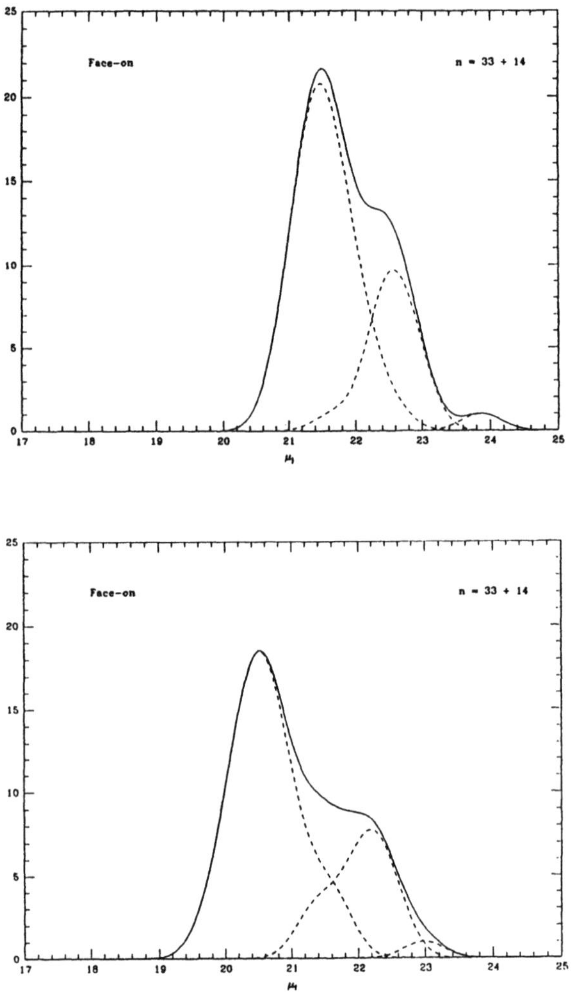

Figure 2: Generalized histograms of the face–on, extrapolated central surface brightness of a complete sample of disk galaxies, selected with a major axis diameter at the 26.5 J–mag arcsec^{-2} isophote larger than 2 arcmin (see van der Kruit 1987a). In both panels, the brighter of the distributions, indicated with dashed lines, contains the systems with morphological type Sc or earlier (so–called non–dwarfs) and the fainter, the remaining galaxies (dwarfs). The single faintest system is NGC 4392, which should probably be deleted from the sample. The top panel shows the data from IIIa–J plates and corresponds to roughly the B–band, while the lower panel comes from IIIa–F plates and is between the V– and R–bands.

The Radial Truncations or Edges in the Disk Light Distributions

Radial cut–offs or truncations in the light distributions of galactic disks were first recognized in photometry of a few edge–on spirals (van der Kruit, 1979; van der Kruit and Searle 1981). It is not surprising that this occurred in those systems because of the longer pathlengths through such disks and the correspondingly higher surface brightnesses. What is somewhat surprising, however, is that the edges were not identified earlier, since these are easily visible to the eye on prints of deep plates, and in fact even on Palomar Sky Survey prints. There appears no record of this in the published literature, except possibly the qualitative remark by Bertola and di Tullio (1975) on the basis of deep plates of two edge–on galaxies, that the "dimensions of the major axis are not getting bigger on the deep plates and the disks tend be sharply bounded."

The truncations or edges are clearly visible in the isophote maps in the publications mentioned. Apart from all the usual checks on background fitting, profiles of stellar images, *etc.* there are two additional arguments in favor of the reality of the edges: (1) in edge–on galaxies, as *e.g.* NGC 4565, the truncations are observed already at levels of 24 or 25 mag arcsec^{-2}, which is at least two magnitudes brighter than the level down to which reliable surface photometry can be done; (2) the sharp truncations in these edge–ons are only observed along the major axes; the bulge profiles along the minor axes from the same data go smoothly down to the limits of the photometry at 27 or 28 mag arcsec^{-2}.

The edges occur in the edge–on systems at about 4 or 5 radial scalelengths from the center, where the radial e–folding drops below 1 kpc. In face–on disks, this radius would correspond to a surface brightness of about 26 or 27 mag arcsec^{-2}, and these sharp declines in surface brightness should also be observable. Yet, there are very few published radial profiles that show evidence for this. The question then is: why is this the case? This has been addressed in van der Kruit (1988) and is related to the manner in which these profiles are determined. It can be illustrated by the example of the almost perfectly face–on galaxy NGC 628. In Figure 3, I show an isophote map of this spiral. Now, the usual way of determining the radial profile is to perform averages over elliptical (in NGC 628, circular) rings centered on the nucleus. The obvious advantage of this is that more solid angle is available for the faint outer parts, where the data are more affected by noise. In NGC 628, the radial profile determined in this way (Shostak and van der Kruit 1984) shows no sharp edge, although a small change of slope is visible at about 4.5 arcmin from the center.

However, notice that the last three contours in Figure 3 are definitely closer together than the brighter ones, and also that the outer contours do deviate clearly from circles. So the problem seems to be that the outer parts of stellar disks have significant deviations from circular symmetry, although sharp declines in surface brightness exist. These then occur at somewhat different radii in

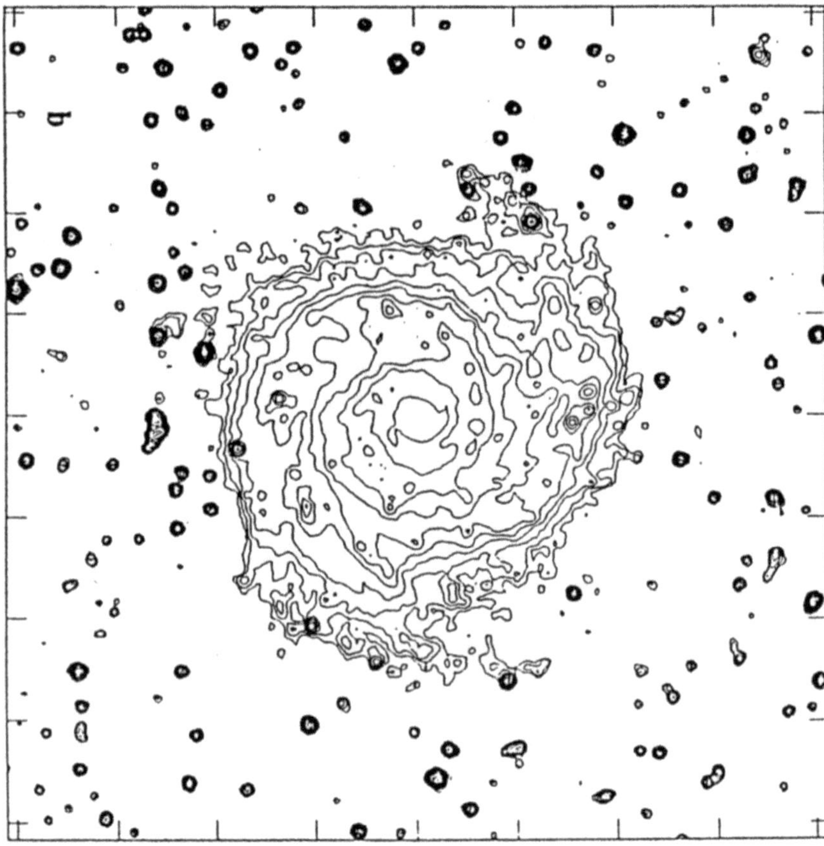

Figure 3: Isophote map of the face–on Sc spiral galaxy NGC 628. The contour interval is 0.5 mag. The outer three or so contours show a smaller spacing than the inner ones, and the outer ones also show significant deviations from circularity. Some star images have been removed from the map.

different position angles, and the procedure of azimuthal averaging then smooths out the edges. Note that this will not apply to edge-on galaxies, where the ends of the major axes refer to small ranges in azimuthal angle in the galaxy's plane. Further, it was noted in the edge-ons that the edges usually occurred at slightly different radii on the two halves of the major axis.

This point is further illustrated in Figure 4, where radial profiles are shown over restricted ranges (30°) of position angle. The profiles that do not end in an asterix (indicating contamination by a stellar image) indeed show sharp declines, but there is a significant variation of the galactocentric distance where this occurs among these profiles. The conclusion then is that the stellar disk of NGC 628 indeed has a relatively sharp edge, but that it varies in galactocentric distance between about 300 and 360 arcsec with azimuthal angle.

A further exercise is to examine the maps of the spiral galaxies in the "Palomar – Westerbork Survey" (Wevers, Allen, and van der Kruit 1986) and determine from the spacing of the outer three isophotes an outer scalelength, then divide these by the scalelength determined from the azimuthally averaged profile. Except for most S0's, which show no obvious effect of truncation, the value of this ratio for the sample then comes out as 0.47 ± 0.17, and the changes of slope occur at 4.5 ± 1.0 inner scalelengths. So, it is possible to identify the disk edges also in moderately inclined spiral galaxies, and these have the same sharpness and radius as in edge-on systems. A simple estimate shows that the small deviations in radius observed can probably survive the shear caused by differential rotation for a Hubble time (van der Kruit 1988).

The Origin of Exponential Disks

In this section, I will discuss the possible origin of the exponential nature of the radial light profiles, but a model for this should of course ideally also explain the constant central surface brightness and the observed edges. However, before this is done, we should first examine the question of whether the mass distribution in galactic disks follows the same exponential distribution. Now, rotation curves cannot be used for this since their shape is, for an important part, determined by the dark matter, which presumably resides in a more or less spherical halo. So, we will have to look first at studies involving vertical dynamics, that is, determinations of disk surface density from the thickness and vertical velocity dispersion of a disk component. Useful components are the old disk stars themselves and the HI gas layer.

As will be discussed in more detail in the next section, the stellar disks have a constant thickness with galactocentric distance, and this property can be used to test whether the surface density distribution is exponential as well. As shown in van der Kruit and Searle (1981), for a constant thickness of a self-gravitating component as the old stellar disk, the velocity dispersion scales as the square root of the surface density. So, if the light and mass distribution have the

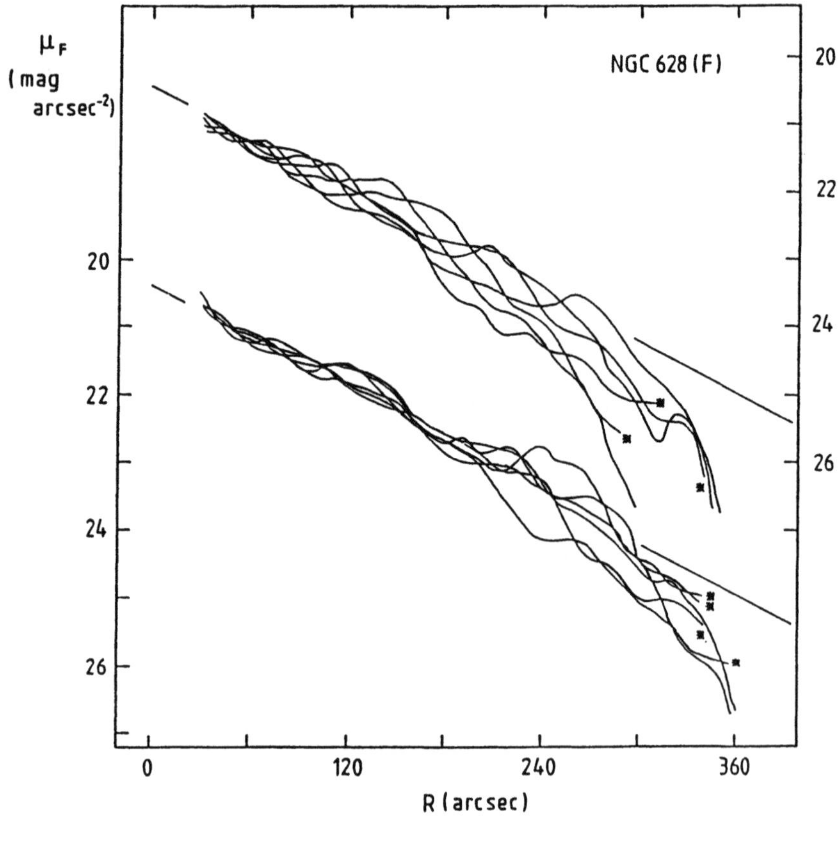

Figure 4: Radial surface brightness profiles of NGC 628 after averaging in intensity in azimuthal sectors of 30°. The profiles that end in an asterix are affected by images of foreground stars and cannot be drawn reliably beyond what is shown. The straight line is the exponential fit to the total profile (averaged over 360° of azimuthal angle) and fits the inner profiles in the figure. There are relatively sharp declines in the profiles between 300 and 360 arcsec in those that can be used, but the precise galactocentric radius where this occurs varies with azimuthal angle.

same radial dependence, the velocity dispersion should fall off with radius in an exponential manner, but with a scalelength that is twice as large as that of the light distribution. This was found to indeed be the case by observing the stellar velocity dispersion in the face-on galaxy NGC 5247 (van der Kruit and Freeman 1986), and has been confirmed for more systems in thesis work by Bottema (in preparation).

For the HI layer, a different picture holds, since there the velocity dispersion has been found to be independent of distance from the center (*e.g.* Shostak and van der Kruit 1984). Then, the prediction is that the thickness of the HI layer increases with galactocentric distance also in an exponential manner with again twice the scalelength of the optical light. This was found to be consistent with HI observations of the edge-on spiral NGC 891 (van der Kruit 1981).

Then there is the question of the central surface density. If constant, it would together with Freeman's law require a constant mass-to-light ratio M/L from disk to disk. This indeed appears to be the case for those galaxies where vertical dynamics can provide an estimate of this property (van der Kruit 1988 and references therein). Even the edges appear to occur in the mass distributions, since these predict a feature in the rotation curve at this radius. Evidence for such a feature can be seen in a few galaxies, most clearly in the HI observations of the edge-on, severely warped (in HI) spiral NGC 4013 (Bottema, Shostak, and van der Kruit 1987).

How can we put this into a picture of disk formation? Of course, we still lack a detailed model for galaxy formation, but a few things can be said anyway. Before doing this, it should be noted that there has been a claim that Freeman's law is not at all an intrinsic property of galactic disks, but derives from dust absorption (Jura 1980). Namely, if disks are optically thick, a constant central surface brightness can arise naturally from the absorption properties. An argument against this is the evidence discussed above, namely that the constancy of surface brightness occurs also at red and near-infrared wavelengths, but more detailed study is necessary to settle this matter definitely.

Below I will discuss a working hypothesis which shows promise of explaining the three features of disks mentioned above, namely the exponential nature, constant central surface brightness, and edges at 4–5 scalelengths. It should be mentioned that a suggestion has been made in the framework of the stochastic, self-propagating star formation (SSPSF) theory that could also explain these features, and it is appropriate that this be discussed here also. Seiden, Schulman, and Elmegreen (1984) suggested a model in which the molecular gas surface density, star formation rate, and surface brightness are all proportional to the total surface density of matter. This results in a constant M/L as mentioned above, which really says that at all radii, there is a similar time integral of the star formation rate, and this may of course occur equally well in other models for the star formation history in disks. It is not entirely clear therefore, that the adoption of SSPSF is essential to this model.

The crucial point is that Seiden *et al.* assume that the disk forms from a falling down of gaseous material from the halo into the disk with conservation of specific angular momentum, and that therefore the surface density profile of the disk is the same as that of a halo that gives rise to a flat rotation curve, namely an 1/r dependence. They argue that over practical ranges of radius this is not crucially different from exponential fits, and Seiden *et al.* continue to show that within SSPSF theory, the edges and constant central surface brightness follow from their model as well. However, the mechanism proposed then implies that the halo must be rotating cylindrically with the same rotation curve as the disk in order to assure that the gas falls down in the disk at the same radius as it originally had in the halo, so that the column density profile between halo and disk is conserved.

My objection against this model is that the halo is assumed to settle in a r^{-2} space density distribution, being supported by its rotation at all positions, without any physical indication of why this should be so. This, after all, puts rather strong and demanding constraints on the initial conditions in the protogalaxy, the collapse conditions, and on the mechanism which provides protogalaxies with angular momentum. Contrary to this, it seems more natural to assume that the flat rotation curves arise because the material (whatever it is) that makes up the halo settles after some relaxation process in a structure that resembles an isothermal sphere, where rotation is not important in supporting the structure. Violent relaxation leads naturally to relaxed and isothermal structures that are to a lesser extent determined by the details of the initial conditions, while the Seiden *et al.* halo has very special properties that need to survive the collapse.

The working hypothesis that I want to describe (see also van der Kruit 1987 and references therein) is based on a few notions. (a) The protogalaxies acquire their angular momentum from tidal interaction with neighbors, so that we may adopt Peebles's (1969) dimensionless parameter λ to describe the overall properties. This is defined as $\lambda = JE^{1/2}G^{-1}M^{-5/2}$ (with J being the total angular momentum, E the total energy, and M the total mass) and should equal about 0.7 from numerical and analytical work. (b) Disks form from uniformly rotating, uniform spheres with conservation of specific angular momentum (Mestel's 1963 hypothesis) in the force field of the dark halo. (c) The specific angular momentum distribution of the protohalo is the same as that of the disk (Fall and Efstathiou 1980), although only the disk is required to conserve this during collapse. Fall and Efstathiou have already shown that halos should contain about 10 times more mass than the disks in order to enable the tidal interaction picture to provide the observed amount of rotation in galaxies.

So, we start out with uniformly rotating, uniform spheres and assume that a major fraction with the same specific angular momentum distribution as the whole protogalaxy collapses to give the dark halo. It is assumed to collapse dissipationlessly by a process such as violent relaxation, and to settle in a structure that resembles an isothermal sphere or at least provides a forcefield that gives a flat rotation curve. The disk (in gaseous form and with dissipation) then settles in this forcefield with conservation of specific angular momentum. The result of such a collapse of a specific angular momentum distribution as in the uniformly rotating, uniform sphere in the forcefield of a flat rotation curve is illustrated in Figure 5. It then follows that the resulting surface brightness distribution is close to that of an exponential disk with a sharp edge (corresponding to the maximum angular momentum in the protogalaxy) at 4.5 scalelengths.

The exponential radial surface density and luminosity profiles, and the edges, thus arise naturally if we assume that galaxies start out as uniformly rotating, uniform spheres; the dark halo forms first and virializes roughly as an isothermal sphere, and the disk material settles according to Mestel's hypothesis. It can be shown then that the constant central surface density and luminosity also follow as long as the dark matter always makes up the same fraction of the material in the protogalaxy (van der Kruit, 1987).

Three comments need to be made to complete this schematic and very crude description. The first concerns the spheroid. Van der Kruit and Searle (1982b) have noted that the spheroid and the disk are two discrete components in the light distributions with different flattenings and argued that this points to two discrete epochs of star formation during galaxy formation, the first occurring early on before the collapse from which a moderately flattened spheroid results after dissipationless virialisation through violent relaxation. Subsequently, the remaining gas settles dissipationally in a flat disk and disk star formation commences. This explains the basic two-component structure of disk galaxies. Note in Figure 5 that there is a pronounced excess in the predicted radial profile at small radii, and it may be that this material, which was originally also at small radii in the protogalaxy, formed stars early on and actually resulted in the spheroid. The second comment concerns the possible "thick disks" (Gilmore and Wyse 1986 and references therein), which, if real, would make up a small fraction of the disk material and which have metal-abundance and kinematics intermediate between spheroid and disk. This may be the result of vigorous star formation in the last stages of initial disk formation. Thirdly, we usually observe gas beyond the edge of the stellar disk and this must then constitute material that fell in over a much longer timescale after disk formation was completed. It must have come from even larger radii than the boundary of the initial protogalaxy and consequently has higher angular momentum, and in general may not be expected to settle in the same plane, so that warped gas layers result. Indeed, warps usually start near the edges of the stellar disks, as is also well-demonstrated by the very large warp in NGC 4013 (Bottema *et al.* 1987).

Z–Profiles and the Constant Thickness

Van der Kruit and Searle (1981) proposed that the vertical density and light distribution of the old disk population could be modelled by that of a locally self–

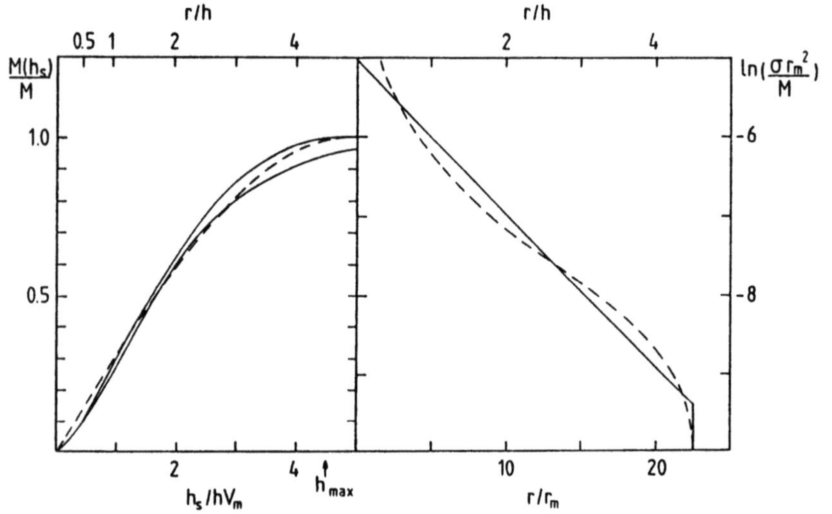

Figure 5: The surface density of a flat disk that results from a uniformly rotating, uniform sphere after it settles with detailed conservation of angular momentum in a forcefield corresponding to a flat rotation curve. The left–hand panel shows the distribution of specific angular momentum in three cases. The lower full–drawn curve is that of an infinite exponential disk with a Fall and Efstathiou (1980; FE) rotation curve, in which the radius r_m, where it reaches the flat part of the curve V_m, equals $0.2h$. The upper full–drawn curve is the same, except that the disk has an abrupt edge at $4.5h$. The dashed curve is that of an uniformly rotating, uniform sphere with maximum specific angular momentum $h_{max} = 4.5h\ V_m$. The right–hand panel displays the surface density distribution in a flat disk with an FE rotation curve for the distribution of specific angular momentum of a uniformly rotating, uniform sphere with $h_{max} = 22.5V_m r_m$ (dashed line). The full–drawn curve is an exponential disk with $h = 5r_m$ and a cut–off at 4.5 scalelengths.

gravitating isothermal sheet, which has the z–distribution $\rho(z) = \rho_o \mathrm{sech}^2(z/z_o)$. This idea is based on the notion that the stars in the solar neighborhood of ages a few Gyrs or more all have a fairly similar velocity dispersion. This function approximates an exponential with scaleheight $z_o/2$ for distances larger than about z_o from the plane. The surprising observation was that the scaleheight z_o turned out from observations to be essentially independent of galactocentric distance.

This remarkable fact has not been explained in detail, but it is known that the velocity dispersion of the stars must result from some kind of secular evolution of the stellar kinematics, invoked by gravitational disturbances of the orbits of the stars (*e.g.* Spitzer and Schwarzschild 1951; Lacey 1984; Carlberg and Sellwood 1985). As noted above, the constant thickness of the old disks implies that the stellar velocity dispersion decreases with distance from the center and it may at first sight not be unreasonable to suppose that any mechanism will be able to provide higher stellar velocities in areas with a higher total (an initially completely gaseous) surface density. In fact, van der Kruit and Searle (1982a) show in a very simple approximation that if the scattering of stellar orbits and the star formation rate are both determined by the same agent (*e.g.* giant molecular clouds), a roughly constant scaleheight may result.

Recently, Wainscoat (1986) has performed near–infrared photometry of a few edge–on galaxies, and found that the extrapolation of the vertical profiles to small values of z does not curve over as the sech^2–function would predict, but seems to be better described by an exponential all the way to the plane. In itself, an excess over the isothermal sheet is not too surprising, since at small z contributions from younger populations with lower velocity dispersions will start to affect the distribution in the sense observed, but one should keep in mind that contributions from red supergiants are expected to the light but not to the density distributions. It is, however, not clear from — in particular — observations in the solar neighborhood, that as sharp a peak as the exponential is expected in the plane (see discussion in van der Kruit 1988). In the last paper, therefore, it is proposed to drop the square from the sech in the original formula and equations for the vertical forcefield; integrated vertical stellar velocity dispersion and responses of isothermal components with negligeable surface densities are derived and compared for the three cases of exponential, sech, and isothermal sheets. These are useful to estimate the effects of the precise distributions at small z on determinations of surface densities from vertical dynamics (see above).

The results of this exercise can be summarized as follows. For the three cases we have:

$$\begin{aligned} \text{isothermal:} \quad & \rho(z) = \rho_o \mathrm{sech}^2(z/2z_e), \\ \text{“sech–distribution:”} \quad & \rho(z) = \rho_o \mathrm{sech}\,(z/z_e), \\ \text{exponential:} \quad & \rho(z) = \rho_o \exp(-z/z_e). \end{aligned}$$

Then, with σ the corresponding surface density, the integrated face–on stellar

de Vaucouleurs, G. 1974, in *I. A. U. Symp. 58, Formation and Dynamics of Galaxies*, ed. J. Shakeshaft (Dordrecht: Reidel), p. 1.
de Vaucouleurs, G. 1979, in *Photometry, Kinematics and Dynamics of Galaxies*, ed. D. S. Evans (Austin: Astron. Dept. Univ. Texas), p. 1.
Wainscoat, R. J. 1986, Ph. D. thesis, Australian National University.
Wevers, B. M. H. R., Allen, R. J., and van der Kruit, P. C. 1986, *Astron. Astrophys. Suppl.* **66**, 505.

Discussion

D. Burstein: I would feel much happier about the reality of truncations in the luminosity profiles of spirals if you did the following test: assume the galaxy has the same luminosity distribution law to infinity (*e.g.* exponential). Take the difference between this assumption and the observed truncation, and see if it yields a constant value for $\mu_B > 24$ mag arcsec^{-2}? If so, that is *prima facie* evidence for sky subtraction errors (1% error in the sky = 0.25 mag error at 26th mag, 0.1 mag at 25th mag, and so on).

van der Kruit: This is a viable test, of course. But we see the effects of disk truncation in edge–on galaxies at such bright levels that incorrect sky subtraction cannot have produced them in any reasonable and conceivable way. Also, the luminosity profiles in these systems along the minor axis or perpendicular to the disk disappear smoothly in the noise as would be expected from a correct treatment of sky level. There is absolutely no doubt that the truncations are real.

M. Capaccioli: Could the truncation in edge–on spirals be due to an increase in scale height at large r, as expected in regions of external gravity?

Now a comment: if Σ_0 is constant and $f(r)$ is exponential, then total mass and angular momentum are both functions of r_e only. This is not easy to understand in the framework of conventional mechanisms to acquire angular momentum.

van der Kruit: There is at most a very minor increase in disk thickness before the truncation radius, and disks probably are self–gravitating up to this radius. Unless a very discrete and sudden increase in thickness occurs at these radii by a considerable factor, we would not have missed it. Your remark is interesting, but could it be that total mass and angular momentum actually determine r_e as it does determine the scale length in my description.

R.–J. Dettmar: Your surface photometry has been used by several groups to model disks with exponentially increasing scale–height near the cut–off. Can you comment on that?

van der Kruit: As I replied to Capaccioli, the increase can be only modest, as has actually been found by some people. Larger increases are only based on extrapolation beyond the truncation radius, and are in any case too gradual to produce an apparent sharp truncation.

J.–C. Pecker: I would hesitate very much to justify the "truncation" by Mestel's theory about the angular momentum: outwards transfer of momentum, through galactic "winds," should make Mestel's prediction obsolete. On the other hand, ionized interstellar medium does exist; and, at that scale, one should perhaps not forget about electromagnetic effects confining the galactic plasma. Near the boundary of a galaxy, this kind of effect might play a role.

van der Kruit: There is indeed little direct evidence for the hypothesis of detailed conservation of angular momentum. Yet the hypothesis does explain the observation. In any case, there is a maximum angular momentum per unit mass in any model that would produce a truncation of some kind, so even if redistribution of angular momentum occurs, a truncation should still result. It is true that we usually see HI beyond the edge of the optical disk in often inclined layers, but then this must be material with high angular momentum that fell in after completion of that part of disk formation that gave rise to the disk observed in the (old) stellar distribution.

P. E. Seiden: A few years ago, on the basis of propagating star formation, we proposed that the disk followed $1/r$–constant instead of an exponential. This function is exponential–like over a large region and fits the observations well. It directly gives a deviation from exponential at a distance of about four times the exponential scale length. Also, it gives a constant extrapolated central surface brightness as a direct consequence of trying to fit an exponential to $1/r$–constant. Do you have an opinion on this explanation?

van der Kruit: As long as your model is consistent with the observations — as you indicate it probably is — it is of course a viable alternative. Many galaxies actually are well-fit by an exponential to small galactocentric distances, and in these cases the extrapolated central surface brightness involves very small or no extrapolations, and the use of the actual fitting function has little effect on the inferred μ_0.

W. T. Sullivan: There appear to be a reasonable number of low surface brightness galaxies that are not dwarfs, *e.g.* higher–redshift UGC galaxies that are not listed in the Zwicky catalog. Have any systematic photometric studies been done on these? Might they substantially widen the μ_0 distribution?

van der Kruit: I do not know of any systematic study of these systems. However, I believe that the only proper approach is to take a well–defined, statistically complete sample as described in my paper. In my sample, all galaxies classified as dwarfs that had measured redshifts turned out to be small systems at small distances. The systems you mention could indeed widen the μ_0 distribution in principle, but it is unclear now whether there are many of them in terms of space density.

A Study of the Sombrero Galaxy (NGC 4594)

Ralf–Jürgen Dettmar

Radioastronomisches Institut der Universität Bonn

Introduction

NGC 4594 (M 104, the "Sombrero") is a southern early type spiral galaxy, classified between S0/a and Sb, which is seen almost edge–on. It is the nearest of a class of massive, luminous early type spirals with very high rotational velocities (Giovanelli *et al.* 1986). It was one of the first galaxies studied by de Vaucouleurs (1948) photometrically and it was proposed as a surface photometric standard by the I. A. U. in 1961. Here we present preliminary results from two–dimensional surface–photometry.

Observations

We have used UK– and ESO–Schmidt plates to obtain surface photometry of NGC 4594 in B_J and in R_F. Photoelectric aperture photometry with the ESO 1–m telescope supplies the absolute photometric scale, and CCD imaging with the ESO/MPI 2.2–m telescope is used to check the linearity and to study the nuclear region.

Results

The bulge of NGC 4594 is detected out to $r \simeq 1000$ arcsec, and is well–described by an $r^{1/4}$ law outside $r = 35$ arcsec. If it is represented as a power law, we obtain $I \sim z^{-2.1}$. The comparison of the minor axis profile with previous work shows sufficient agreement at intermediate brightness levels, while obvious discrepancies appear in the outer and inner parts, *e.g.* the bulge is slightly

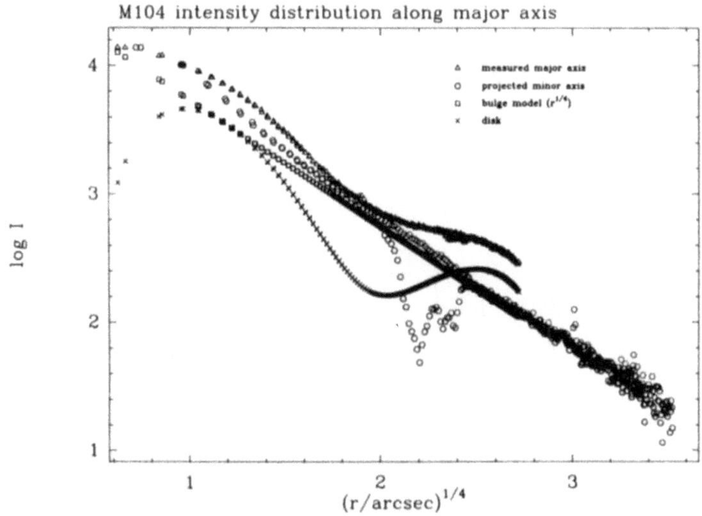

Figure 1: Intensity distribution, the bulge models, and the resulting disk along the major axis from CCD photometry. The inner disk is visible up to 15 arcsec where the main disk starts to become dominant.

steeper and less extended that found by Burkhead (1986). The ellipticity of the bulge becomes rapidly circular at $r = 300$ arcsec, and the colour gradient of the bulge is found to be less than 0.15 magnitudes in $B - V$ for $r < 10$ arcmin. To study the characteristics of the disk, we have modelled the bulge by extrapolating an $r^{1/4}$ law into the nuclear region and by projecting the minor axis intensity distribution according to ellipticity. With both models, a hole of the disk becomes visible and the inner disk described by Burkhead (1986) shows up (Figure 1). This inner disk is very thin and elongated. It is red with $(g-i) = 2.5$ (Figure 2), and also a second inner dust lane ($r \sim 40$ arcsec) inclined to the disk is visible in the two-dimensional colour-distribution. This unique configuration meight be related to the active nucleus which is very unusual for spiral galaxies (Bajaja *et al.* 1988).

References

Bajaja, E., Dettmar, R.-J., Hummel, E., and Wielebinski, R. 1988, *Astron. Astrophys.*, in press.

Burkhead, M. S. 1986, *Astron. J.* **91**, 777.

Giovanelli, R., Haynes, M. P., Rubin, V. C., and Ford, W. K. 1986, *Astrophys. J. Lett.* **301**, L7.

de Vaucouleurs, G. 1948, *Ann. d'Astrophys.* **11**, 247.

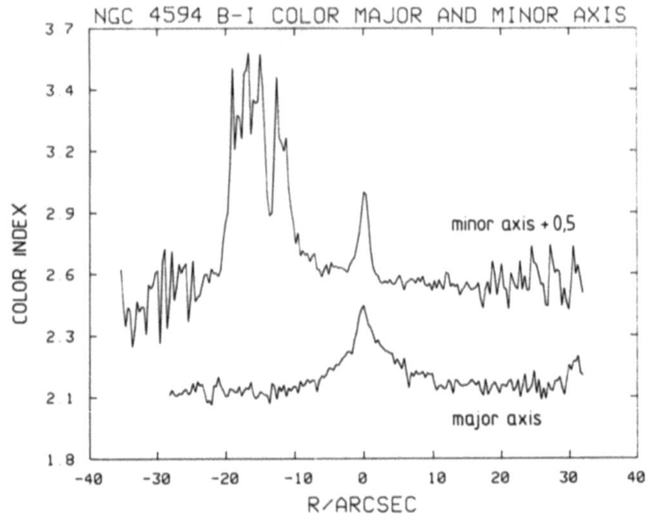

Figure 2: The inner disk is easily visible in the colour distributions along the minor and major axis. It is a red and elongated feature with a scale height of only 1/10 of the main disk which possesses a hole in the nuclear region.

Visible and Near Infrared Photometry of NGC 4736

M. Prieto, C. Muñoz-Tuñón, J. Beckman, A. Campos, and J. Cepa

Instituto de Astrofisica de Canaries, Universidad de La Laguna, Tenerife, Spain

NGC 4736 is classified by de Vaucouleurs (RC2, 1976) as SA(r)ab. It has tightly wound arms, and it has a ring shape with an extended envelope.

Principal features of the gaseous component (Bosma *et al.* 1977) are gas out to 3 arcmin galactocentric distance as detected in radio continuum emission, an inner ring of HII regions at a radius between 40 to 60 arcsec — where there is also a ring of neutral gas — and a weaker outer ring of neutral gas at 4 arcmin from the nucleus. The centre of the galaxy is marked by an absence of gas.

In order to obtain some characteristics of the stellar population, we have carried out visual and infrared observations, which permit us to obtain information about the stellar population.

The visible observations were obtained on the nights 15–18 June 1986, at the f/15 cassegrain focus of the 1–m Jacobus Kapteyn Telescope at the Observatorio del Roque de los Muchachos on La Palma. The photometry was carried out with a cooled EMI 9658AM photomultiplier, in a series of stepped rasters using an aperture of 28 arcsec. We covered an area of 600x180 arcsec in right ascension and declination, respectively. Details of the signal to noise ratio and of the reduction are given in Beckman *et al.* (1987).

Data were reduced to "face on" taking the position angle and the inclination plane of the galaxy to be 35° (Bosma *et al.* 1977) and 112° (van der Kruit 1974), respectively. Optical data have been corrected for the internal extinction given the distribution in H I (Bosma *et al.* 1977) and H_2 (Garman and Young 1986). The correction method is described in detail in Battaner *et al.* (1986) and in Prieto *et al.* (1985).

The near infrared observations were obtained the nights 19–20 March 1988 with an infrared photometer at the Cassagrain focus of the 1.5–m Carlos Sanchez Telescope of the Instituto de Astrofisica de Canarias, at the Observatorio del Teide, Tenerife. The detector used was an indium antimonide photoconductor.

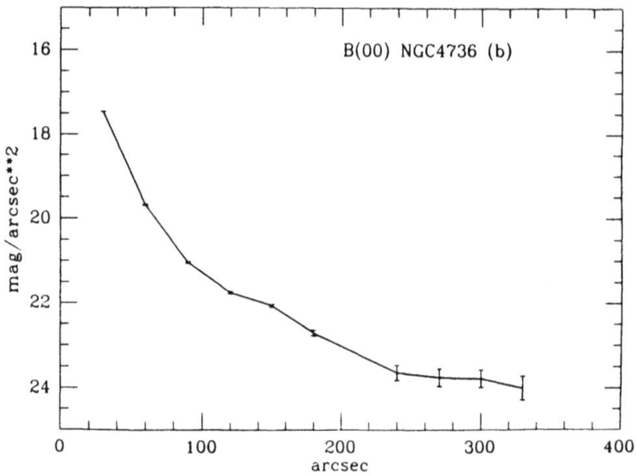

Figure 1: Luminosity profile in the B–band. It has been averaged over azimuth, corrected for internal extinction, extinction in our Galaxy, and deprojected to face–on.

The near infrared J filter has an effective wavelength of 1.3μm with 0.24μm half–band width. When pumped to operate at 63 K, the system was sky–noise limited. A speed of 10 arcsec/sec throughout, a sampling rate of 5 data/sec, and an aperture of 20 arcsec was used, so the signal was somewhat oversampled. We averaged every six data points to reduce the noise. Scans were taken at 10 arcmin outwards from the center of NGC 4736 towards the north and south.

Figure 1 shows the B–band luminosity profile. Figure 2 shows the near–infrared "J" profile, and Figure 3 (a and b), the $U - B$ and $B - V$ colour index profiles.

In most of the profiles, we can distinguish five sections: a) an inner intense region (out to 40 arcsec); b) an inner region of lower flux gradient (from 40 to about 100 arcsec); c) a region of spiral structure (from 100 to 200 arcsec); d) a gap at about 250 arcseconds; and, finally e) a faint emission ring centered at 300 arcsec galactocentric radius. These galactic components are most easily seen in the "J" profiles due the higher spatial resolution. From the index colour profiles, we note that a bluer region is observed at about 50 arcseconds radius; this is associated with the inner ring of H II regions and a zone of recent star formation.

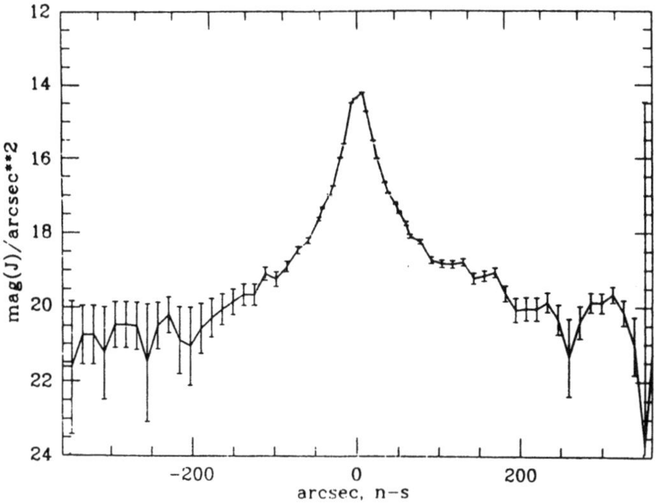

Figure 2: Luminosity profile in the near-infrared "J"-band along the north–south direction, deprojected to face-on.

References

Battaner, E., Beckman, J. E., Mediavilla, E., Prieto, M., Sanchez-Magro, C., Muñoz–Tuñón, C., and Sanchez-Saavedra, M. L. 1986, *Astron. Astrophys.* **161**, 70.

Beckman, J. E., Cepa, J., Prieto, M., and Muñoz–Tuñón, C. 1987, *Rev. Mexicana Astron. Astrophys.* **14**, 134.

Bosma, A., van der Hulst, J. M., and Sullivan, W. T. 1977, *Astron. Astrophys.* **57**, 373.

Garman, L. E. and Young, J. S. 1986, *Astron. Astrophys.* **154**, 8.

van der Kruit, P. C. 1974, *Astrophys. J.* **188**, 3.

Prieto, M., Battaner, E., Sanchez, C., and Beckman, J. 1985, *Astron. Astrophys.* **146**, 297.

de Vaucouleurs, G., de Vaucoulerus, A., and Corwin, H. G. 1976, *Second Reference Catalogue of Bright Galaxies*, (Austin: Univ. Texas Press).

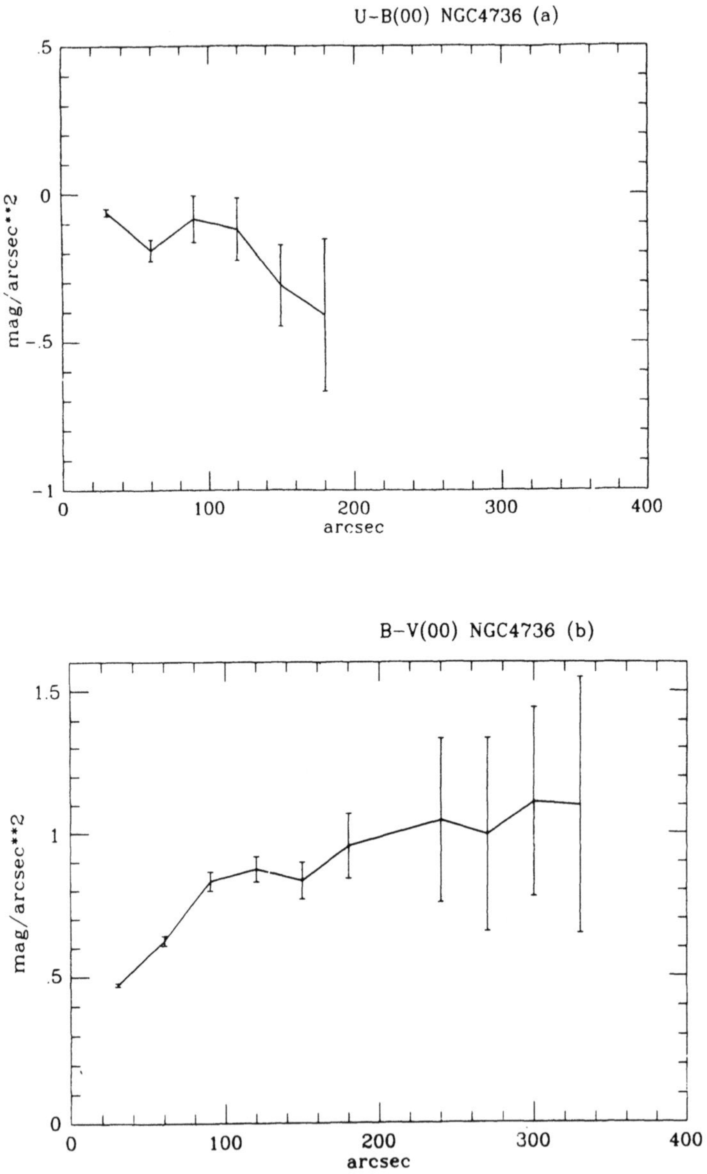

Figure 3: $U - B$ (top) and $B - V$ (bottom) colour index profiles.

Computer Processed Color Images of Spiral Galaxies

Philip E. Seiden, Debra Meloy Elmegreen, and Bruce G. Elmegreen

IBM Thomas J. Watson Research Center, Yorktown Heights, New York, U.S.A.

The painstaking and extensive work on spiral galaxies carried out by Antoinette and Gérard de Vaucouleurs is at the same time impressive and intimidating. It is clear from their work, and the work of other fine investigators in the field, that the job of collecting the extensive data necessary to understand the structure and evolution of spiral galaxies is not easy. The reduction of the astronomical plates to extract the appropriate data is difficult and time consuming. For that reason, we began a program to investigate how the addition of color and digital processing could provide more informative images. Computerized color imagery using multiple passbands may be an important new tool for analyzing galaxies. Computers can superimpose the various images of a galaxy, and then realign the result to a face-on orientation for a better view of the spirals, or subtract the underlying disk light to emphasize only the azimuthal variations. The computer processed images show the shapes and color distributions of spiral structure better then conventional photographs. Spiral arms can often be traced for more than a complete turn around the nucleus, inner ovals are sometimes revealed, and blue patterns of star formation can be identified.

Most of our work up to the present has concentrated on using B and I images. These two images are aligned and normalized with respect to each other, and a fake G image is produced from their average. The resulting B, G, and I images are combined on an RGB video monitor to produce a color composite, using B for blue, G for green, and I for red.

The normalization of the blue and red intensities was done by equating the total light in the two passbands that comes from the region between 1/3 and 2/3 of the optical radius at 25th magnitude per square arcsecond. This gives an attractive color image of the galaxy, with the blue enhanced just enough to show most of the large regions of star formation.

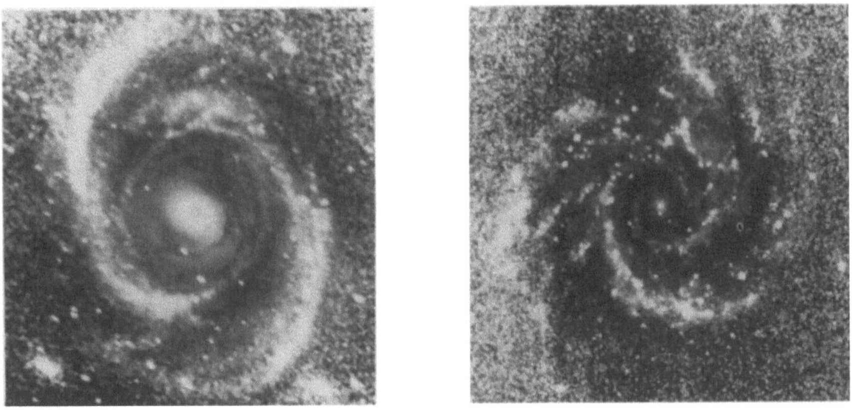

Figure 1: Face–on galactic images with the average radial surface brightness subtracted (B band). The left–hand image is NGC 3031 (M 81) and the right–hand image is NGC 628 (M 74) (the figures are in black and white because color processing was not available).

Except for the enhancement of the blue image the images are still in the normal form. Two problems remain; firstly, all galaxies are not face–on and it is harder to see the structure of a galaxy inclined to the field of view. Therefore, we rectify the images to their face–on position by stretching them by the inverse of the cosine of their inclination angle.

The second problem is that the surface brightness of a galaxy falls off strongly with radius. The dynamic range of photographic paper is not great enough to show the whole luminous range of a galaxy. To see the dim outer regions the center would have to be strongly saturated. Conversely, if the picture were exposed lightly enough to see the details of the galactic center, then the outer arms would be completely invisible. To eliminate this difficulty we remove the radial gradient of surface brightness. The first step is to calculate the average radial dependence of the surface brightness from the face–on image. We then smooth this curve and subtract it from the face–on image. That is, we subtract from every pixel the value appropriate to its galactic radius. The resulting image has an average surface brightness independent of radius, and it is possible to see the central regions and the far edges with ease on the same picture. Two examples are shown in Figure 1.

NGC 3031 is an example of a spiral pattern that is continuous and symmetric over much of the optical disk. It is a good example of a spiral density wave. Figure 1, however, clearly shows that the spiral structure in NGC 3031 covers only the outer half of the galaxy; it terminates near the bulge. This may be an example where a wave excited in the outer part of the disk, possibly by the companion galaxy M 82, is absorbed by an inner Lindblad resonance or reflected by a Q–barrier produced by the bulge.

NGC 628 is a grand–design multiple arm spiral with at least one long blue arm. In contrast to NGC 3031, however, this galaxy appears to be an example of pure star formation with no underlying wave. In Figure 1, the long arm can be seen starting at the nucleus and can be traced for as much as 540 degrees toward the outer edge of the galaxy. This is evident on radially subtracted images because they show structure in both the inner and outer regions of a galaxy without saturation of the nucleus.

Computer processed color images immediately reveal the features discussed above as well as many others. Each processed image of a spiral galaxy seems to show something new. For example, there are occasionally weak oval–like structures in the center. The distributions of spiral arms in the outer regions of multiple arm galaxies are often asymmetric. And, as noted for NGC 628 the spiral arms in some grand design galaxies can be traced for long distances. All these features are more easily seen after the images have been rectified and after the average radial profiles have been subtracted.

Modelling the Luminosity Distributions of Edge–On Spiral and Lenticular Galaxies

Martin Shaw[1] and Gerard Gilmore[2]

[1]Department of Astronomy, University of Manchester, England, U.K.
[2]Institute of Astronomy, Cambridge, England, U.K.

Motivation

An increasing proportion of external galaxies show substantial deviations from the luminosity distributions predicted by standard fitting functions (see *e.g.* van der Kruit and Searle 1982, Michard 1985, and Capaccioli *et al.* 1987).

However, it is currently unclear whether such discrepancies are a result of an inappropriate choice of fitting function (or set thereof), or because of an inappropriate modelling procedure, since *objective* (*i.e.* statistical) selection criteria are rarely used in an attempt to differentiate between respective fitting functions or between a variety of model combinations. It is also unclear whether the discrepancies are themselves systematic in nature (pertinent to a search for intermediate or "thick–disc" components).

Present Work and Results Thus Far

To study the nature of the galaxian non–thin disc light in general, and to address the above questions in particular, we have constructed a nonlinear, least–squares modelling algorithm. The principle is to search for an unconstrained minimum in the sum of the squares of the residuals and the reduced χ^2 between the observed two–dimensional surface brightness distribution, and that predicted on the basis of the chosen model components and values for the parameters that define them.

The technique is an improvement on previous schemes not only for its objectivity, but because of the wide range of possible model combinations tested, and because the complete data array is modelled at each iteration. It has been tested on numerous model galaxies (both with and without the addition of random and systematic errors) as well as to existing literature data for NGC 891 and NGC 4565, before being subsequently applied to CCD photometry for 11 large ($\geq$ 5 arcmin), normal, edge-on ($i \geq 75°$) spirals and lenticulars *not* displaying "box/peanut" bulge morphology.

The principal result of our tests has been to show that it is indeed possible to define the most appropriate model combination applicable to each galaxy *solely* on the basis of the respective goodness-of-fit estimators (contrary to previous expectations in the literature). We find the following additional results:

1. In only two systems does the adoption of a three-component combination yield a statistically significant improvement over a simple two-component set.
2. There is *no* evidence of a correlation between the number of components required to describe the observations and the dominance of the "bulge/spheroid".
3. For all galaxies in which a disc component can be clearly defined, the scale height is constant to better than $\pm$4% of the derived mean value (*i.e.* there is no evidence for disc "flaring" to ~5 disc scale lengths for galaxies ranging from S0 to Sc).
4. For all galaxies in the present sample for which clear non-disc populations are isolated in the modelling, we find no evidence of a correlation between the shape of the non-disc light and the degree of disc dominance: the gravitational field of the disc appears insufficient to determine the flattening of the non-thin disc population (as suggested by Barnes and White 1984). The shape of this component thus appears primordial in origin.

Departures From the Best-Fit Models

Almost all galaxies in the present sample show substantial (*i.e.* $\geq$ 0.4 mag arcsec^{-2}) departures from the best-fitting model combinations (see Table 1). Because of the wide range of model combinations tested, it is clear that such discrepancies are not sufficiently systematic to be accounted for merely by the presence of an additional model component. The discrepancies take two forms:

- An underestimate of the light at large Z-heights. Although an incorrect estimate of the sky background is a possible source, one requires the sky estimates to be in error by ~4 times the assigned uncertainties (which are typically $\pm$0.5% of the local mean sky). This hypothesis could be checked in future using larger format imaging devices.

Galaxy	Type	exp. time (secs.) & passband	"best-fit" model combination	SUMSQ ratio	red. χ^2 ratio	consistency of disc scale parameters between passbands	consistency of non-disc scale parameters between passbands	rms residual (mag. $arcsec^{-2}$)
NGC 2295	Sab	2000 (R)	"exp. + $r^{1/4}$ law"	0.97	0.97	—	—	± 0.25
NGC 3115	S0	220 (B) 1500 (R)	"exp. + $r^{1/4}$ law"	0.98	0.86	—	73 %	± 0.2 ± 0.2
NGC 3573	S0	2000 (B) 2x2000 (R)	"exp. + $r^{1/4}$ law" "$sech^2$ + $r^{1/4}$ law"	0.82 0.70	0.83 0.76	60 %	85 %	± 0.3 ± 0.25
NGC 4289	Scd	2x1000 (R)	"exp. + $r^{1/4}$ law"	0.81	0.79	—	—	± 0.3
NGC 5078	Sa	3x2000 + 1000 (R)	"$r^{1/4}$ law"	0.08	0.08	—	—	± 0.2
NGC 5170	Sc	3x2000 (B) 2x2000 (R)	"exp. + exp. + $r^{1/4}$ law"	0.62	0.90	93 %	94 %	± 0.3 ± 0.2
IC 4351	Sb	300 (R)	"exp. + $r^{1/4}$ law"	0.94	0.86	—	—	± 0.3
A0919–33	Sbc	2000 (B) 2000 (R)	"exp. + $r^{1/4}$ law"	0.94	0.90	66 %	75 %	± 0.2 ± 0.2
A0931–32	Scd	2000 (R)	"exponential"	0.27	0.31	—	—	± 0.4
A1611–00	Scd	2000 (R)	"$sech^2$ + $r^{1/4}$ law"	0.98	1.03	—	—	± 0.35
UGC 7170	Sc	2000 (B) 2000 (R)	"exponential"	0.09	0.08	97 %	—	± 0.15 ± 0.2
NGC 891	Sb	literature (J) (U′) (F)	"$sech^2$ + $sech^2$"	1.00	0.86	87 %	76 %	± 0.2
NGC 4565	Sb	literature (B) (R)	"$sech^2$ + exp. + exp."	1.03	0.95	77 %	85 %	± 0.25

Table 1 : SUMSQ and red. χ^2 ratios are averaged sums of squares of residuals and reduced χ^2 for the "best-fit" model combinations compared to the next best fit.

- Factor ~2 inconsistencies between the scale parameters in the B and R passbands for a number of candidates. These may be due to the nature of the error distributions adopted in the respective data sets, and can be accounted for by imposing "minimum errors" in both passbands of ± 0.2 mag arcsec^{-2}.

After correction for the above effects, we obtain final solutions which fit the observations to within a typical r.m.s. residual of ± 0.3 mag arcsec^{-2} and yield model parameters consistent with colour to ~30%.

With the intention of gleaning more information concerning the intrinsic properties of galactic bulges, work is currently underway to examine the two dimensional distribution of residuals from the fit in each galaxy to ascertain the extent to which the discrepancies may also be correlated with morphological characteristics such as the degree of bulge dominance. There are already suggestions that the non-thin disc light in seemingly simple lenticular galaxies is not adequately described by standard fitting functions (see above references).

References

Barnes, J. and White, S. D. M. 1984, *Mon. Not. Roy. Astron. Soc.* **211**, 753.
Capaccioli, M., Held, E. V., and Nieto, J.-L. 1987, *Astron. J.* **94**, 1519.
van der Kruit, P. C. and Searle, L. 1982, *Astron. Astrophys.* **110**, 79.
Michard, R. 1985, *Astron. Astrophys. Suppl.* **59**, 205.

Near–Infrared Imaging of Edge–On Galaxies

Richard J. Wainscoat

NASA Ames Research Center, Moffett Field, California, U. S. A.

Summary

Images in the near-infrared passbands J, H, and K of nine edge-on spiral galaxies have been obtained using two different techniques: a) raster scanning a single detector across the galaxy using the Anglo-Australian Telescope (AAT) and b) using an infrared array camera on the United Kingdom Infrared Telescope (UKIRT). The results for one of the galaxies studied with the AAT — IC 2531 — are summarised and a simple model for the stellar and dust distributions in IC 2531 is described.

Introduction

Edge-on galaxies are an excellent source of information about galactic structure. They allow us to study the vertical (z) distribution of stars, and also yield information about the radial distribution of light. This information is of particular value in understanding the formation and dynamical evolution of spiral galaxies.

The light from the old disk is dominated by K-giants — consequently a substantial part of the stellar radiation is emitted at near-infrared wavelengths. The distribution of old stars can therefore be best studied in the near-infrared, since the contribution from the bluer young stellar population is smaller. More importantly, extinction is much less than at optical wavelengths. In edge-on spiral galaxies, a strong dust lane usually hides the region near $z = 0$, meaning that study of this part of an edge-on galaxy is impossible at optical wavelengths. Studies in our own Galaxy can determine stellar distributions, kinematics, and abundances close to the plane. Near–infrared wavelengths represent the only means of obtaining analogous measurements in external galaxies.

Until recently, technological limitations and the bright near–infrared background (telescope + sky = 12–12.5 mag arcsec^{-2} at K) have made near–infrared imaging very difficult. The observations of IC 2531 discussed here were made using a single detector scanned across the galaxy to form an image. This technique relies heavily upon the precise pointing and control possible with the AAT and requires a large amount of telescope time, but nevertheless produces excellent data, reaching to about 20 mag arcsec^{-2} at K. The newly–developed infrared array cameras provide the opportunity to make observations with a resolution close to that possible optically, with good signal–to–noise ratios.

The above techniques using near–infrared imaging to probe galactic structure are practical only as red as the K passband (2.2μm). At K, the light is still dominated by starlight; longward of K, hot dust starts to make a significant contribution, and the thermal background level is very high relative to the surface brightness of the galaxies being studied.

AAT Imaging

The AAT has been used to make near–infrared images at J, H, and K of six edge–on galaxies — NGC 55, NGC 4594, NGC 7123, NGC 7814, IC 2531, and A 0106–8034 (Wainscoat 1986). IC 2531 will be discussed here since it is the closest in type to our own Galaxy.

IC 2531 is a southern Sc galaxy having a nearly exactly edge–on orientation, making it an ideal candidate for the study of the vertical distribution of the disk stars. It has a very low far–infrared flux density — its light distribution is therefore likely to be dominated by the old population in the redder passbands. Oversampled J, H, and K images were obtained with the AAT using a 5″ aperture. *UBVRI* CCD images were obtained for comparison.

Vertical cuts were made through all images. The cuts through the optical data show a highly exponential decay in the z direction, especially in the redder optical passbands. This exponential nature persists into very small z until the dust lane is reached. A clear excess over the isothermal sech^2 distribution proposed by van der Kruit and Searle (1981a,b, 1982) is evident. Vertical cuts through the near–infrared images, in particular at K, are also highly exponential, with their exponential shape persisting to $z = 0$. The K light distribution in IC 2531 therefore appears to be well fitted by:

$$L(R,z) = \begin{cases} L_0 \exp(-R/h - |z|/h_z) & R < R_{max} \\ 0 & R > R_{max} \end{cases} \quad (1)$$

where h_z is the vertical scale height. The excess of stars that this distribution has relative to the isothermal case implies that the vertical velocity dispersion should increase with z (approaching a constant value as z increases).

A simple model of the light and dust distributions in IC 2531 has been constructed. This model consists of three components: the old disk, a young disk, and a disk of dust. Colour images constructed from the optical CCD data, in particular the $U - V$ image, show that young stars are present in IC 2531 and must be considered in the modelling — the dust lane is *blue* in $U - V$. In view of the exponential structure found in the K image and in the redder optical images, Equation (1) was used as the functional form of the old disk. The same distribution, with the same radial scale length, but different scale heights, was used for the young disk and the dust. Values of $h_{z_y} = h_z/8$ for the young disk and $h_{z_d} = h_z/4$ for the dust were found to produce good fits (h_z is the scale height of the old disk).

This model produces a good fit to the light distribution observed in IC 2531 from U through K. The general shape of the $UBVRI$ profiles is well reproduced, and the shape and the depth of the dust lane match the observations very well. The JHK profiles also match well, but have lower resolution, so are not as good a test.

The observations and modelling of IC 2531 summarised here are discussed in more detail by Wainscoat *et al.* (1988).

Infrared Array Camera Images

Near-infrared images of several well-known edge-on galaxies (NGC 891, NGC 4565, NGC 5907 and IC 2531) were recently obtained using the infrared array camera on UKIRT. A 62 × 58 element InSb array with a pixel size of 1.2″ was used. Each galaxy was observed as a composite of many overlapping frames. Further image processing needs to be performed before these images can be properly analysed, but the images already graphically illustrate the difference in appearance that these galaxies undergo between optical and near-infrared wavelengths, and demonstrate the value of the K wavelength for studies of galactic structure.

References

van der Kruit, P. C. and Searle, L. 1981a, *Astron. Astrophys.* **95**, 105.
van der Kruit, P. C. and Searle, L. 1981b, *Astron. Astrophys.* **95**, 116.
van der Kruit, P. C. and Searle, L. 1982, *Astron. Astrophys.* **110**, 61.
Wainscoat, R. J. 1986, Ph. D. Thesis, Australian National University.
Wainscoat, R. J., Freeman, K. C., and Hyland, A. R. 1988, *Astrophys. J.*, submitted.

Photometric Decomposition of Galaxies

F.Simien

Observatoire de Lyon, France

I. Introduction

The concept of photometric decomposition probably had its origin in Hubble's (1926) classic work on galaxy morphology, where he situated the different types along a sequence of decreasing importance of the central concentration. Later, Baade (1944) showed that this could be considered in terms of the two major stellar components and that this sequence was actually one of decreasing relative importance of Population I with respect to Population II. This increased the interest of a quantitative analysis of the photometric contributions. Search for luminosity laws typical of each component can be traced back to 1913, when Reynolds studied the center of M 31 and proposed a formula subsequently applied to ellipticals by Hubble (1930). De Vaucouleurs (1948, 1953) introduced the $r^{1/4}$ law as a new standard for spheroidal populations. The exponential luminosity decrease in the disks was first noticed by Patterson (1940), and later by de Vaucouleurs (1956). These results provided the basis for subsequent photometric decompositions. The first complete one was presented by de Vaucouleurs (1959) who showed that the brightness profile along the major axis of M 31 resulted from the superposition of two components with the above-mentioned characteristics. Actually, this analysis was made in 1957; in 1958, using better data, he made a more detailed decomposition, on both main axes, and the disk turned out to be more complicated than just exponential, with, among other features, a vanishing brightness at the center. This provided early evidence that photometric decomposition is not always a straightforward process. Afterwards, in many more cases the model turned out to be partially inadequate. Nevertheless, the great majority of the results obtained so far come from the original scheme, and, as developed later, many of their conclusions cannot be questioned.

Table 1: The standard photometric model

	Bulge	Disk
Geometry	Oblate, similar, ellipsoidal isodensity surfaces	
True axial ratio	~0.5 - ~0.9	~0.1 - ~0.3
Isophotes	Similar ellipses	
Apparent ratio	q_I	q_{II}
Effective radius	$r_{e,I}$	$r_{e,II}$
Effective luminosity	$\mu_{e,I}$	$\mu_{e,II}$
Profile	$\mu = \mu_e + 8.325(a^{1/4} - 1)$	$\mu = \mu_e + 1.823(a - 1)$
Integrated magnitude	m_I	m_{II}

Notes: r_e is the semi–major axis of the effective isophote containing half the total light of a component, and μ_e is the effective brightness of this isophote (in mag arcsec^{-2}); $a = r/r_e$. The customary use of I and II for bulge and disk, respectively, is not related to their stellar populations (II and I).

Much work has been devoted to photometric separation of bulge and disk contributions (with far less attention to other components). Early studies, like those of van Houten (1961) and Freeman (1970) have become classic papers. Yoshizawa and Wakamatsu (1975) were the first to systematically fit both components. Many papers followed and, lately, large samples of homogeneous data have been analysed (*e.g.* Kent 1985, hereafter K85; Kodaira *et al.* 1987, hereafter KWO).

The outline of this paper is as follows: Section II deals with the "standard" photometric model, reviews its limitations and presents a few alternatives. Section III describes most of the decomposition methods found in the literature. Section IV presents a selection of published results based mostly on the standard model. The main applications are discussed in Section V. Section VI presents a few conclusions.

II. The Photometric Models

A. The Standard Model

Let us consider a model galaxy made of a spheroid (or bulge) and a disk, whose relevant characteristics are summarized in Table 1. The model is fully described by a set of 6 parameters: $r_{e,I}, \mu_{e,I}, q_I, r_{e,II}, \mu_{e,II}$, and q_{II}. Among the subsequently derived quantities, a particularly important one is the fractional luminosity of the bulge, which may be expressed either as the ratio of bulge and disk luminosities, or as the ratio of bulge to total luminosities, or even as a magnitude difference, $\Delta m_I = m_I - m_{galaxy}$.

This standard model, as first proposed by de Vaucouleurs, is able to generate a family which mimics the Hubble sequence fairly well, with only a limited number of free parameters. The major point of this similarity is the unambiguous separation of the two main stellar populations. This represents an appropriate level for the *a priori* information needed to perform photometric decompositions in a systematic way, at least to a first approximation.

In many cases, departures from the model are obvious, and they fall into two different classes: those coming from features which are superimposed on the (hopefully) unperturbed model, and those which reflect more fundamental model inadequacies. Actually, these two classes may not be completely separated, but this addresses a complicated problem which is beyond the scope of the present paper.

B. Superimposed Features

The most common are: nuclei, lenses, bars, rings, and spiral arms, but this list is not exhaustive.

Nuclei seem to be quite common among spheroidal components (Kormendy 1985, and references therein). When an accurate measurement of the point-spread function is available, comparison between the convolved model and the data often reveals an excess luminosity at the very center: in favourable cases, this can be accounted for by a star-like nucleus, but the problem is sometimes more complex (see below).

The lens can be a source of important perturbation in the inner profile, and it sometimes prevents the bulge from showing any dominating contribution except at the very center. A bar may be a little easier to separate, specially if it is tilted. These features have not been studied systematically enough for standard profiles or shapes to be found (Duval and Athanassoula 1983; Duval and Monnet 1985). The rings are usually characterized by local bumps in the brightness distributions, with a gaussian residual profile (Buta 1984), and they are often relatively easy to separate. This author, however, points out that a ringed galaxy can exhibit a complex structure.

The spiral arms locally enhance the brightness of the flat component, and it is the underlying old disk which is supposed to be exponential (Schweizer 1976). The removal of these arms can noticeably modify the parameters of the disk, and ensure greater homogeneity within a sample of objects.

The presence of the above perturbing features can significantly contribute to the uncertainty in decomposition results. The luminosity of these features is often small with respect to the whole galaxy (typically a few percent, and up to about 20%, but this estimate may be very color-dependent); locally, however, they can alter the brightness distribution significantly. With the hypothesis that underlying standard components are present, model fittings can be attempted, provided that a) the perturbed regions are avoided, or b) the features have been previously erased from the data.

C. More Fundamental Model Inadequacies

In many cases the standard model proves to be only a rough approximation, even in the absence of the above-discussed perturbing features. Actually, all the main characteristics outlined in Table 1 can be contradicted by particular cases.

The geometry and luminosity gradient in bulges and, more generally, spher oidal systems are complex questions (Capaccioli, this conference; Nieto 1988, and references therein); here, only the main points have to be pointed out, as follows.

In many cases, the bulge departs markedly from the $r^{1/4}$ law, for spirals as well as S0's: see, *e.g.* Jensen and Thuan (1982), van der Kruit and Searle (1981a, 1982a; hereafter KSI, KSIII), Dettmar and Wielebinski (1986), Capaccioli *et al.* (1985, 1987), Michard (1985), and Simien and Michard (1985). This effect is noticed as an upward concavity in the bulge profile plotted in $r^{1/4}$ scale. Very near the center, the behavior of the bulge may be very different from a galaxy to another (Kormendy 1985). For very late-type galaxies, there is an additional source of uncertainty: a possible confusion between a very small bulge and a nucleus.

Several galaxies show evidence for important variations in the flattening of the bulge (Burkhead 1979; KSI–II). The outer isophotes may have axial ratios intermediate between those typical of disks and bulges. The hypothesis of elliptical isophotes is sometimes irrelevant, as box–and peanut–shaped contours have been observed in many spirals and lenticulars (Jarvis 1986). Pointed isophotes have been noticed even after removing the disk light (KSI–II).

The disks also exhibit a wide variety of characteristics. Many do not reach the center of the galaxy (de Vaucouleurs 1958; Freeman 1970: Type II profiles). In several S0's, the disk profile is nowhere exponential (Kormendy 1977). Edge–on disks show an outer cut–off at a few effective radii (KSI–IV), and this has also been identified in face–on and moderately inclined galaxies (van der Kruit 1988, and references therein). Cuts perpendicular to the major axis of edge–on disks are not consistent with the exponential model near the plane (Burstein 1979; KSI–IV).

The above list of problems and inadequacies, for both bulges and disks, appears impressive, and one may ask what is left in the reliability of the standard method. Fortunately, there are often some compensations between different factors, and significant characteristics can be highlighted (see Section IV). The uncertainty caused by the use of the standard model is difficult to estimate, partly because the errors have often been dominated by the effects of poor or incomplete data, and/or summary analysis.

D. Alternative Models

For edge–on disks, KSI–IV proposed the "sech^2" model (see also van der Kruit, this conference). Face–on projection is exponential, and the standard disk remains a good approximation at moderate inclinations. Lately, van der Kruit (1988) has considered an intermediate model.

For the bulges, no alternative to the $r^{1/4}$ law has yet emerged as a new standard. In the case of NGC 891, an edge–on Sb, Bahcall and Kylafis (1985) found that state–of–the–art data were unable to distinguish between the $r^{1/4}$ model and a thick disk with the same sech^2 law as the thin disk. This is a result with many consequences, but it has recently been challenged by Shaw and Gilmore (1988a, hereafter SG).

In their highly–detailed study of NGC 891 and NGC 4565, SG do not rule out exponential disks nor $r^{1/4}$ bulges, but test a wide variety of model combinations including, in addition, sech^2 components; thick disks as well as thin ones are considered. Although presented specifically for edge–on galaxies, their method is not restricted to these.

The nature of the thick disk as an intermediate component is still a matter of debate (Burstein 1979; KSI; Wyse and Gilmore 1988, and references therein).

III. Decomposition Techniques

A. Data Presentation

The form in which the surface brightness of a galaxy is available is of major importance for the decomposition technique. Long restricted to one or two profiles, the photometric data actually involved in the process are now often representing a much larger amount of information, up to the full two–dimensional array of pixel luminosities. Still, the bulk of the published results rely on limited data. Several cases deserve a few comments:

a) a profile along a single direction, *e.g.* the major axis, can bring out the main features and components of a galaxy and, in favourable cases, determine several parameters (in nearly face–on objects, for example); but it can also prove to be misleading. The same is true for the profile as a function of the equivalent radius, if the variation of axial ratio is not taken into account. Data on both major and minor axes allow a more detailed analysis, provided that the isophotes are reasonably elliptical. The projected profile (Watanabe *et al.* 1982) is independent of inclination and yields the generalized profile, a powerful tool, but the non–axisymmetric features should be first removed from the raw data.

b) azimuthal averages along ellipses with a fixed axial ratio (usually set by an outer isophote) enhance the S/N ratio; however, some information is lost in the inner regions if a rounder bulge dominates the luminosity there (Boroson 1981). The effect can be important with highly-inclined galaxies.

c) parameters of ellipses fitted to the isophotes, together with their brightness level, have been widely used. In the most favourable cases, they take into account most of the available photometric information. Sometimes, however, this representation is limited to a first-order approximation, due to the presence of, for example, perturbing features (as spiral arms), transitions between bulge- and disk-dominated regions (see below), or environmental effects.

d) isophote fitting with parametrized measurement of the departure from perfect ellipse shape probably represents the next-to-last step toward the full mapping of a galaxy. In early-type galaxies, Fourier coefficients of order 4 and up in azimuthal angle conveniently approximate the pointed isophotes thought to be caused by the presence of a flat component superposed on a spheroid (Carter 1978, 1987; Jedrzejewski 1987; Bender and Möllenhoff 1987; Michard and Simien 1988). An equivalent parametrization would be far more complicated for later-type galaxies with isophotes distorted by, *e.g.* spiral arms, although theoretically feasible (Kuhl and Giardina 1982). It should be stressed that, in many cases, the choice of a red or infrared color band can significantly reduce isophote distortions (Boroson *et al.* 1983).

e) the full array of pixel luminosities in a galaxy frame is sometimes considered as the ultimate arbiter in the model fitting (SG). This should increasingly be the case: the handling of such a large amount of data is no more a technical burden, and its freedom from model-dependent processes is often (however, not always) a major advantage.

B. Methods and Algorithms

Most of the classical methods for decomposing the light distribution in a galaxy rely on one or more of the following points: the fundamental difference between the brightness gradients of the two components, the difference in their intrinsic flattening, and the separation in the ranges where each one dominates the light.

Looking for a linear fit to a photometric profile within one of these ranges (in r scale for the disk and $r^{1/4}$ for the bulge), allows one to subtract the contribution of the corresponding component; then, determination of the other one is made by a second fit within its own range, in the appropriate scale. Kormendy (1977) presented this scheme as the first step of an iterative process. This allows one to compare the resultant profile with the data outside the fitting ranges as a test of overall quality-of-fit. The regions with perturbing features can easily be avoided.

Kormendy (1977) also proposed a slightly different method in which a standard, non–linear least squares fit is performed simultaneously on both fitting ranges, and solved directly for the r_e and μ_e parameters of bulge and disk components.

Both of these methods have been used by many other authors since then, sometimes with specific modifications (Boroson 1981; Burstein 1979; Simien and de Vaucouleurs 1983, 1986, hereafter SdV; K85). When a single profile is used, the coupling between the parameters makes the global fit little sensitive to large variations in their values. Profiles on both main axes impose many more constraints and significantly reduce the uncertainties.

In the method of KWO, there is no initial guess on the bulge and disk parameters, but a model grid in a three–dimensional space of global photometric parameters: diameter, mean surface brightness, and concentration index. The 10 best–approximating models are selected, and then tested for goodness–of–fit with the generalized profile, a process which ultimately defines the adopted model. This method has the advantage of being fully automatic and free of any subjective choice. Perturbing features, however, were not taken into account by KWO.

Schombert and Bothun (1987) used initial estimates of bulge and disk parameters within a model grid, and a χ^2 minimization routine to fit a single (and unspecified) photometric profile. Boundary values were set to avoid unrealistic solutions. They showed that the typical S/N ratio in the photometric data obtained at a small telescope with a CCD detector is not a major factor in limiting the accuracy of the decomposition; nor are the expected errors in the sky brightness determinations. They found that departures of the data from the standard model is the main point. They estimated at 2.0 the upper limit in the χ^2 beyond which the standard decomposition becomes meaningless. For "good" galaxies, by contrast, the accuracy in the bulge–to–disk ratio determined by their method is typically 20% (presumably, several other methods would give the same estimate).

For their model combinations, SG use both the sum of the square of the residuals and a reduced χ^2, calculated over the two-dimensional data, as estimators of the quality–of–fit. One of their main arguments is that these estimators are able to differentiate between the number of components required. Their expected analysis of many more galaxies may bring out common characteristics, which would be of major interest (Shaw and Gilmore 1988b, and this conference).

Kent (1986) presented a completely different approach: he made no assumption on the fitting laws for either component. He just assumed that each one is characterized by elliptical isophotes of constant, and essentially different, flattening. Then, an iterative process calculated the bulge and disk profiles. Not surprisingly, the former was found to be quite different from the $r^{1/4}$ model. This method presents many advantages, but also obvious limitations: a) it does not work with face–on galaxies or when the bulge and disk have roughly the

Table 2: A selection of photometric decomposition results.

Authors	Mags.	Types	N	Data	Outline of method
Freeman (1970)	*B*	L^-–Im	36	Equiv. pr.	fD, mag. subtract.
Yoshizawa, Wakamatsu (1975)	*B*	L^-–Sd	24	Equiv. pr.	fD, fB
Kormendy (1977)	*B*	L^-/L	8	Major axis	fD, fB, iteration (hereafter K–iter), stand. non–lin. least squares (ls)
Burstein (1979)	*B*	L^-–L^+	12	Major axis	K–iter; "Rx" meth.
Boroson (1981)	*B*	L^+–Sc	26	Ell. av. pr. azm. pr.	K–iter
Whitmore, Kirshner (1981)	*B*	L^-–Scd	26	Equiv. pr.	K–iter
Simien, de Vaucouleurs (1983; 1986: SdV)	*B*	L^-–Sd	32+66	Main axes	Modified K–iter
van der Kruit (1987: vdK)	"*J*"	L–Im	51	Ell. av. pr.	fD
Kodaira *et al.* (1987: KWO)	V	L^-–Im	167	Gen. pr.	Model grid
Kent (1985: K85)	*r*	L^-–Sc	105	Main axes	ls, K–iter
Kent (1986)	*r*	Sb–Sc	37	Main axes	No *a priori* fitting laws; iteration
Boroson *et al.* (1983)	*I*	L–Sc	62	Ell. av. pr. asm. pr.	K–iter, disk parameters only

Notations: *Profiles* (pr.): equiv. = equivalent; ell. av. = elliptically averaged; azm. = azimuthal; gen. = generalized. *Method*: fD = exponential fit to disk; fB = $r^{1/4}$ fit to bulge.

same flattening, b) it is sensitive to the presence of non–axisymmetric features and, c) its reliability depends on the determination of the bulge flattening.

To summarize, many different methods have been used to determine the standard model which best describes an observed luminosity distribution. Since many, sometimes fundamental departures have been pointed out, however, new methods are being tested which rely on different sets of *a priori* information. Still, a large number of galaxies are approximated adequately enough by the standard model, and a sample of results is discussed in the next Section.

IV. Results of the Standard Decomposition.

A. Selection of the Samples

Table 2 collects papers presenting photometric decompositions, and selected for their statistical importance. Almost all of them rely on the standard model. The samples are restricted to studies which attempted to fit both components, with only a few exceptions. These include the work by Freeman (1970), who fitted the disk only, but pioneered the systematic determination of parameters along the Hubble sequence; and the large samples of disk parameters by Boroson *et al.* (1983), and van der Kruit (1987, hereafter vdK).

SdV analysed 32 galaxies with published and unpublished data, and collected results on these and 66 additional objects from the five preceding entries in Table 2 (Yoshikawa and Wakamatsu 1975, Kormendy 1977, Boroson 1981, Burstein 1979, Whitmore and Kirshner 1981). The merged set was partially homogeneized. Below, these results will be compared with those of K85, KWO, and vdK.

B. The Fractional Luminosity of the Bulge

The trend in the mean magnitude difference Δm_I along the Hubble sequence is shown in Figure 1. The scatter in each individual distribution is large. The dispersion of determinations within a single stage is about 0.6 mag; it is thought to be due mainly to errors in the photometry and the decomposition and, to a lesser extent, to classification errors (about 0.7 T units for most well–determined types).

Broad agreement is noticed between SdV and K85, despite the waveband difference. Early–type galaxies from KWO are also in agreement, but large systematic discrepancies appear at stages later than Sbc.

C. Search for Biases

At each stage T, the residuals to the mean, $\Delta m_I - \overline{\Delta m_I}$, can be correlated with their equivalents in, for example, apparent and absolute magnitude, inclination, and dimension of the galaxies. SdV did not find significant correlations. Still, selection effects are present: a) the measurement of a very small bulge in a highly–inclined late–type galaxy is seldom possible, and attention has to be paid to a possible bias toward small values of Δm_I for T greater than 5 or 6; b) on the contrary, the early–type sample may be biased toward large Δm_I values, just because many lenticulars with very faint disks were probably misclassified as ellipticals and rejected from the selection. Recently, a more consistent attention to all early–type objects has fortunately begun to reduce this bias (Carter 1987). In Figure 1, the mean value of Δm_I for $T \leq 0$ is likely to be too high.

Correlation with color residuals reveals a loose effect (Whitmore 1984, SdV), the galaxies with brighter–than–average bulges tending to be redder than average, as expected.

D. Photometric Parameters of Bulge and Disk Components

Mean values of r_e and μ_e (corrected to face-on) *versus* T are displayed in Figures 2 and 3. All dimensions have been converted to the short extragalactic distance

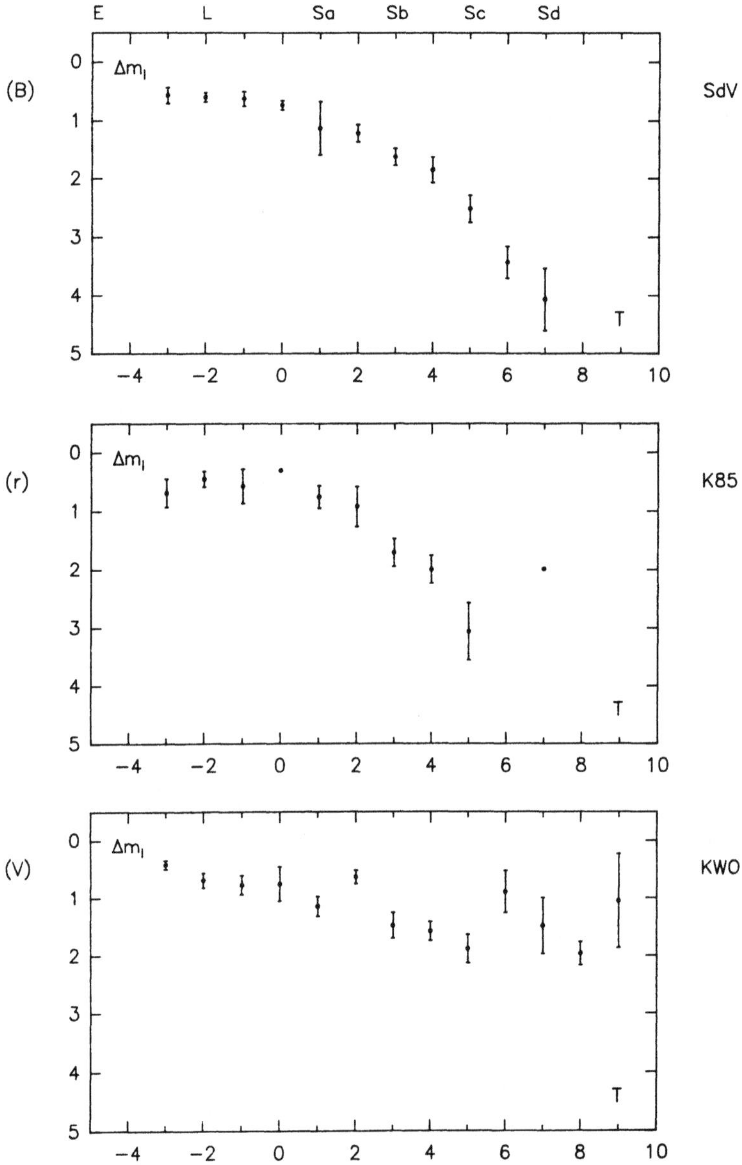

Figure 1: Mean value of fractional luminosity of the bulge, Δm_I, *versus* stage T. Ordinates are in B, r, and V mags, respectively, for the SdV, K85, and KWO samples. Notes for K85 data: T mostly from RC2; discrepant point is NGC 5996, for which T is very uncertain.

scale ($H_0 = 100$). It should be borne in mind that the different samples refer to the B, "J," V, and r bands, and that intercomparisons should remain cautious.

The bulge parameters show a very large scatter, which is not quite surprising. The effective radii $r_{e,I}$ (Figure 2) from SdV and K85 are in rough agreement from Sa to Scd. However, there is a systematic discrepancy for earlier types, which reflects selection effects or, more likely, differences in the decomposition techniques. For later types, the decrease in the bulge dimensions is more apparent with the SdV sample. By contrast, data from KWO show a surprising increase of $r_{e,I}$ with T.

The bulge effective brightness $m_{e,I}$ (Figure 3) becomes fainter along the Hubble sequence. This is much more pronounced for the KWO sample and it may be related, in this case, to the increase in $r_{e,I}$. An unexplained discrepancy of the order of one magnitude appears between the SdV and K85 data after the latter have been tentatively converted to B magnitudes ($B \sim r + 1.3$).

For the disk parameters the B–band results of SdV have been complemented by those of vdK obtained in the nearby "J" band (van der Kruit 1979), after correction by the mean relation $B = J + 0.15$. The scatter in the individual $r_{e,II}$ and $\mu_{e,II}$ distributions is much smaller than for the corresponding bulge parameters, and the agreement between the different samples is better. This is an obvious consequence of a) the shallower luminosity gradient and, b) the usually wider range in which the disk can be fitted.

The K85 and KWO samples show no definite trend in $r_{e,II}$ along the Hubble sequence (Figure 2). In the SdV results, on the contrary, a small systematic variation is noticed. A significant correlation exits between the $(r_{e,II} - \overline{r_{e,II}})$ and $(M_t - \overline{M_t})$ residuals at each T, which reflects the well-known diameter–luminosity relation (Heidmann 1969, Paturel 1979), restricted to the disks. Presumably, corrections according to this relation would partially compensate for selection effects, and would yield a better–defined mean $r_{e,II}$ depending on M_t.

The distribution of $\mu_{e,II}$ (Figure 3) shows overall agreement between the samples, after allowing for color differences ($B - V \sim 0.7$, $B - r \sim 0.9$). The scatter is lower than in $\mu_{e,I}$. No systematic variation is apparent as a function of T, and the mean value ($\mu_B \sim 23.5$) is in close agreement with the Freeman (1970) rule.

Absolute magnitudes show that bulges are, on average, brightest at T $\sim$ 0, and then decrease monotonically along the Hubble sequence (Meisel and Ostriker 1984; SdV). The maximum brightness of the disks occurs at T $\sim$ 3. There is evidence for a mean decrease in the brightness of both components at T $<$ 0, but this may be influenced by selection effects.

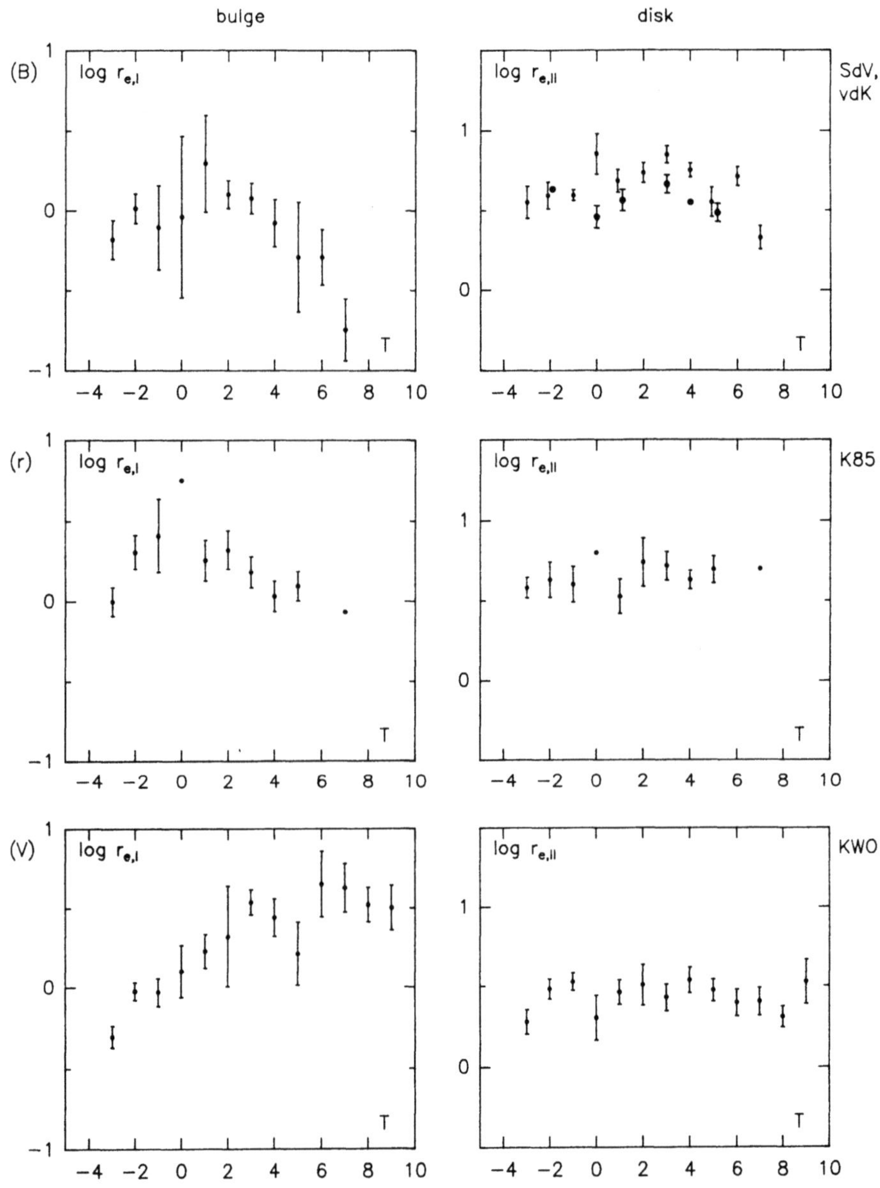

Figure 2: Mean effective radii, log r_e (kpc), *versus* T, for bulges and disks. For the disks, *B*–band data (SdV) and "*J*"–band data (vdK: larger dots) are displayed on the same graph.

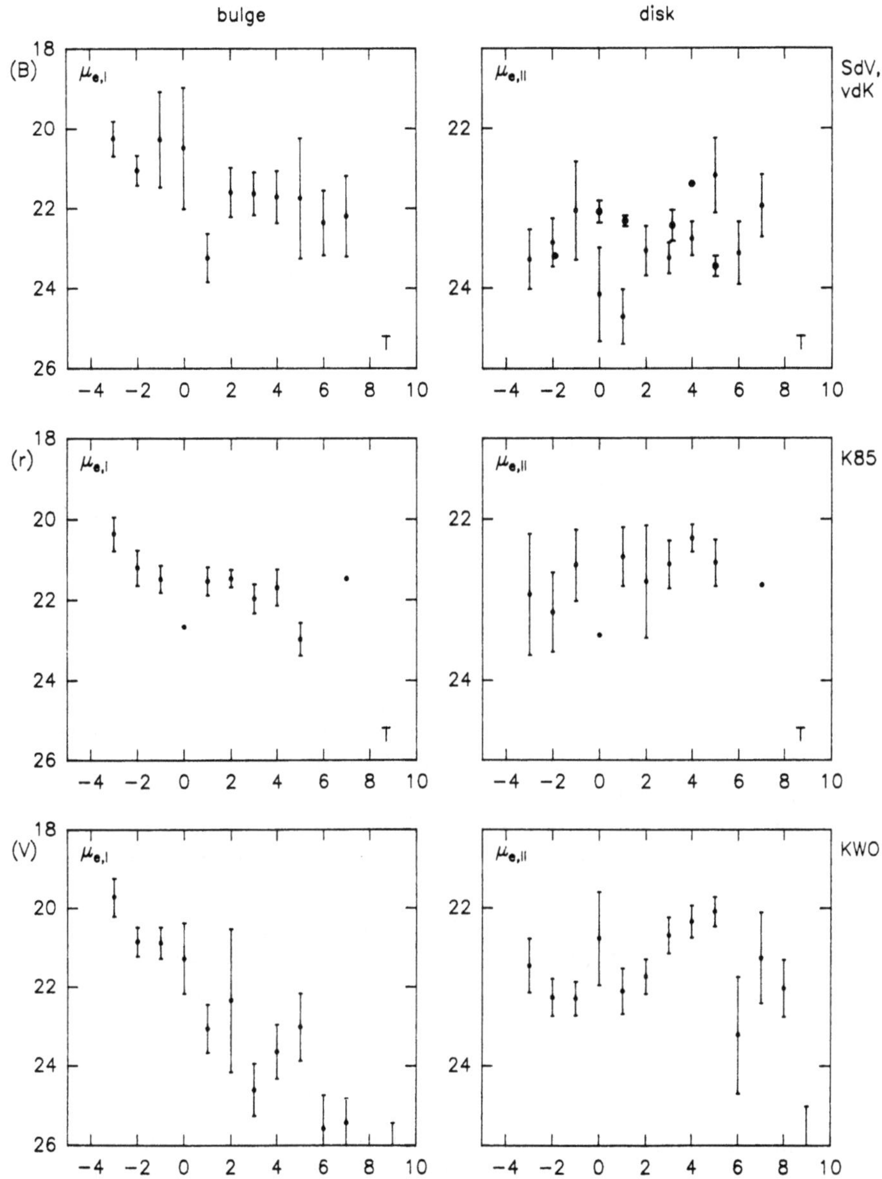

Figure 3: Mean effective luminosities μ_e, corrected to face–on values (in B, r, and V mags), *versus* T for bulges and disks. For the disks, "J" mags from vdK (larger dots) have been corrected to B ($B \sim J + 0.15$).

V. Applications

A. The Revised Morphological Type T

The smooth variation of $\overline{m_I}$(T) confirms quantitatively the statistical validity of the T scale in RC2 (de Vaucouleurs *et al.* 1976). But Δm_I alone does not allow a good classification, mostly because of the inaccuracy in its determination (concentration indices, far easier to determine, are more appropriate).

B. Formation and Evolution of Galaxies

Statistics on bulge and disk parameters provide useful data for topics in galaxy formation (Barnes and White 1984; other references in K85) and galaxy evolution (Guiderdoni 1987; Ferrini and Galli 1988). A single example: it has been suggested that S0's form a gas–poor sequence parallel to the gas–rich sequence of spirals (van der Bergh 1976), but this is contradicted by the photometric decompositions, as no lenticular shows a bulge–to–disk ratio similar to those of late–type spirals. This gives support to the "intrinsic" theory of S0 formation.

C. Comparison between Bulges and Elliptical Galaxies

This can be made through the following two relations:

a) the Faber and Jackson (1976) relation, $L \propto \sigma^n$; This was investigated by Whitmore *et al.* (1979), Whitmore and Kirshner (1981), and Kormendy and Illingworth (1983), with the result that unbarred bulges of lenticulars and spirals do not differ significantly from ellipticals in terms of central dynamics and stellar content. De Vaucouleurs and Olson (1982) analysed the above relation as a distance indicator for early–type galaxies; their results could be slightly corrected for S0's by taking into account the disk luminosity.

b) the relation between log $r_{e,I}$ and $\mu_{e,I}$ (Kormendy 1982, and references therein); SdV and K85 confirmed that E's and S0 bulges share the same regression line, and that bulges of spirals are significantly shifted. KWO also addressed this question.

D. Light Mixing in Absorption Spectroscopy

Once limited to its innermost part, long–slit absorption spectra of the bulge now reach regions where mixing with disk light can be significant (see *e.g.* Illingworth and Schechter 1982, McElroy 1983, and Whitmore *et al.* 1984). The apparent rotation velocity of the bulge, for example, has to be corrected for the effect of line–of–sight integration through the disk, and this requires knowledge of the relative luminosity of both components. Another case of mixing: spectra of the nucleus contaminated by light from the bulge (Whitmore 1980).

E. Flat Rotation Curves and Dark Matter

Much work has been devoted to infer halo characteristics from the rotation of spirals (see Freeman 1987 for a review). One of the classical procedures consists of: a) calculating the kinematical contribution predicted by both visible components [Monnet and Simien (1977) dealt with the standard case], b) adjusting their amplitude by assigning an M/L ratio to each one, and then, c) looking at what remains after removing the observed rotation.

Point b) above can cause large uncertainty, as the constraints on the M/L ratios are rather loose. The classical view of a dominant contribution of the visible matter in the inner regions has even been challenged (Burstein and Rubin 1985). Nevertheless, several works made use of the bulge and disk rotation curves. Kent (1986) applied his special decomposition technique to 37 Sb and Sc galaxies. Athanassoula *et al.* (1987) studied 48 spirals from Sa to Sd; their bulge determinations, needed for only a part of the sample, are approximate, but they found this not to be a critical point for the accuracy of their results.

F. Dynamics of Spheroidal Systems

Detailed dynamical models of the bulge component require knowledge of the mass distribution, as can be derived from the photometric decomposition (Simien *et al.* 1979, Bacon 1985, Jarvis and Freeman 1985).

Stellar dynamics may also provide another approach to the (bulge + disk + halo) models, for galaxies with significant bulge, with both stellar rotation and velocity dispersion measurements (Whitmore and Kirshner 1981, Bacon and Martinet 1985).

VI. Conclusion

The standard bulge/disk decomposition has yielded a fair amount of results (a couple of hundred galaxies), and has brought out interesting characteristics. But the scatter in these results is very large and, so far, no clear evaluation has been made of the following three contributions: a) the uncertainties in data handling and model fitting, b) the limitations of the model itself, and c) the intrinsic (or cosmic) scatter. The answer may eventually come from more elaborate models, if a higher degree of *a priori* information can be established statistically. Still, more has yet to be learned from the standard picture; but this will require great care in model fitting and rigorous attention to all peculiarities of each galaxy to be analysed.

Statistical or individual, applications are many. Prior to decomposition of a galaxy, selection of the model depends on the ultimate purpose; the following two examples are extreme cases: a) if an approximate decomposition is needed, as for the statistics on some integrated parameters, then a simple model relying on limited *a priori* information is suitable; b) if high accuracy is wanted (and allowed by the data), previous knowledge may be insufficient, and trial–and–error is likely to be the best tool; then, quality–of–fit criteria must not neglect a basic requirement: the decomposition has to be physically meaningful.

Generalization of two dimensional data should contribute to future progress. For large samples, one can imagine expert systems able to: a) detect (almost) all the peculiarities of a galaxy, b) make the right choice in a library of models and methods and, c) eliminate unsatisfactory solutions — a critical point, as misjudgement in setting the rules may lead to biased results.

References

Athanassoula, E., Bosma, A., and Papaioannou, S. 1987, *Astron. Astrophys.* **179**, 23.

Baade, W. 1944, *Astrophys. J.* **100**, 137.

Bacon, R. 1985, *Astron. Astrophys. (Lett.)* **147**, L16.

Bacon, R. and Martinet, L. 1985, in *I. A. U. Symposium 117, Dark Matter in the Universe*, ed. J. Kormendy and G. R. Knapp (Dordrecht: Reidel), p. 134.

Bahcall, J. N. and Kylafis, N. D. 1985, *Astrophys. J.* **288**, 252.

Barnes, J. and White, S. D. M. 1984, *Mon. Not. Roy. Astron. Soc.* **211**, 753.

Bender, R. and Möllenhoff, C. 1987, *Astron. Astrophys.* **177**, 71.

van den Bergh, S. 1976, *Astrophys. J.* **206**, 883.

Boroson, T. A. 1981, *Astrophys. J. Suppl.* **46**, 177.

Boroson, T. A., Strom, K. M., and Strom, S. E. 1983, *Astrophys. J.* **274**, 39.

Burkhead, M. S. 1979, in *Photometry, Kinematics and Dynamics of Galaxies*, ed. D. S. Evans (Austin: Astron. Dept., Univ. of Texas), p. 143.

Burstein, D. 1979, *Astrophys. J.* **234**, 435.

Burstein, D. and Rubin, V. C. 1985, *Astrophys. J.* **297**, 423.

Buta, R. J. 1984, Ph. D. thesis, University of Texas (Austin)

Capaccioli, M., Held, E. V., and Nieto, J.-L. 1985, *Lect. Notes Phys.* **232**, 265.

Capaccioli, M., Held, E. V., and Nieto, J.-L. 1987, *Astron. J.* **94**, 1519.

Carter, D. 1978, *Mon. Not. Roy. Astron. Soc.* **182**, 797.

Carter, D. 1987, *Astrophys. J.* **312**, 514.

Dettmar, R.-J. and Wielebinski, R. 1986, *Astron. Astrophys. (Lett.)* **167**, L21.

Duval, M. F. and Athanassoula, E. 1983, *Astron. Astrophys.* **121**, 297.

Duval, M. F. and Monnet, G. 1985, *Astron. Astrophys. Suppl.* **61**, 141.

Faber, S. M. and Jackson, R. E. 1976, *Astrophys. J.* **204**, 668.

Ferrini, F. and Galli, D. 1988, *Astron. Astrophys.* **195**, 27.

Freeman, K. C. 1970, *Astrophys. J.* **160**, 811.

Freeman, K. C. 1987, in *I. A. U. Symposium 117, Dark Matter in the Universe*, ed. J. Kormendy and G. R. Knapp (Dordrecht: Reidel), p. 119.

Guiderdoni, B. 1987, *Astron. Astrophys.* **172**, 27.

Heidmann, J. 1969, *Astrophys. Letters* **3**, 19.

van Houten, C. J. 1961, *Bull. Astron. Inst. Netherlands*, **16**, 1.
Hubble, E. P. 1926, *Astrophys. J.* **64**, 321.
Hubble, E. P. 1930, *Astrophys. J.* **71**, 231.
Illingworth, G. and Schechter, P. L. 1983, *Astrophys. J.* **256**, 481.
Jarvis, B. J. 1986, *Astron. J.* **91**, 65.
Jarvis, B. J. and Freeman, K. C. 1985, *Astrophys. J.* **295**, 324.
Jedrzejewski, R. I. 1987, *Mon. Not. Roy. Astron. Soc.* **226**, 747.
Jensen, E. B. and Thuan, T. X. 1982, *Astrophys. J. Suppl.* **50**, 421.
Kent, S. M. 1985, *Astrophys. J. Suppl.* **59**, 115 (K85).
Kent, S. M. 1986, *Astrophys. J.* **91**, 1301.
Kodaira, K., Watanabe, M., and Okamura, S. 1987, *Astrophys. J.* **62**, 703 (KWO).
Kormendy, J. 1977, *Astrophys. J.* **217**, 406.
Kormendy, J. 1982, in *Morphology and Dynamics of Galaxies*, ed. L. Martinet and M. Mayor (Sauverny: Geneva Observatory), p. 113.
Kormendy, J. 1985, *Astrophys. J.* **292**, L9.
Kormendy, J. and Illingworth, G. 1983, *Astrophys. J.* **265**, 632.
van der Kruit, P. C. 1979, *Astron. Astrophys. Suppl.* **38**, 15.
van der Kruit, P. C. 1987, *Astron. Astrophys.* **173**, 59 (vdK).
van der Kruit, P. C. 1988, *Astron. Astrophys.* **192**, 117.
van der Kruit, P. C. and Searle, L. 1981, *Astron. Astrophys.* **95**, 105 (KSI).
van der Kruit, P. C. and Searle, L. 1981, *Astron. Astrophys.* **95**, 116 (KSII).
van der Kruit, P. C. and Searle, L. 1982, *Astron. Astrophys.* **110**, 61 (KSIII).
van der Kruit, P. C. and Searle, L. 1982, *Astron. Astrophys.* **110**, 79 (KSIV).
Kuhl, F. P. and Giardina, C. R. 1982, *Comput. Graph. and Image Proc.* **18**, 236.
McElroy, D. B. 1983, *Astrophys. J.* **270**, 485.
Meisel, A. and Ostriker, J. P. 1984, *Astron. J.* **89**, 1451.
Michard, R. 1985, *Astron. Astrophys. Suppl.* **59**, 205.
Michard, R. and Simien, F. 1988, *Astron. Astrophys. Suppl.* **74**, 25.
Monnet, G. and Simien, F. 1977, *Astron. Astrophys.* **56**, 173.
Nieto, J.-L. 1988, in *Proceedings* of the Second Extragalactic Regional Meeting (held in Cordoba, November 30 - December 5, 1987).
Patterson, F. S. 1940, *Harvard Bull.* No. **914**, 9.
Paturel, G. 1979, *Astron. Astrophys.* **71**, 19.
Reynolds, R. H. 1913, *Mon. Not. Roy. Astron. Soc.* **74**, 132.
Schombert, J. M. and Bothun, G. D. 1987, *Astron. J.* **93**, 60.
Schweizer, F. 1976, *Astrophys. J. Suppl.* **31**, 313.
Shaw, M. A. and Gilmore, G. F. 1988a, in press (SG).
Shaw, M. A. and Gilmore, G. F. 1988b, in preparation.
Simien, F. and de Vaucouleurs, G. 1983, in *I. A. U. Symposium 100, Internal Kinematics and Dynamics of Galaxies*, ed. E. Athanassoula (Dordrecht: Reidel), p. 375.
Simien, F. and de Vaucouleurs, G. 1986, *Astrophys. J.* **302**, 564 (SdV).

Simien, F. and Michard, R. 1985, *Lect. Notes Phys.* **232**, 345.
Simien, F., Pellet, A., and Monnet, G. 1979, *Astron. Astrophys.* **72**, 12.
de Vaucouleurs, G. 1948, *Ann. d'Astrophys.* **11**, 247.
de Vaucouleurs, G. 1953, *Mon. Not. Roy. Astron. Soc.* **113**, 134.
de Vaucouleurs, G. 1956, *Occasional Notes Roy. Astron. Soc.* **3**, 129.
de Vaucouleurs, G. 1958, *Astrophys. J.* **128**, 465.
de Vaucouleurs, G. 1959, in *Handbuch der Physik* **53**, ed. S. Flügge (Berlin: Spinger–Verlag), p. 275.
de Vaucouleurs, G., de Vaucouleurs, A., and Corwin, H. G. 1976, *Second Reference Catalogue of Bright Galaxies* (Austin: Univ. of Texas Press) (RC2).
de Vaucouleurs, G. and Olson, D. W. 1982, *Astrophys. J.* **256**, 346.
Watanabe, M., Kodaira, K., and Okamura, S. 1982, *Astrophys. J. Suppl.* **50**, 1.
Whitmore, B. C. 1980, *Astrophys. J.* **242**, 53.
Whitmore, B. C. and Kirshner, R. P. 1981, *Astrophys. J.* **250**, 43.
Whitmore, B. C., Kirshner, R. P. and Schechter, P. L. 1979, *Astrophys. J.* **234**, 68.
Whitmore, B. C., Rubin, V. C., and Ford, W. K. 1984, *Astrophys. J.* **287**, 66.
Whitmore, B. C. 1984, *Astrophys. J.* **278**, 61.
Wyse, R. F. G. and Gilmore, G. 1988, *Astron. J.* **95**, 1404.
Yoshizawa, M. and Wakamatsu, K. 1975, *Astron. Astrophys.* **44**, 363.

Discussion

B. C. Whitmore: Having played this game of decomposition of the luminosity profile, I can attest that it can be a frustrating experience, due to many of the problems you mentioned. With this in mind, I should like to suggest that another type of observation that can give information about the bulge–to–disk ratio (B/D) (besides the luminosity profile and the two–dimensional shape) is the color of a galaxy, since there is a good correlation between B/D and color. For example, $B - V$ for a bulge is ≈ 0.9, while $B - V \approx 0.5$ for a disk. If $B - V$ (total) $= 0.7$, this immediately suggests B/D ≈ 1. A more sophisticated treatment would measure B/D (bulge) and B/D (disk) for each individual galaxy by using regions dominated by the bulge and disk.

Simien: Such a color method would probably not work for early– or late–type galaxies.

Whitmore: That is certainly true, but still leaves a lot of galaxies. I might note that the other two methods also only work for certain galaxies (the luminosity profiles method does not work well for bulge dominated; two–dimensional shape does not work at all for face–on galaxies). I think the best approach will be to use all three methods simultaneously.

S. Okamura: I cannot tell easily whether the discrepancy of bulge magnitude between Kodaira *et al.* and other two studies is due to numerous differences in samples, color bands, and data reduction procedures; or to some artifact such as systematic errors. The tendency we see in our data is that brightness scale becomes fainter while length scale becomes larger in the bulges of later–type galaxies. This implies that bulges of very late–type galaxies are like icebergs in the sea. We can see only a small fraction of it. When we compute the *total* magnitude of such bulges using the parameters, we get rather bright magnitudes. However, in the case of galaxies, we are not sure how much matter actually lies beneath sea level. There is also the possibility that this tendency is to some extent an artifact due to the lack of spatial resolution of data in the central region. It is true that bulge parameters are much more sensitive to spatial resolution than disk parameters. However, we believe that the presence of such a tendency, apart from the range of variation, is real, because the parameter correlation diagrams (log r_0 *versus* μ_0) of all three studies show the same tendency.

R. J. Buta: When dealing with decompositions, one must be careful not to try to force–fit objects which are clearly outside the scope of applicability of the usual fitting functions. For example, in the case of NGC 1433, there might be a temptation to obtain the disk scale length by drawing a straight line beneath the bumps caused by the inner and outer rings. However, the rings are probably not just enhancements within an exponential disk, but represent material which has been gathered at the expense of another region, such as co–rotation. In barred galaxies, the disks are probably so restructured that decomposition techniques applied to them will be misleading.

Simien: I agree. But lacking proper modelling, I consider this kind of guided fit as likely to provide a closer approximation to the old disk than the blind fit. I admit that the process is crude, and I just hope that the difference is statistically significant (and in the right direction!).

The Kinematics of Early Type Galaxies

Roger L. Davies

Kitt Peak National Observatory, Tucson, Arizona, U. S. A.

Abstract

Recent observations of the kinematics of the central regions of M 31, M 32 and NGC 4594 at high spatial resolution have revealed steep central velocity gradients and high velocity dispersions that have been interpreted as indicating the presence of large central masses. Axisymmetric, constant mass–to–light ratio models can account for the steep rise in velocity dispersion found in these galaxies if the stellar orbits become increasingly radial towards the center, but cannot simultaneously reproduce the steep velocity gradients observed. When central point masses of $10^{6-9} M_{\odot}$ are added to these models, both the rotation and dispersion profile can be fit simultaneously. Alternative explainations for the observed behavior need to be investigated. For example, in the case of M 31, a nuclear bar viewed end–on has been suggested to account for the observations. In addition, we need to understand how the presence of a black hole would effect the distribution of the orbits of the stars, and what observable effects might be generated.

A substantial fraction of ellipticals appear to contain extended central stellar components that have kinematics independent from those of the bulk of the galaxy. Recent data indicate that 10–30% of ellipticals have such central components. The kinematics of the interstellar medium, when one is present in ellipticals, is also frequently decoupled from the kinematics of the stars. These phenomena appear to be direct evidence that the accretion of other galaxies was a common occurance during the evolution of ellipticals.

Preliminary results of the application of Binney's statistical test to determine the true shapes of elliptical galaxies indicate that a typical galaxy is between oblate and maximally triaxial. In triaxial galaxies stars populate the box orbits which can transport material from the outer parts of galaxies to the center. It now seems likely that typical elliptical galaxies possess the ingredients necessary to produce an active nuclues once a source of fuel is encountered, namely a central engine and a mechanism for transporting material from the outside into the center of the galaxy.

The modified Faber–Jackson relation, which is now well established in elliptical galaxies, appears to apply equally well to the bulge components of early type disk galaxies in the Virgo and Coma clusters. This suggests a considerable commonality in the formation process of ellipticals and bulges. In addition, if this relation is found to hold more universally, it can be used to improve the precision of relative distance estimates to aggregates of early type galaxies, and removes the uncertainties that arise from ambiguities in the Hubble classification of distant galaxies.

1. Introduction

Modern observations of the kinematics of elliptical and lenticular galaxies began in the mid–to–late 1970s when it became apparent that the rotation velocities of the most luminous elliptical galaxies provide insufficient support to account for their flattened figures (Bertola and Capaccioli 1975, Illingworth 1977). It now appears that the flattening of the figures of ellipticals results from an anisotropic distribution of random velocities rather than from rotation, so that elliptical galaxies do not have to be oblate spheroids, but could have ellipsoidal figures (Binney 1978).

This picture was augmented in the early '80s by studies of low luminosity ellipticals and of the bulges of spiral galaxies. These systems, which have photometric properties that form a continuum with those of the giant ellipticals, have rotation velocities that are consistent with them being oblate figures flattened by rotation (Kormendy and Illingworth 1982, Davies *et al.* 1983). Statistically, there is a relation between luminosity and the degree of support provided by rotation rather than random motions. Early type systems with low luminosity (ellipticals and bulges) appear to be flattened by rotation and are consistent with having oblate figures; giant ellipticals tend to be supported by anisotropic dispersions, although they exhibit a large range in v/σ.

This situation is summarized in Figure 1 which shows the distribution of galaxies of different luminosities and morphological types in the v/σ *versus* ϵ diagram. Best estimates of the shape of the bulge of The Galaxy and the bulge of M 31 (Lindblad 1956, Stark 1977, Gerhard and Vietri 1986) indicate that spiral bulges may also be triaxial.

In this review, I will emphasize recent developments in this field; reviews of earlier work can be found in Davies (1987) and Binney (1982). In the following, I discuss (1) the existance of kinematic subsystems in ellipticals, (2) the preliminary results of the application of statistical tests to indicate whether or not

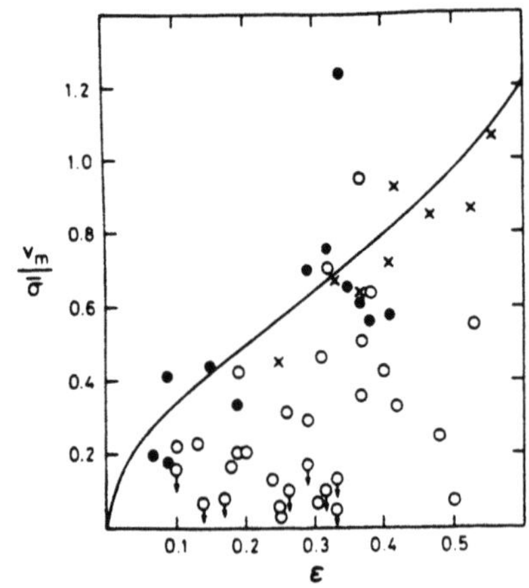

Figure 1: The peak rotation velocity divided by velocity dispersion plotted against ellipticity, for a sample of luminous ellipticals (open circles), low luminosity ellipticals ($M_B > -20.5, H_0 = 50\,\mathrm{km}\,sec^{-1}$ Mpc^{-1}) and the bulges of spirals (crosses). The mean line for oblate isotropic galaxies galaxies is shown.

ellipticals are triaxial, and (3) the modified Faber–Jackson relation for elliptical galaxies and, in particular, for early type disk galaxies.

2. Kinematic Subsystems

a. The Nuclear Regions of Nearby Bulges

In recent studies of the kinematics of the Local Group galaxies M 31 and M 32, Kormendy (1988a) and Dressler and Richstone (1988) have drawn attention to the sharp increase in the stellar velocity dispersion in the center of these galaxies and the steep central gradient in the rotation curve. By subtracting the light of the bulge, Kormendy determined a peak rotation velocity for the nucleus of 149 km sec^{-1} and a central velocity dispersion of 245 km sec^{-1}, rising from 145 km sec^{-1} in the bulge. Kormendy's data are reproduced in Figure 2. Both authors were able to account for the velocity dispersion profiles in axisymmetric models of constant mass–to–light ratio by including more radial orbits in the central regions. These radial orbits have low angular momentum making it impossible to account simultaneously for the rise in velocity dispersion and the steep gradient in velocity. Both papers demonstrated that the addition of a central point mass of zero luminosity enabled the observed central kinematics to be reproduced; masses in the range 10^{6-7} and $10^{7-8} M_\odot$ were required for M 32 and M 31, respectively. The possibility that a dramatic change in the character of the stellar population, that would generate a large number of high

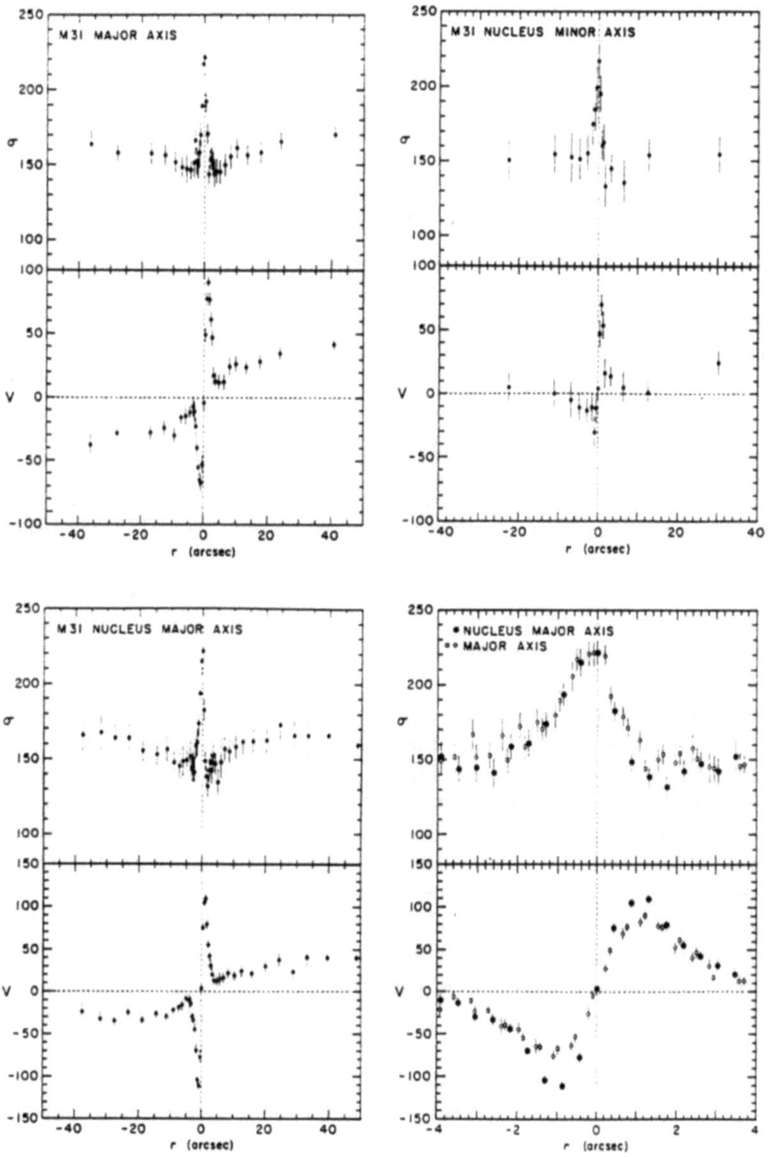

Figure 2: Reproduced from Kormendy (1988a) with permission. The rotation and velocity dispersion profiles for the center of M 31. The central points are spaced at intervals of 0.44 arc-seconds.

M/L stellar remnants in the nuclei, seemed unlikely to these authors because of the similarity of the spectral energy distributions of the nuclei and bulges. They prefered the hypothesis that both galaxies possess a central black hole.

By subtracting the light of the bulge from that of the nucleus of M 31, Kormendy determined the velocity dispersion of the "outer nucleus" to be $100\,\mathrm{km\,sec^{-1}}$, which is lower than that of the bulge at the same radius (2 arc–seconds = 6 pc); however, this result was not confirmed by Dressler and Richstone. The high v/σ found by Kormendy for the outer nucleus led him to suggest that the nuclear component might be a disk. Gerhard (1988) has investigated this possibility, suggesting that such a disk would form by dissipation into the plane of the inner bulge. The results of the Stratoscope II experiment (Light, Danielson and Schwarzschild 1974) indicate that the nucleus is misaligned with the inner bulge by 10 degrees. A similar misalignment between the bulge and disk of M 31 has been known for some time (Lindblad 1956). This misalignment has been interpreted as arising from the projection of the triaxial figure of the bulge of M 31 onto the sky, rather than as a physical misalignment which would not be expected to persist. Gerhard assumed that these three components share a common equatorial plane so that the bulge is prolate–triaxial (Stark 1977) and the nuclear disk is a flat bar viewed end–on. In this picture, the large observed rotation velocities arise as a result of large non–circular streaming velocities in the bar. This work indicates that a black hole may not be required and demonstrates that a nuclear bar viewed end–on provides an alternative interpretation of the M 31 data.

Kormendy (1988b) has made high spatial resolution measurements of the central kinematics of NGC 4594, the Sombrero galaxy, an Sa with one of the most luminous bulge components known. He finds the kinematics to be similar to those in M 31, exhibiting a steep central velocity gradient and a peaked central velocity dispersion. In addition, a clearer signature of a nuclear disk of stars is present as the velocity dispersion dips to well below that of the bulge at radii of 3–6 arc–seconds ($\sigma = 180\,\mathrm{km\,sec^{-1}}$ compared to $240\,\mathrm{km\,sec^{-1}}$ in the bulge). These data are interpreted as showing that the V–band mass–to–light ratio increases by a factor of 10 at radii less than 1 arc–second, suggesting that a dark central mass of order $10^9 M_\odot$ is present.

The analysis used to infer the presence of a large central mass relies on the fact that bulges exhibit substantial rotation velocities that limit the degree of radial anisotropy that can be invoked to account for the rising velocity dispersion. There is no reason to suspect that such mass concentrations might not be present in the slow rotating giant ellipticals. If early type systems frequently possess large central masses, postulated to be black holes, there are important implications for models of galaxy formation and the onset of activity in galaxies. These results need to be scutinized carefully and alternative explainations sought. For example, we need to understand how the presence of a black hole would cause the distribution of orbits of the stars to be modified and what observable effects that evolution might generate (see Norman, May, and van

Albada 1985).

b. Extended Stellar Sub–Structures

NGC 5813 was the first elliptical galaxy for which a central component with independent kinematics was discovered. Efstathiou, Ellis, and Carter (1982) carried out photographic photometry and longslit spectroscopy on this galaxy, and discovered a central component extending to radii of 6 – 7 arc–seconds superimposed on a normal elliptical galaxy luminosity profile. The velocity dispersion of the central component was low, around 210 km sec^{-1}, and its rotation velocity high, around 90 km sec^{-1}, whereas the bulk of the galaxy exhibited slow rotation, 20 km sec^{-1}, and a higher velocity dispersion, 250 km sec^{-1}. Kormendy (1984) pointed out that this "core–within–a–core" structure and unusual kinematics would be expected if the central component were the tidal remnant of a dense, low luminosity elliptical that had been captured by a more luminous galaxy, and therefor suggested a merger origin for this system.

Until 1987, NGC 5813 was the only elliptical galaxy known to have this two component stellar kinematic structure. The recent advent of high signal–to–noise, high spatial resolution kinematic mapping of ellipticals has resulted in many galaxies with similar features being discovered. Franx and Illingworth (1988) discovered that approximately one quarter of the stars in the core of IC 1459 out to a radius of 5 arc-seconds rotate in the opposite sense to those in the bulk of the galaxy. Again the core stars have a much greater rotation velocity than the main part of the galaxy. The higher v/σ of the central component led them to suggest that it could be a disk. Franx and Illingworth used the cross correlation method to determine the central absorption line profiles and discovered asymmetries of opposite senses on opposite sides of the center of the galaxy that are typical of those produced along lines of sight through a disk embedded in a hot component. They estimated that the mass of the counter–rotating component is 10^{10} $M_{\odot}$, and suggested that it was formed in an accretion event, either as a tidally disrupted remnant or as a starburst. In addition, they found that the ionized gas that is coextensive to the counter–rotating core stars, has the same sense of rotation as the bulk of the galaxy, perhaps indicating that the galaxy has experienced more than one accretion event.

Jedrzejewski and Schechter (1988) have mapped a sample of 14 E2 galaxies chosen to have little if any isophote twisting and no morphological peculiarities; of these, 3 have kinematically distinct cores. NGC 4494 and NGC 7626 exhibit cores that rotate faster than the bulk of the galaxy and, in NGC 3608, the core is seen to counter–rotate. Davies and Birkinshaw (1988) mapped 14 galaxies at 4 or more position angles; while the data were not of uniform quality, one well studied galaxy, NGC 4472, was found have an independent central component, consistent with counter–rotation. Bender (1988) finds 4 galaxies with kinematically distinct stellar cores out of 7 galaxies he studied; he confirms the fast rotation of the core of NGC 4494 and finds a similar fast rotating core in NGC 4406.

He also reports the discovery of two counter–rotationg cores in NGC 4365 and NGC 5322. He notes that in NGC 5322, where the counter–rotation has a high amplitude (80 km sec^{-1}), the velocity dispersion falls by 10–20% in the counter–rotating region. This is not seen in IC 1459, but may be present in the dispersion profiles of NGC 4494 and NGC 7626. In an unpublished study of 28 ellipticals, Tonry has found one–third of the galaxies to show features in their rotation curves that could be interpreted as evidence for distinct kinematic components. In Figure 3, I have collected the data on counter–rotating cores discussed above to illustrate the amplitude and scale of the phenomenon.

The galaxies in which these features have been found are usually photometrically and morphologically normal. In most cases, they were chosen for detailed study to carry out Binney's test for triaxiality (see section 3). The most dramatic examples of central kinematic substructures in ellipticals are surely evidence for past accretion events. These galaxies show the signature of a past merger in their internal kinematics long after they appear to be relatively normal galaxies in other respects. Approximately 10–30% of ellipticals appear to exhibit central regions with independent kinematics. If all of them can be attributed to past accretion events, then they are direct evidence that mergers were an important physical process in the evolution of a large fraction of early type galaxies. It would be interesting to investigate whether similar structures in rotation curves might arise in models of formation by a single collapse, and whether the conditions required to set up such structures are realistic.

c. Counter–Rotating Gas

The rotation axis of the neutral or ionized gas in several elliptical galaxies has been shown to be different from the rotation axis of the stars, *e.g.* NGC 1052: Davies and Illingworth (1986), van Gorkom *et al.* 1986, and NGC 5128: Wilkinson *et al.* (1986). These misalignments have been used to suggest that the gas was accreted after the galaxy formed. Gas disks with rotation axes anti–parallel to the stellar component have been found in both elliptical galaxies (NGC 7097: Caldwell, Kirshner, and Richstone 1986, NGC 5898: Bertola and Bettoni 1988) and a lenticular galaxy (NGC 4546: Galletta 1987). Bertola, Buson, and Zeilinger (1988) have kinematically mapped the ionised gas and stars in four ellipticals with major axis dust lanes, and find two of the gas disks to rotate in the same sense as the stars, and two with rotation axes that are antiparallel. In all these cases, the detailed three dimensional geometry remains uncertain because of the degeneracy introduced by projection onto the sky. However, it is clear that the decoupling of the kinematics of the interstellar medium and stellar components of early type galaxies is common; this strengthens the hypothesis that the interstellar medium in these galaxies is often accreted.

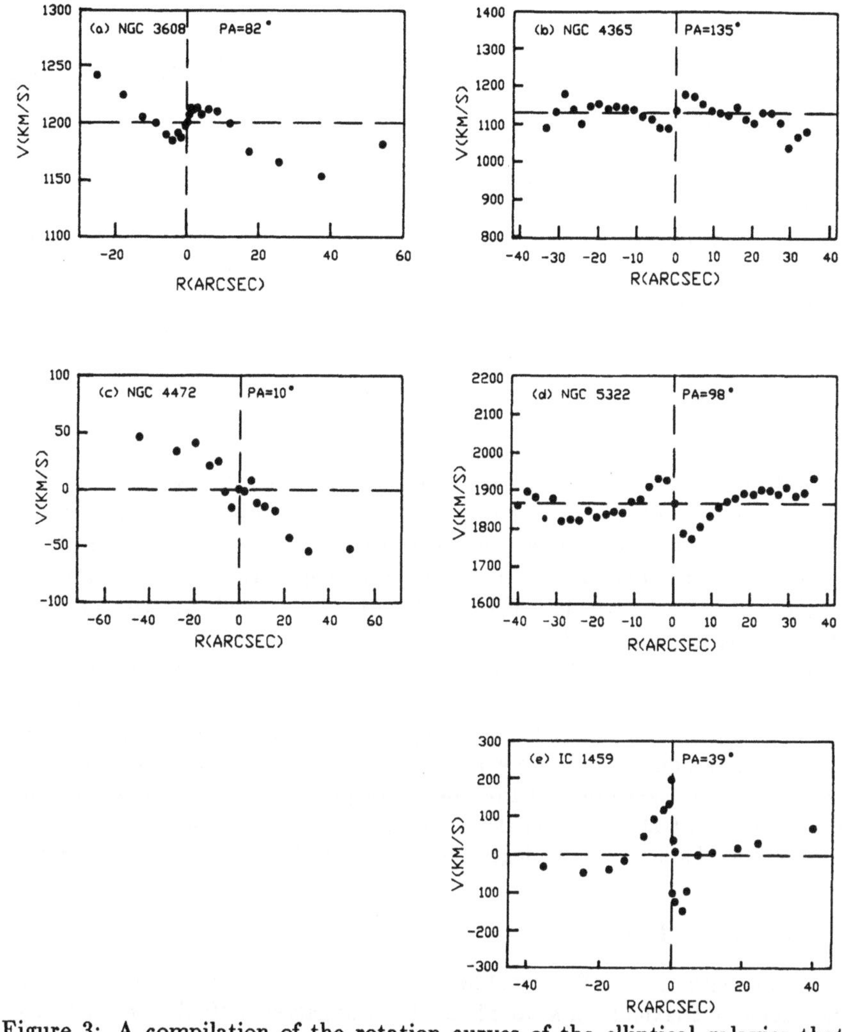

Figure 3: A compilation of the rotation curves of the elliptical galaxies that have been discovered to have counter-rotating stellar cores. Not all galaxies have been observed at enough position angles to establish that the core and the bulk of the galaxy have antiparallel angular momentum vectors, but clearly, the central kinematics are in all cases dramatically independent of those of the bulk of the galaxy. (a) NGC 3608 observed by Jedrzejewski and Schechter (1988); (b) NGC 4365 observed by Bender (1988); (c) NGC 4472 observed by Davies and Birkinshaw (1988); (d) NGC 5322 observed by Bender (1988); and (e) IC 1459 observed by Franx and Illingworth (1988). Note that in (a) and (c), the velocity scale extends $\pm$ 100 km sec^{-1} from the origin; whereas in (b), (d), and (e), it extends $\pm$ 300 km sec^{-1}. In (c) and (e), velocities are referred to the center of the galaxy, whereas in (a), (b), and (d), they are absolute.

3. Statistical Tests for the Intrinsic Shapes of Elliptical Galaxies

The low rotation velocities of luminous elliptical galaxies indicate that they need not be oblate spheroids flattened by rotation, but are likely to be ellipsoidal and supported by an anisotropic distribution of the random component of stellar velocities. A commonly used indicator of triaxiality has been the apparent twisting of the position angle of the major axis that would arise as a result of the projection onto the sky of a triaxial galaxy with an ellipticity that chnages with radius. Thus, the presence of isophote twisting at small radii in a substantial fraction of ellipticals reported in a number of recent papers (Djorgovski 1985, Jedrzejewski 1987, and Peletier *et al.* 1988) has be interpreted as evidence for intrinsic triaxiality. Further, Franx (1988) has shown that triaxial models that follow Stackel potentials can show an ellipticity gradient, but no isophote twist when projected onto the sky. Thus, for galaxies exhibiting ellipticity changes, the absence of an isophote twist cannot be used to infer an axisymmetric figure.

A variety of individual galaxies show particular features that indicate that they are likely to be triaxial. These include the large iosphote twist and complex kinematics of NGC 596 (Williams 1981), and the minor axis rotation in NGC 4261 (Davies and Birkinshaw 1986). In addition, dust lanes that are not oriented along either the projected major or minor axes (Hawarden *et al.* 1981 and Sparkes *et al.* 1985) also suggest a triaxial morphology. However, these examples do not indicate what fraction of ellipticals are detectably non–axisymmetric or how different from oblate or prolate a typical galaxy may be. Two statistical tests based on the internal kinematics of ellipticals have been proposed to characterize the degree of triaxiality of a typical elliptical galaxy.

a. Binney's Test

In an influential paper, Binney (1985) proposed a statistical test to determine the degree of triaxiality of a sample of elliptical galaxies. He developed a kinematic description of triaxial ellipticals which followed a particular velocity law, and allowed streaming motions about the axis that the figure tumbles about. He investigated two orientations of the figure with respect to the angular momentum vector, the tumbling bar, with the angular momentum vector parallel to the short body axis; and the spindle, with the angular momentum vector parallel to the long body axis. He showed that if elliptical galaxies are triaxial, then streaming motions along the apparent major axes are accompanied by similar motions along the minor axes. He parameterised the triaxiality, Z, such that $b = Zc + (1 - Z)a$, so that $Z = 1$ is a prolate figure, and $Z = 0$ is an oblate figure; and the minor axis rotation by $\mu = V_{min}/\sqrt{V_{min}^2 + V_{maj}^2}$. Using the tumbling bar model, he then derived the probability that a galaxy of a given triaxialty, Z, and apparent axial ratio, q, will have a minor axis rotation

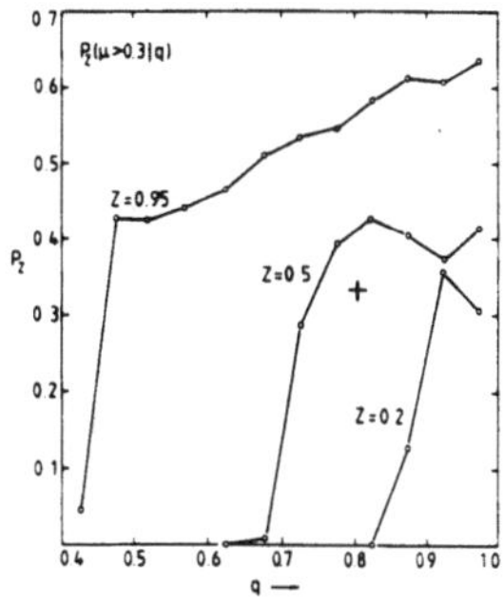

Figure 4: Adapted from Binney (1985). The probability that a galaxy will have $\mu > 0.3$ plotted as a function of apparent axial ratio q. Three lines are shown corresponding to the expected behavior for samples of galaxies that are oblate–triaxial ($Z = 0.2$), maximally triaxial ($b = (a + c)/2, Z = 0.5$), and prolate–triaxial ($Z = 0.95$). Of those galaxies with adequate published data (12), one third have $\mu > 0.3$ and the average axial ratio is 0.8; this point is marked +.

greater than a given μ for a range of μ's. He repeated this calulation for Z's covering the range of close–to–oblate and close–to–prolate. Figure 4 reproduces his figure which shows the probability of observing galaxies with measurable minor axis rotation, taken to be $\mu > 0.3$, as a function of apparent axial ratio for galaxies with $Z = 0.2$, 0.5, and 0.95. This shows that for close–to–oblate galaxies, it is those galaxies that are round (i.e. viewed pole–on) that are expected to show the largest μ values. Observationally, these galaxies are those that show the smallest rotation velocities. To determine μ values, rotation on both the major and minor axes needs to be detected. The successful application of this test called for high signal–to–noise, high spatial resolution rotation curves for elliptical galaxies with moderate flattenings on both their major and minor axes.

Most of the published velocity maps that include both the photometric major and minor axes are not adequate to generate meaningful values of μ, primarily because of uninterestingly high upper limits on the rotation along the minor axis. For the purposes of illustration, I have assembled the relevant published data and determined the range of values for both the major and minor axis rotation velocities. I used 8 galaxies from Davies and Birkinshaw (1988), together with NGC 5128 (Wilkinson *et al.* 1986), NGC 1316 (Bosma, Smith, and Wellington 1984), NGC 596 (Williams 1981), and IC 1459 (Franx and Illingworth 1988). Using the measured values for the rotation velocities, the mean value of μ derived for this sample is 0.36. (Taking the extremes to produce the largest values of μ, i.e. the upper limit on the minor axis rotation velocity and the lower limit on the major axis rotation velocity, the median value of μ is 0.45). The sample has a mean ellipticity of 0.20, with a dispersion of 0.07. One third of the galaxies have $\mu > 0.3$; I have plotted this point on Figure 4. These data are hardly adequate to the task, but suggest that the sample can be characterised as being a little on the oblate side of perfectly triaxial, a conclusion in agreement with Binney's original result. Several groups have collected data specifically to carry out this test: Franx, Illingworth, and Heckman (1989), Jedrzejewski and Schechter (1989), and Bender and colleagues; a better characterization of the distribution of figure types will be forthcomming when these studies are completed. Preliminary indications from these authors are that minor axis rotation is detected in 15–30% of ellipticals with a mean ellipticity of 0.2, which is consistent with ellipticals being significantly triaxial and closer to oblate than prolate. The statistical methods developed to analyse these new data will hopefully indicate with what certainty the hypothesis that all ellipticals are oblate or prolate can be rejected. In addition, these new data will indicate whether the model used by Binney is an appropriate one. For example, if the distribution of galaxies with ellipticity 0.2 indicates that a typical elliptical is triaxial, but closer to oblate than prolate, then there should be a very small number of galaxies with ellipticities > 0.35 that have detected minor axis rotation. If more galaxies like NGC 4261 are discovered, with $V_{min} >> V_{maj}$, it may indicate that a small fraction of ellipticals have the spindle geometry with the long axis being parallel to the rotation axis.

b. Levison's Test for Round Galaxies

Levison (1987) has developed a test for triaxiality based on the models of triaxial E5–6 elliptical galaxies constructed by Levison and Richstone (1987). These models were constructed from libraries of orbit families using the method of linear programming (Schwarzschild 1979). The potentials are logarithmic and scale free, so that the rotation velocity and velocity dispersion are a function of angle only. Levison considered two models, one nearly oblate with axis ratios 0.53:0.82:1.00 and one nearly prolate with 0.56:0.66:1.00. He demonstrated that when viewed so as to appear round, both models possess substantial streaming motions having high v/σ. In a separate argument, he inverted the observed distribution of axial ratios of ellipticals under the assumption that the intrinsic distribution is triaxial. From this, he concluded that between 3–23% of apparently round galaxies will have an intrinsic axial ratio flatter than 0.73, and thus will exhibit measurable rotation velocities at some position angle.

These models do not make the assumptions explicit in Binney's development, that there is a specific velocity law and that the galaxy is a tumbling bar with streaming about the short axis only. However, they do require (1) that $M/L \propto r$, to match elliptical galaxy light profiles, and (2) that half the X–tube orbits (those with net angular momentum along the long axis) are counter–rotating, to avoid producing galaxies with fast rotation along the minor axis. Levison's analysis applies only to galaxies that are round in projection, and the method requires a more complete mapping of the velocity field because the "major" and "minor" axes are not defined for round galaxies. Gonzalez, Levison, and Richstone (1989) are pursuing a sample of round ellipticals to search for rotation; they expect to find a small fraction that do possess rotation if ellipticals are generically triaxial. While this method is less generally applicable than that proposed by Binney, a significant result could be forthcomming if a sample of, say, 50 round ellipticals can be assembled.

The two tests described above are completely independent and, of course, Binney's method can be applied to the data collected for the test proposed by Levison. Preliminary indications from the groups investigating Binney's test indicate that ellipticals are usually significantly triaxial. In any case, it seems likely that a more definitive view of the figure types of ellipticals will emerge soon.

4. The Modified Faber–Jackson Relation for Elliptical and Lenticular Galaxies

Faber and Jackson (1975) discovered a relationship between the central velocity dispersion (σ) of elliptical galaxies and their luminosities (L), such that $L \propto \sigma^4$. This relationship was used by Tonry and Davis (1981) to map the local velocity field and estimate the velocity with which the Local Group is falling into the Virgo cluster. The dispersion in luminosity at fixed velocity dispersion implied an uncertainty in their distance estimator of 32% per galaxy.

A second parameter in the Faber–Jackson relation for ellipticals has now been identfied that can be used to improve its precision as a distance estimator. The derivation of a similar relationship for lenticular galaxies (and even early type spiral galaxies) would improve our ability to predict the distances to aggregates of early type galaxies. In addition, the existence of such a relationship that includes later Hubble types would suggest a substantial commonality in the modes of formation of the early type components of galaxies. I will briefly discuss the now well–developed description of the global properties of ellipticals, and go on to review attempts to apply the same methods to later type galaxies.

Faber *et al.* (1987) and Djorgovski and Davis (1987) have concluded that the global properties of elliptical galaxies populate a plane in the 3–space of $\log R_e$, $\log \sigma$, and $\log I_e$. Earlier, Lauer (1985) found a similar relationship between these quantities determined for the cores of ellipticals. Lucey (1986) has independently derived a distance indicator based on the same philosophy. Fuller discussions of this topic can be found in the above references. The mean relationship between the global properties of ellipticals obtained from these studies is

$$R_e \propto \sigma_e^{1.37} \times I_e^{-0.87}.$$

As discussed by Faber *et al.* (1987), the fact that ellipticals populate such a plane is expected if the only constraint on galaxy formation is the Virial Theorem; then M/L is a unique function of position in the plane. The same group, Dressler *et al.* 1987, have formulated a distance indicator which incorperates the improvement in precision due to the second parameter that is based on the relationship between the diameter enclosing a blue surface brightness of 20.75 (D_n) and the velocity dispersion, $D_n \propto \sigma^{1.2}$. The dispersion in diameter at fixed velocity dispersion is equivalent to an uncertainty in distance of 23% per galaxy. This method has been used to map the velocity field out to the Coma cluster (Lynden–Bell *et al.* 1988). Using a similar method, Lucey and Carter (1988) have estimated the peculiar velocities of different groups in the direction of Centaurus.

The first investigation of the Faber–Jackson relation for lenticular and early–type spiral galaxies was made by Whitmore, Kirshner, and Schechter (1979). They generated magnitudes for the bulge components of these composite systems by fitting an exponential disk profile to the outer parts of the galaxy profiles and subtracting the disk component from the total; this was possible for about half of their galaxies. For the remainder, they assumed a mean disk surface brightness of $B = 21.42\,\mathrm{mag\,arcsec}^{-2}$ and subtracted that from the total. They determined absolute magnitudes by assuming a smooth Hubble flow. They found a similar $L \propto \sigma^4$ relation for bulges, but also found that the dispersions were 15% lower than those of ellipticals of the same luminosity. The scatter in luminosity at fixed σ implied a distance uncertainty of 38% per galaxy.

The situation was further investigated by de Vaucouleurs and Olson (1982). They established the distance moduli to a sample of elliptical and lenticular galaxies from optical teritiary indicators, from a revised Tully–Fisher relation for those early type galaxies associated with spirals, and from group recession velocities assuming a uniform Hubble Flow. They made no attempt to estimate the bulge magnitudes, using total magnitudes throughout. The velocity dispersions were compiled from available sources. They noted that $U - V$ color was a significant second parameter in the blue L, σ relation for both elliptical and lenticular galaxies. They investigated the effect of waveband on the slope and zero point of the Faber–Jackson relation for E's and S0's and showed, that in the V band, both E and S0 galaxies obey the same relation. The dispersion about the mean relation was substantially larger for lenticulars than for ellipticals (0.77 mags compared to 0.55 mags; corresponding to 42% and 29% in distance). Dressler and Sandage (1983) measured rotation curves and central velocity dispersions for 32 S0 galaxies. They used total blue absolute magnitudes taken from the Revised Shapley–Ames Catalog (Sandage and Tammann 1981) and estimated the bulge magnitudes by eye. They found that the blue L, σ relation for the bulges of disk galaxies was the same as that for ellipticals. They confirmed, however, that the bulges scatter more about the mean relation than do the ellipticals.

Kormendy and Illingworth (1983) compared a sample of ellipticals (43) with a sample of S0's (33) and spirals (15) that included barred and unbarred galaxies. They derived absolute magnitudes using both a uniform Hubble flow model and a Virgocentric infall model, but their results were not sensitive to this choice. They were particularly careful to use only disk galaxies that had published decompositions of the disk and bulge components, so that the most accurate bulge magnitudes could be determined. In addition, they examined edge-on disk galaxies separately, and excluded galaxies that showed morphological evidence that the bulges were contaminated by disk material. They found that the bulges of unbarred spiral and S0 galaxies obeyed the same blue L, σ relation as ellipticals. There were two caveats to this result: (1) One-third of barred galaxies had dispersions significantly lower than non-barred galaxies of the same luminosity, which Kormendy and Illingworth attributed to contamination of the bulges of barred galaxies by disk material transported inward by the bar. (2) The bulges of edge-on galaxies have larger velocity dispersions than those that are less inclined. They suggested that this could be due to an anisotropy in the velocity dispersion, or to a central mass that generates rapid central rotation.

More recently, Dressler (1987) has studied the lenticular and early type spiral galaxies in the Virgo and Coma clusters. Using direct frames taken with the 4-Shooter on the 200-inch telescope, he determined an equivalent to D_n for these galaxies and used previous determinations of velocity dispersions. He found the "D_n, σ" relation for these cluster disk galaxies to be remarkably tight and identical to that of the ellipticals in the same clusters. In the cluster environment, it appears that there was a substantial commonality in the processes of formation and evolution experienced by elliptical galaxies and the bulges of spiral galaxies. The dispersion in D_n at fixed σ for the disk galaxies was found to be equivalent to an uncertainty in distance per galaxy of 15% and 17% for Virgo and Coma, respectively, compared to 17% and 18% for the ellipticals. Thus, improved relative distance estimates to aggreagates of early type galaxies can be made by using both types of galaxies.

Dressler also investigated the V-band L, σ relation for both bulges and elliptical galaxies in Coma. He used the photographic isophotal magnitudes from Godwin and Peach (1977) together with estimates of the bulge-to-disk ratios made from large scale plates taken on the DuPont telescope. He found that the L, σ relation for Coma bulge galaxies has a scatter of 0.59 magnitudes (31%), comparable to the scatter for the ellipticals in the Coma cluster (Dressler *et al.* 1987), suggesting that the Faber-Jackson relation for bulges in clusters is tighter than that for "field" bulges. However the L, σ relation for bulges in Coma is offset from that for the ellipticals, such that the bulges are under luminous by 0.4 magnitudes compared to the ellipticals. It seems possible that if total rather than bulge magnitudes were used, this result would be consistent with that of deVaucouleurs and Olson (1982). However, it is inconsistent with the results of Kormendy and Illingworth (1982) and Dressler and Sandage (1983) that indicated that the Faber-Jackson relation was the same for ellipticals and bulges when bulge magnitudes were used.

5. Summary

Dramatic new data from Kormendy (1988 a,b) and Dressler and Richstone (1988) on the kinematics of the central regions of M 31, M 32, and NGC 4594, have shown both very steep rotation curves and peaked velocity dispersion profiles. Constant mass–to–light ratio models can reproduce the observed velocity dispersion profiles by invoking radial anisotropy, but cannot simultaneously produce the fast rotation observed. These data have been interpreted in terms of models that contain a central dark mass ranging from $10^6 - 10^9 M_\odot$.

The recent discovery of stellar subsystems with independent kinematics, in the centers of several elliptical galaxies by Illingworth and Franx (1988), Jedrzjewski and Schechter (1988), Davies and Birkinshaw (1988), and Bender *et al.* (1988), appears to be direct evidence that these galaxies have in the past accreted others. Initial indications are that these subsystems are quite common, occurring in 10–30% of galaxies studied. The galaxies exhibiting this behavior often do not show obvious morphological signs of peculiarity, so this kinematic signature may be the longest lived evidence of a past merger. In addition, the kinematics of the interstellar medium in ellipticals is often found to be different from that of the underlying galaxy. Together, these two phenomena suggest that mergers were common during the evolution of elliptical galaxies.

Binney (1985) and Levison (1987) have presented statistical tests to determine the true shape of elliptical galaxies. These tests are based on the general property of triaxial galaxies that, when viewed in projection on the sky, streaming motions along the major axis can be associated with streaming motions of comparable magnitude about the minor axis. Preliminary results indicate that 15–30% of E2 galaxies have detected minor axis rotation; using Binney's test, this corresponds to a typical galaxy having a figure that is significantly triaxial, being between oblate and maximally triaxial. If a typical elliptical galaxy is triaxial, then stars can populate the box orbits which allow stars from the outer parts of a galaxy to pass close to the center.

If elliptical galaxies are triaxial, and if it is common for them to have accreted other galaxies during their evolution, then a source of fuel for an active nucleus, and a mechanism for getting that fuel from the outside of the galaxy into the center, is available. If, in addition, ellipticals generally have central mass concentrations that could act as engines for the activity, then all the necessary ingredients for generating an active nucleus in an elliptical galaxy appear to be present. This suggests that a typical elliptical galaxy that might normally be in the quiescent state could become active by undergoing an injection of fuel resulting from a merger or interaction.

In the Faber–Jackson (L, σ) relation for elliptical galaxies, the scatter at fixed velocity dispersion was typically 0.55–0.7 magnitudes corresponding to an uncertainty in distance of 29–38% per galaxy. More recently, this relation has been modified to include the effects of surface brightness as a second parameter to produce a distance indicator with a scatter of 0.45 magnitudes (23%). In extending the second parameter formulation to disk systems, Dressler (1987) has shown that the D_n, σ relation for bulges in the Coma and Virgo clusters is identical to that of the elliptical galaxies. This suggests a considerable degree of commonality in the formation mechanisms of the early type components of galaxies, and also that these disk systems can be used to improve the precision of distance estimates to aggregates of early type galaxies.

Acknowledgments

This review was written at a time when a number of new results of the kinematics of elliptical galaxies were emerging. It would have been impossible for me to have presented an up-to-date picture without a number of groups making their preliminary results available to me prior to publication. I am particularly indebted to Marijn Franx, Garth Illingworth, Robert Jedrzejewski, Paul Schechter, Ralf Bender, Stefan Wegner, Francesco Bertola, and Reynier Peletier in this regard. In the course of preparing this review, I enjoyed productive discussions with James Binney, Garth Illingworth, Hal Levison, Colin Norman, Tim de Zeeuw, John Kormendy, and Alan Dressler. I would like to thank Francesco Bertola, Lucio Buson, and Werner Zeilinger for their generous hospitality in Padova immediately prior to this meeting. I would like to thank Joyce DuHammel and Carl Wetzel for producing the figures.

References

Bender, R., 1988, preprint.
Bertola, F. and Capaccioli, M. 1975, *Astrophys. J.* **200**, 439.
Bertola, F. and Bettoni D. 1988, *Astrophys. J.* **329**, 102.
Bertola, F., Buson, L., and Zeilinger, W. W. 1988, in preparation.
Binney, J. J. 1978, *Mon. Not. Roy. Astr. Soc.* **183**, 501.
Binney, J. J. 1982, *Ann. Rev. Astr. Astrophys.* **20**, 399.
Binney, J. J. 1985, *Mon. Not. Roy. Astr. Soc.* **212**, 767.
Bosma, A., Smith, R. M., and Wellington, K. 1984, *Mon. Not. Roy. Astr. Soc.* **212**, 301.
Caldwell, N., Kirshner, R. P., and Richstone, D. O. 1986, *Astrophys. J.* **305**, 136.
Davies, R. L., Efstathiou, G. P., Fall, S. M., Schechter, P. L., and Illingworth, G. D. 1983, *Astrophys. J.* **266**, 41.
Davies, R. L. and Birkinshaw, M. 1986, *Astrophys. J. Lett.* **303**, L45.
Davies, R. L. and Illingworth, G. D. 1986, *Astrophys. J.* **302**, 234.
Davies, R. L. 1987, in *Proc.* of I. A. U. Symp. **127**, *The Structure and Dynamics of Elliptical Galaxies*, ed. T. de Zeeuw, (Dordrecht: Reidel) p. 63.
Davies, R. L. and Birkinshaw, M. 1988, *Astrophys. J. Suppl.* **68**, in press.
de Vaucouleurs, G. and Olson, D. W. 1982, *Astrophys. J.* **256**, 346.

Djorgovski, S. 1985, Ph. D. thesis, University of California, Berkeley.
Djorgovski, S. and Davis, M. M. 1987, *Astrophys. J.* **313**, 59.
Dressler, A. 1987, *Astrophys. J.* **317**, 1.
Dressler, A., Lynden–Bell, D., Burstein, D., Davies, R. L., Faber, S. M., Terlevich, R. J., and Wegner G. 1987, *Astrophys. J.* **313**, 42.
Dressler, A. and Richstone, D. O. 1988, *Astrophys. J.* **324**, 701.
Dressler, A. and Sandage, A. 1983, *Astrophys. J.* **265**, 664.
Efstathiou, G. P., Ellis, R. S., and Carter, D. 1982, *Mon. Not. Roy. Astr. Soc.* **201**, 975.
Faber, S. M. and Jackson, R. 1976, *Astrophys. J.* **204**, 668.
Faber, S. M., Dressler, A., Davies, R. L., Burstein, D., Lynden–Bell, D., Terlevich, R. J., and Wegner G. 1987, in *Nearly Normal Galaxies from the Planck Time to the Present,* ed. S. M. Faber, (New York: Springer–Verlag) p. 175.
Franx, M. 1988, *Mon. Not. Roy. Astr. Soc.* **231**, 285.
Franx, M. and Illingworth, G. D. 1988, *Astrophys. J. Lett.* **327**, L55.
Franx, M., Illingworth, G. D., and Heckman, T. 1989, in preparation.
Galletta, G. 1987, *Astrophys. J.* **318**, 531.
Gerhard, O. E. and Vietri, N. 1986, *Mon. Not. Roy. Astr. Soc.* **223**, 377.
Gerhard, O. E. 1988, *Mon. Not. Roy. Astr. Soc.* **232**, 13P.
Godwin, J. G. and Peach, J. V. 1977, *Mon. Not. Roy. Astr. Soc.* **181**, 323.
Gonzalez, J. J., Levison, H., and Richstone, D. O. 1989, in preparation.
Hawarden, T. G., Elson, R. A. W., Longmore, A. J., Tritton, S. B., and Corwin, H. G. 1981, *Mon. Not. Roy. Astr. Soc.* **196**, 747.
Illingworth, G. D. 1977, *Astrophys. J. Lett.* **218**, L43.
Jedrzejewski, R. I. 1987, *Mon. Not. Roy. Astr. Soc.* **226**, 747.
Jedrzejewski, R. I. and Schechter, P. L. 1988, *Astrophys. J. Lett.* **330**, L87.
Jedrzejewski, R. I. and Schechter, P. L. 1989, in preparation.
Kormendy, J. 1984, *Astrophys. J.* **287**, 577.
Kormendy, J. 1988a, *Astrophys. J.* **325**, 128.
Kormendy, J. 1988b, preprint.
Kormendy, J. and Illingworth, G. D. 1982, *Astrophys. J.* **256**, 460.
Kormendy, J. and Illingworth, G. D. 1983, *Astrophys. J.* **265**, 632.
Lauer, T. 1985, *Astrophys. J.* **292**, 104.
Levison, H. 1987, *Astrophys. J. Lett.* **320**, L93.
Levison, H. and Richstone, D. O. 1987, *Astrophys. J.* **314**, 476.
Light, E. S., Danielson, R. E., and Schwarzschild, M. 1974, *Astrophys. J.* **194**, 257.
Lindblad, P. 1956, *Stockholm Obs. Annals* **19**, No. 2.
Lynden–Bell, D., Faber, S. M., Burstein, D., Davies, R. L., Dressler, A., Terlevich, R. J., and Wegner G. 1988, *Astrophys. J.* **326**, 19.
Lucey, J. 1986, *Mon. Not. Roy. Astr. Soc.* **222**, 417.
Lucey, J. and Carter, D. 1988, preprint.
Norman, C. A., May, A., and van Albada, T. S. 1985, *Astrophys. J.* **296**, 20.

Peletier, R. F., Davies, R. L., Illingworth, G. D., Davis, L. E., and Cawson, M. 1988, in preparation.

Sandage, A. and Tammann, G. A. 1981, *A Revised Shapely–Ames Catalog of Bright Galaxies*, (Washington, D. C.: Carnegie Institution of Washington).

Schwarzscild, M. 1979, *Astrophys. J.* **232**, 236.

Sparkes, W. B., Wall, J. V., Thorne, D. J., Jorden, P. R., Van Breda, I. G., Rudd, P. J., and Jorgenson, H. E. 1985, *Mon. Not. Roy. Astr. Soc.* **217**, 87.

Stark, A. 1977, *Astrophys. J.* **213**, 368.

Tonry, J. and Davis, M. M. 1981, *Astrophys. J.* **246**, 680.

van Gorkom, J. H., Knapp, G. R., Raimond, E., Faber, S. M., and Gallagher, J. S. 1986, *Astr. J.* **91**, 791.

Whitmore, B., Kirshner, R. P., Schechter, P. L. 1979, *Astrophys. J.* **234**, 68.

Wilkinson, A., Sharples, R. M., Fosbury, R. A. E., and Wallace, P. T. 1986, *Mon. Not. Roy. Astr. Soc.* **218**, 297.

Williams, T. B. 1981, *Astrophys. J.* **244**, 458.

Discussion

E. M. Burbidge: These are beautiful and very fascinating new data. I have a comment and two questions. I find it surprising that mergers of galaxies should be frequent enough to account for the number of anomolous velocities. Why should such mergers occur especially in counter–rotating systems? And what would be the relaxation time, for what would appear to be a dynamically unstable situation, to settle into a co–rotating system?

Davies: The past frequency of mergers and their role in galaxy evolution are hotly debated. The most dramatic examples of counter–rotating cores seem to be direct evidence of past mergers. If the galaxies showing central "features" in their rotation curves, are also interpreted as cores with independent stellar kinematics that have resulted from mergers, then a substantial fraction of ellipticals appear to have experienced such events during their evolution. I would be interested to hear what other mechanisms, if any, might be candidates for producing such phenomena. As far as the relaxation time is concerned, the features could be very long lived as the two body relaxation time is very much longer than the Hubble time for these systems.

J. Binney: In relation to the points raised by Dr. Burbidge, I would say that insofar as I understand the situation, counter–rotation, and possibly rotation at right angles, would be able to persist for a two–body relaxation time which Dr. Davies has pointed out is many Hubble times. But rotation about axes not aligned in such special ways would, I believe, phase mix away until one of the prefered orientations was achieved.

K. C. Freeman to Binney: What kind of satellite orbit would be required to produce a counter–rotating core: does it need to be close to radial?

J. Binney: No, I don't think the orbit need be radial. If a counter–rotating satellite were to spiral into the center of a giant, the counter–rotation of the satellite would tend to limit tidal damage. This would also allow it to arrive intact — and still counter–rotating — at the center.

F. Bertola: I would like to point out the extreme usefulness of the presence of gaseous disks in elliptical galaxies in order to derive their intrinsic shape. Constraints such as the apparent ellipticity of the stellar body, the axial ratio of the outer distribution of gas, and the difference between the rotation axis of the gas and the projected minor axis of the starlight, allow us to derive the intrinsic axial ratios b/a and c/a and the viewing angles θ and ϕ. This method has recently been applied to NGC 5077 which turns out to be a triaxial, but almost oblate, galaxy. (Bertola, Bettini, Danziger, Sadler, and de Zeeuw, in preparation).

J.–L. Nieto: I would like to comment about disks, boxiness, and merging. There is a class of E galaxies that may have escaped formation by a merging process: disk Es that appear to be isotropic rotators, in the continuity of the S0s in the Hubble sequence. The analysis of the available data shows that two types of boxiness may coexist that reflect two possible formation processes, either merging (Binney and Petrou 1985) or three dimensional bar instabilities (Combes and Sanders 1981). This is briefly discussed in a poster paper presented at this meeting.

Davies: I am not sure that all ellipticals that have positive amplitudes of the $\cos 4\theta$ term in the Fourier expansion of their isophotes, the "disk Es," are consistent with being isotropic rotators.

W. T. Sullivan: Have any correlations been found between the presence of the unusual central kinematic components and the presence of active galactic nuclei?

Davies: There are very few systems with which to judge that, and active galaxies are often studied in more detail so that such features are more likely to become apparent. A carefully selected sample of active galaxies and a control sample would have to be studied to answer your question. Bender, Döbereiner, and Möllenhoff (Astron. Astrophys. 1987, 177, L53) have suggested that the amplitude of the $\cos 4\theta$ term in the Fourier expansion of the isophote shape is correlated with the presence of radio activity in the sense that galaxies exhibiting boxy isophotes are more frequently radio sources.

R. Bender: We find correlations between X–ray, radio emission, and box–shaped isophotes. Therefore, it might be that we have a multiple correlation between X–ray emission, radio activity, boxy isophotes, and kinematic peculiarities (*e.g.* minor axis rotation, counterrotating cores).

K. C. Freeman: Do galaxies showing inner kinematic subsystems also show outer Malin–Carter shells?

Davies: The galaxies included in the studies that discovered the independent stellar kinematics were chosen to be broadly normal, not possessing any particular peculiarities. However, IC 1459 has outer wisps similar to the Malin–Carter shells, and NGC 7626 clearly exhibits those features; in addition, both are radio sources. The authors of these studies have always taken direct frames before undertaking the spectroscopy. Once their studies are complete, the fraction of galaxies with independent core kinematics that exhibit shells should be evaluated and compared with the fraction for galaxies with no evidence of independent core kinematics. I hope they will address this issue specifically.

Bender: As far as I know, there do exist distinct morphological differences between galaxies showing counter–rotating cores and minor axis rotation, and galaxies which do not show these features. At least in our sample of galaxies, those having significant minor axis rotation show boxy isophotes. In contrast, those ellipticals which contain weak disks are rapid rotators; these most likely have isotropic velocity dispersions. In a similar way, I suppose that counter–rotating cores are solely associated with galaxies which show boxy isophotes or merging residues (like dust, etc.). Therefore, these results confirm that we have two types of ellipticals: (1) objects with box-shaped isophotes, most likely being merger remnants, and (2) objects containing weak disks forming a continuous sequence in morphological and kinematical properties with S0 galaxies.

M. Capaccioli: Are there realistic alternatives to merging which may account for counter–rotation?

Davies: Nobody has suggested any to me! But given that such a configuration is stable, perhaps a more reasonable question would be to ask whether such systems are likely to have been set up at the time of formation (perhaps by mergers, but perhaps not) and whether we have any hope of dating the event that gives rise to them. Franx and Illingworth point out that there is no difference in color between the central counter–rotating component and the bulk of the galaxy, so at least it is not a very young phenomenon in IC 1459.

Models of the $r^{1/4}$ Law

James Binney

Department of Theoretical Physics, University of Oxford, England, U. K.

Summary

N–body models show that a system obeying the $r^{1/4}$ law forms when a cloud of stars relaxes from a cold, clumpy initial configuration. Bertin and Stiavelli (1984) have suggested a simple analytical fit to the distribution function of such a system. Jaffe (1987) and Tremaine (1987) have explained the structure of such models at large radius in terms of a simple picture of violent relaxation. A similar analysis of the population of the most tightly bound orbits suggests that violently relaxed systems should have central density cusps in which $\rho \propto r^{-3/2}$.

1. $r^{1/4}$ Model Versus the Isothermal Sphere

Elliptical galaxies are smooth, apparently relaxed systems. So it is natural to ask whether they are maximum entropy configurations. Unfortunately, if one extremizes the entropy of a self–gravitating stellar system subject to given mass and energy (Ogorodnikov 1965; Lynden–Bell 1967), one recovers the distribution function of the isothermal sphere, which has little in common with the system which in projection obeys the $r^{1/4}$ law. Indeed, Young (1976) found that the density of the latter goes as

$$\rho(r) \quad \propto \quad \begin{cases} \exp(-7.669r^{1/4})/r^{3/4} & \text{for } r \lesssim 10^{-3}r_e \\ \exp(-7.669r^{1/4})/r^{7/8} & \text{for } r \gtrsim 10^{-1}r_e \end{cases} \tag{1a}$$

rather than

$$\rho(r) \quad \propto \quad \begin{cases} \text{constant} & \text{for } r \ll r_c \\ r^{-2} & \text{for } r \gg r_c \end{cases} \tag{1b}$$

as in an isothermal sphere. Why are relaxed galaxies not isothermal? There are two reasons. (i) Since an unbounded isothermal sphere has infinite mass, the distribution function to which entropy extremization leads is inconsistent with the constraint of finite mass. Hence, what the standard derivation of the isothermal sphere shows is that an unbounded stellar system of finite mass admits no configuration of stationary entropy. (ii) In the light of this remark one might be tempted to place the system in a large but finite box.[1] However, this still does not lead to the prediction that elliptical galaxies should be isothermal, since even confined isothermal spheres of high central concentration represent merely saddle points of the entropy (Katz 1978).[2] Thus, we should not be surprised that elliptical galaxies are thoroughly nonisothermal. The methods of classical statistical mechanics are just not powerful enough to handle this interesting case.

2. $r^{1/4}$ Models from N–Body Simulations

What we can learn about the endpoints of violent relaxation from n–body models? Simulations of the virialization of clouds of stars were first undertaken in the early 1970's (for a review see Gott 1977), but failed to produce any very convincing representation of an $r^{1/4}$–galaxy. The breakthrough came in 1982 when van Albada showed that an excellent representation of an $r^{1/4}$–galaxy forms when stars relax from a very cold and clumpy initial configuration. Villumsen (1982), McGlynn (1984), and May and van Albada (1984) confirmed that the earlier n–body simulations had failed to generate $r^{1/4}$–like systems because they had lacked the spatial resolution and/or the ability to handle asymmetric configurations required of a simulation of collapse from very cold initial configurations.

In van Albada's $r^{1/4}$–model, the velocity anisotropy parameter is given by

$$\beta \equiv 1 - \frac{\sigma_\theta^2}{\sigma_r^2} \simeq 0.5\left[1 + \log_{10}\left(\frac{r}{r_e}\right)\right] \quad (0.1 < \frac{r}{r_e} < 10). \tag{2}$$

Thus, the velocity ellipsoid is isotropic well within r_e, and at large r becomes significantly elongated in the radial direction.

In van Albada's initial configuration, the energies of particles cover a narrow range, while the final energy distribution is much broader. Binney (1982) showed

[1] There is a useful analogue in the Saha equation; an isolated hydrogen atom is guaranteed to be ionized even at arbitrarily low temperature because it has an infinite number of ionized states only a finite distance above its ground state. We get sensible predictions by immersing the atom in a dilute cloud of free electrons, or, equivalently, by placing the atom in a box.

[2] Core collapse is a physical manifestation of this result; entropy–generating two–body encounters cause a highly concentrated isothermal sphere to evolve *away* from isothermality.

that the energy distribution of an isotropic $r^{1/4}$ model is remarkably well fit by the simple formula

$$\left.\frac{\mathrm{d}M}{\mathrm{d}E}\right|_{\text{final}} \propto e^{-\beta E} \quad (\beta < 0) \tag{3}$$

and argued that this result should hold also for anisotropic models. Van Albada's energy distribution is consistent with this formula.

3. Understanding the Formation of an $r^{1/4}$ System

Bertin and Stiavelli (1984, 1987) have sought a simple distribution function which (i) yields $\rho(r) \propto r^{-4}$ as $r \to \infty$ and (ii) yields $\beta(0) = 0$ and $\beta \to 1$ as $r \to \infty$. A distribution function that satisfies these criteria is

$$f = F_0|E|^{3/2} \exp\left[-\left(\frac{E + \frac{1}{2}L^2/r_a^2}{\sigma^2}\right)\right] \tag{4}$$

In this expression, F_0 is a constant, E and L are the energy and total angular momentum of an individual star, σ is a free parameter, and r_a is an eigenvalue used to ensure that the density vanishes at infinity. The Bertin and Stiavelli models are parametrised by their central potentials $\Phi(0)$. $r^{1/4}$–like models are obtained for a suitable value $|\Phi(0)|/\sigma^2$.

Numerical simulations of the formation of $r^{1/4}$ models leave one with the impression that the structure of the distribution function is determined by the exchange of energy and angular momentum that is effected by the system's violently fluctuating gravitational potential close to the moment $t_{\min}$ of maximum collapse. Jaffe (1987) pointed out that $\rho \propto r^{-4}$ for $r \gg r_e$ is a natural consequence of the fluctuating central potential (the "reactor") having no way of knowing the precise value of the escape energy. Consequently, the central reactor sets up an energy distribution that is continuous across $E = 0$, and $\mathrm{d}M/\mathrm{d}E \simeq$ constant for $E \approx 0$. For an isotropic model, the phase–space density f is related to $\mathrm{d}M/\mathrm{d}E$ by [*e.g.* Binney and Tremaine (1987), equation (4-157b)]

$$f = g^{-1}\frac{\mathrm{d}M}{\mathrm{d}E} \quad \text{where} \quad g(E) \equiv 16\pi^2 \int_0^{\Phi(r)=E} \sqrt{2[E - \Phi(r)]}\, r^2 \mathrm{d}r. \tag{5}$$

In the envelope $\Phi \simeq -GM/r$, so $g(E) \propto |E|^{-5/2}$, and we have $f \propto |E|^{5/2}$ for $E \simeq 0$. The density in the envelope is therefore

$$\rho \propto \int f(E)\sqrt{E - \Phi}\, \mathrm{d}E \propto |\Phi|^4 \propto r^{-4}. \tag{6}$$

Tremaine (1987) obtains a similar result by a slightly different route. He argues that at $t_{\min}$ all parts of the reactor's velocity space (up to the escape velocity) are fairly uniformly populated. If velocity space were approximately uniformly populated at every radius, we would have $f \sim \exp(-\beta E)$, with β a small number of indeterminate sign. However, high-energy orbits will not attain populations $\propto \exp(-\beta E)$ because (i) only orbits that pass through the reactor are significantly populated, and (ii) the fluctuations of the central potential are too short-lived to fully populate even these orbits. The actual population achieved by a reactor-penetrating orbit is reduced by the ratio of the orbit's period T_r to the duration $t_{\rm eff}$ of the maximum collapse phase. Thus, the density of stars on orbits that pass through the core is

$$\begin{aligned} f(E) &\propto \min(1, t_{\rm eff}/T_r) \times \exp(-\beta E) \\ &\propto |E|^{3/2} \exp(-\beta E), \end{aligned} \tag{7}$$

where $T_r \propto |E|^{-3/2}$ has been used. Finally, we must take into account the fact that high-angular momentum orbits cannot be populated by the central reactor because they don't pass through the core. Thus, Tremaine argues that violent relaxation leads to a distribution function of the form

$$f \propto |E|^{3/2} \times \begin{pmatrix} \text{decreasing} \\ \text{function L} \end{pmatrix} \times \begin{pmatrix} \text{decreasing function} \\ \text{of } |E| \text{ which is} \\ \text{non-zero at } E=0 \end{pmatrix} \tag{8}$$

The Bertin and Stiavelli distribution function (4) is about the simplest distribution function satisfying these criteria. Bertin and Stiavelli assume that $\beta > 0$, but Merritt, Tremaine, and Johnstone (1988) have shown that closer approximations to the $r^{1/4}$ model are obtained with $\beta < 0$ as one would expect given that $\mathrm{d}M/\mathrm{d}E$ for the $r^{1/4}$ model is of the form (3).[3]

Thus, in broad terms, we understand why violent relaxation leads to the asymptotic density profile characteristic of elliptical galaxies at large r. Do similar arguments explain why giant galaxies have central density cusps rather than isothermal cores? I think the answer to this question is "yes." My argument will hinge on the use of $\mathrm{d}M/\mathrm{d}E$ as the natural determinant of $\rho(r)$ independent of uncertainty about the anisotropy profile $\beta(r)$. To build confidence in this method, let me quickly rework Jaffe's argument along the lines I shall apply to the core.

Stars spend most of their time near apocentre, that is, at the radius at which $\Phi(r) = E$. Hence, ρ and $\mathrm{d}M/\mathrm{d}E$ are approximately related by

$$\frac{\mathrm{d}M}{\mathrm{d}E} \simeq 4\pi r^2 \rho(r) \left(\frac{\mathrm{d}r}{\mathrm{d}\Phi}\right)_{\Phi=E}. \tag{9}$$

Consider a star that is expelled from the central reactor with a speed v_0 that is

[3]One obtains $\mathrm{d}M/\mathrm{d}E$ for the Bertin and Stiavelli distribution function by integrating over the volume $(2\pi)^{-1}T_r L\mathrm{d}L$ of phase space with E in the range $(E+\mathrm{d}E, E)$. It is easy to see that for (4), $\mathrm{d}M/\mathrm{d}E \sim$ constant at $E \sim 0$ in accordance with Jaffe's argument.

close to the escape speed. Since in the relaxed galaxy $\Phi \simeq -GM/r$ for large r, this star's final apocentric radius $r_{\rm fi}$ is related to v_0 by

$$-\frac{GM(v_0)}{r_{\rm fi}} = \tfrac{1}{2}v_0^2 + \Phi(0, t_{\rm min}), \tag{10}$$

where $M(v_0)$ is the mass in stars with speeds smaller than v_0. Solving for $r_{\rm fi}$, making the approximation $M(v_0) \simeq$ constant and differentiating, we find $\mathrm{d}r_{\rm fi} \propto r_{\rm fi}^2 v_0 \mathrm{d}v_0$. We now have

$$\begin{aligned} \rho(r) &\simeq \frac{\mathrm{d}M}{4\pi r_{\rm fi}^2 \mathrm{d}r_{\rm fi}} \propto \frac{(\mathrm{d}M/\mathrm{d}v_0)\,\mathrm{d}v_0}{r_{\rm fi}^2 \mathrm{d}r_{\rm fi}} \\ &\propto \frac{(\mathrm{d}M/\mathrm{d}v_0)\,\mathrm{d}v_0}{r_{\rm fi}^4 v_0 \mathrm{d}v_0} \propto \frac{1}{r_{\rm fi}^4}, \end{aligned} \tag{11}$$

where the last proportionality assumes that $(\mathrm{d}M/\mathrm{d}v_0)v_0^{-1}$ is approximately constant for speeds near v_0. Thus, this argument recovers Jaffe's result $\rho \propto r^{-4}$ without assuming that the distribution function is of the form $f(E)$.

Now consider the situation at small r. Suppose that in the relaxed configuration we have $\rho \propto r^{-\alpha}$ $(\alpha < 2)$. Then, at small r, the relaxed potential satisfies

$$\Phi(r, \infty) = \Phi(0, \infty) + \frac{r^{2-\alpha}}{2-\alpha}, \tag{12}$$

so $\mathrm{d}\Phi \propto r^{1-\alpha}\mathrm{d}r$. The relaxed energy distribution is therefore

$$\begin{aligned} \frac{\mathrm{d}M}{\mathrm{d}E} &\simeq \frac{\mathrm{d}M}{\mathrm{d}\Phi} \propto \frac{r^{-\alpha}r^2\mathrm{d}r}{r^{1-\alpha}\mathrm{d}r} = r \\ &\propto \mathcal{E}^{1/(2-\alpha)} \quad [\mathcal{E} \equiv E - \Phi(0,\infty)]. \end{aligned} \tag{13}$$

In the reactor, the phase-space density is essentially constant over the low-E, low-L portion of phase space into which are scattered the stars which will wind up in the core of the relaxed system. Hence, in the reactor at the smallest values of E

$$\begin{aligned} \left.\frac{\mathrm{d}M}{\mathrm{d}E}\right|_{t_{\rm min}} \propto \frac{g(E, t_{\rm min})}{16\pi^2} &= \int_0^{r_{\rm max}(E)} v r^2 \mathrm{d}r \\ &= \int_0^{r_{\rm max}} \sqrt{2[E - (\Phi(0, t_{\rm min}) + \tfrac{1}{2}\Omega^2 r^2)]}\, r^2 \mathrm{d}r \\ &= \frac{\pi}{4\Omega^3}\big[E - \Phi(0, t_{\rm min})\big]^2. \end{aligned} \tag{14}$$

Here, Ω is the circular frequency at the centre of the minimum configuration. The energies of these stars change considerably during the ensuing violent relaxtion, but mostly as a consequence of more energetic stars moving away from them as the system reexpands. So, we won't be much in error if we suppose that for core stars $\mathcal{E} = E - \Phi(r = 0, t)$ is approximately constant during the reexpansion. We may then equate equation (13) for $\mathrm{d}M/\mathrm{d}E$ in the final configuration with equation (14), expressed as a function of $\mathcal{E}$, for the energy distribution at in the reactor, to find

$$\mathcal{E}^2 \propto \mathcal{E}^{1/(2-\alpha)} \tag{15}$$

which clearly requires $\alpha = \frac{3}{2}$. Thus, we *do* expect galaxies to have singular central densities $\rho \propto r^{-3/2}$, indeed densities more singular than the $r^{1/4}$ law predicts. As Jaffe (1983) has pointed out, some galaxies are well fitted by central profiles as steep as $\rho \propto r^{-2}$.

References

Bertin, G. and Stiavelli, M. 1984, *Astron. Astrophys.* **137**, 26.

Binney, J. J. 1982, *Mon. Not. Roy. Astron. Soc.* **200**, 951.

Binney, J. J. and Tremaine, S. D. 1987, *Galactic Dynamics*, (Princeton: Princeton University Press).

Gott, J. R. 1977, *Ann. Rev. Astron. Astrophys.* **15**, 235.

Jaffe, W. 1983, *Mon. Not. Roy. Astron. Soc.* **202**, 995.

Jaffe, W. 1987, in *Structure and Dynamics of Elliptical Galaxies*, ed. T. de Zeeuw, (Dordrecht: Reidel) p. 511.

Katz, J. 1978, *Mon. Not. Roy. Astron. Soc.* **183**, 765.

Lynden–Bell, D. 1967, *Mon. Not. Roy. Astron. Soc.* **136**, 101.

May, A. and van Albada, T. S. 1984, *Mon. Not. Roy. Astron. Soc.* **209**, 15.

McGlynn, T. 1984, *Astrophys. J.* **281**, 13.

Merritt, D., Tremaine, S. D., and Johnstone, D. 1988, *Mon. Not. Roy. Astron. Soc.*, to be submitted.

Ogorodnikov, K. F. 1965, *Dynamics of Stellar Systems*, (Oxford: Pergamon).

Stiavelli, M. and Bertin, G. 1987, *Mon. Not. Roy. Astron. Soc.* **229**, 61.

Tremaine, S. D. 1987, in *Structure and Dynamics of Elliptical Galaxies*, ed. T. de Zeeuw, (Dordrecht: Reidel) p. 367.

Young, P. J. 1976, *Astron. J.* **81**, 807.

van Albada, T. S. 1982, *Mon. Not. Roy. Astron. Soc.* **201**, 939.

Villumsen, J. V. 1982, *Mon. Not. Roy. Astron. Soc.* **199**, 493.

Discussion

G. Bertin: Thank you for mentioning our work with M. Stiavelli. I have three comments to make: (i) We did *not* impose the $\rho \sim r^{-4}$ behavior at large radii in order to get the $r^{1/4}$ law, but argued on the basis of phase space arguments (Bertin and Stiavelli 1984). The $r^{1/4}$ behavior came as a surprise. Actually, $\rho \sim r^{-4}$ goes against a common belief that the $r^{1/4}$ law is associated with $\rho \sim r^{-3}$ behavior. (ii) I still see Tremaine's argument (IAU Symposium 127) as qualitatively consistent with our choice of distribution functions, but not really able to imply the $|E|^{3/2}$ factor we have used. (iii) About the distribution N(E): Stiavelli and I feel that the exponential behavior is related to the inner singularity of the potential (expected to be roughly logarithmic) rather than to the overall mass distribution.

Binney: On your first point, the point is that you require r^{-4}, which as Jaffe (1982) points out, is the best power–law approximation to the asymptotic structure of the $r^{1/4}$ model. Young (1976) shows that the true structure is more complex than a power law, but of course the mass interior to r converges, which for $\rho \sim r^{-3}$, it would not. On your second point, I would say that Tremaine's argument, while only heuristic, is persuasive. On your third point, I would say that the exponential behavior $dN/dE \propto e^{-\beta|E|}$ holds over energies that correspond to the observationally interesting domain $0.05 < r/r_e < 5$.

Rotating Cores in Elliptical Galaxies*

Ralf Bender

Landessternwarte Königstuhl, Heidelberg, F.R.G.

The structure of cores in elliptical galaxies can provide considerable insight into the formation and evolution of these galaxies. Compared to the rather detailed photometric data on cores (*e.g.* Lauer 1985, Kormendy 1985), up to now little is known about their kinematics. The findings of Kormendy (1984), who found a core–within–a–core in NGC 5813, and Franx and Illingworth (1988), who observed a core rotating in a sense opposite to the main body in IC 1459, indicate that the core structure of elliptical galaxies might be strongly influenced by merging.

Here I present the results of the observations of four luminous elliptical galaxies, which were found to exhibit significant rotation in their cores (NGC 4365, NGC 4406, NGC 4494, and NGC 5322).

The rotation curve of the central part of NGC 4494 largely resembles that of the bulge of M 31 (Kormendy 1988). However, the scales are very different. The core and main body of NGC 4494 are rotating in the same sense. The rotation velocity has a minimum at 7 arcsec and increases again towards the nucleus. The main body of NGC 4494 most likely is rotationally flattened (Bender 1988a). The most plausible explanation of the observed kinematics in NGC 4494 are variations in velocity anisotropy as discussed by *e.g.* Vietri (1986) or Gerhard (1988).

In NGC 4365 and NGC 4406, the rotation axes of cores and main bodies of the galaxies are misaligned, *i.e.* their kinematics are decoupled. This follows from the fact that the main bodies of these galaxies rotate nearly along their apparent minor axes (Wagner *et al.* 1988), while the cores seem to rotate along an axis near the major axes of the galaxies (for a more detailed discussion see Bender 1988b). The cores dominate the rotation and velocity dispersion curves inside of roughly three photometric core radii (Lauer 1985).

---

*Based on observations carried out at the German–Spanish Astronomical Center, Calar Alto, Spain

NGC 5322 shows strong counter rotation between the core and the outer parts. Surface photometry of this galaxy suggests the presence of a nearly edge-on disk embedded in the core or even dominating the core (Bender *et al.* 1988). Obviously, the core structure of NGC 5322 is similar to the one of IC 1459, for which Franx and Illingworth (1988) inferred the existence of a cold disk-like component by purely spectroscopic arguments.

A morphological feature which NGC 4365, NGC 4406 and NGC 5322 have in common are box-shaped isophotes (Bender *et al.* 1988). Theoretical as well as observational arguments in favor of an origin of boxiness in interaction or merging are given by several authors (Binney and Petrou 1985, Nieto 1988). Signs of interaction and merging are also present in IC 1459 (see Franx and Illingworth 1988). Therefore, it seems plausible to assume that these four objects, which exhibit counter-rotating cores or which show kinematic misalignment between their cores and their main bodies, are merger remnants. The most stable core of the progenitors survived the merging process and settled in the center of the potential well. If interaction indeed is the cause of the kinematic peculiarities reported on here, it is likely that a large fraction of box-shaped or disturbed ellipticals shows similar properties.

A more detailed presentation of the data will be given in Bender (1988b).

References

Bender, R. 1988a, *Astron. Astrophys. Letters* **193**, L7.
Bender, R. 1988b, *Astron. Astrophys. Letters*, in press.
Bender, R. Döbereiner, S., and Möllenhoff, C. 1988, *Astron. Astrophys. Suppl.*, in press.
Binney, J. and Petrou, M. 1985, *Mon. Not. Roy. Astron. Soc.* **214**, 449.
Franx, M. and Illingworth, G. D. 1988, *Astrophys. J. Letters* **327**, L55.
Gerhard, O.E. 1988, Private communication.
Kormendy, J. 1984, *Astrophys. J.* **287**, 577.
Kormendy, J. 1985, *Astrophys. J. Letters* **292**, L9.
Kormendy, J. 1988, *Astrophys. J.* **325**, 128.
Lauer, T. R. 1985, *Astrophys. J.* **292**, 104.
Nieto, J.-L. 1988, in 2*da* *Reunion de Astronomica Extragalactica*, Nat. Academy of Sci. Cordoba, in press.
Vietri, M. 1986, *Astrophys. J.* **306**, 48.
Wagner, S. J., Bender, R., and Möllenhoff, C. 1988, *Astron. Astrophys. Letters* **195**, L5.

NGC 4546, the Double–Spin SB0

Daniela Bettoni[1] and Giuseppe Galletta[2]

[1]Osservatorio Astronomico di Padova, Italy
[2]Dipartimento di Astronomia, Università di Padova, Italy

NGC 4546 is a barred S0 seen almost edge–on whose stellar disc is cut on the equatorial plane by a disc of gas rotating in the opposite direction and with relative circular velocities as great as 400 km sec^{-1}. Due to the high rotational velocities involved, the counterrotation between gas and stars is evident from a simple inspection of the spectra (Galletta 1986, 1987). Despite to this peculiarity, the morphology of NGC 4546 is quite normal: CCD images in the V band taken at the 1.52–m telescope at La Silla indicate the presence of a bright bulge, including ~50% of the total light, perfectly aligned with a stellar disc of constant flattening and with 71° of inclination with respect to the plane of the sky. The bar is quite small, extending $\sim 10''$ on either side of the nucleus, and ends with two pseudo–spiral arms. A knotted dark lane crosses the bar and the pseudo–arms. In Hα, the image of the galaxy is also quite normal: a slightly irregular disc of gas is visible, extending for $\sim 50''$ from the nucleus and with about the same ellipticity as the stellar disc.

Spectroscopy of the galaxy performed at the 2.2–m ESO/MPI telescope at La Silla indicates quite circular stellar motions within the disc with the exception of the region dominated by the bar, where orbits appear to be elongated parallel to the bar axis (Figure 1b). Gas streaming is always retrograde with respect to the stellar motions and appears to be more irregular (Figure 1d). In contrast to the stellar orbits, however, the gas motions seem to be elongated perpendicularly to the major axis of the bar. The stellar velocity dispersion σ_s and the equivalent quantity for the gas σ_g = FWHM/2.35 (obtained by deconvolving the instrumental profile from the FWHM of the [O III] line) both appear peaked toward the galaxy nucleus (Figures 1c,e), with local irregularities for the gas. The intensity ratios between [N II], [S II], and [O III] lines along the major axis indicate values intermediate between those of the nuclei of galaxies and those of galactic HII regions.

It is probable that NGC 4546 belongs to the same class of S0's with polar rings discussed by Schweizer *et al.* (1983), where the gas is known to have an angular momentum perpendicular to the axis of the stellar disc. Similarly to these cases, many elliptical galaxies with dust lanes (Bertola and Galletta 1978) show the same kind of decoupling between gas and star kinematics (see Bertola 1987 for a review). Among these, the dust lane E0.5 galaxy NGC 5898 (Bettoni 1984, Bertola and Bettoni 1988) and the E5 NGC 7097 (Caldwell *et al.* 1986) have gas counterrotating with respect to the stars, like NGC 4546. The kinematic properties of all these galaxies suggest that the gas observed at present has been recently accreted from outside, by means of collision and disruption of a *small* dark cloud or a gas–rich dwarf system. Deep imaging of the field shows that no connection is present with the galaxy ZWG 014.074, an apparently neighboring galaxy of 15th magnitude, and suggests that the original galaxy or cloud colliding with NGC 4546 has been completely disrupted.

In addition to the case of NGC 4546, another barred S0 has been very recently detected with gas and stars apparently counterrotating: this is the almost face–on galaxy NGC 2217 (Bettoni and Galletta, in preparation).

References

Bettoni, D. 1984, *The Messenger* **37**, 17.

Bertola, F. 1987, in *Structure and Dynamics of Elliptical Galaxies*, I.A.U. Symp. No. **127**, ed. T. de Zeeuw (Dordrecht: Reidel), p. 135.

Bertola, F. and Bettoni, D. 1988, *Astrophys. J.* **329**, in press.

Bertola, F. and Galletta, G. 1978, *Astrophys. J. Lett.* **226**, L115.

Caldwell, N., Kirshner, R. P., and Richstone, D. 1986, *Astrophys. J.* **305**, 136.

Galletta, G. 1986, The Messenger **45**, 18.

Galletta, G. 1987, *Astrophys. J.* **318**, 531.

Schweizer, F., Whitmore, B. C., and Rubin, V.C. 1983, *Astron. J.* **88**, 909.

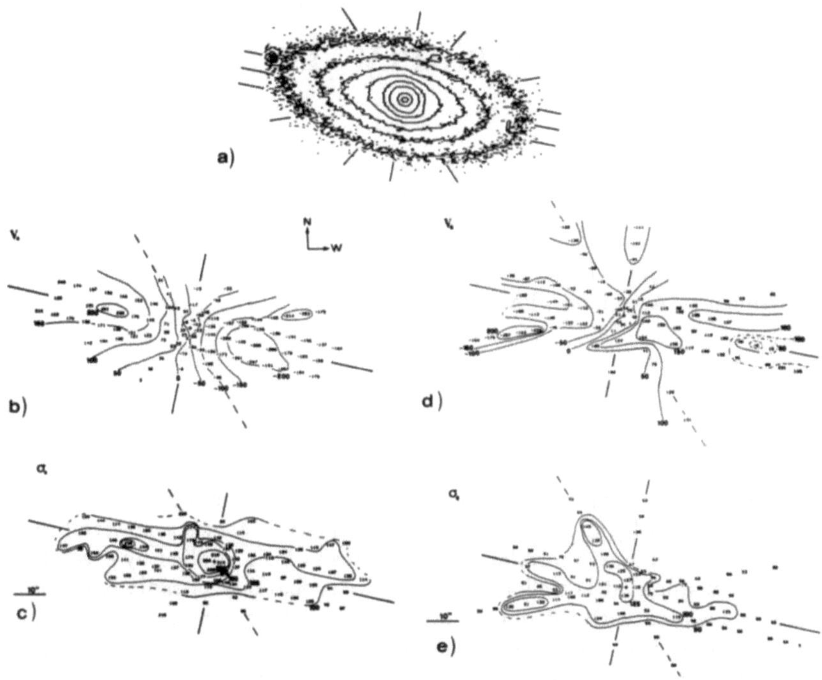

Figure 1: (a) Inner isophotes of NGC 4546, with the positions of the spectrograph slit drawn at the boundary. (b) Measured velocities for stars, rebinned in intervals of 5" each and referred to the systemic velocity. Hand fits of the lines of equal velocity are also shown. (c) The same as in b, but for the velocity dispersion. (d) The same as in b, but for the gas velocities. (e) The same as in b, but for the gas velocity dispersion, defined as σ_g = FWHM/2.35.

Radio Emission and Optical Properties of Early Type Galaxies

M. Calvani[1], G. Fasano[2], and A. Franceschini[3]

[1]Sissa–Isas, Trieste, Italy
[2]Padova Astronomical Observatory, Italy
[3]Department of Astronomy, Padova University, Italy

The Context

Statistical analyses of optical parameters of radio galaxies may help us to understand the reason why an ordinary early type galaxy becomes a powerful radio emitter. Some important correlations have already been studied by several authors. For example, it is well known that the probability of radio emission is strongly correlated with the optical luminosity of the galaxy (*e.g.* Colla *et al.* 1975, Auriemma *et al.* 1977). Strong radio emission seems also to be more common in ellipticals than in S0 or Sa galaxies (*e.g.* Heeschen 1970, Dressel 1981, Corday 1986).

A more controversial issue concerns a possible dependence of the apparent flattening on radio emission for early type galaxies, such that powerful radio ellipticals would appear rounder on average than the radio–quiet objects. This possibility was discussed, with contradictory results, by Heeschen (1970), Hummel (1980), Dressel (1981), Disney *et al.* (1984) and Sparks *et al.* (1984). In this work, we perform a thorough analysis of the relationships between radio emission and optical properties of early type galaxies.

Two independent works by Djorgowski *et al.* (1987) and Dressler *et al.* (1987) have recently suggested that the phenomenology of elliptical galaxies in the optical band can be completely described by a set of three basic parameters, namely the absolute magnitude, M (or, equivalently, the galaxy radius), the mean surface brightness, μ, and the central velocity dispersion σ. We have looked for possible correlations between such parameters and the radio luminosity. In addition, we have re–analyzed the relationship between radio emission and apparent axial ratios.

The Data Set

The sample used for our analysis is based on three independent radio surveys of optically selected galaxies, namely:

- the sample of Disney and Wall (1977) with the amendments discussed in Sparks *et al.* (1984); this gives 175 galaxies (E's and S0's) with radio flux density above 12 mJy at 5 GHz.
- The sample by Sadler (1984a,b,c). This has 248 objects with radio data at 2.7 GHz above 30 mJy. This sample partially overlaps that by Disney and Wall (74 objects in common).
- The sample of Dressel and Condon (1978), which provides 493 early type objects with radio fluxes at 2.38 GHz above 15 mJy.

Our sample therefore includes 842 early type galaxies.

It is worth stressing that:

i) our statistical analysis was carried out using ***survival analysis*** techniques, which allow us to properly exploit the information content of the upper limits to the radio flux as well as of the radio detections;

ii) in order to have more reliable morphological classifications and apparent axial ratios, we derived these quantities by averaging data from all available catalogues;

iii) at variance with most of the previous studies, we have extended the analysis also to the S0 galaxies.

Results

Dependence of $logP_5$ on M_B

Some of the results of previous papers concerning the dependence of radio emission on the optical luminosity in early type galaxies have been confirmed and strengthened by our analysis. The correlation between radio power and optical absolute magnitude, discussed by *e.g.* Fanti *et al.* (1973), Colla *et al.* (1975), Auriemma *et al.* (1977), and Dressel (1981), is confirmed here to a high level of significance. Also, we confirm the results by Auriemma *et al.* (1977) that the radio–optical luminosity correlation is *not linear*, but breaks up at $P_5 \sim 10^{22}$ (W/Hz/sr) or, equivalently, at $M_B \sim -21$.

Moreover, we find that ellipticals of high optical luminosity tend to be relatively more luminous in the radio band than S0 galaxies of the same absolute B magnitude. This result, supported by a detailed regression analysis of $logP_5$ on M_B, applies to galaxies brighter than $M_B \sim -21$. On the other hand, objects of lower optical luminosity seem to be characterized by more similar distributions of radio to optical emission, as pointed out by Franceschini *et al.* (1988).

Correlations Between $logP_5$ and σ or μ

Among the various parameters characterizing elliptical galaxies, the central velocity dispersion σ and the average surface brightness μ, are those most tightly related to the depth of the potential well inside the galaxy. We find a clear correlation between σ and radio power for E's in our sample. However, this turns out to be a second–order effect induced by the well known $M_B - \sigma$ relation. In addition, no correlation of the radio power with the mean surface brightness is seen.

These facts seem to imply that, although there is a dependence of the radio power on the total mass of the systems — as revealed by the $P_5 - M_B$ correlation — the radio power from an elliptical galaxy may not be related to the depth of the potential well. Similar results were found for the S0 galaxies in our sample, although any dependences of σ on M_B or P_5 turned out to be less pronounced than those shown by the E galaxies.

Correlation between P_5 and (b/a)

Our analysis confirms the tendency of the brightest radio E's ($logP_5$ (W/Hz/sr) ≥ 22.5) to be less flattened in comparison with the overall population. Indeed, a Kolmogorov–Smirnov test applied to the marginal distributions of apparent axial ratios for the two subsamples of E galaxies with $logP_5$ larger or smaller than 22.2 shows that the null hypothesis of a common parent population is rejected at the 99.9% confidence level. Survival analysis techniques confirm this result (correlation coefficient $CC = 0.22 \pm 0.05$).

It is important to stress that the optical selection of our sample is not responsible for such effect: we have found no dependence of the axial ratios on the absolute magnitude, in agreement with previous results (see Disney *et al.* (1984) and references therein). We refer the reader to Disney *et al.* (1984) and Sparks *et al.* (1984) for a discussion of possible interpretations of the correlation between radio power and flattening in E's.

An analysis of the dependence of P_5 on b/a for the S0 galaxies in our sample leads to the same qualitative result previously discussed for the E's, although with a considerably smaller significance ($\sim$ 95%). Of course, the existence for S0's of such correlation is not easily understood, unless we invoke effects of anisotropic radio emission. A more reasonable explanation is that high radio luminosity S0 galaxies are often misclassified objects (see Ekers and Ekers 1973).

In fact, by looking only at the eleven S0's in our sample with $P_5 > 10^{23}$, we found that, after excluding the dubious cases, the $P_5 - (a/b)$ correlation disappears.

References

Auriemma, C., Perola, G. C., Ekers, R., Fanti, R., Lari, C., Jaffe, W. J., and Ulrich, M. H. 1977, *Astron. Astrophys.* **57**, 41.

Colla, G., Fanti, C., Gioia, I., Lari, C., Lequeux, J., Lucas, R., and Ulrich, M. H. 1975, *Astron. Astrophys.* **38**, 209.

Corday, R. A. 1986, *Mon. Not. Roy. Astron. Soc.* **219**, 575.

Disney, M. J., Sparks, W. B. and Wall, J. V. 1984, *Mon. Not. Roy. Astron. Soc.* **206**, 899.

Disney, M. J. and Wall, J. V. 1977, *Mon. Not. Roy. Astron. Soc.* **179**, 234.

Djorgowsky, S. and Davis, M. 1987, *Astrophys. J.* **313**, 59.

Dressel, L. L. 1981, *Astrophys. J.* **245**, 25.

Dressel, L. L. and Condon, J. J. 1978, *Astrophys. J. Suppl.* **36**, 53.

Dressler, A., Lynden–Bell, D., Burstein, D., Davies R. L., Faber, S. M., Terlevich, R. J., and Wegner, G. 1987, *Astrophys. J.* **313**, 42.

Ekers, R. D. and Ekers, J. A. 1973, *Astron. Astrophys.* **24**, 247.

Fanti, R., Gioia, I., Lari, C., Lequeux, J. and Lucas, R. 1973, *Astron. Astrophys.* **24**, 69.

Franceschini, A. 1988, *Mon. Not. Roy. Astron. Soc.*, in press.

Heeschen, D. S. 1970, *Astron. J.* **75**, 523.

Hummel, E. 1980, Ph. D. Thesis, University of Groningen.

Sadler, E. M. 1984a, *Astrophys. J.* **89**, 23.

Sadler, E. M. 1984b, *Astrophys. J.* **89**, 34.

Sadler, E. M. 1984c, *Astrophys. J.* **89**, 53.

Sparks, W. B., Disney, M. J., Wall, J. V. and Rodgers, A. W. 1984, *Mon. Not. Roy. Astron. Soc.* **207**, 445.

Interacting Pairs of Elliptical Galaxies

E. Davoust[1], Ph. Prugniel[1], and J. Arnaud[2]

[1]Observatoire Midi–Pyrénées, Toulouse, France
[2]CFHT Corporation, Kamuela, Hawaii, U. S. A.

Summary

We have found a morphological signature of gravitational interaction in a sample of 50 close pairs of elliptical galaxies. The halos of both galaxies in five, and possibly more, pairs are offcentered and this offcentering is symmetric with respect to the center of the pair. The displacement of the nucleus with respect to the halo in each galaxy, or the formation of an asymmetric tidal bulge, are two possible causes for this effect.

We have undertaken a detailed morphological and dynamical study of a large sample of close pairs of elliptical galaxies, in order to investigate the hypothesis that they represent early stages of a merger event, which may occur after repeated close encounters. Images of 50 pairs were obtained with the 2–m telescope of Pic du Midi and with the 3.6–m CFH telescope, and spectra of 19 pairs were obtained at the 1.93–m telescope of Observatoire de Haute–Provence.

In the course of the photometric reduction, we were struck by asymmetric distortions in the isophote maps which turned up in a significant proportion of pairs. The shape of the distortions is common to all these pairs: the outer isophotes of each galaxy are displaced with respect to the inner ones in opposite senses along a direction nearly perpendicular to the line joining the centers of the two galaxies. The deformation thus appears symmetric with respect to the center of the pair.

We have found these distortions on the maps of eleven pairs. The detailed image analysis confirms that the offcenterings are intrinsic to each galaxy in five

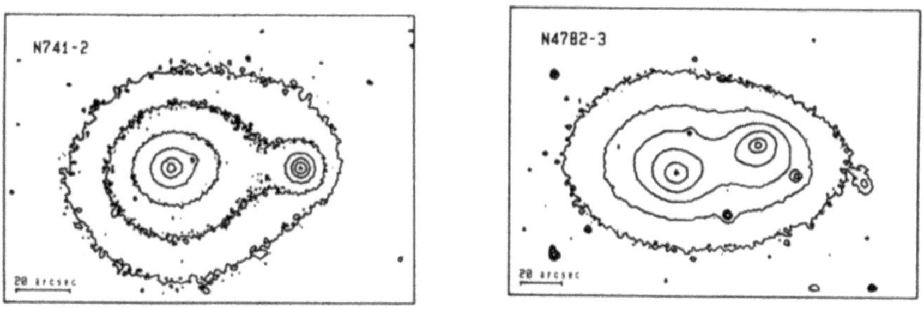

Figure 1: Isophotal maps of CCD images in the V band of two pairs of elliptical galaxies. The step is one magnitude and the lowest level is at 23 mag arcsec^{-2}.

pairs. The isophotal maps of two pairs are presented in Figure 1. The centers of the outer isophotes are indicated by crosses. The outer isophotes of both components of each pair are offcentered by about 10% in radius.

Data on the individual pairs are gathered in Table 1. The velocity differences ΔV, available for four pairs, are quite large, over 300 km sec^{-1}, the present pairs are thus experiencing near–parabolic encounters.

Table 1.

Name	ΔV	Separation		Lum.
	km s^{-1}	arcsec	kpc	ratio
NGC 741–2	369	50	11	0.09
NGC 1587–8	305	60	12	0.3
NGC 2672–3	574	33	7	0.08
NGC 4782–3	647	42	8	0.95
UGC 12064		36	12	0.25

In six additional pairs (NGC 545–7, 750–1, 5654, 5860, 7778–9 and K 399), the offcenterings might be real, but less convincing, as they could also be explained by the superimposition of two elongated ellipticals with almost parallel major axes.

Three kinds of gravitational phenomena can be invoked to explain the observed effect:

- gravitational wakes due to dynamical friction,
- differential deflections in an interpenetrating encounter,
- tidal bulges.

A dynamical wake is caused by the deflection of the particles of a galaxy in the gravitational field of another galaxy. Such wakes have been found in numerical experiments (White 1983) and predicted analytically (Weinberg 1986).

Although there is no fully self-consistent study that describes two wakes in an uneven pair, it is unlikely that a wake of the observed amplitude can be produced behind the large galaxy by the particles of the small one.

Numerical experiments by Aguilar and White (1986) have shown that the center of a galaxy is displaced after it experiences a close interpenetrating encounter with another galaxy. The dynamical time scale in the center is much shorter than the duration of the encounter, hence the core responds like a solid body and is pulled off its position. The outer halo experiences an impulsive symmetric perturbation, and its response is delayed and symmetric.

Tidal bulges are the response of galaxies to a tidal field. This field is asymmetric in close encounters: the stress is larger on the sides of the two galaxies oriented toward the barycenter of the pair. This qualitative effect has been found in numerical experiments by Dekel *et al.* (1980): the halo of the perturbed galaxy is populated by stars in elongated rather than circular orbits.

One may wonder why this morphological effect is not evident in the other close pairs of our sample. There are at least two circumstances in which this effect does not exist. Some pairs may not be physical, and others may be seen before or long after the encounter. Moreover, this effect may very well exist, but remain below the detection threshold, if the pairs are on a highly inclined orbit with respect to the line of sight, or if the galaxies are very close and of comparable mass.

We have found a remarkable morphological effect common to five pairs of elliptical galaxies, and possibly six others. The morphological distortion reported here is very useful for understanding both the internal dynamics of individual galaxies and the dynamical state of pairs, because its presence provides constraints on the structural and dynamical parameters of the systems.

References

Aguilar, L. A. and White, S. D. M. 1986, *Astrophys. J.* **307**, 97.
Dekel, A., Lecar, M., and Shaham, J. 1980, *Astrophys. J.* **241**, 946.
Weinberg, M. D. 1986, *Astrophys. J.* **300**, 93.
White, S. D. M. 1983, *Astrophys. J.* **274**, 53.

Isophotal Twisting in Isolated Elliptical Galaxies

Giovanni Fasano and Carlotta Bonoli

Padova Astronomical Observatory, Italy

Introduction

The dramatic burst of interest in early-type galaxies which took place in the late seventies was triggered both by kinematical observations (Betola and Capaccioli 1975, Illingworth 1977) and by good quality imaging revealing a degree of complexity so far unexpected in these objects. In particular, the discover of isophotal twisting gave support to the triaxial models developed by Binney (1978). Measures of twisting in large samples have been published by a number of authors (*e.g.* King 1978, Carter 1978, Strom and Strom 1978, Leach 1981, Kent 1984, Lauer 1985, Michard 1985, Djorgovsky 1986, and Jedrzejewski 1987) who obtained percentages of significantly twisted objects up to 60%.

However, Kormendy (1982), analyzing the sample used by di Tullio (1979), suggested that isophotal twisting can result mostly from tidal interactions and found a correlation between tidal class and the amount of twist in the class. If so, isolated objects should not exhibit considerable twisting. To verify this, we decided to observe, with the Asiago's CCD camera, a sample of field E's taken from the UGC catalogue.

Observations and Data Reduction

The selection of the galaxy sample was automatically made on the basis of the angular distance from the nearest object in the catalogue (not less than 8 times the galaxy's diameter) and was improved by examining each candidate object on the POSS.

Accurate sky level determinations and flat fielding checks are crucial to avoid systematic errors in the position angle (PA) and ellipticity measurements. For

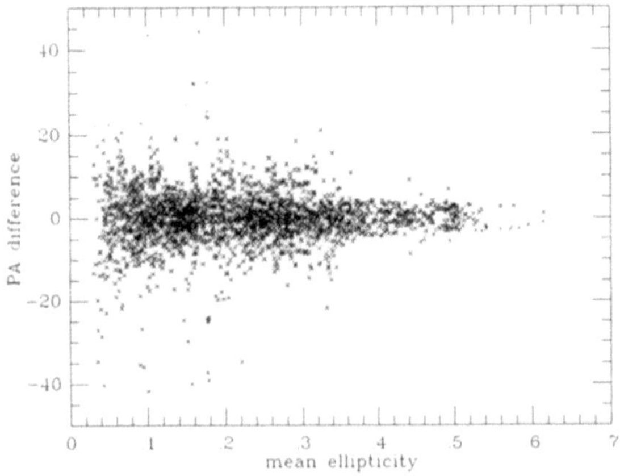

Figure 1: Differences between the PA estimates given in the literature are plotted versus the corresponding mean ellipticities. The sample consists of 57 galaxies studied by more than one author. Each point represents the difference between the PA estimates given by two different authors for the same radius in the same galaxy. Reprinted from Fasano and Bonoli (1988a) with permission.

this reason, we have considered only objects with angular dimensions small enough to secure a sufficient part of the frame to be filled with the sky and, at the same time, big enough to secure a sufficient resolution. The resulting sample consists of 45 objects, which have been observed in the R band. Only two of the galaxies in the sample were previously observed. The images were pre-reduced with the standard procedures and then analysed by INMP technique (Capaccioli *et al.* 1984) in order to fit elliptical contours to the true isophotes. In this way we have obtained the luminosity, ellipticity and PA profiles.

Systematic errors in the ellipticity and PA measurements are usually poorly considered. Indeed, Fasano and Bonoli (1988a), by comparing all the published data for objects which have been observed by more than one author, found a disagreement in the PA measurements considerably larger than quoted errors, especially for small ellipticity isophotes (see Figure 1). Then, to attach realistic error bars to our measurements, we have analysed by INMP artificial frames reproducing the observed galaxies and mimicking the estimated standard deviations in the flat fielding procedure and background subtraction.

The detailed luminosity, ellipticity and PA profiles for the entire sample are presented in Fasano and Bonoli (1988b).

Results

After the analysis of the CCD frames, some galaxies in our sample of *isolated* ellipticals turned out to be not properly isolated and/or not purely ellipticals. In particular, four objects are suspected to be the brightest members of distant clusters. We also found eight objects showing pointed isophotes or bar like profiles. That is not surprising, however, for it reflects the intrinsic heterogeneity of the so-called ellipticals (see Nieto 1988 for a comprehensive review).

We now give a brief description of the profiles:

a) Luminosity profiles. About 1/3 of the galaxies in the sample follow rather faithfully the $r^{1/4}$ law on both major and minor axes, the others showing (especially on the major axis) more or less relevant departures from it. In some cases, the luminosity profiles seem to be well described by an exponential law.

b) Ellipticity profiles. Within a large variety of behaviours, we found a predominance of outwards increasing ellipticities, with a more or less extended *plateau* in the intermediate region. This kind of profile is particularly pronounced in highly flattened objects. However, 15% of the galaxies in the sample show a decrease of the ellipticity in the outer regions.

c) Position angle profiles. We found that about half of the galaxies in the sample have PA variations larger than the estimated errors. Thus, isophotal twisting occurs in isolated galaxies roughly with the same frequency as for objects in randomly selected samples.

Discussion

We draw attention to the possible correlations between twisting and other parameters. First, our sample seems to reproduce fairly well the correlation obtained by Galletta (1980) between flattening and isophotal twisting (see Figure 2). Moreover, a first look at the data would suggest that the most evident twistings are found in those galaxies which show luminosity profiles more complex than the $r^{1/4}$ law. To investigate how systematic this behaviour can be, we estimated the departures of our luminosity profiles from the $r^{1/4}$ law by means of a $r^{1/4}$ + exponential decomposition. The resulting exponential/$r^{1/4}$ luminosity ratios (D/B) range between -1.2 and 0.5 in decimal logarithm. In Figure 3, we have plotted the maximum twistings versus $\log(D/B)$. This figure confirms our first impression that purely $r^{1/4}$ galaxies show no twisting within the errors (apart from U10103, which is an extremely round object whose outer isophotes are strongly disturbed by the presence of a small companion), whereas, when the exponential component becomes important, galaxies can exhibit high values of the twisting.

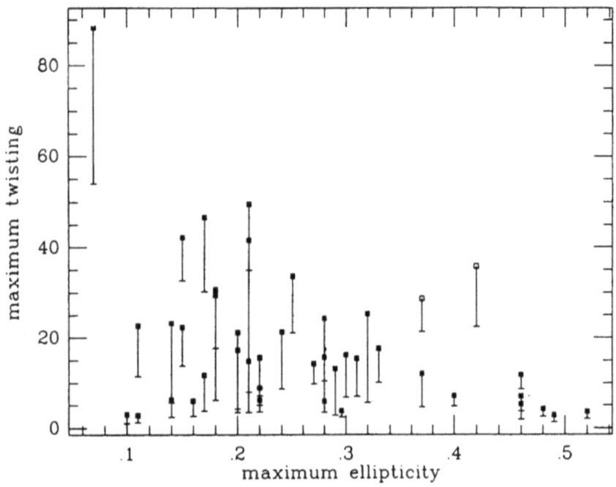

Figure 2: Galletta's (1980) correlation for our galaxy sample. The two open squares represent two suspected barred S0's (UGC 1043 and UGC 2656).

The commonly proposed mechanisms able to generate isophotal twisting in galaxies are 1) triaxiality, 2) tidal interactions, 3) dust lanes, 4) bars, and 5) change of the orientation of the principal axes. The last mechanism could be important in a scenario where elliptical galaxies mainly owe their structure to merging events (Gerhard 1983, Nieto 1988). Such a mechanism should be able to explain the correlation shown in Figure 3, if we consider elliptical galaxies as objects with two axisymmetric components ($r^{1/4}$ + exponential) whose principal axes are not aligned [a similar model was proposed by Ruiz (1976) for the central region of M 31]. Alternatively, a triaxial component with an exponential profile could be invoked, as suggested by Galletta (private communication). Both models would also reproduce Galletta's correlation fairly well, the first one requiring the reasonable assumption that more flattened $r^{1/4}$ spheroids contain exponential components less tilted than the rounder ones.

In conclusion, we would stress the need for analyzing the large samples existing in the literature in order to verify the correlation shown in Figure 3, in particular when tidal effects cannot be neglected.

References

Bertola, F. and Capaccioli, M. 1975, *Astrophys. J.* **200**, 439.

Binney, J. 1978, *Mon. Not. Roy. Astron. Soc.* **183**, 501.

Capaccioli, M., Held, E. V., and Rampazzo, R. 1984, *Astron. Astrophys.* **135**, 89.

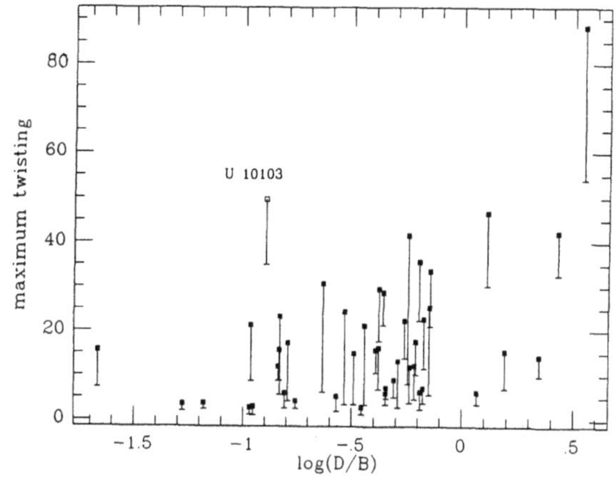

Figure 3: Maximum twisting versus $\log(D/B)$ luminosity ratio for galaxies in our sample. High values of twisting are found only for $log(D/B) > -0.35$, apart from UGC 10103, which is an almost round object whose outer isophotes are disturbed by the presence of a small companion.

Carter, D. 1978, *Mon. Not. Roy. Astron. Soc.* **182**, 797.
Djorgovski, S. 1986, Ph. D. Dissertation, University of California, Berkeley,.
Fasano, G. and Bonoli, C. 1988a, in preparation .
Fasano, G. and Bonoli, C. 1988b, in preparation .
Galletta, G. 1980, *Astron. Astrophys.* **81**, 179.
Gerhard, O. E. 1983, *Mon. Not. Roy. Astron. Soc.* **203**, 19p.
Illingworth, G. 1977, *Astrophys. J. Lett.* **218**, 43.
Jedrzejewski, R. I. 1987, *Mon. Not. Roy. Astron. Soc.* **226**, 747.
Kent, S. M. 1984, *Astrophys. J. Suppl.* **56**, 105.
King, I. R. 1978, *Astrophys. J.* **222**, 1.
Kormendy, J. 1982, in *Morphology and Dynamics of Galaxies*, Twelfth Advanced Course of the Swiss Society of Astronomy and Astrophysics, ed. L. Martinet and M. Mayor (Sauverny: Geneva Observatory), p. 115.
Lauer, T. R. 1985, *Astrophys. J. Suppl.* **57**, 473.
Leach, R. 1981, *Astrophys. J.* **248**, 485.
Michard, R. 1985, *Astron. Astrophys. Suppl.* **59**, 205.
Nieto, J. L. 1988, in *Second Extragalactic Astronomy Regional Meeting*, Cordoba, Argentina, in press .
Ruiz, M. T. 1976, *Astrophys. J.* **207**, 382.
Strom, K. M. and Strom, S. E. 1978, *Astron. J.* **83**, 73.
di Tullio, G. A. 1979, *Astron. Astrophys. Suppl.* **37**, 591.

The Formation of Elliptical Galaxies

Jean-Luc Nieto

Observatoire Midi-Pyrénées, Toulouse, France

Summary

We summarize recent results from the analysis of isophote shapes in E galaxies and present a scenario in which E galaxies are made of two classes: S0–like objects, at the low end of the L_D/L_T sequence of (early-type) galaxies, and merger products. There is a smooth transition between these two classes.

1. E's With Pointed Isophotes

E's with pointed isophotes are discussed in detail in Nieto *et al.* (1988). They present the same photometric and kinematical characteristics as bulge–dominated S0's (type S0$^-$), whose prototype is NGC 3115: they are S0–like objects with a faint disk, at the low end of the L_D/L_T sequence of (early–type) galaxies, and a rotationally flattened spheroid. Their percentage may be as high as 30 to 50% of the conventionally called E galaxies, but identifying them is difficult because of the faintness of the disk and a not always favorable orientation of the system with respect to the line of sight. In this last respect, several features in the "bona fide" E NGC 3379 suggest that this galaxy is in fact a face–on S0.

The trend with luminosity of kinematical properties of E's found by Davies *et al.* (1983) may be explained by the fact that disk–E's, as spirals or S0s, are preferentially less massive objects than large E's (Sandage *et al.* 1985).

2. E's With Boxy Isophotes

Some properties of boxy E's or S0 bulges suggest that these objects are the result of merging processes. However, boxy E's and boxy S0's present dramatically different kinematical and nuclear properties telling us that boxiness in both systems cannot be due to the same formation process. A recent analysis (Bender and Nieto 1988) of the profiles of the a_4 coefficient in the Fourier analysis of the deviations from pure elliptical isophotes in a large sample of E galaxies led to the identification of two general types of boxy E's: (a) purely boxy E's, *e.g.* E's where isophotes are boxy throughout the galaxy, and (b) boxy disk–E's, *e.g.* E's where the two types of deviations (disk and boxiness) coexist on the same isophotal contours, making them look like boxy lemons. Typical cases of the latter type are NGC 3610 or NGC 4697.

If we except a few atypical cases, boxiness in purely boxy E's is correlated with nuclear activity, anisotropy of the velocity dispersion, and conventional signatures of merging processes (see below). Boxy disk–E's, like standard boxy S0's, do not show these properties: they are isotropic systems with no particular nuclear activity.

3. Two classes of E's

It follows that we identify two main classes of "ellipticals:" disk–E's or S0–like objects, whether boxy or not, and another class that appears to result from merging processes because the members show features or properties well–recognized as signatures of merging processes (or accretion of all kinds): decoupling of gas and stars, counterrotating cores, dust–lanes, shells, polar rings, pure boxiness, nuclear activity, (possibly also anisotropy and triaxiality), etc...

However, there must be a smooth transition between these two classes, since recent observations have shown that disk–galaxies have experienced merging processes, suggesting that the disk structure resists merging processes better than believed before. It now appears that mergers more often involve one small–mass object (and a normal–mass galaxy) than two spirals of comparable mass. Although they may have quite spectacular signatures, these small mergings may not have strongly affected the original disk structure.

It is therefore quite conceivable that these two classes overlap. While merger products would be second–generation systems, disk–E's, although subjected in the past to several merging processes, still bear their original structure. The presence today of a disk in a galaxy is probably related to the history of the galaxy, and notably to the way the successive mergings have affected the original disk structure.

4. What does it explain?

We conclude that the morphological class of E galaxies is made in fact of two classes of objects having very different physical properties, in spite of an apparently similar morphology: S0–like objects and merger products. This would explain several puzzling features in E's. Notably:

(a) statistical tests on their three–dimensional shapes are inconclusive because the samples investigated are a mixture of two different classes of objects,

(b) the "remarkable" regularity (Djorgovski and Davis 1987) shown by the fundamental parameters is the result of quite frequent small merging processes that produce a smooth transition between the physical characteristics of the two classes,

(c) disk–E's (as well as spirals or S0's) have a lower mass than that of standard E's because they have experienced fewer (or less efficient) merging processes.

This would also reconcile the apparent contradiction between the continuity suggested by the Hubble sequence and the hypothesis of catastrophic merging processes as a discontinuous way of forming E's. This also suggests that (normal–mass) galaxies were all formed with disks and that the presently observed Hubble sequence is the result of continuous merging processes which have affected the original Hubble sequence. A full account of this paper is in Nieto (1988).

References

Bender, R. and Nieto, J.–L. 1988, in preparation.

Djorgovski, S. and Davis, M. 1987, *Astrophys. J.* **319**, 59.

Davies, R. L., Efstathiou, G., Fall, S. M., Illingworth, G., and Schechter, P. L. 1983, *Astrophys. J.* **266**, 41.

Nieto, J.–L., Capaccioli, M., and Held, E. V. 1988, *Astron. Astrophys.* **195**, L1.

Nieto, J.–L. 1988, "Early-Type Galaxies," in *2da Reunion Regional de Astronomia Extragalactica* (Cordoba: Academia Nacional de Ciencias de Cordoba), in press.

Sandage, A., Binggeli, B., and Tammann, G. A. 1985, in The Virgo Cluster, ed. O.–G. Richter and B. Binggeli (Garching bei München: ESO), p. 239.

Imaging Fabry–Perot Interferometry in Extragalactic Astronomy

William D. Pence

Space Telescope Science Institute, Baltimore, Maryland, U. S. A.

Summary

This paper reviews the impact that imaging Fabry–Perot interferometry has made on studies of the velocity fields in external galaxies over the past 20 years. A bibliography of all published papers is given, and some of the main scientific results are summarized. The future role of Fabry–Perot interferometry in extragalactic studies is briefly assessed.

Instrument Description

A typical Fabry–Perot interferometer consists of a reducing camera (collimator plus camera lens) which serves to provide a region of collimated (parallel) light in which the interference pre–filter and the étalon can be placed. The reducing camera also serves to decrease the focal ratio of the telescope beam to match the final image scale with the detector pixel size and to improve the efficiency of detecting faint extended sources. The detector can range from a simple photographic plate to a more complex two–dimensional photon counting device or CCD.

Without the étalon in the optical path, the instrument is simply a fast direct imaging camera. Most existing instruments are used in this configuration to take narrow–band interference filter images. When the étalon, which consists of two highly reflective parallel surfaces, is inserted in the beam the spatial image is modulated by a circular interference pattern. The flux measured at each position in the field corresponds to the emitted monochromatic radiation whose wavelength depends on the angle of incidence to the étalon and on the optical spacing between the two reflective surfaces.

Further details on the optical theory and practical design considerations of Fabry–Perot interferometry can be found in many previous reviews. See Hernandez (1986) for a mathematical analysis of the theory and practice of Fabry–Perot interferometry; and Courtès (1960, 1973), Vaughan (1967), Taylor and Atherton (1980), de Vaucouleurs and Pence (1980), Boulesteix *et al.* (1983), and Bland and Tully (1988b) for reviews of Fabry–Perot interferometers specifically designed for astronomical applications.

Comparison with Spectroscopy

The unique advantage of Fabry–Perot interferometry is that it allows a small spectral range (typical 10 – 100 Å free spectral range for galaxy kinematics) to be scanned simultaneously over a two–dimensional field of view. Thus, one sacrifices the broad spectral coverage available with a long–slit spectrograph in favor of increased spatial coverage. For applications such as mapping the velocity field of a galaxy where only a single emission line (*e.g.* Hα) needs to be measured, a Fabry–Perot interferometer is vastly more efficient than a conventional spectrograph.

Disadvantages of a Fabry–Perot interferometer include the susceptibility to ghost images caused by internal reflections between the various optical elements. In practice, this often limits the faintest emission that can be measured since it becomes uncertain whether the apparent image is real or is a reflection from brighter emission elsewhere in the field. Future optical designs should devote special attention to reduce the amount of reflected light which reaches the detector. Fabry–Perot interferometers also tend to have a brighter sky background because of overlapping orders of interference from the adjacent continuum which can be a serious problem for absorption line studies.

Another common problem with Fabry–Perot interferometers is that the narrow free spectral range that is typically used for galaxy kinematics is often insufficient to record the broader emission from the nuclear region of the galaxy. This could be overcome by using an étalon with a larger free spectral range, but at the cost of decreased velocity resolution. In practice, it is often necessary to make separate observations of the nuclear region, either with a different étalon or with a conventional spectrograph.

Static Fabry–Perot Interferometers

The first–generation Fabry–Perot devices employed an étalon with a rigidly maintained optical spacing between the two reflective surfaces. The interferograms that are produced consist of a concentric pattern of interference fringes superposed on the continuum image of the galaxy. There are typically 5 or

10 measurable fringes crossing the galaxy image, and the radial velocities are derived by measuring the radius to points along each fringe. The interferograms are analogous to a multislit spectrograph which provides simultaneous measurements at several cuts across the galaxy. In order to obtain velocity measurements at other locations in the galaxy additional exposures are necessary, after moving either the étalon or the telescope itself. This technique is most suited for galaxies which contain broad regions of relatively uniform surface brightness line emission, and not for galaxies containing only small discrete H II regions since distorted velocities will be derived if the H II region happens to lie in the wings of an interference fringe.

Marseilles Group

Following the discovery of Fabry–Perot interferometry in the late 1890s, it was first applied to the study of an astronomical source (the Orion Nebula) by Buisson, Fabry, and Bourget (1914) at Marseilles Observatory. Imaging Fabry–Perot interferometry was not used in extragalactic studies until the technique was resurrected by Courtès (1960), also at Marseilles Observatory. Courtès's first instruments required very fast, short focal length optical systems in order to record the interference fringes on unaided photographic plates. Due to the resulting small image scale the early studies tended to be of large, nearby galaxies. Over the years their instrumentation has steadily improved with the addition of image tube intensifiers, use of "insect–eye" lenses to improve the spatial coverage of the fringes, and, more recently, development of a scanning Fabry–Perot system, called CIGALE (similar to the TAURUS instrument to be discussed later).

Over the past twenty years, about 40 papers on the velocity fields of two dozen galaxies have been published by this group (Table 1) as a result of using Fabry–Perot interferometers on a variety of telescopes around the world. For most of these galaxies, the papers present catalogues of H II regions, derived rotation curves, fits to various mass models, and in some cases, discussions of deviations from circular rotation. Studies of several Sc galaxies revealed the existence of a new class of very extended H II regions (Monnet 1971) which could be detected over the entire disk of the galaxies and proved invaluable for mapping their velocity field. The paper on M 33 by Boulesteix, Dubout–Crillon, and Monnet (1981) is notable for the novel use of Fabry–Perot interferograms to map the [O III]/Hβ line ratio over the galaxy to measure the physical conditions in the ionized disk gas. Recent studies with their scanning instrument include the central region of M 31 (Boulesteix *et al.* 1987), and the interacting pair of galaxies NGC 7752–7753 (Marcelin *et al.* 1987).

Table 1.
Bibliography of Fabry-Perot Papers on Galaxies

	Marseilles Group
M 33	Carranza *et al.* 1968
M 31	Deharveng and Pellet 1969
NGC 4449	Crillon and Monnet 1969a
M 51	Carranza, Crillon, and Monnet 1969
NGC 4631	Crillon and Monnet 1969b
Cen A	Sersic 1969
NGC 2403	Deharveng and Pellet 1970a
NGC 4490-85	Boulesteix, Dubout-Crillon, and Monnet 1970
M 33	Deharveng and Pellet 1970b
M 33	Boulesteix and Monnet 1970
LMC	Georgelin and Monnet 1970
SMC, LMC	Carranza, Monnet, and Cheriguene 1971
M 33	Monnet 1971
LMC	Cheriguene and Monnet 1972
M 33,31,51, N 4631	Courtès 1973
M 33	Comte and Monnet 1974
M 31	Deharveng and Pellet 1975
NGC 628	Monnet and Deharveng 1977
NGC 7793	Carranza and Aguero 1977
M 33	Boulesteix *et al.* 1979
M 101	Comte, Monnet, and Rosado 1979
NGC 2997	Milliard and Marcelin 1981
LMC	Rosado *et al.* 1981
M 33	Boulesteix, Dubout-Crillon, and Monnet 1981
M 83	Comte 1981
NGC 1313	Marcelin and Athanassoula 1982
NGC 925	Marcelin, Boulesteix, and Courtès 1982
Cen A	Marcelin *et al.* 1982
NGC 1566	Comte and Duquennoy 1982
NGC 1566	Comte 1983
NGC 7741	Duval 1983
Cen A	Marcelin 1983
NGC 2903	Marcelin, Boulesteix, and Georgelin 1983
NGC 300	Marcelin, Boulesteix, and Georgelin 1985a
NGC 7741	Duval and Monnet 1985
LMC, SMC	Marcelin, Boulesteix, and Georgelin 1985b
LMC	Lavel *et al.* 1987
M 31	Boulesteix *et al.* 1987
NGC 7752-53	Marcelin *et al.* 1987
NGC 6946	Bonnarel *et al.* 1988
NGC 4321	Arsenault *et al.* 1988

Table 1 (continued).

	Texas Group
NGC 1569	de Vaucouleurs, de Vaucouleurs, and Pence 1974
NGC 253	Pence 1978
NGC 7793	Davoust and de Vaucouleurs 1979
NGC 2537	de Vaucouleurs and Pence 1980
NGC 7793	Davoust and de Vaucouleurs 1980
NGC 253	Pence 1981
NGC 6503	de Vaucouleurs and Caulet 1982
M 83	de Vaucouleurs, Pence, and Davoust 1983
N3351,4725,4736	Buta 1988
	University of Maryland
M 51	Tully 1974
	TAURUS
M 82	Taylor and Atherton 1980
NGC 5253	Atherton *et al.* 1982
Vela Ring Galaxy	Taylor and Atherton 1982
M 83	Allen *et al.* 1983
Cen A	Taylor and Atherton 1983
Cen A	Phillips *et al.* 1984
NGC 1068	Atherton, Reay, and Taylor 1985
NGC 5643, 7582	Morris *et al.* 1985
NGC 247, 3109	Carignan and Freeman 1985
NGC 3109	Carignan 1985
NGC 1365	Teuben *et al.* 1986
NGC 1433	Buta 1986
NGC 7531	Buta 1987
NGC 2685, 4650A	Nicholson *et al.* 1987
Cen A	Bland, Taylor, and Atherton 1987a
Cen A	Bland, Taylor, and Atherton 1987b
NGC 4650A	Sparke, Taylor, and Nicholson 1987
NGC 4027	Pence *et al.* 1988
NGC 1365	Edmunds, Taylor, and Turtle 1988
NGC 1512	Buta 1988
Cen A	Bland and Taylor 1988
Cen A	Sparke *et al.* 1988

Table 1 (concluded).

	Rutgers/CTIO
M 82	Williams, Caldwell, and Schommer 1984
NGC 4449	Malumuth, Williams, and Schommer 1986
Cen A	Hayes, Schommer, and Williams 1987
NGC 5728	Schommer *et al.* 1988
	Hawaii
NGC 4151	Cecil, Bland, and Tully 1987
M 82	Bland and Tully 1988
M 51	Cecil 1988
NGC 1068	Cecil, Bland, and Tully 1988
NGC 1068	Bland *et al.* 1988
NGC 1068	Widemann *et al.* 1988

Texas Group

In the early 1970s, Gérard de Vaucouleurs at the University of Texas developed a Fabry–Perot interferometer, called the Galaxymeter, which was modeled on the design of the French instrument. The main improvement in this instrument was the incorporation of an image tube intensifier to increase the speed of the system. (Tully, 1974, was actually the first to use an image intensifier on a Fabry–Perot interferometer; however, it was only used to study one galaxy). Because of its greater sensitivity, the Galaxymeter could afford to use a slower, longer focal length camera lens, with a resulting significant increase in image size and angular resolution. For this reason, the Galaxymeter could be used to study smaller and more distant galaxies than were previously accessible.

This instrument has been used by de Vaucouleurs and his collaborators to continue their studies of late–type spiral galaxies and, in particular, the Sculptor group of galaxies. The papers on the two Sculptor Group spiral galaxies NGC 253 (Pence 1981) and NGC 7793 (Davoust and de Vaucouleurs 1980) both present very extensive photometric and kinematic observations and derived mass models. More recently, Buta (1988) used the Galaxymeter with an improved larger image tube to study the dynamics of several ringed spiral galaxies.

Scanning Fabry–Perot Interferometers

In order to take full advantage of the capabilities offered by Fabry–Perot interferometry, one needs to be able to scan across the free spectral range of the étalon at each position in the two–dimensional image by varying the optical path length between the two reflecting surfaces of the étalon. In this way one can measure the line profile, and hence the radial velocity at each image pixel, and eliminate the ambiguity present in fixed–étalon instruments caused by the convolution of the spatial intensity variations with the instrumental interference pattern. Two techniques have been used to modulate the optical path length of the étalon: pressure variation, and piezo–electric variation of the physical separation of the reflective surfaces.

University of Maryland

Tully (1974) at the University of Maryland was the first to study an external galaxy using a scanning Fabry–Perot interferometer. The étalon was contained in a pressure chamber which could be regulated to vary the index of refraction of the gas medium between the reflective surfaces. A series of 15 image tube intensified photographic interferograms of M 51 were obtained at different pressure settings which covered the free spectral range of the étalon. The plates were digitized and the Hα line profile then was reconstructed from the measured interferogram intensities at each point in the field. This study of M 51 produced the most detailed velocity map of any previously studied external galaxy, and it remains as one of the few studies to find direct kinematic evidence for velocity streaming across the arms as predicted by the spiral density wave theory.

TAURUS

Tully's pioneering study of M 51 demonstrated the power of scanning Fabry–Perot interferometry, but two key technical advances were necessary to improve its reliability and convenience before the technique would become generally accessible. The first improvement was to replace the pressure scanning mechanism, which is limited in range and is difficult to regulate, by étalons in which the gap is precisely controlled by piezo-electric spacers. These étalons are currently marketed by Queensgate Instruments, Ltd. and have been used in most scanning Fabry–Perot instruments. The second improvement was to replace the photographic plate with a photon counting detector or CCD to increase the quantum efficiency and provide a more convenient direct digital (and linear) measurement.

These technical advances came together in the late 1970s with the development of the TAURUS Fabry–Perot interferometer (Taylor and Atherton 1980) used at the Anglo–Australian Telescope. This instrument combined a variable gap étalon with the Image Photon Counting System (IPCS) already in use at the AAO to provide a major leap forward in control and versatility. Under microcomputer control, the étalon can be rapidly stepped across the free spectral range, and the photon events recorded at each étalon setting are co–added into a three–dimensional data cube (two spatial and one spectral dimension) stored in computer memory. Repeated scanning over the spectral range minimizes any variations in the observing conditions (seeing, or transparency) between the different frames in the data cube during the integration period.

The use of a linear detector and fine sampling of the free spectral range provides information on the entire line profile, not just the centroid, or radial velocity. Thus, TAURUS has been effectively used in the study of complicated dynamical systems with multiple emission line components or asymmetric profiles such as in the central regions of NGC 1068 (Atherton, Reay, and Taylor 1985) and NGC 1365 (Teuben *et al.* 1986; Edmunds, Taylor, and Turtle 1988). Arguably, the most extensive analysis to date of any external galaxy has been by Bland and collaborators on the peculiar galaxy Centaurus A (Bland, Taylor, and Atherton 1987, Bland and Taylor 1988; Sparke, Nicholson, Taylor, and Bland 1988). This series of papers is based on some 17500 Hα and 5300 [N II] TAURUS spectra and concludes that the prominent dust lane is a projection of a severely warped, thin disk of gas and dust. This study also illustrates what is perhaps an increasing trend in this field: a few nights of observing with a Fabry–Perot interferometer can provide an almost overwhelming amount of data which can take months, or even years to interpret.

The success of the original TAURUS instrument on the AAT during 1981 – 1983, and on the Issac Newton Telescope at La Palma since 1984, has lead to the development of a second generation version, called TAURUS II, which has recently been commissioned on both the AAT and the William Herschel Telescope. TAURUS II was designed as a common–user facility for these observatories and is now being used during a significant fraction of the total observing time on these telescopes.

Rutgers/CTIO

At about the same time that TAURUS was being built, a group at Rutgers University (Roesler *et al.* 1982) developed their own pressure scanned Fabry–Perot instrument using a CCD detector. This was used for studies of two galaxies: M 82 (Williams, Caldwell, and Schommer 1984) and NGC 4449 (Malumuth, Williams, and Schommer 1986). A later version of the instrument, using a Queensgate piezo-electric controlled étalon, has recently been on extended loan at CTIO, and has been used on a number of interesting projects, most of which still in the data reduction and analysis stage.

Hawaii

Scanning Fabry–Perot systems are now available on both the CFH 3.6-m and the University of Hawaii 2.2-m telescopes. The French supplied CFHT system can be used either with a photon–counting detector or a CCD, while the U. H. system (Bland and Tully 1988b) uses only CCD detectors. The advantages of a CCD over photon counting systems are the small geometric distortion (which simplifies the data reduction process) and the fact that CCDs can be used on bright, large dynamic range objects such as galactic nuclei. CCDs have the disadvantage that the read–out noise and relatively long read–out times limit the scanning rate of the system, which makes the data more susceptible to atmospheric variations during the observation.

The limited results published so far with these instruments are already providing some interesting new discoveries. For instance, Bland and Tully (1988a) used the CFHT Fabry–Perot interferometer to demonstrate that the emission from the optical halo and filaments around M 82 consists of two distinct components: bright narrow–line filaments in a bipolar outflow, and a diffuse slowly rotating broad–line halo. The study of NGC 1068 (Cecil, Bland and Tully 1988) using the U.H. system should also be mentioned as the first case in which a polarizing filter was combined with a scanning Fabry–Perot interferometer to obtain spectropolarimetric measurements in the narrow line region near the Seyfert nucleus.

Summary Statistics and Future Trends

As of the date that this review was written (April 1988) there were a total of 83 papers in the literature reporting the results of imaging Fabry–Perot interferometry on some 48 different galaxies. This is an average of only 4 papers per year over the past 20 years, but the number of published papers has been increasing very rapidly over the past few years. There were 9 Fabry–Perot papers published in 1987, and 15 papers have already been published, or are in press, in the first 4 months of 1988. It seems clear that the number of Fabry–Perot studies of galaxies will continue to grow as new instruments proliferate to other major observatories.

It is also apparent that only a fraction of the world's extragalactic astronomers currently have had much experience with Fabry–Perot interferometers as compared with conventional spectrographs or photometers. This is demonstrated by the fact that there is a total of only 69 different authors on the Fabry–Perot papers of which only 40 of the names appear on more than one paper. This limited access to Fabry–Perot data is understandable due to the complexity of operating the early instruments. In many cases, only the original developers of the instrument were qualified to observe with them. This situation has changed substantially with the new generation of instruments, exemplified by TAURUS II, which are much simpler to operate and are specifically designed as common–user instruments. At some observatories, Fabry–Perot observing has become so routine that it is offered as a remote service to astronomers; in this way, several different proposals requiring a similar instrumental configuration can be executed more efficiently in a single run by an observatory scientist, and the resulting data are then sent to the requesting astronomer. In addition, the initial data reduction procedure for Fabry–Perot data has been automated to the point where it is not unreasonable to expect that in the near future astronomers will receive calibrated velocity, linewidth, and line intensity maps without having to do any of the observing or processing themselves. They can then concentrate instead on the more interesting problem of analyzing and interpreting the resulting maps.

The main bottleneck in the process of publishing results obtained with imaging Fabry–Perot interferometry is in the analysis and interpretation stage. There can be so much complex detail in the final velocity field and line profile maps that it overwhelms our current ability to understand and model the data. It is often necessary, at least initially, to simply concentrate on the overall features and trends in the data. There is a great need for more powerful computer software and hardware to deal with the large n–body and hydrodynamical codes that are necessary to model the observed galaxy kinematics. Use of powerful interactive imaging microcomputer work stations, with links to supercomputers for large numerical simulations, seems to offer the best hope for improving this situation. The growth of standard data analysis packages (*e.g.* STARLINK, IRAF, MIDAS) should also make the reduction and analysis software more generally available.

In the future, we can expect to see Fabry–Perot interferometry used in several exciting new ways. It should be possible to map stellar velocity fields by measurements of absorption lines profiles using an imaging Fabry–Perot interferometer; several different groups are currently working on projects of this type. It remains to be seen how sensitive this technique will be (continuum light from neighboring orders of interference will tend to reduce the contrast of the absorption lines) but if it is successful, it would be a very useful technique for mapping the velocity field in the center of triaxial elliptical galaxies, for instance, which in the past have only been sampled by long slit spectroscopy at a few position angles. Fabry–Perot interferometry should also be an efficient method of measuring the rotation curves in field spiral galaxies as a substitute for the H I line–width used in the Tully–Fisher distance indicator relation. The additional information available from knowing the shape of the rotation curve, as well as its amplitude, may enable more precise distance determinations to be made.

The most exciting new development in the use of Fabry–Perot interferometry is probably the extension to the infrared wavelength region. Dramatic new observations in the 1 – 5 micron infrared region by T.R. Geballe, I.S. McLean, I. Gatley, and G. Wright have been reported (Gatley 1988) using a 58 x 62 InSb array mounted behind a Fabry–Perot interferometer at UKIRT. While it is still too early to report specific results, it is evident that this technique will lead to many exciting new discoveries in objects which are inaccessible at visible wavelengths.

References

Allen, R. J., Atherton, P. D., Oosterloo, T. A., and Taylor, K. 1983, in *I. A. U. Symp. 100, Internal Kinematics and Dynamics of Galaxies*, ed. E. Athanassoula (Dordrecht: Reidel), p. 147.

Arsenault, R., Boulesteix, J., Georgelin, Y., and Roy, J. 1988, *Astron. Astrophys.***200**, 29.

Atherton, P. D., Reay, N. K., and Taylor, K. 1985, *Mon. Not. Roy. Astron. Soc.* **216**, 17p.

Atherton, P. D., Taylor, K., Pike, C. D., Harmer, C. F. W., Parker, N. M., and Hook, R. N. 1982, *Mon. Not. Roy. Astron. Soc.* **201**, 661.

Bland, J., Cecil G. N., Tully R. B., and Widemann, Th. 1988, preprint.

Bland, J. and Taylor, K. 1988, *Mon. Not. Roy. Astron. Soc.*, submitted.

Bland, J., Taylor, K., and Atherton, P. D. 1987a, in *I. A. U. Symposium 127, Structure and Dynamics of Elliptical Galaxies*, ed. T. de Zeeuw (Dordrecht: Reidel), p. 417.

Bland, J., Taylor, K., and Atherton, P. D. 1987b, *Mon. Not. Roy. Astron. Soc.* **228**, 595.

Bland, J. and Tully, R. B. 1988a, *Nature*, in press.

Bland, J. and Tully, R. B. 1988b, in *New Directions in Spectrophotometry*, ed. A. G. Davis Philip, D. S. Hayes, and S. J. Adelman, (Schenectady: L. Davis Press).

Bonnarel, F., Boulesteix, J., Georgelin, Y. P., Lecoarer, E., Marcelin, M., Bacon, R., and Monnet, G. 1988, *Astron. Astrophys.* **189**, 59.

Boulesteix, J., Colin, J., Athanassoula, E., and Monnet, G. 1979, *Photometry, Kinematics, and Dynamics of Galaxies*, ed. D. Evans (Austin: Univ. Texas Astron. Dept.), p. 271.

Boulesteix, J., Dubout–Crillon, R., and Monnet, G. 1970, *Astron. Astrophys.* **8**, 204.

Boulesteix, J., Dubout–Crillon, R., and Monnet, G. 1981, *Astron. Astrophys.* **104**, 15.

Boulesteix, J., Georgelin, Y. P., Lecoarer, E., Marcelin, M., and Monnet, G. 1987, *Astron. Astrophys.* **178**, 91.

Boulesteix, J., Georgelin, Y. P., Marcelin, M., and Monnet, G. 1983, *SPIE Conf. Inst. Astron.* **V., 445**, 37.

Boulesteix, J. and Monnet, G. 1970, *Astron. Astrophys.* **9**, 350.

Buisson, H., Fabry, Ch., and Bourget, H. 1914, *Astrophys. J.* **40**, 241.

Buta, R. 1986, *Astrophys. J. Suppl.* **61**, 631.

Buta, R. 1987, *Astrophys. J. Suppl.* **64**, 1.

Buta, R. 1988, *Astrophys. J. Suppl.* **66**, 233.

Carignan, C. 1985, *Astrophys. J.* **299**, 59.

Carignan, C. and Freeman, K. C. 1985, *Astrophys. J.* **294**, 494.

Carranza, G., Courtès, G., Georgelin, Y. P., Monnet, G., and Pourcelot, A. 1968, *Ann. d'Astrophys.* **31**, 63.

Carranza, G., Crillon, R., and Monnet, G. 1969, *Astron. Astrophys.* **1**, 479.
Carranza, G., Monnet, G., and Cheriguene, M. F. 1971, *Astron. Astrophys.* **10**, 467.
Carranza, G. J. and Aguero, E. L. 1977, *Astrophys. Sp. Sci.*, **47**, 397.
Cecil, G. N. 1988, *Astrophys. J.* **329**, 38.
Cecil, G. N., Bland, J., and Tully, R. B. 1987, in *Active Galactic Nuclei*, ed. P. J. Wiita (Atlanta: Georgia St. Univ).
Cecil, G. N., Bland, J., and Tully, R. B. 1988, *Astrophys. J.*, submitted.
Cheriguene, M. F. and Monnet, G. 1972, *Astron. Astrophys.* **16**, 28.
Comte, G. 1981, *Astron. Astrophys. Suppl.* **44**, 441.
Comte, G. 1983, in *I. A. U. Symp. 100, Internal Kinematics and Dynamics of Galaxies*, ed. E. Athanassoula (Dordrecht: Reidel), p. 151.
Comte, G., Monnet, G., and Rosado, M. 1979, *Astron. Astrophys.* **72**, 73.
Comte, G. and Duquennoy, A. 1982, *Astron. Astrophys.* **114**, 7.
Comte, G. and Monnet, G. 1974, *Astron. Astrophys.* **33**, 161.
Courtès, G. 1960, *Ann. d'Astrophys.* **23**, 115.
Courtès, G. 1973, *Vistas in Astron.* **14**, 81.
Crillon, R. and Monnet, G. 1969a, *Astron. Astrophys.* **1**, 449
Crillon, R. and Monnet, G. 1969b, *Astron. Astrophys.* **2**, 1.
Davoust, E. and de Vaucouleurs, G. 1979, in *Photometry, Kinematics and Dynamics of Galaxies,* ed. D. Evans (Austin: Univ. Texas Astron. Dept.), p. 255.
Davoust, E. and de Vaucouleurs, G. 1980, *Astrophys. J.* **242**, 30.
Deharveng, J. M. and Pellet, A. 1969, *Astron. Astrophys.* **1**, 208.
Deharveng, J. M. and Pellet, A. 1970a, *Astron. Astrophys.* **7**, 210.
Deharveng, J. M. and Pellet, A. 1970b, *Astron. Astrophys.* **9**, 181.
Deharveng, J. M. and Pellet, A. 1975, *Astron. Astrophys.* **38**, 15.
Duval, M. F. 1983, in *I. A. U. Symp. 100, Internal Kinematics and Dynamics of Galaxies*, ed. E. Athanassoula, (Dordrecht: Reidel), p. 237.
Duval, M. F. and Monnet, G. 1985, *Astron. Astrophys. Suppl.* **61**, 141.
Edmunds, M. G., Taylor, K., and Turtle, A. J. 1988, *Mon. Not. Roy. Astron. Soc.*, in press.
Gatley, I. 1988, *Bul. Am. Astron. Soc.* **19**, 1079.
Georgelin, Y. and Monnet, G. 1970, *Astrophys. Lett.* **5**, 213.
Hayes, J., Schommer, R. A., and Williams, T. B. 1987, in *I. A. U. Symp. 127, Structure and Dynamics of Elliptical Galaxies*, ed. T. de Zeeuw (Dordrecht: Reidel), p. 419.
Hernandez, G. 1986, *Fabry–Perot Interferometers* (Cambridge: Cambridge Univ. Press).
Lavel, A., Boulesteix, J., Georgelin, Y. P., Georgelin, Y. M., and Marcelin, M. 1987, *Astron. Astrophys.* **175**, 199.
Malumuth, E. M., Williams, T. B., and Schommer, R. A. 1986, *Astron. J.* **91**, 1295.

Marcelin, M. 1983, in *I. A. U. Symp. 100, Internal Kinematics and Dynamics of Galaxies*, ed. E. Athanassoula (Dordrecht: Reidel), p. 335.
Marcelin, M. and Athanassoula, E. 1982, *Astron. Astrophys.* **105**, 76.
Marcelin, M., Boulesteix, J., Courtès, G., and Milliard, B. 1982, *Nature* **297**, 38.
Marcelin, M., Boulesteix, J., and Courtès, G. 1982, *Astron. Astrophys.* **108**, 134.
Marcelin, M., Boulesteix, J., and Georgelin, Y. 1983, *Astron. Astrophys.* **128**, 140.
Marcelin, M., Boulesteix, J., and Georgelin, Y. P. 1985a, *Astron. Astrophys.* **151**, 144.
Marcelin, M., Boulesteix, J., and Georgelin, Y. 1985b, *Nature* **316**, 705.
Marcelin, M., Lecoarer, E., Boulesteix, J., Georgelin, Y., and Monnet, G. 1987, *Astron. Astrophys.* **179**, 101.
Milliard, B. and Marcelin, M. 1981, *Astron. Astrophys.* **95**, 59.
Monnet, G. 1971, *Astron. Astrophys.* **12**, 379.
Monnet, G. and Deharveng, J. M. 1977, *Astron. Astrophys.* **58**, L1.
Morris, S. L., Ward, M. J., Whittle, D. M., Wilson, A. S., and Taylor, K. 1985, *Mon. Not. Roy. Astron. Soc.* **216**, 193.
Nicholson, R. A., Taylor, K., Sparks, W. B., and Bland, J. 1987, in *I. A. U. Symp. 127, Structure and Dynamics of Elliptical Galaxies*, ed. T. de Zeeuw (Dordrecht: Reidel), p. 415.
Pence, W. 1978, *Univ. Texas Publ. in Astron.* No. **14**.
Pence, W. 1981, *Astrophys. J.* **247**, 473.
Pence, W. D., Taylor, K., Freeman, K. C., de Vaucouleurs, G., and Atherton, P. 1988, *Astrophys. J.* **326**, 564.
Phillips, M. M., Taylor, K., Axon, D. J., Atherton, P. D. and Hook, R. N. 1984, *Nature* **310**, 554.
Roesler, F. L., Oliversen, R. J., Scherb, F., Lattis, J., Williams, T. B., York, D. G., Jenkins, E. B., Lowrance, E. B., Zucchino, P., and Long, D. 1982. *Astrophys. J.* **259**, 900.
Rosado, M., Georgelin, Y. P., Georgelin, Y. M., Lavel, A., and Monnet, G. 1981, *Astron. Astrophys.* **97**, 342.
Schommer, R. A., Caldwell, N., Wilson, A. S., Baldwin, J. A., Phillips, M. M., Williams, T. B., and Turtle, A. J. 1988, *Astrophys. J.* **324**, 154.
Sersic, J. L. 1969, *Nature* **224**, 253.
Sparke, L. S., Nicholson, R. A., Taylor, K., and Bland, J. 1988, preprint.
Sparke, L. S., Taylor, K., and Nicholson, R. A. 1987, in *I. A. U. Coll. 96, The Few Body Problem*, ed. M. J. Valtonen (Dordrecht: Reidel).
Taylor, K. and Atherton, P. D. 1980, *Mon. Not. Roy. Astron. Soc.* **191**, 675.
Taylor, K. and Atherton, P. D. 1982, *Mon. Not. Roy. Astron. Soc.* **208**, 601.
Taylor, K. and Atherton, P. D. 1983, 1983, in *I. A. U. Symp. 100, Internal Kinematics and Dynamics of Galaxies*, ed. E. Athanassoula (Dordrecht: Reidel), p. 331.

Teuben, P. J., Sanders, R. H., Atherton, P. D., and van Albada, G. D. 1986, *Mon. Not. Roy. Astron. Soc.* **221**, 1.
Tully, R. B. 1974, *Astrophys. J. Suppl.* **27**, 415.
de Vaucouleurs, G. and Caulet, A. 1982, *Astrophys. J. Suppl.* **49**, 515.
de Vaucouleurs, G., de Vaucouleurs, A., and Pence, W. 1974, *Astrophys. J. Lett.* **194**, L119.
de Vaucouleurs, G. and Pence, W. 1980, *Astrophys. J.* **242**, 18.
de Vaucouleurs, G., Pence, W., and Davoust, E. 1983, *Astrophys. J. Suppl.* **53**, 17.
Vaughan, A. H. 1967, *Ann. Rev. Astron. Astrophys.* **5**, 139.
Widemann, Th., Bland, J., Tully, R. B., and Cecil, G. N. 1988, preprint.
Williams, T. B., Caldwell, N., and Schommer, R. A. 1984, *Astrophys. J.* **281**, 579.

Discussion

J.–C. Pecker: The beautiful data you have shown concerning such objects as Centaurus A or M 82 lead me to ask whether or not systematic studies of "active" galaxies (say Seyfert I or Seyfert II galaxies, or the jet of M 87) are under way in a realistic scheme? It would be of the highest interest.

Pence: The relatively narrow free spectral range (FSR) of the étalons which were typically used in the first generation Fabry–Perot interferometers was not suitable for studying very broadline emission objects such as Seyfert galaxies. Recently, however, the University of Hawaii instrument has been used with a wide FSR étalon to study several Seyfert galaxies.

R. B. Tully: At Hawaii, we are concentrating on the active galactic nuclei problem. We have a Fabry–Perot étalon that combines large free spectral range and high finesse, hence reasonable spectral resolution. This also avoids the problem of spectral confusion. It is turning out that there are frequently extended regions about nearby active nuclei with anomalous velocity fields that must be rlated to the energetic nuclear events.

R. J. Buta: One of the great values of the Fabry–Perot approach is, of course, the two–dimensional velocity fields that are derived. This is especially important for barred galaxies, which have generally been avoided in large–scale rotation curve studies owing to difficulties with non–circular motions and derivation of orientation parameters. With good velocity coverage, we can attempt to extract rotation curves of barred galaxies and thereby achieve a better balance among the galaxy family dimension for questions of rotation properties and dark matter.

A Circumnuclear Ring of Enhanced Star Formation in the Spiral Galaxy NGC 4321*

Robin Arsenault[1], Jacques Boulesteix[2], Yvon Georgelin[2], and Jean-Rene Roy[3]

[1] European Southern Observatory, Garching bei München, F.R.G.
[2] Observatoire de Marseille, Marseille, France
[3] Département de Physique, Université Laval, Québec, Canada

Observations

Thirty-two $H\alpha$ interferograms (1.5 arcsec/pixel) have been obtained on the spiral galaxy NGC 4321 with CIGALE, the photon counting camera of the Observatoire de Marseille, coupled with a scanning Fabry-Perot interferometer. The interferograms have been transformed into 31 monochromatic images of 0.037 nm bandwidth, covering a spectral range centered on the redshifted $H\alpha$ line. An $H\alpha$ image (continuum free; see Figure 1) and a 656.3 nm continuum image have also been obtained from those interferograms. Finally, the radial velocity of the ionized gas has been measured for more than 10,000 points across NGC 4321 (Arsenault *et al.* 1988).

The $H\alpha$ image shows an unusual double-lobe structure in the nuclear region. The 656.3 nm continuum image does not show any corresponding feature. The $H\alpha$ velocity field shows a high velocity gradient in the central part as well as two local maximas following the central rapid velocity rise. Non-circular motions are seen associated with the spiral arms and the bar.

*Based on observations taken at the Canada-France-Hawaii Telescope

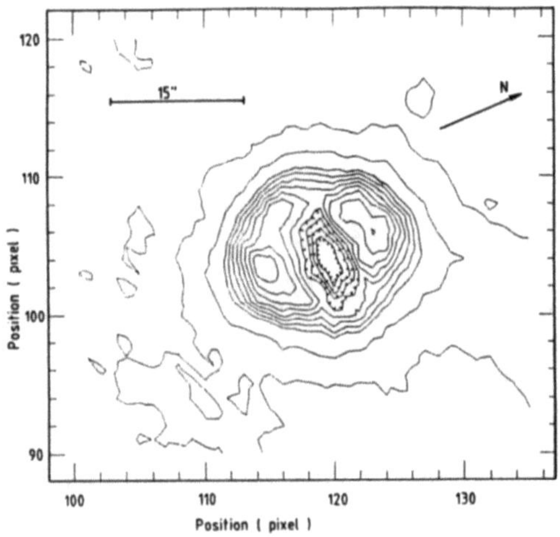

Figure 1: Hα image of the nuclear region of NGC 4321. Ticked contours indicate a depression (hole).

Interpretation

The Hα velocity field allows one to determine the rotation curve and therefore the position of the Inner Lindblad Resonances (ILR) in NGC 4321. This is illustrated in Figure 2 where the plot shows the function $\omega - \kappa/2$, ω being the angular velocity and κ the epicyclic frequency. The horizontal line represents the angular velocity of the spiral pattern. We assume that corotation occurs at the end of the bar that is at a radius of 4700 pc, adopting a distance of 13.8 Mpc (de Vaucouleurs 1984). The ILRs are found at the intersection of the spiral pattern angular velocity with the function $\omega - \kappa/2$. NGC 4321 has two ILRs located at 350 and 1100 pc. These radii correspond almost perfectly with the physical extent of the Hα nuclear ring.

A clear scenario appears where the Hα gas clouds orbiting in the gravitational potential of NGC 4321 loose angular momentum at the crossing of the dust lanes along the bar. Therefore, gas is continuously fed to the nucleus. The orbits of the gas clouds are strongly perturbed at the ILRs. Models show that between the galactic center and the first ILR (inner ILR), stable orbits are elongated parallel to the bar; between the two ILRs (inner ILR and outer ILR) they are perpendicular to the bar; and again parallel to the bar outside the outer ILR (Contopoulos and Papayannopoulos 1980). Those perturbations of orbits increase the collision rate between the clouds which in turn increases the star formation rate at these radii.

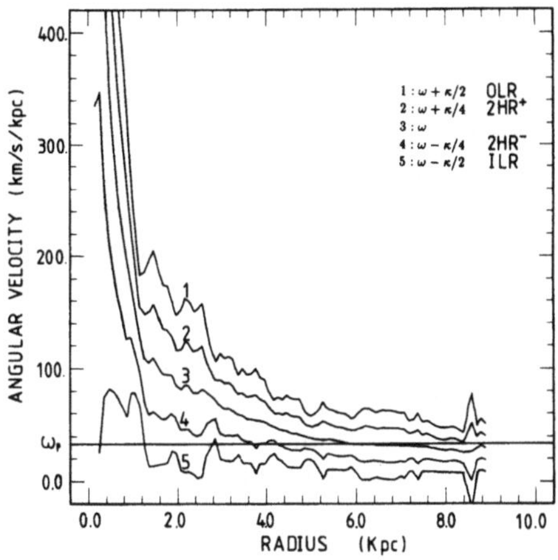

Figure 2: There is a resonance when the expression $\omega \pm \frac{\kappa}{N}$ equals the pattern rotation speed.

Further results

A morphological comparison between NGC 4321 and the the computer simulations of Schwarz (1981) is very instructive. Schwarz's simulations tend to form rings associated with the resonances. On a deep blue plate of NGC 4321 (Sandage and Tammann 1981) one can easily identify an outer ring or pseudo-ring (diameter = 9.9 kpc) of the type R'_1 (see Buta, this conference), located slightly inside the OLR, as well as an inner broken ring in 4 sections (diameter = 4.1 kpc) limited by the bar and elongated parallel to the bar major axis. Although the R'_1 outer rings are rather unstable they will form under the condition that the disk does not extend passed the OLR. This condition is very likely to be fulfilled in NGC 4321 since Warmels (1986) has shown that the disk of spirals have been stripped by the intergalactic medium in the Virgo cluster of galaxy. The ring identifications make sense since the ratio of the outer ring diameter on the inner ring diameter is 2.4 which is the value expected for a flat rotation curve between the rings. If one choose the pattern speed (of the bar in the model of Schwarz) which would intersect the resonance curves at the ring diameters, one find a pattern speed of 33 $\mathrm{km\,s^{-1}\,kpc^{-1}}$, which is slightly lower than what we first assumed. However, this last value does not change much the position of the ILRs and gives a perfect match with the inner ring position (associated with the second harmonic resonance 2HR−) and a good match for the outer ring which would be inside the OLR. Furthermore, corotation occurs in a regions depleted of matter.

Conclusion

The CIGALE observations are a powerful tool to investigate the internal kinematics of galaxies. In the particular case of NGC 4321, the combination of Hα image, 656.3 nm continuum image and Hα velocity field give access to morphological and dynamical information which allows to infer the nature of the nuclear ring. The computer simulations of Schwarz are consistent with the behavior shown by NGC 4321, where 3 rings are identified and corresponds to the positions of the ILRs (nuclear ring 450 pc), the 2HR– resonance (inner ring 4.1 kpc) and the OLR (outer ring 9.9 kpc). Recent observations have revealed nuclear rings in NGC 1097 (Hummel *et al.* 1987) and NGC 1433 (Buta 1986), showing that NGC 4321 is not an isolated case. It is tempting to suggest that nuclear star formation activity in galaxies could be link to the dynamical behavior of such galaxies and in particular to the existence of the ILRs.

References

Arsenault, R., Boulesteix, J., Georgelin, Y., and Roy, J.-R. 1988, *Astron. Astrophys.*, accepted.

Buta, R. 1986, *Astrophys. J. Suppl.* **61**, 631.

Contopoulos, G. and Papayannopoulos, Th. 1980, *Astron. Astrophys.* **92**, 33.

Hummel, E., van der Hulst, J. M., and Keel, W. C. 1987, *Astron. Astrophys.* **172**, 32.

Sandage, A. and Tammann, G.A. 1981, *A Revised Shapley-Ames Catalog of Bright Galaxies* (Washington: Carnegie Institution of Washington).

Schwarz, M.P. 1981, *Astrophys. J.* **247**, 77.

de Vaucouleurs, G. 1984, In *The Virgo Cluster*, eds. O.-G. Richter and B. Binggeli (Garching bei München: ESO).

Warmels, R. H. 1986, Ph. D. thesis, Rijksuniversiteit te Groningen.

An Intensive Study of M 81

Frank Bash
Astronomy Department, University of Texas at Austin, U. S. A.

Introduction

Michele Kaufman (at Ohio State University) and I are conducting an extensive observational study of M 81 in order to understand where the various components of the spiral arms lie, how they relate to one another and how they agree with theoretical predictions. The work includes several additional collaborators which depend on the particular study. In the first paper (Bash and Kaufman 1986), we investigate the radio continuum structure of M 81 by means of λ 6–cm and λ 20–cm VLA continuum maps. In paper II (Kaufman *et al.* 1987), we examine M 81's H II regions. Paper III (in preparation) investigates the dust lanes and molecular gas. Paper IV will examine the relative location of all of these species with respect to one another and with respect to theoretical predictions while Paper V will examine M 81's nucleus.

The period 1907–1939 can be regarded as the golden age for understanding the structure and internal workings of stars. Beginning with Emden and continuing with such notables as Eddington, Jeans, and Bethe, the equilibrium structure and energy sources of main sequence stars were worked out. We may now be in a similar period for understanding the spiral galaxies. The Density Wave theory of Lin and Shu (1964) followed by the Two–Armed Spiral–Shock picture of Fujimoto (1968) and Roberts (1969) gave hope that we could understand the physical basis for the spiral arms and how that spiral structure is connected to the process of star formation. Maybe we could understand what determines how "two–armed" a galaxy will be, what sets the rate of star formation along the arms, what kind of star is born in the arms, what orbits they start with and how molecular clouds are involved.

Of course, problems have developed with our over–optimism. The work of Gerola and Seiden (1978) has shown us that the density wave theory is not needed to describe some, perhaps many, spiral galaxies. Indeed, they have shown that these galaxies seem adequately to be described by a stochastic picture. There is also the likely possibility that density waves are transient. Finally,

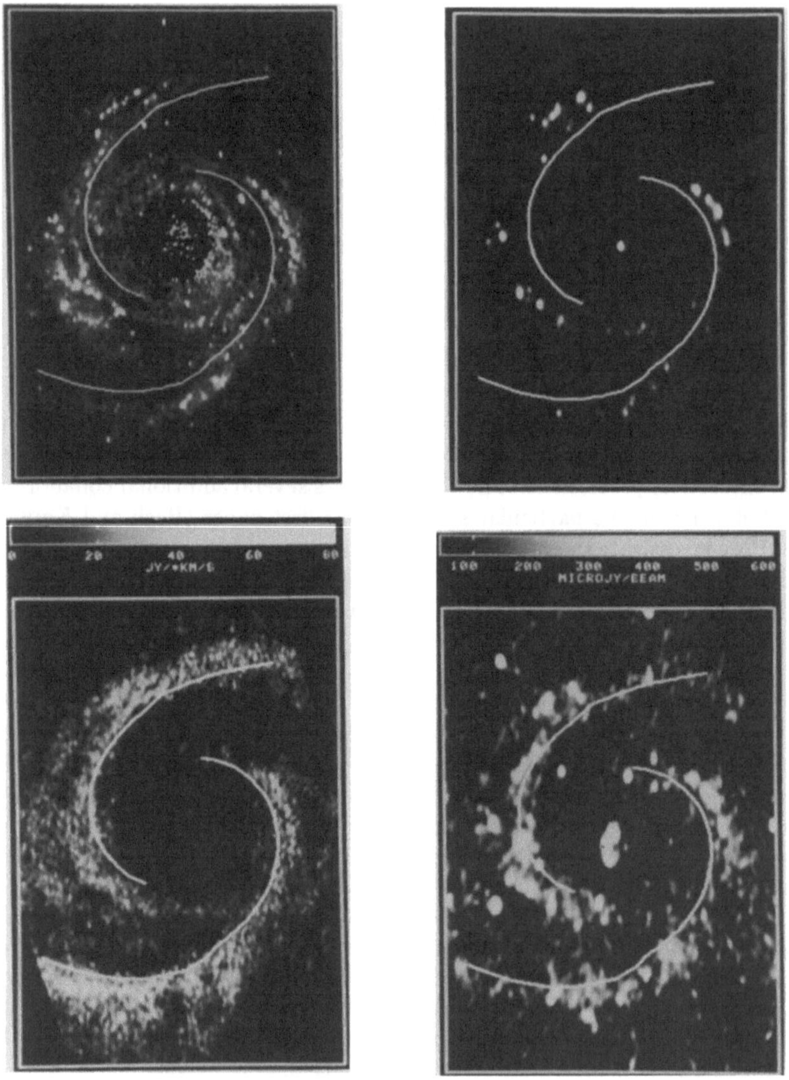

Figure 1. Four face-on pictures of M 81. Each panel shows the observed H I velocity shock. The upper-left panel shows young, blue stars, the upper-right panel shows H II regions, the lower left panel shows the H I column density and the lower right panel shows the λ 20-cm continuum radiation.

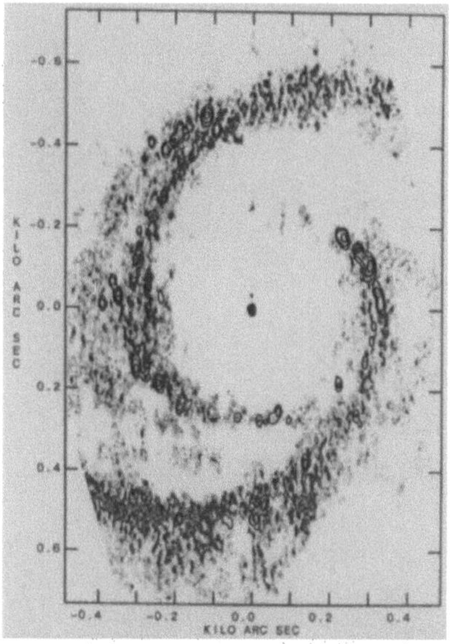

Figure 2. A face–on picture of M 81 with the major axis horizontal and North on the left. The greyscale shows the H I and the superimposed contours show the H II regions.

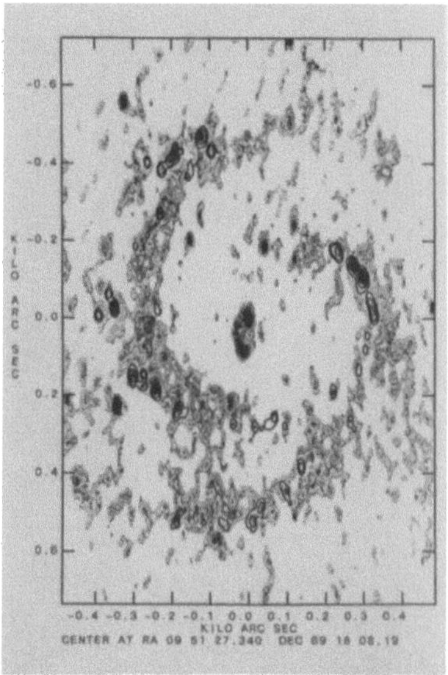

Figure 3. Same as Figure 2, but with λ 20–cm continuum radiation shown by the greyscale. Again, the H II regions are shown by the superimposed contours.

Elmegreen (1987) has recently shown us that the degree of organization of a spiral galaxy (or how "two–armed") is unrelated to its rate of star formation as measured by its Hα flux.

It seemed to us that the time was ripe to observe extensively a simple, two–armed spiral galaxy.

The Choice of M 81

The galaxy M 81 was chosen for an extensive observational study since it is near, large, well–oriented, and likely to contain a density wave. We adopt a distance of 3.3 Mpc for M 81. Its major spiral structure is ~15 arc–minutes across which is ≈ 100 beam diameters given the synthesized beamwidth of the VLA. At this resolution, M 81's spiral arms are well–defined. M 81 is inclined by 59° to the plane of the sky which gives a very effective combination of seeing the location and structure of the spiral arms, plus good measurements of velocities in the plane of the galaxy like those predicted peculiar velocities caused by density waves. Finally, M 81 is simple, clearly two–armed, and is the best example of a galaxy in which density waves actually exist, thanks to the work of Visser (1980 a,b).

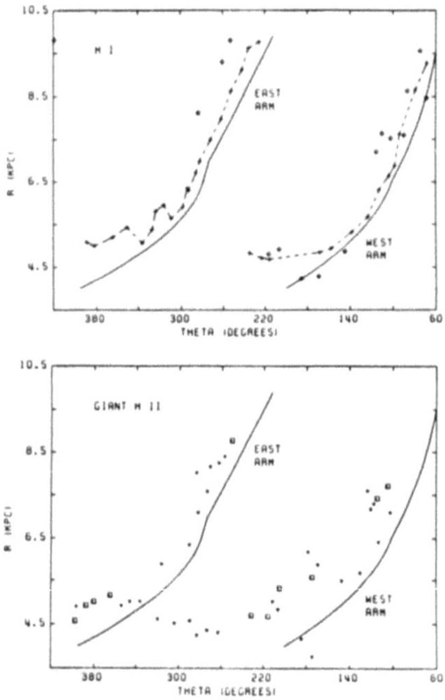

Figure 4. The location of H II regions and H I in M 81. The face–on coordinate system is polar with the galactocentric azimuth (THETA) increasing in the direction of rotation. The distance from the center assumes M 81 is 3.3 Mpc away. The observed location of the H I "shocks" are shown by the heavy solid lines. In the top figure, the supergiant H II regions are marked by a box. In the bottom figure, the dashed line connecting the "+" symbols traces the main H I ridge while the diamond symbols trace spurs of H I.

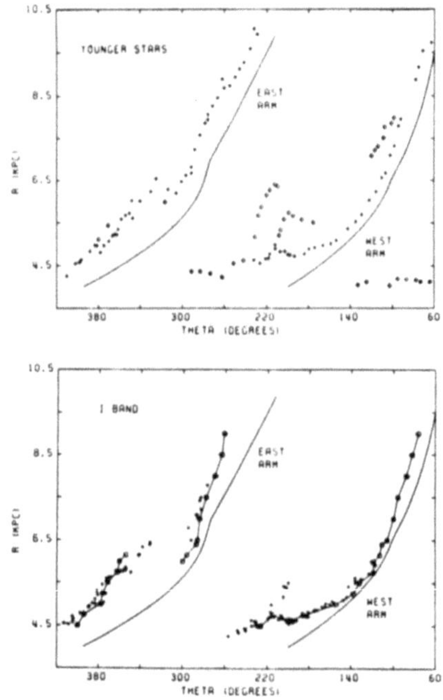

Figure 5. Like Figure 4, but for young stars and the I–band light. The top figure shows ridges of young, blue stars marked with a dashed line connecting "+" symbols. Spurs are marked with diamond symbols. The bottom figure shows the I–band light. The "+" signs show the main ridge of I–band light along the arms. The solid line which connects the octagons shows our adopted potential minimum.

The Aims

We have two aims in this research.

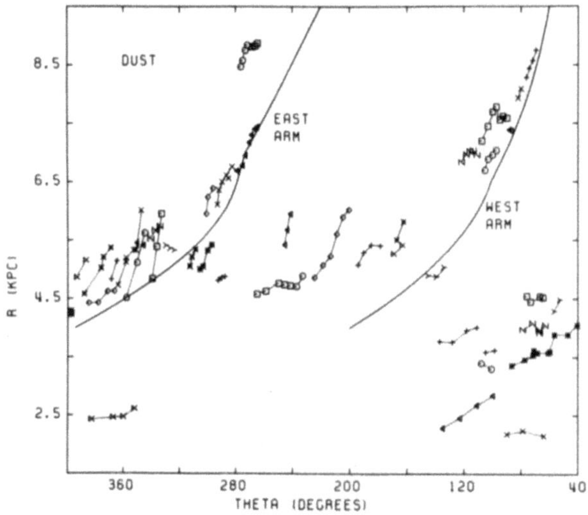

Figure 6: Like Figures 4 and 5 but depicting dust lanes.

1. We wish to measure, independently of any theory, the actual locations of *e.g.* H I, H II regions, dust lanes, and the observed H I velocity discontinuity ridge (which is identified as a shock wave in the Two–Armed Spiral Shock Theory) along the arms. We can locate, for example, the H II regions with respect to the velocity discontinuity in order to see whether they all lie "downstream" from it and how far.

2. We can then compare these relative locations to the predictions of theory, *e.g.* the Density Wave Theory.

The result of this work will be much more than just a detailed description of M 81 or the confirmation or rejection of a theory. For example, if the H II regions all lie downstream from the velocity discontinuity, and if the spiral shock somehow triggers star formation, then the distance of an H II region from the shock is related to how long the process takes.

The Data

The following data, much of it having at least 10 arc–second resolution, have been gathered by Kaufman and me, and by various collaborators. In each case, the collaborator is noted.

1. VLA λ 6–cm and λ 20–cm continuum maps to locate the thermal radiation from the H II regions and the non–thermal radiation from supernova remnants and the spiral arms.

2. VLA 21–cm H I maps with measured radial velocities (Hine and Rots)

3. Hα images (Kennicutt and Hodge)

4. Dust lane maps (Elmegreen)

5. Digitized B and I–band images (Elmegreen)

6. Radial velocities of 17 giant H II regions (Levreault)

7. Einstein X–ray map (Fabbiano)

8. CO observations using the Owens Valley Millimeter Array (Vogel, Kutner) and the IRAM 30–m telescope (Combes, Brouillet, Baudry)

The Results

The study of M 81 is in progress. We will mention, briefly, some of the results so far.

The H I data clearly reveals the sharp velocity discontinuity found by Visser. It runs along the inside edges of the spiral arms approximately like logarithmic spirals, but the two arms differ in pitch angle (Bash and Kaufman 1986). The arm nearer M 82 seems to be distorted.

Giant H II regions, young, blue stars along the arms, the centers of the smooth, red arms on the I–band plate, most of the dust lanes, most of the arm non–thermal emission, and the H I surface density maximum all lie "downstream" from the velocity discontinuity ridge.

Figure 1 shows four pictures of M 81, turned face–on. The major axis is horizontal with North on the left. The observed location of the velocity discontinuity ridge is shown on all four panels. The upper left panel shows the young stars (the Elmegreen B–band image with her red I–band plate subtracted in order to remove the smooth underlying disk). The upper right panel shows the giant H II regions from the VLA λ 20–cm map, the lower left shows the integrated H I surface density from the Hine and Rots VLA map and the lower right shows the faint λ 20–cm non–thermal radiation as shown on our VLA map.

Figures 2 and 3 show where H II regions, H I, and the non–thermal radio radiation along the arms lie with respect to each other. On Figure 2, we show the H I column density from the Hine and Rots data plus our data on the location of the H II regions. Note that the H II regions generally lie along the H I ridge. Figure 3 shows the faint 20–cm continuum along the spiral arms with the H II regions. Again, the H II regions lie along the non–thermal arms. Both the H I

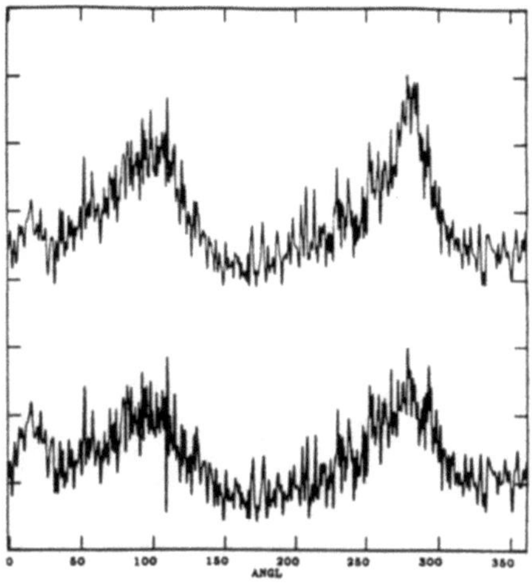

Figure 7: The upper panel shows the I–band surface brightness of M 81 around a circle, for the face–on galaxy, whose radius is 7.5 kpc. The lower panel shows the I–band trace with blue light removed (from a B–band image). The lower trace is more sinusoidal.

arms and the non–thermal arms seem to terminate in a ring whose radius is about 4.5 kpc and which we have suggested is the inner Lindblad Resonance.

Figures 4, 5, and 6 show details of the location of these things with respect to the observed velocity discontinuity. Each figure shows a polar coordinate system with a face–on galactocentric azimuth versus distance from the center. The azimuth (THETA) increases in the direction of rotation. We assume M 81 to be at 3.3 Mpc. The observed velocity discontinuity is shown as the pair of solid lines labeled as "East Arm" and "West Arm" in each figure. The ring which was mentioned above is seen near R = 4.5 kpc where objects (*e.g.* H II regions) are spread over a range of azimuth. Otherwise each species seems to lie downstream (greater theta) from the velocity discontinuity.

Figure 4 (top) shows the ridge of the maximum of H I surface density where spurs from the main ridge also are shown. Figure 4 (bottom) shows the giant H II regions. The ring at R $\approx$ 4.5 kpc is clearly seen.

Figure 5 (top) shows the ridges of young (blue) stars along the arms. Again, spurs off of the ring and arms are shown. Figure 5 (bottom) shows the location of the maximum surface brightness of the arms from Elmegreen's I plate. A solid line connects those points which seem to be along a smooth, continuous, arm–like feature.

Finally, Figure 6, on a slightly different scale, shows the locations of the dust lanes as traced by Elmegreen. Continuous features are connected by thin

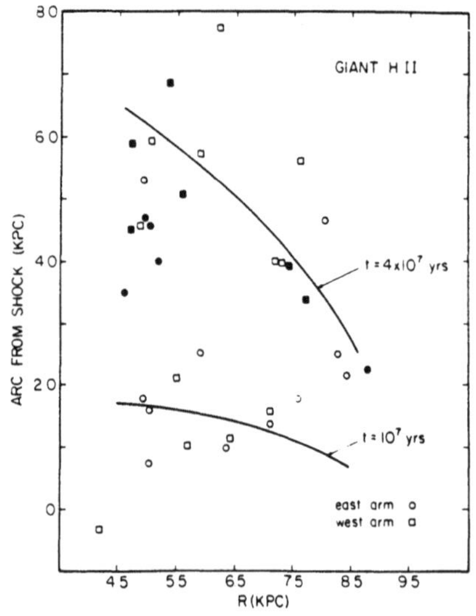

Figure 8: The location of H II regions in M 81 with the supergiant ones shown as filled symbols. Each H II region is plotted at its distance from the center of M 81 and distance downstream from the H I shock. The two curves show the distance downstream attained by an object moving at the local circular velocity and assuming that the shock moves with a pattern speed of $\Omega_p = 18\,\mathrm{km\,sec^{-1}\,kpc^{-1}}$.

lines. The ring, near R = 4.5 kpc, has many dust–lane spurs extending from it. Outside of R ≈ 6 kpc, the dust lanes seem to lie roughly parallel to the velocity discontinuity, but not on it.

The data above are a part of the results of our attempt to find the relative locations of the spiral arm tracers independent of any theory for spiral structure. In the context of the predictions of the density wave theory, we present Figures 7 and 8. Again, these are only examples of what can be done with these data.

Figure 7 shows the surface brightness of M 81, measured around a circle of R = 7.5 kpc in a face–on projection of the galaxy. The upper trace shows the *I*–band surface brightness. The two peaks come from the crossings of the two arms. This kind of data presentation was first done by Schweizer (1976). The *I*–band light especially emphasizes the red, low–mass stars which fill the galaxy's disk including the interarm regions. However, the blue, young spiral arm stars also produce *I*–band light. The density wave theory suggests that the underlying stellar disk should show a sinusoidal trace in Figure 7 and that the gas, responding to the potential well of the old disk stars, should be compressed in a narrow shock near the potential minimum. When this gas produces young, blue, stars they should lie in a lane which is narrow compared to the old, underlying stars. The lower trace in Figure 7 shows that when the light from the B–plate is subtracted from the I–plate in an attempt to remove the young stars, the trace becomes much more sinusoidal. This is in general accord with what the density wave theory predicts.

Figure 8 shows how these data can give an estimate for the time scales relevant to spiral arm star formation. The figure shows the giant H II regions in M 81 with the most luminous ones shown by filled symbols. The coordinates are the length of the circular arc downstream from the velocity discontinuity (the density wave shock) against distance from the center. The circle is centered on the center of a face–on M 81. We can compute the distance expected for circular orbits, the rotation curve assumed by Visser (1980 a, b) and a spiral pattern speed of 18 km sec^{-1} kpc^{-1} which Visser also assumed. We show how far an object on a circular orbit should move out in front of the shock in 10 and 40 million years. (The 10 and 40 million year curves cross at corotation, here = 11.3 kpc.)

On Figure 8, one sees that the bulk of the H II regions seem to lie between 10 and 40 million years downstream from the spiral shock. The most luminous ones seem to be nearer 40 million than 10 million years downstream.

Summary and Future Plans

We have collected a detailed set of diverse data on spiral arm tracers in a simple, two–armed spiral galaxy, M 81. These data give us precise relative locations of these tracers along the spiral arms so that we can see, independent of any theory, the structure of the spiral arms in a simple spiral galaxy and, in the context of the density wave theory, see if the theory fits the observations and use its implications to deduce information on time scales for star formation along the spiral arms. The work seems to have some promise in continuing these beginning stages of attempting a physical understanding of the operation of spiral galaxies.

I wish to acknowledge the vital importance of Michele Kaufman's contributions and together we thank all of our other collaborators for their generous assistance. I also wish to acknowledge the support of the National Science Foundation through grant AST86–11784.

References

Bash, F. N. and Kaufman, M. 1986, *Astrophys. J.* **310**, 621.

Elmegreen, B. G. 1987, in , I. A. U. Symposium No. **115**, *Star Forming Regions*, ed. M. Peimbert and J. Jugaku, (Dordrecht: Reidel) p. 457.

Fujimoto, M. 1968, in, I. A. U. Symposium No. **29**, *Non–Stable Phenomena in Galaxies*, ed. V. A. Ambartsumian, (Yerevan: Armenian Acad. Sci.) p.

453.
Gerola, H. and Seiden, P. E. 1978, *Astrophys. J.* **223**, 129.
Kaufman, M., Bash, F. N., Kennicutt, R. C., and Hodge, P. W. 1987, *Astrophys. J.* **319**, 61.
Lin, C. C. and Shu, F. H. 1964, *Astrophys. J.* **140**, 646.
Roberts, W. W. 1969, *Astrophys. J.* **158**, 123.
Schweizer, F. 1976, *Astrophys. J. Suppl.* **31**, 313.
Visser, H. C. D. 1980a, *Astron. Astrophys.* **88**, 149.
Visser, H. C. D. 1980b, *Astron. Astrophys.* **88**. 159.

Discussion

M. Capaccioli: Did you take into account the different absorption in the near and far sides of M 81?

Bash: Yes. In Kaufman *et al.* (1987, *Astrophys. J.* **319**, 61), we measured the visual extinction in the directions of each of 42 giant H II regions. They are distributed over both the east and west sides (far and near sides, respectively) of the galaxy. We find no significant difference in the average visual extinction between the giant H II regions along the east and west arms.

G. Courtes: I prefer to think of "velocity gradient" rather than "velocity discontinuity," but anyway, what "radial velocity discontinuity" did you find?

Bash: We see about 30 km sec^{-1}.

Courtès: In our study of the southern arm of M 33, we found less (5 to 15 km sec^{-1}), but the M 33 arms are not as short as the arms of M 81.

J.–C. Pecker: You have shown a magnificent set of coherent views of the various components of M 81. You said that the CO data are badly needed. I agree, but the "proto–stellar" objects or star–forming regions may also be observed in the far infrared with the IRAS satellite. Has this been done? If so, what are the results?

Bash: The IRAS satellite did observe M 81. Unfortunately, the angular resolution is insufficient for our purposes. We are planning 2–5μm observations using the new infrared array camera just coming into use.

H I and CO Emission in the Hot–Spot Barred Spiral NGC 4314

J. A. Garcia–Barreto[1], F. Combes[2], and C. Magri[3]
[1]Instituto de Astronomia, Universidad Nacional de Mexico
[2]Observatoire de Meudon and École Normale Supérieure, Paris, France
[3]Astronomy Department, Cornell University, New York, U.S.A.

NGC 4314 is a strongly barred spiral galaxy with an inner ring, having the peculiarity of a conspicuous nuclear spiral structure inside the ring [see the Hubble Atlas (Sandage 1961) p. 44]. The ring is optically sprinkled with hot spots. Radio–continuum emission has been mapped at 2–, 6–, and 20–cm with the VLA by Garcia–Barreto and Pişmiş (1985): the nuclear ring is obvious (Figure 1) and highly contrasted, revealing non–thermal and thermal emissions. H I emission was observed at Arecibo (3.3 arc–minute resolution), but only a very weak H I mass was derived of about a few 10^6 $M_\odot$ (Figure 2). This is probably related to the smooth and featureless appearance of the bar and outer arms of this galaxy, classified as anemic by van den Bergh (1976).

We mapped the center of the galaxy in CO(1–0) and CO(2–1), with 23 arcsec and 14 arcsec resolution, using the IRAM 30–m millimetric telescope, at Pico Veleta, Spain. The molecular mass derived towards the center is a 3–4 $\times$ 10^8 $M_\odot$, which is very high for this early–type SBa galaxy, and not expected from the far infrared IRAS fluxes (7.3 Jy at 100μm). The derived dust temperature, 35°K, suggests, however, that massive star formation is presently going on.

Figure 3 displays the CO(2–1) map, made at 10 arcsec sampling on a grid aligned on the major axis of NGC 4314. Spectra reveal two velocity components that can be interpreted as coming from a rotating ring, that would be coincident with the Hα ring (or tightly wound spiral) observed by Wakamatsu and Nishida (1980). The CO(2–1)/CO(1–0) ratio varies across the map, between 0.5 and 2.

The high concentration of stellar mass towards the nuclear region (2 $\times$ 10^9 $M_\odot$ from the Hα rotation curve), suggests the presence of an inner Lindblad resonance. The location of the resonant region is coincident with the ring, if corotation is taken at the end of the bar. The concentration of molecular

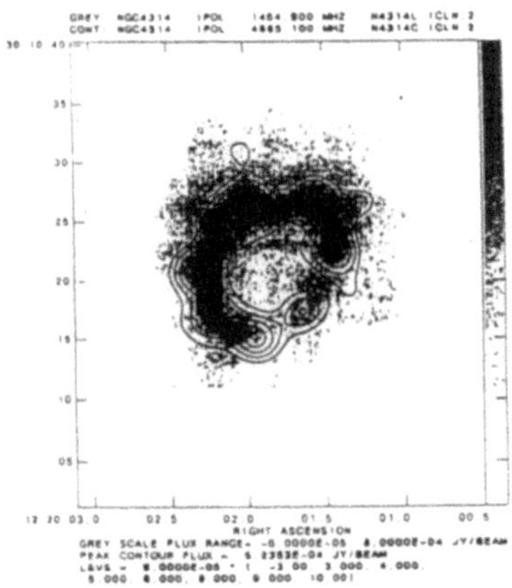

Figure 1. Radio continuum emission at 20–cm (grey scale) and 6–cm (contours) from the VLA observations of Garcia–Barreto and Pismis (1987).

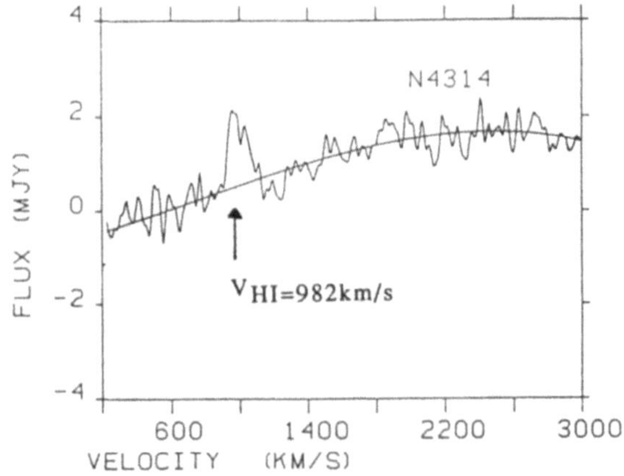

Figure 2. H I spectrum obtained at Arecibo (3.3 arc–minute resolution). Mean radio position is $\alpha = 12^h\ 20^m\ 01.^s8$, $\delta = 30^\circ\ 10'\ 21''$

clouds (and consequent star formation) in the nuclear ring could then be naturally explained, from the angular momentum transfer induced by the strong bar (Combes and Gerin 1985).

J. A. G.–B. acknowledges partial financial support from Conacyt (Mexico) and a travel grant from Aeromexico (Mexico).

References

van den Bergh, S. 1976, *Astrophys. J.* **206**, 883.
Combes, F. and Gérin, M. 1985, *Astron. Astrophys.* **150**, 327.
Garcia–Barreto, J. A. and Pişmiş, P. 1985, *Bul. Am. Astron. Soc.* **17**, 893.
Sandage, A. 1961, *The Hubble Atlas of Galaxies*, Carnegie Publication No. **618** (Washington, D. C.: Carnegie Institution of Washington).
Wakamatsu, K. and Nishida, M. T. 1980, *Publ. Astron. Soc. Japan* **32**, 389.

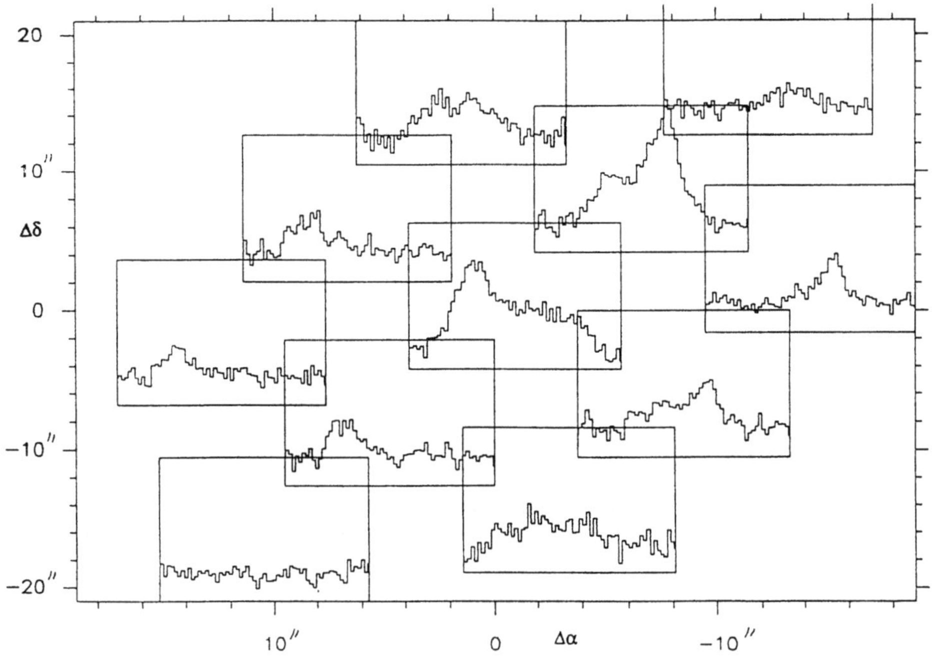

Figure 3. CO(2–1) map obtained with the IRAM 30–m telescope, on a grid aligned along the major axis. The temperature scale is in T_r^*, from −0.05 to 0.2 °K. The velocity scale is from 840 to 1160 km sec^{-1} heliocentric.

Optical and Radio Properties of Nearby, Non–Cluster Spirals

G. Giuricin, F. Mardirossian, and M. Mezzetti

Dipartimento di Astronomia, Trieste, Italy

Summary

Inspecting the radio continuum properties of about 200 nearby, non–cluster spirals, we have found that the radio continuum emission (per unit light) of these spirals is slightly correlated with their colour indices $(U-B)_o^T$, and more closely to their Hα emission strengths, but not to indices of H I content.

The Data Sample

The study of the relations between the radio continuum emission of spirals and their other global properties (*e.g.* H I content, colours, Hα emission strengths, infrared emission) which are known to be probably different in different environments, can be of aid in understanding the nature of the radio continuum emission, and in clarifying to what extent the radio properties of spirals can be considered as sensitive probes of their environment. In the present study, we examine a sample of nearby non–cluster spirals, for which effects of environment are likely to be small in comparison with those acting on cluster members.

As a sample of (nearby) non–cluster galaxies with known H I and radio continuum data, we have considered: i) the spiral and irregular members of Geller and Huchra's (1983) galaxy groups, excluding the 34 groups which, according to the two authors, probably contain interlopers or are parts of rich clusters, and excluding the distant groups (*i.e.* , those with corrected galactocentric velocities $V_o \geq 3000\,\mathrm{km\,sec^{-1}}$); ii) nearby ($V_o < 3000\,\mathrm{km\,sec^{-1}}$) spirals of the list of 324 "isolated" galaxies whose H I properties have been surveyed by Haynes and Giovanelli (1984), and from the compilation of Giuricin *et al.* (1985), who relied essentially on the reference sources cited in Huchtmeier *et al.* (1983). We have characterized the H I content of a galaxy by evaluating the parameters M_H/L [ratio of the H I mass to the corrected total blue luminosity expressed in the RC2 system (de Vaucouleurs *et al.* 1976)] and σ_H (mean apparent surface density) defined as

$$\sigma_H = 4 \cdot M_H/(\pi \cdot D^2) \quad (1)$$

where D is the absolute linear diameter (reduced to face-on value at the galactic pole) at the 25.0 mag arcsec^{-2} brightness level in the RC2 system. The H I deficiency (or excess) of a galaxy is then evaluated by means of the residuals $\Delta \log \sigma_H = \log \sigma_H - < \log \sigma_H >_T$ and $\Delta \log M_H/L = \log M_H/L - < \log M_H/L >_T$, where the standard values $< \log s_H >_T$ and $< \log M_H/L >_T$ for each galaxy morphological type refer to a reference sample, as tabulated in Guiderdoni and Rocca-Volmerange (1985).

The radio continuum fluxes (or their upper limits) of our galaxies have been mostly taken from the Arecibo survey of Dressel and Condon (1978) and in some cases from Hummel (1980), Kotanyi (1980), Harnett (1982), Hummel *et al.* (1985), and Sulentic (1976). We have transformed the fluxes observed at different frequencies to 2.4 GHz using the power law spectrum $f_v \sim r^{-0.8}$. We have evaluated the common logarithm of the ratio R between the radio continuum (at a frequency of 2.4 GHz) and the optical luminosity defined as

$$\log R = \log f + 0.4(B_T^o - 12.5) \quad (2)$$

where B_T^o is the corrected blue total magnitude in the RC2 system, and f is the flux density at 2.4 GHz in mJy.

Results

There appear to be no significant correlations between $\log R$ and the H I deficiency parameters $\Delta \log \sigma_H$ and $\Delta \log M_H/L$. On the other hand, $\log R$ turns out to correlate negatively with the corrected colour index $(U-B)_o^T$; in fact, for N = 38 detected galaxies, we have obtained the Spearman correlation coefficient $r_s = -0.28$ (which is significant at the $\sim$ 95% confidence level). But $\log R$ does not correlate significantly with the corrected colour index $(B-V)_o^T$. For our galaxies, we have taken the colour indices from the RC2 and from Davis and Seaquist (1983). We have found a good correlation between $\log R$ and $\log W$ [logarithm of the Hα + [NII] emission-line equivalent widths, which are taken directly from Kennicutt and Kent (1983)]; in fact, for N = 56 detected spirals we have obtained $r_s = 0.61$, which is significant at the > 99.99% confidence level.

Hence, our analysis of a large sample of non-cluster spirals, in which the presence of upper limits is considered, indicates that the radio continuum emission of spirals is slightly linked to their integrated star formation histories (to which the colours $(U-B)_o^T$ are related); it is more closely linked to their current star formation rates (which are essentially represented by Hα emission strengths). The absence of a significant correlation between $\log R$ and $(B-V)_o^T$ may be due to the fact that $(B-V)_o^T$ is less sensitive that $(U-B)_o^T$ to the level of star formation. The absence of a relation between the H I content and the radio continuum emission in our sample may be related to the fact that star formation activity may be more associated with molecular gas rather than with H I gas.

In conclusion, an inspection of the radio and optical properties of nearby spirals yields further support to the (controversial) view that the radio continuum emission of spiral galaxies is certainly connected with their young stellar population, although a contribution from the old disk population can not be excluded.

References

Davis, L. E. and Seaquist, E. R. 1983, *Astrophys. J. Suppl.* **53**, 269.
Dressel, L. L. and Condon, J. J. 1978, *Astrophys. J. Suppl.* **36**, 53.
Geller, M. and Huchra, J. P. 1983, *Astrophys. J. Suppl.* **52**, 61.
Giuricin, G., Mardirossian, F., and Mezzetti, M. 1985, *Astron. Astrophys.* **146**, 317.
Guiderdoni, B. and Rocca–Volmerange, B. 1985, *Astron. Astrophys.* **151**, 108.
Harnett, J. I. 1982, *Australian J. Phys.* **35**, 321.
Haynes, M. P. and Giovanelli, R. 1984, *Astron. J.* **89**, 758.
Huchtmeier, W. K., Richter, O.–G., Bohnenstegel, H.–D., and Hauschildt, M. 1983, *ESO Preprint* No. **250**.
Hummel, E. 1980, *Astron. Astrophys. Suppl.* **41**, 151.
Hummel, E., Pedlar, A., van der Hulst, J. M., and Davies, R. D. 1985, *Astron. Astrophys. Suppl.* **60**, 293.
Kennicutt, R. C. and Kent, S. M. 1983, *Astron. J.* **88**, 1094.
Kotanyi, C. G. 1980, *Astron. Astrophys. Suppl.* **41**, 421.
Sulentic, J. W. 1976, *Astrophys. J. Suppl.* **32**, 171.
de Vaucouleurs, G., de Vaucouleurs, A., Corwin, H. G. 1976, *Second Reference Catalogue of Bright Galaxies* (RC2) (Austin: University of Texas Press).

A Physical Model of the Gaseous Dust Band of Centaurus A

Richard A. Nicholson[1], Keith Taylor[2], Joss Bland[3], and Linda S. Sparke[4]

[1]Sussex University, Brighton, U.K.
[2]Anglo-Australian Observatory, Epping, N.S.W., Australia and Royal Greenwich Observatory, Sussex, U.K.
[3]University of Hawaii, Honolulu, Hawaii, U.S.A.
[4]Kapteyn Laboratory, Groningen University, Netherlands.

Introduction

We present a physical model of Centaurus A which satisfactorily explains the morphology and kinematics of the gaseous dust band, primarily revealed through the comprehensive TAURUS data from Bland *et al.* (1987). At the same time, we take care to make the model consistent with what is known of the stellar component and radio structure.

Kinematic Model

The velocity field of the ionized HII gas has been analysed through fitting an optimised set of differentially rotating annuli whose inclination, β, and orientation, α, change with radius. This process is secure for regions of the disc beyond the region ($r \sim 120$ arcsec) of line–splitting. Clearly any such analysis within this radius becomes unreliable. Beyond $r \sim 250$ arcsec, where ionized gas is no longer observed, we have attempted to extrapolate the warp to rproduce qualitatively the general appearance of the dust band as observed in broad–band optical images of the galaxy.

The resultant warped disk model is shown in Figure 1. Given a dust–to–gas ratio which increases linearly with radius, the predicted dust lane morphology is shown in Figure 2. Several of the spatial characteristics of the dust lane are

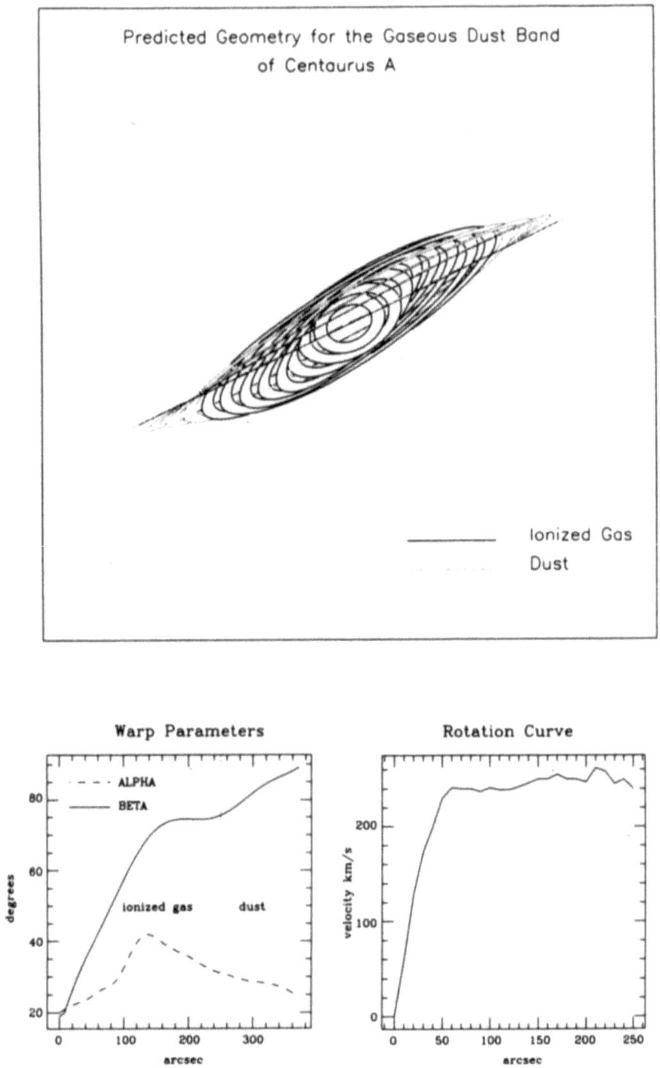

Figure 1: (top) Predicted geometry for the gaseous dust band of Centaurus A. Solid lines show the predicted distribution of ionized gas, while dotted lines show the dust. (bottom left) Warp parameters (inclination β and orientation α) as a function of radius, and (bottom right) the predicted rotation curve.

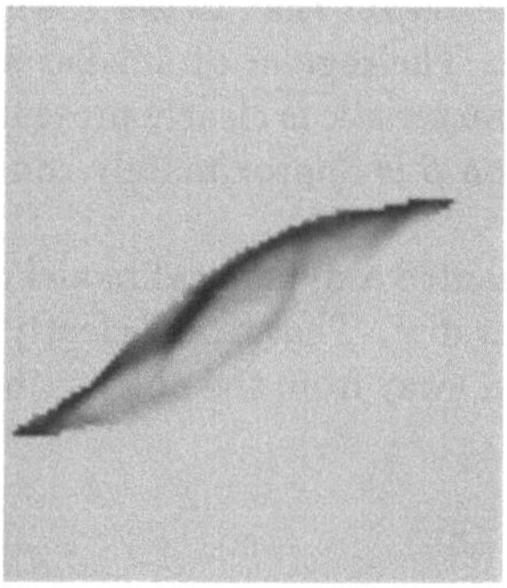

Figure 2: Modelled appearance of the dust lane of Centaurus A.

qualitatively reproduced with this model: 1) the prominent dust lane to the north–east of the nucleus, 2) the rounded feature seen to the south–west and its merger with a less prominent band of dust to the south–east, and 3) the general shape and complexity of the bands to the west. The model also reproduces the two symmetric line–split regions, as well as the twisted iso–velocity tongues seen on Hα velocity maps.

Implications of the Warp Geometry

The monotonic variation of the inclination angle, β, (see Figure 1b) implied by the model, strongly suggests that this feature is the result of differential precession about an axis close to the plane of the sky. The precession axis cannot lie exactly along the photometric minor axis as this would result in material between 100 and 200 arcsec precessing in the opposite sense (relative to its orbital motion) to the rest of the disc. However, if the precession axis is at $40°$ or above, the precession sense is constant, and material at small radii lags behind material at larger radii. This immediately implies that the gas disc is precessing in an oblate rather than a prolate potential and hence forms a polar ring about the stars.

We can test this hypothesis by examining the variation of position angle, α, on the sky, which can now be interpreted as an inclination angle to the pole of the system. A massless gas disc, through a process of dissipative differential precession, will tend to settle into the preferred plane of an oblate system, so that the major axis of the disc should twist towards the major axis of the stellar ellipsoid with decreasing radius. This is clearly true for the disc at radii, $r \leq 120$ arcsec. However, at larger radii, the gas disc twists towards the pole with decreasing radii; this is opposite to the direction expected. In S0 polar ring galaxies, however, this effect may be caused by the self–gravity of the gas disc. In such systems, the region that twists towards the pole precesses as a single entity and is stable. The regions at smaller and larger radii still precess as test particles. This characteristic is clearly present in Figure 1b where, within the HII disc, the inclination β is approximately constant in the region where α twists towards the pole.

Sparke (1986) has presented a dynamical model of Centaurus A as an oblate stellar ellipsoid with a polar disc. That model clearly reproduces the characteristic twist first towards, then away from the pole, with a monotonically increasing twist.

Conclusions

The kinematic model presented successfully reproduces the observed dust band morphology, velocity field, and line–width maps. Further, the warp geometry of the disc is consistent with the idea that Centaurus A is a nearly oblate elliptical galaxy with a self–gravitating polar disc.

References

Bland, J., Taylor, K., and Atherton, P. D. 1987, *Mon. Not. Roy. Astron. Soc.* **228**, 595.

Sparke, L. S. 1986, *Mon. Not. Roy. Astron. Soc.* **219**, 657.

CCD Observations of Gas and Dust in NGC 4696: Implications for Cooling Flows?

W. B. Sparks and F. Macchetto

Space Telescope Science Institute, Baltimore, Maryland, U.S.A.

Introduction

NGC 4696 is a typical, nearby radio elliptical galaxy, occupying a central dominant location in the Centaurus cluster, and is a prime candidate for a galaxy hosting a cooling flow (Fabian *et al.* 1982, Canizares 1988). It contains an asymmetric dust lane coincident with which are optical emission lines, strong in [N II] relative to Hα (*e.g.* Jørgensen *et al.* 1983), and the system is also a source of far infrared emission (Jura *et al.* 1987).

In this and other cooling–flow galaxies, the optical emission lines have been taken as evidence that the hot X–ray gas is indeed cooling, and that we are seeing the thermally unstable condensations. However, dust is not easily accounted for in such a picture. We find using continuum and narrow–band direct CCD images that the dust is apparently quite normal in its properties, which favours an external infall origin.

We consider the thermal interaction between infalling material and hot gas, and find consistency between the observed energetic output of the dust lane balanced by energy input from the X–ray gas by electron heat conduction.

Properties of the Dust Lane

The dust lane forms a one–sided arc suggestive of infall to the nucleus. A similar distribution of line emission implies that the one–sidedness is intrinsic and not due to projection effects. Extinction was estimated at each point for both the V and R filters, by modelling the underlying galaxy light using ellipse fitting plus interpolation. The derived absorption at these two wavelengths shows no significant difference in relative amplitude to that of our galaxy (Figure 1). These data also imply that the dust covering factor is high.

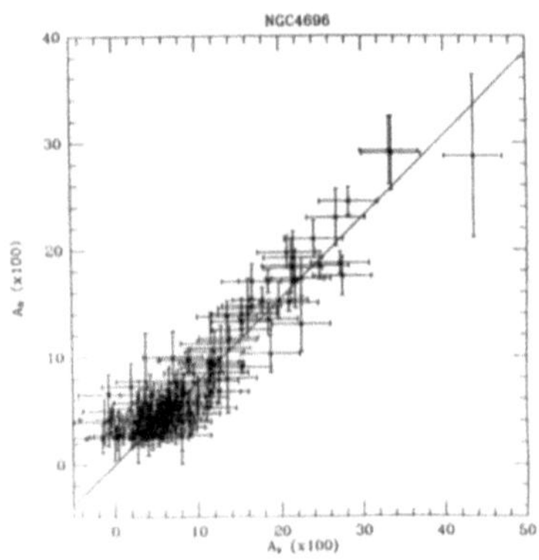

Figure 1. Comparison of extinction at R to extinction at V. The line has a slope appropriate to dust in *our* galaxy.

For a normal gas–to–dust ratio, the extinction implies a total gas mass of $\approx 3.5 \times 10^{7}_{\odot}$. We find a total radiated flux in Hα + [N II] of 9×10^{40} erg sec^{-1} implying a total emission from line radiation of $\approx 1 \times 10^{43}$ erg sec^{-1}. The infrared flux is also $\approx 1 \times 10^{43}$ erg sec^{-1}. Correlating emission line luminosity with optical depth, we infer typical gas densities which, to within a factor of two or three, lead to pressure balance with the ambient X–ray gas (Matilsky *et al.* 1985).

Heat Conduction

Comparison of the heat input by saturated conduction (Cowie and McKee 1977) to the energy input by shocks at typical dynamical velocities, suggests that conduction dominates. The saturated heat flux is $\approx 3 \times 10^{43}$ erg sec^{-1}, which compares well with the inferred total emission from the dust lane of $\approx 2 \times 10^{43}$ erg sec^{-1}.

Saturated heat flow is expected to be valid only as an initial condition (hot medium adjacent to cold), so we have also investigated steady–state solutions, generalizing the Cowie and McKee (1977) solution to non–zero heat inflow. Heat inflow and mass loss rate are related by the outer boundary condition (that the temperature approaches a constant value)

$$Q_0 = Q_s \frac{2}{7}(1 - w)$$

where Q_0 is the heat inflow, w is the mass loss rate normalized to the classical Cowie and McKee (1977) evaporative mass loss rate and $Q_s = 4\pi r_c \kappa_\infty T_\infty$. The steady–state heat flow into the dust lane depends on the details of the mass loss or mass condensation rate but in general would be somewhat lower than the observed output. Radiative losses have been ignored here, but are not expected to affect the result significantly (Böhringer and Hartquist 1987).

Summary and Speculation

We find a normal wavelength dependence of extinction in the dust lane of NGC 4696, suggesting an external origin for the gas and dust. The energy radiated by the dust lane via emission lines plus infrared radiation may be balanced by heating from the X–ray gas to the dust lane by electron conduction. Direct heat tranfer from the hot gas to the cool could therefore be the emission line excitation mechanism.

If valid generally, such a model would impact on cooling flow theories: (i) by modifying interpretation of the correlation between optical emission and X–ray emission (Hu *et al.* 1985) (ii) forced cooling of the hot halo may give rise to cooler X–ray gas, often seen in other similar systems, and previously interpreted as strong supporting evidence for the existence of cooling flows (iii) the structure of some thermal conduction models of X–ray halos is amended by the imposition of the inner cold boundary condition.

That is, a number of the signatures of a cooling flow would be produced inevitably by the infall. Given the simplicity and inevitability of the process, we suggest that many of the optical emission systems in strong X–ray emitting regions are caused by the infall of cold material into a hot environment.

Acknowledgment

STScI is operated by the Association of Universities for Research in Astronomy, Inc., under contract with the National Aeronautics and Space Administration.

References

Böhringer, H. and Hartquist, T. W. 1987, *Mon. Not. Roy. Astron. Soc.* **228**, 915.

Canizares, C. R. 1988, in *Cooling Flows in Clusters and Galaxies*, ed. A. C. Fabian, in press.

Cowie, L. L. and McKee, C. F. 1977, *Astrophys. J.* **211**, 135.

Fabian, A. C., Atherton, P. D., Taylor, K., and Nulsen, P.E.J. 1982, *Mon. Not. Roy. Astron. Soc.* **201**, 17P.

Hu, E. M., Cowie, L. L., and Wang, Z. 1985, *Astrophys. J. Suppl.* **59**, 447.

Jørgensen, H. E., Nørgaard–Nielsen, H. U., Pederson, H., and Schnopper, H. 1983, *Astron. Astrophys.* **122**, 301.

Jura, M., Kim, D. W, Knapp, G. R., and Guhathakurta, P. 1987, *Astrophys. J. Lett.* **312**, L11.
Matilsky, T., Jones, C., and Forman, W. 1985, *Astrophys. J.* **291**, 621.

NGC 2777: Amorphous — and Young?

Jack W. Sulentic

Department of Physics and Astronomy, University of Alabama, Tuscaloosa, U.S.A.

1. Introduction

We report imaging and spectroscopy of the irregular galaxy NGC 2777. The galaxy was observed as part of a pilot program to study the properties of companion galaxies. NGC 2777 (m_{pg} = 13.9; $V_\odot$ = 1319 km sec^{-1}) is a companion of the Sa galaxy NGC 2775 (m_{pg} = 11.4; $V_\odot$ = 1180 km sec^{-1}). NGC 2777 exhibits a high surface brightness core surrounded by an extended halo (0.8 × 0.6 arcmin). The core (0.6 × 0.4 arcmin) is broken up into numerous knots.

Spectroscopic observations made with the Palomar 5–m telescope in the blue ($\lambda\lambda$3700–4300Å) reveal an early A–type absorption spectrum with the Balmer series detected out to H 12 and weak Ca K. Red sensitive observations with the KPNO 4–m reflector reveal a rich emission line spectrum with weak Mg I (λ5175Å) and Na D (λ5892Å) absorption. Comparison of the absorption spectra with stellar population synthesis spectra (Keel 1983) suggest that they are consistent with an age t$\leq$ 10^9 yr. We do not see, however, WR–type features that might be expected if the spectrum were simply dominated by a burst of O–B star formation (see *e.g.* Keel 1987).

2. A Blue Wing

An extensive wing was found on the blue side of the Hα line in NGC 2777. Allowing for the presence of [N II] λ6548Å, this feature extends approximately 10^3 km sec^{-1} blueward of the Hα centroid. This wing is visible across at least the entire high surface brightness core of NGC 2777. A possible interpretations for this feature involves the outflow of ionized gas from the galaxy (see *e.g.*

Ulrich 1972). The size (about 2 kpc, if at 12 Mpc distance) of the galaxy and its morphology make a nuclear origin for this gas unlikely. Instead, an galaxy–wide outflow related to massive ongoing star formation is more attractive. The lack of evidence for O–B stars in the spectra of NGC 2777 is perhaps surprising if a strong superwind model is invoked.

3. The Amorphous Class

NGC 2777 is found to possess all of the defining characteristics of the amorphous class of galaxies (Sandage and Brucato 1979). These properties include: 1) "irregular" morphology (in the sense that the objects are clearly not E, S0, or Spiral in form), 2) early type (A) absorption spectra, 3) large IR $60\mu m/100\mu m$ ratios indicating a warm dust component, presumably due to active star formation (usually ≥ 0.5). Examination of existing galaxies that are classified amorphous, reveals that a significant number of such objects are companions to brighter galaxies. Well–known examples include M 82 and NGC 3077 associated with M 81; NGC 404 associated with M 31; NGC 5195 associated with M 51; NGC 1510 associated with NGC 1512; and NGC 5253 associated NGC 5128. This observation led Cottrell (1978) to propose that this class of galaxies (then called Irr II) were the ephemeral products of very close encounters. It would be useful to see a more complete survey of the numbers and spectroscopic properties of the amorphous class along with their associative properties. Fourteen galaxies listed as "amorphous" (or "amorphous?") in the *Revised Shapley-Ames Catalogue* suggests a frequency of about 0.01 for this class.

4. The Missing Link?

The slightly higher, and possibly quantized, redshift ($\Delta V = +139$ km sec^{-1}) of NGC 2777 (relative to NGC 2775) has been found to be characteristic of companion galaxies (Sulentic 1984; Arp and Sulentic 1985). Such companion galaxies are hypothesized to be quite young within the context of the discordant redshift hypothesis. It has been suggested that they originate in an ejection process from the nuclei of larger and older galaxies. In this context, the amorphous class might represent a missing link between the discordant redshift quasars and normal Doppler redshifted galaxies. We argue that the predisposition for such galaxies to 1) be companions of bright spiral galaxies in loose groups (see *e.g.* Cottrell 1978), 2) to exhibit systematically higher redshifts than the largest galaxy in the group, and 3) to have characteristically early–type spectra are consistent with the unconventional origin that is proposed for them. All of these characteristics require further study in well–defined surveys. The single most important observation, however, involves the demonstration that these galaxies are really young.

References

Arp, H. C. and Sulentic, J. 1985, *Astrophys. J.* **291**, 88.
Cottrell, G. 1978, *Mon. Not. Roy. Astron. Soc.* **184**, 259.
Keel, W 1983, *Astrophys. J.* **269**, 466.
Keel, W. 1987, *Astron. Astrophys.* **172**, 43.
Sandage, A. and Brucato, R. 1979, *Astron. J.* **84**, 472.
Sulentic, J. 1984, *Astrophys. J.* **286**, 442.
Ulrich, M.-H. 1972, *Astrophys. J.* **178**, 113.

VLA Observations of Unusual H I Distributions for Coma Cluster Spirals

Woodruff T. Sullivan, III*

Institute of Astronomy, Cambridge, England, U.K. and
Observatoire de Meudon, France

Introduction

The influence of the cluster environment on the neutral hydrogen contents of spiral galaxies in clusters has been well established over the past decade. It is found that spirals in the central regions of those clusters having the most intracluster gas and the lowest percentage of spirals exhibit H I deficiencies (when compared with field spirals of the same size and luminosity and morphological type) ranging from a factor of 2 to 10. The Virgo cluster is the best–studied example of the moderately deficient class, and the Coma cluster is the extreme case of the very deficient type. Until recently, however, it has not been possible to study the details of the distribution of the H I in these deficient cluster galaxies. The first detailed studies with synthesis telescopes were carried out (on the Virgo cluster) on the WSRT by Warmels (1986) and on the VLA by van Gorkom *et al.* (1984). These showed that the H I in the deficient spirals was primarily missing from the outer parts of the galaxies; that is, that the size of the H I disks was much smaller than the optical disks, the opposite of the usual situation in the field. Furthermore, van Gorkom *et al.* (1984) and Cayette *et al.* (1988) have several cases of very asymmetric H I distributions, which can reasonably be attributed to either tidal interactions with other cluster members, or to interactions (such as ram–pressure stripping) with the intracluster medium. With these interesting results in hand, I have observed the H I distributions

*On leave from Department of Astronomy, University of Washington, Seattle, Washington, U.S.A.

in several deficient spirals in the Coma cluster, where one might expect even stronger effects. Here I give a preliminary account of these observations.

Observations

Observations of eight spirals in three separate fields were made with the C and D arrays of the Very Large Array. The extreme faintness of the H I signals, and the relatively poor ratio of synthesized beam size to optical galaxy size, mean that the results reported here are not at all as detailed as those for the five–times–closer Virgo cluster. Furthermore, in order to gain sufficient signal–to–noise, almost all velocity resolution was forfeited, and I will here discuss only the velocity–integrated, total hydrogen maps. The galaxies were chosen to include several of the most deficient observed at Arecibo (Sullivan *et al.* 1981, Bothun *et al.* 1985) with the constraints, however, of a 35 arcmin primary beam for the VLA antennas and a limited redshift range observable at any one time. Two spirals of normal gas content and located in the outer part of the cluster (IC 842 and IC 4088) were observed for comparison and, as Figure 1 shows, their H I distribution is indeed normal — centered on the optical galaxies and of considerably greater extent than the UGC sizes (*i.e.* to the 25.5 mag arcsec^{-2} isophote). One surprise, however, was the discovery that almost as much neutral hydrogen is associated with two much fainter companions to the north of IC 4088 as with the main galaxy itself.

The H I distributions of the six inner spirals are also shown in Figure 1. Two of the galaxies, despite having positions near the cluster center, have normal hydrogen contents, and indeed their H I distributions are much larger in extent than the optical size. The centroid of the H I emission for Zw 160–106 is not significantly offset from the optical center, but for Zw 160–076 the map does appear to show an offset. The VLA observations for this galaxy's field, however, were plagued by humanmade radio interference, with the result of a loss of sensitivity and, more seriously, systematic biases evident in the overall map. This means that the H I distributions derived from the faint signals of Zw 160–076 and NGC 4848 are less reliable than others and that the apparent offset in Zw 160–076 must be treated as tentative.

Discussion

The four central galaxies that are H I-deficient by factors of 2 to 10 as a group, show much smaller H I extents than optical, in a manner similar to the earlier results for the Virgo cluster. The two brightest and largest spirals in the Coma cluster, NGC 4911 and NGC 4921, are of particular interest because each exhibits significant offsets between its H I centroid and optical center. The formal values of the offsets are 18" and 20", respectively, relative to their major axis sizes of 80" and 150". The signal levels and synthesized beam sizes are such

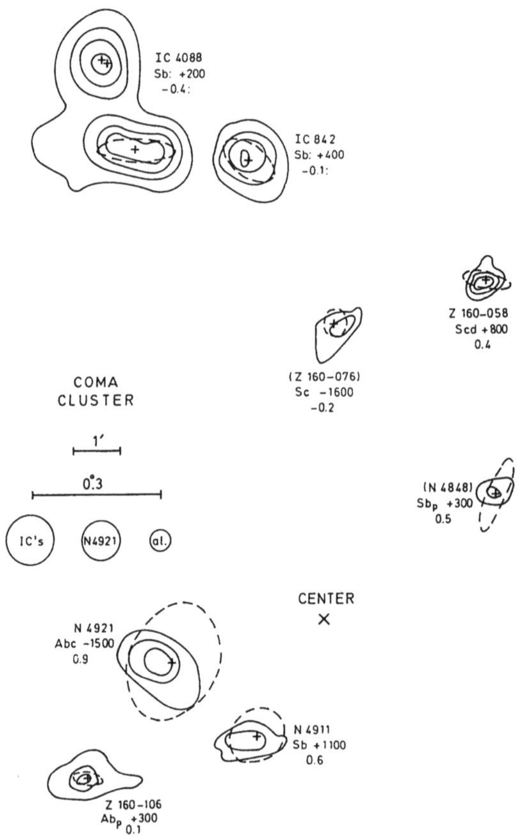

Figure 1: Maps of the total column density of H I (solid lines) for eight spiral galaxies in the Coma cluster. The position of each galaxy's optical center (cross) and each galaxy's shape (dashed lines for the UGC major and minor axes) are shown with respect to the center of the cluster (note the two angular scales indicated). The synthesized beam sizes for the various galaxies are indicated left–center, being 65" for IC 842 and IC 4088, 50" for NGC 4921, and 28" for all others. The outermost contours for each galaxy are about 1.5 times the r.m.s. map noise and correspond to H I column densities of 5×10^{19} atoms cm^{-2} for NGC 4921 and the two IC galaxies, and 2.5×10^{20} atoms cm^{-2} for all others. Parentheses around the names Zw 160–076 and NGC 4848 indicate possible systematic map defects. Each galaxy has indicated its morphological type (A = anemic), radial velocity with respect to that of the cluster mean (6900 km sec^{-1}), and deficiency factor (Sullivan *et al.* 1981), defined as the (logarithmic) difference between the galaxy's observed M_H/L, and that expected for a field spiral of the same morphological type (positive values mean H I deficiency).

that these specific values should be taken as preliminary, but the existence of the offsets seems clear. Such large offsets for field spirals are extremely rare in the *overall* H I centroid (we are not speaking here of faint outer regions), although this remains only a qualitative statement because a thorough study of H I centroids in galaxies has never been conducted. Notice also how the position angles of the H I distributions for these two galaxies do not agree with the optical position angles. The past orbital history of either galaxy may have brought it into a gravitational encounter, but the offsets would seem more likely to result from the asymmetric effects of ram-pressure stripping.

In conclusion, this first rough look at the H I distributions in several of the highly deficient spirals in the central regions of the Coma cluster has indicated very unusual distributions, appropriate to the extreme conditions present in this rich cluster.

I thank the Institute of Astronomy and the Observatoire de Meudon for their hospitality during part of this work, and the Graduate School Research Fund of the University of Washington for its support.

References

Bothun, G. D., Aaronson, M., Schommer, R. A., Mould, J., Huchra, J., and Sullivan, W. T. 1985 , *Astrophys. J. Suppl.* **57**, 423.

Cayatte, V., van Gorkom, J. H., Balkowski, C., and Kotanyi, C. 1988, *Astron. J.*, submitted.

van Gorkom, J. H., Balkowski, C., and Kotanyi, C. 1984, in *Clusters and Groups of Galaxies*, ed. F. Mardirossian, G. Giuricin, and M. Mezzetti (Dordrecht: Reidel), p. 261.

Sullivan, W. T., Bothun, G. D., Bates, B., and Schommer, R. A. 1981, *Astron. J.* **86**, 919.

Warmels, R. H. 1986, Ph. D. thesis, University of Groningen.

Support for Three Controversial Claims Made by Gérard de Vaucouleurs

R. Brent Tully

Institute for Astronomy, University of Hawaii

Summary

Evidence is presented in support of views long held by de Vaucouleurs concerning the distribution of nearby galaxies, the distance scale, and the existence of peculiar streaming velocities.

I. The Plane of the Local Supercluster

In the early 1950s, de Vaucouleurs suggested the existence of a flattened "Local Supercluster" (de Vaucouleurs 1956, 1975 and references therein). The plane delineated by this structure defined the equator of the supergalactic coordinate system (de Vaucouleurs *et al.* 1976). Today, with growing knowledge of the existence of filamentary large–scale structure (Giovanelli *et al.* 1986; de Lapparent *et al.* 1986), de Vaucouleurs's basic viewpoint is probably generally accepted, though the situation was different only a decade ago (Bahcall and Joss 1976).

If there are still reservations about the reality of the concentration of nearby galaxies to a plane, perhaps the following figures will allay those concerns. Figures 1 and 2 illustrate the distribution of galaxies within the conventional Local Supercluster. In each case, galaxies are found to concentrate to a horizontal band — the supergalactic equator. Figure 3 is a smoothed histogram of galaxy counts as a function of distance from the central plane.

Actually, the display in Figure 3 was derived subsequent to an attempt to "fine–tune" the supergalactic coordinate system in the light of modern data and the results presented there involve a 4° tilt of the standard plane. That fine–tuning was considered a waste of time in the end. The plane is corrugated in detail and over a scale of 3000 km s^{-1} centered on the Virgo Cluster, could be described as slightly saddle-shaped: in the line–of–sight toward the cluster, the plane rises then falls, while on an orthogonal cut through the cluster, the plane falls then rises. Overall, de Vaucouleurs's original definition of the plane cannot yet be significantly improved upon. Globally, the plane has FWHM $\simeq$300 km sec^{-1}, a small value in spite of the corrugation, and it is typically two–thirds this value in local regions.

The big surprise has been the linear extent of the plane. It now appears that rich clusters within $0.1c$ in the south galactic hemisphere are concentrated toward this same plane (Tully 1986). A one–dimensional two–point correlation analysis provides the most convincing evidence for the reality of this claim (Tully 1987). Figure 4 is a display of the structure on a scale of $0.1c$ — the Pisces–Cetus Supercluster Complex.

If it turns out to be true that the two planes are real, then the alignment implies a remarkable dynamic connection over a scale of 20,000 km sec^{-1}.

II. Extragalactic Distances: The Short Scale

The debate between Sandage and Tammann (1975, 1984) as proponents of the long distance scale and de Vaucouleurs (1986; see also de Vaucouleurs and Bollinger 1979) as a proponent of the short scale is well–known. With recent improvements in detector technology and more complete H I surveys it has become possible to improve the calibration of luminosity–line width distance–estimator relationships (Tully and Fisher 1977). With CCD R– and I–band photometry, the dispersion in the intrinsic relationships seems to be only 0.25 mag, or 12% uncertainty in distance, to a single object (Pierce and Tully 1988). The local calibration seems to be much improved recently, now that CCD observations of RR Lyrae stars (Prichett and van den Bergh 1987) and Cepheids (Freedman and Madore 1988) have been made in several nearby galaxies. Perhaps the most accurate estimates to date of the distances to the Virgo and Ursa Major clusters are 15.6 and 15.5 Mpc (Pierce and Tully 1988). The combined cluster data superimposed on the local calibration is shown in Figure 5. If the Virgocentric retardation of the Galaxy is 300 km sec^{-1}, then $H_0 \simeq 85$ km sec^{-1} Mpc^{-1}.

A fair consistency is emerging. With H–band estimates of the distances to distant clusters, after adjustment to a common zero–point, the Hubble Constant is determined to be 88 km sec^{-1} Mpc^{-1} (Aaronson *et al.* 1986). This estimate is expected to be free of uncertainties due to streaming motions. Then an estimate of H_0, based on bias–free distance measurements of field galaxies in the vicinity of the Local Supercluster, is $85 < H_0 < 95$ (Tully 1988). This latter estimate *is* sensitive to streaming motions, as will be discussed in Section III. Together, these results provide reasonable confirmation of the short distance scale.

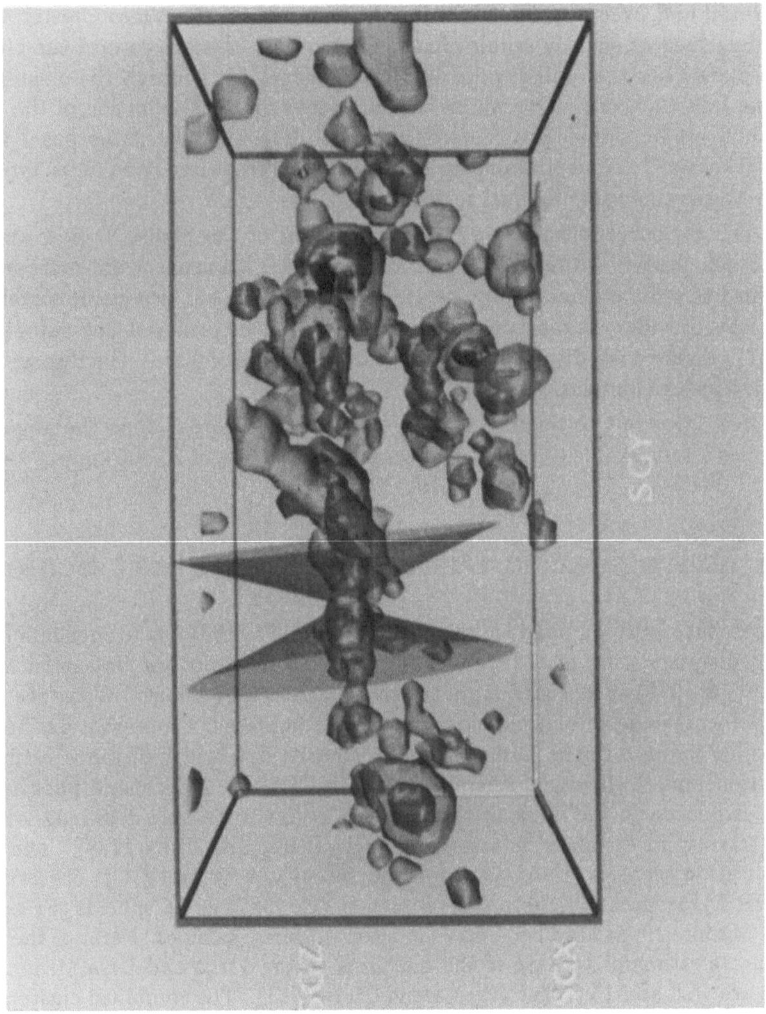

Figure 1: Density contour map of nearby galaxies: edge–on view of part of the plane of the Local Supercluster. The long axis corresponds to 1500 $\mathrm{km\,sec^{-1}}$ = 17 Mpc if $H_0 = 90\,\mathrm{km\,sec^{-1}\,Mpc^{-1}}$. Outer contour: 1.7 galaxies $\mathrm{Mpc^{-3}}$; inner contour; 7 galaxies $\mathrm{Mpc^{-3}}$. The Galaxy is located at the apex of the two cones which outline $b = \pm 20°$. The region between the two cones is largely obscured. The supergalactic equator runs horizontally through the position of the Galaxy. The main concentration of galaxies on the equator is the Coma–Sculptor Cloud. The concentration below the equator is the Leo Spur.

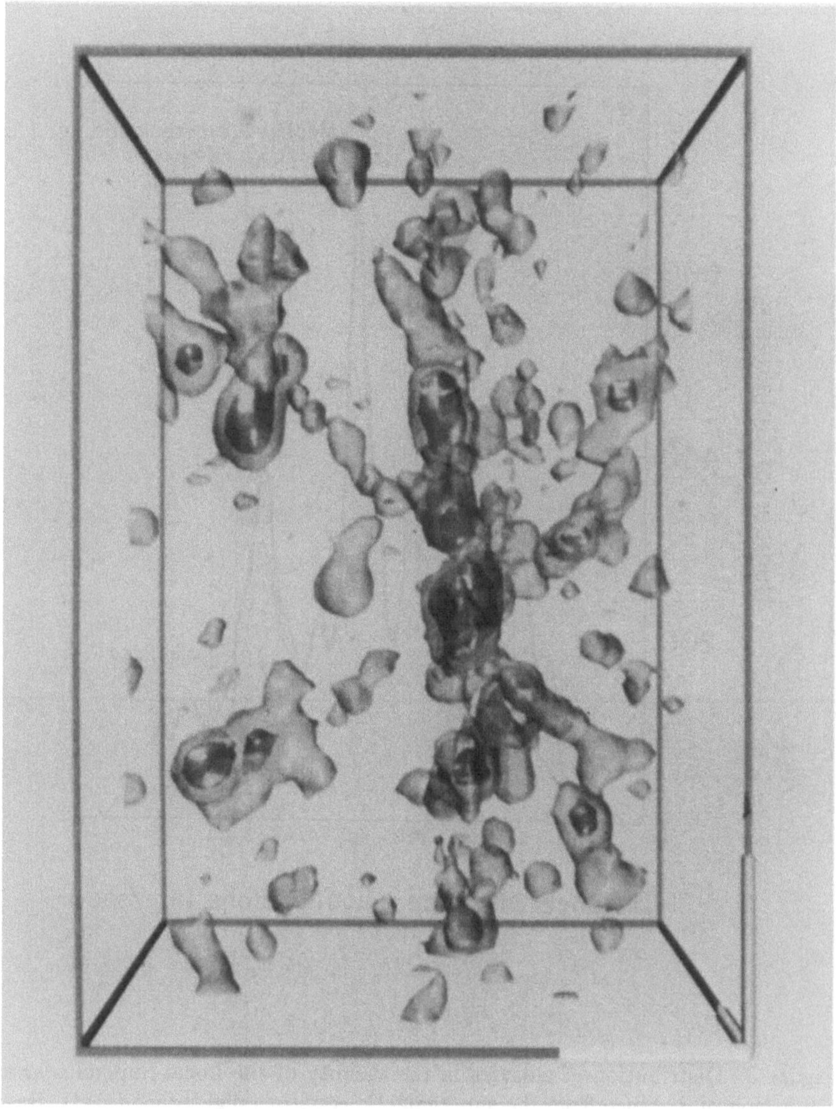

Figure 2: Density contour map of the Local Supercluster: edge–on view orthogonal to Figure 1. The long axis corresponds to 3750 km sec^{-1} = 42 Mpc if H_0 = 90 km sec^{-1} Mpc^{-1}. Outer contour: 0.9 galaxies Mpc^{-3}; inner contour 2.6 galaxies Mpc^{-3}. The Galaxy is near the center on the front surface of the cube. The horizontal concentration of galaxies lies along the supergalactic equator. The Virgo Cluster lies within this concentration near the middle of the cube.

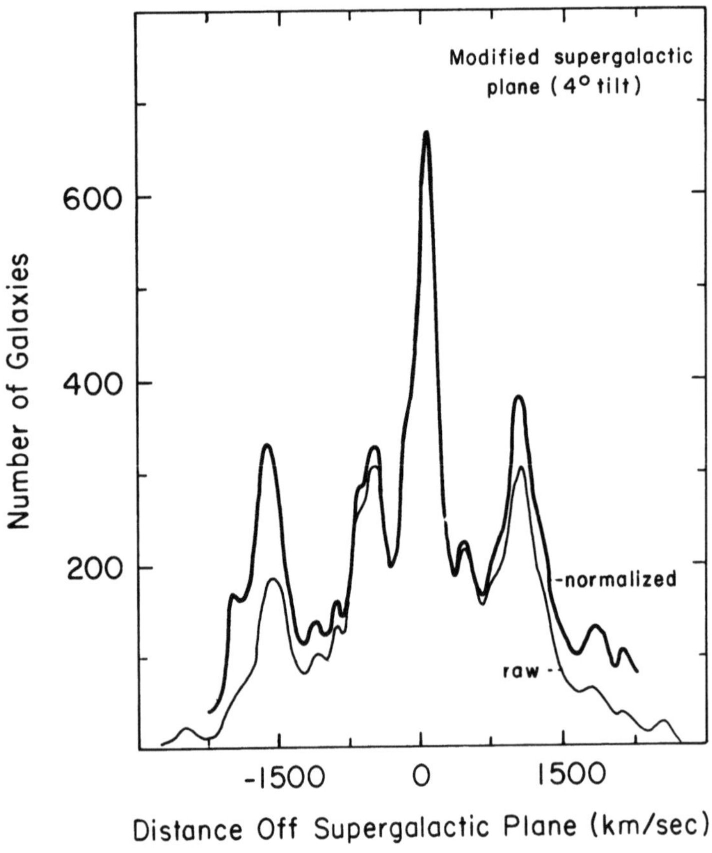

Figure 3: Distribution of galaxies in the vicinity of the Local Supercluster as a function of distance from the supergalactic equator (slightly modified). Raw: observed counts; normalized: counts adjusted for geometric and obscuration biases. The Virgo Cluster contributes 130 counts to the central peak.

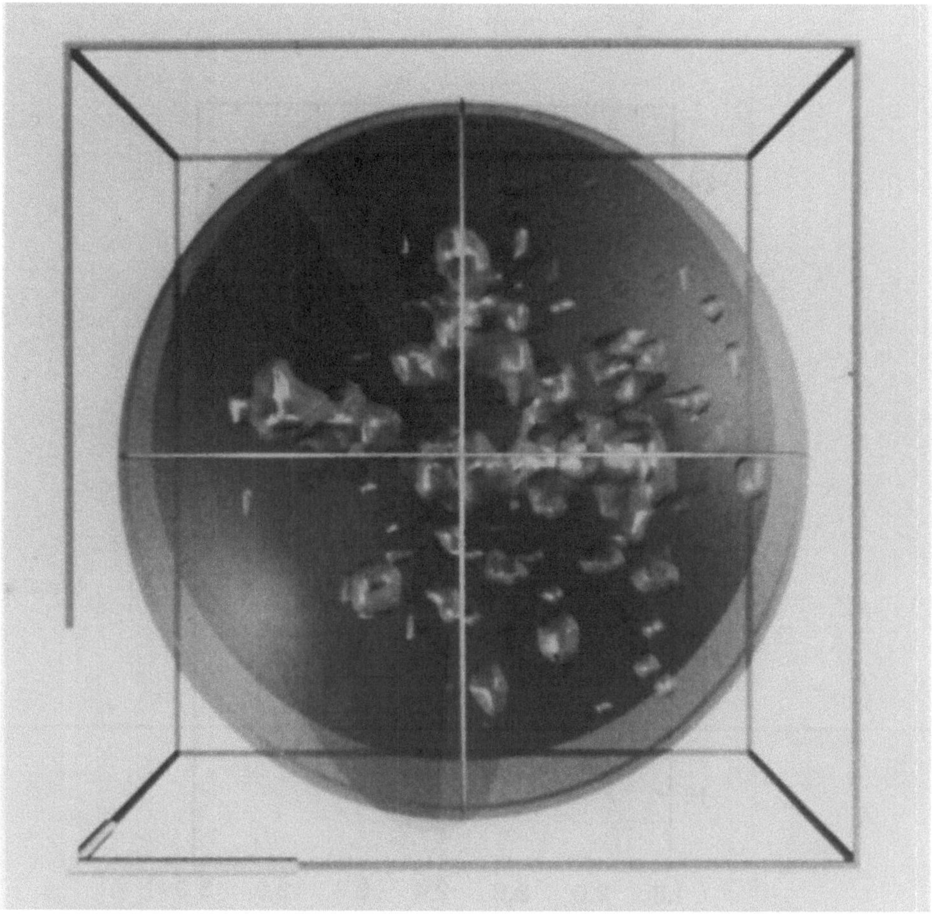

Figure 4: Density contour map of rich clusters of galaxies: from the same viewing direction as Figure 2. The diameter of the spherical region is 0.2 $c =$ 620 Mpc if $H_0 = 90\,\mathrm{km\,sec^{-1}\,Mpc^{-1}}$ (proper distance). Outer contour: 3×10^{-5} clusters $\mathrm{Mpc^{-3}}$; inner contour: 9×10^{-5} clusters $\mathrm{Mpc^{-3}}$. This view is of clusters in the south galactic hemisphere and is essentially looking down the axis of the cone that delineates galactic obscuration. The concentration of clusters to the supergalactic equator in the center of the field and extending toward the right is the Pisces–Cetus Supercluster Complex. The foreground part of the Aquarius Supercluster Complex is seen at high supergalactic latitudes. The flair of clusters upward from the supergalactic equator to the left of center is in the Indus region, but the left hemisphere on this map lies in the south and is incompletely surveyed.

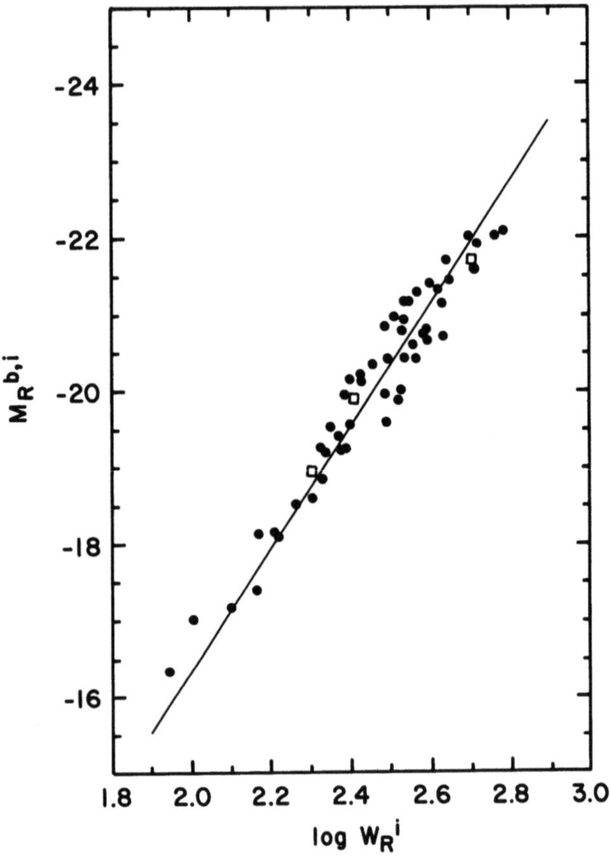

Figure 5: R–band luminosity–line width relation for galaxies in the Virgo and Ursa Major clusters. The absolute magnitude scale, $M_R^{b,i}$, is established by a fit to the three local calibrators indicated by open symbols. The straight line is the least squares fit with uncertainties in line width, W_R^i. This line is the bias–free distance estimator relationship.

III. The Local Velocity Anomaly

Apparently it was Rubin (1951) who first saw evidence for what will be called here the "local velocity anomaly," and de Vaucouleurs certainly saw the evidence a few years later (de Vaucouleurs 1958, de Vaucouleurs and Bollinger 1979). The assertion is that galaxies in the nearest clouds have low velocities for their distances. This claim is based on distance estimates using luminosity–line width relations in a formulation that should avoid Malmquist bias (Tully 1988).

Malmquist bias arises because of a propensity to select galaxies from the field in a magnitude–limited sample that lie *above* the mean relationship in a complete sample. The galaxies are assumed to be less luminous than they really are on average, distances are underestimated, and the Hubble Constant is overestimated. However, if line widths are unbiased, then the relationship that bisects a complete sample in line width provides an unbiased distance estimator in the sense that randomly selected galaxies in a magnitude–limited sample are no more likely to lie on one side of the formulated relationship than on the other.

In the present case, the cluster sample shown in Figure 5 was used to generate the fit to a complete sample. Unbiased distances could then be derived for galaxies in the field. The Hubble ratio (velocity/distance) can be determined for each galaxy with a measured distance, and mean values of the ratio can be found for members of a common cloud. In Figure 6, mean values of the Hubble ratio are shown for all clouds with at least five distance estimates. Open circles show results assuming uniform expansion of the universe, and closed circles show results modified by a specific Virgocentric perturbation.

It seems that if only the cloud that we live in (the Coma–Sculptor Cloud) and the nearest adjacent cloud (the Leo Spur) are considered, then one would conclude $H_0 \simeq 63\,\mathrm{km\,sec^{-1}\,Mpc^{-1}}$. By contrast, if only the remaining clouds are considered, then H_0 is 80 or 95 depending on the assumed kinematic model.

The dependency of the Hubble Constant on a choice of kinematic model follows because of the general expectation that H_0 will be underestimated by an observer inside an overdense region who makes measurements confined to that overdense region (Tully 1988). In the present case, the correct value of H_0 should be larger than the value of 80 found in the uniform Hubble flow case and a tentative estimate is $85 < H_0 < 95\,\mathrm{km\,sec^{-1}\,Mpc^{-1}}$.

An interpretation of the lower value of the Hubble ratio found in the two nearby clouds is that expansion motions have been retarded by a local concentration of matter. The effect would be explained by roughly $10^{14}\,M_\odot$ associated with the Coma–Sculptor Cloud, whence $M/L \sim 300 M_\odot/L_\odot$, over a dimension of 15 Mpc.

The trend from low values of H_0 seen in the nearby clouds to high values of H_0 at larger distances probably lies at the core of the controversy over the value of the Hubble Constant. One school of thought (Bottinelli *et al.* 1986, Tammann 1987, Sandage 1988) would have it that the apparent increase in H_0 is due to Malmquist bias and holds that the correct value is on the low side. On

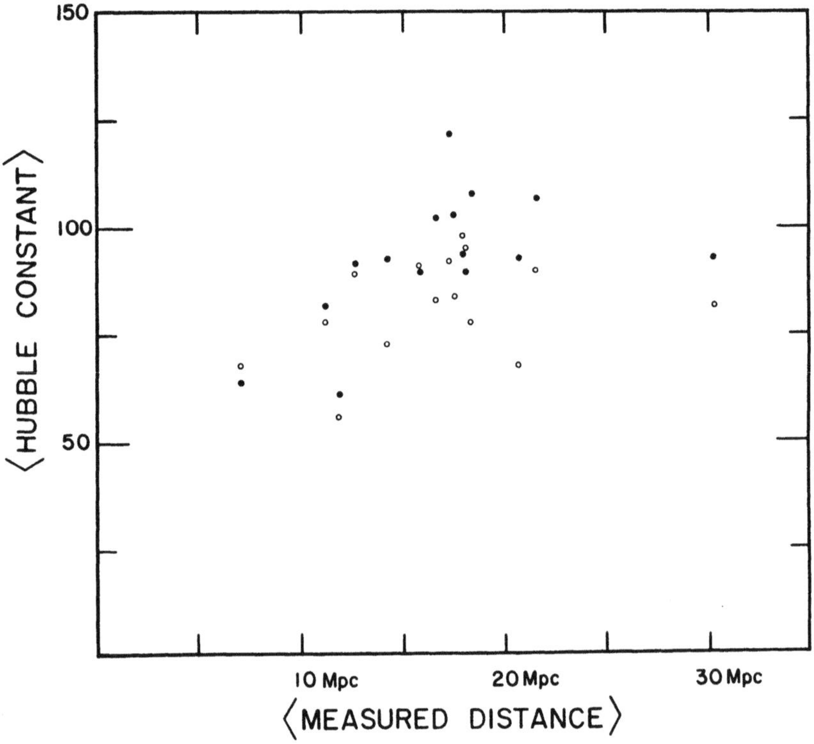

Figure 6: Mean Hubble ratio for clouds of nearby galaxies. Open circles: undisturbed Hubble flow model; filled circles: specific Virgocentric mass model. The Hubble ratio is low for the two clouds illustrated in Figure 1.

the other hand, if it is true that Malmquist bias has been avoided in the present study (Tully 1988), then a physical interpretation for the increase in H_0 is in order, perhaps even the explanation offered here.

References

Aaronson, M., Bothun, G., Mould, J. R., Huchra, J. P., Schommer, R. A., and Cornell, M. E. 1986, *Astrophys. J.* **302**, 536.

Bahcall, J. N. and Joss, P. C. 1976, *Astrophys. J.* **203**, 23.

Bottinelli, L., Gouguenheim, L. Paturel, and Teerikorpi, P. 1986, *Astron. Astrophys.* **156**, 157.

Freedman, W. L. and Madore, B. F. 1988, *Astrophys. J.*, submitted.

Giovanelli, R., Haynes, M. P., and Chincarini, G. L. 1986, *Astrophys. J.* **300**, 77.

de Lapparent, V., Geller, M. J., and Huchra, J. P. 1986, *Astrophys. J. Lett.* **302**, L1.

Pierce, M. J. and Tully, R. B. 1988, *Astrophys. J.* **330**, 579.

Pritchet, C. J. and van den Bergh, S. 1987, *Astrophys. J.* **316**, 517.

Rubin, V. 1951, *Astron. J.* **56**, 47.

Sandage, A. 1988, *Astrophys. J.*, in press.

Sandage, A. and Tammann, G. A. 1975, *Astrophys. J.* **197**, 265.

Sandage, A. and Tammann, G. A. 1984, *Nature* **307**, 326.

Tammann, G. A. 1987, in *I. A. U. Symp. 124, Observational Cosmology*, ed. A. Hewitt, G. Burbidge, and L. Z. Fang (Dordrecht: Reidel), p. 151.

Tully, R. B. 1986, *Astrophys. J.* **303**, 25.

Tully, R. B. 1987, *Astrophys. J.* **323**, 1.

Tully, R. B. 1988, *Nature*, in press.

Tully, R. B. and Fisher, J. R. 1977, *Astron. Astrophys.* **54**, 661.

de Vaucouleurs, G. 1956, *Vistas in Astron.* **2**, 1584.

de Vaucouleurs, G. 1958, *Astron. J.* **63**, 253.

de Vaucouleurs, G. 1975, *Astrophys. J.* **202**, 610.

de Vaucouleurs, G. 1986, in *Galaxy Distances and Deviations from Universal Expansion*, ed. B. F. Madore and R. B. Tully (Dordrecht: Reidel), p. 1.

de Vaucouleurs, G. and Bollinger, G. 1979, *Astrophys. J.* **233**, 433.

de Vaucouleurs, G., de Vaucouleurs, A., and Corwin, H. G. 1976, *Second Reference Catalogue of Bright Galaxies* (Austin: Univ. of Texas Press).

Discussion

M. Burbidge: In the Boötes void, for example, people have been searching for anomalously small or faint galaxies — do you know what is the status of that search?

T. Thuan: Schneider and I have just finished a 21–cm redshift survey of all 1800 dwarfs and Magellanic–type galaxies in the Nilson catalogue that covers the whole northern sky. Detailed comparison of the spatial distribution of bright and dwarf galaxies (in the CfA slice: *Astrophys. J. Lett.* **315**, L93, 1987; and several other regions of space) shows that H I–rich dwarf galaxies ($M_B < -13$) do not fill in the voids and are just as narrowly confined to the bright galaxy structures as the bright galaxies themselves.

Tully: I agree with Thuan, but can comment on a subtlety. It is known that high–luminosity early–type systems tend to lie in higher density regions, but now I have explored the trend to quite low luminosities and find a steady progression in the sense that toward the lower luminosities and later types, the distances between neighbors continues to increase, though these systems are still confined to filamentary structures (*Astron. J.*, July 1988).

H. Corwin: Can you talk a bit more about the effect of the Malmquist bias on your distance scale?

Tully: The amplitude of Malmquist bias is sensitively dependent on the *slope* of your distance–estimator relationship. It is possible to assume a steep enough slope to generate *anti–Malmquist bias.* I claim to have a recipe for the slope that produces zero bias at any distance.

P. Brosche: Don't you feel the necessity to introduce a second parameter (besides the internal velocity) for estimating the absolute luminosity? Although the residual dispersion of the latter may be small, the averages at a fixed velocity width of the distant galaxies and of calibrators could be different due to selection effects acting on the second parameter.

Tully: There could be a second parameter. I am always on the lookout for such a thing, and we have seen marginal evidence for segregation by morphological type. However, the scatter is already so low as to be comparable with the observational uncertainties, and there are advantages to using the simplest possible relationship. With our present formulation, if a candidate galaxy is any nonpathological spiral with a suitable inclination, we work to get a suitable H I profile and include the system in our analysis.

D. Burstein: As I think is well–known by now, our group investigates large–scale motion in a scale–invariant manner, using velocity as our relative measure of distances. As such, I would be very interested in the reanalysis of your data in terms of a great–attractor–plus–Virgo–infall–plus–local–anomaly model.

Tully: You and I have been talking about a local anomaly with effects in the same region of space. Yet they may not be quite the same phenomenon. There could be one thing causing a bulk motion with respect to the microwave background that you see (a push from the Local Void?) and another thing causing the fluctuation in the Hubble ratio that I see (mass in the local Coma–Sculptor Cloud?).

Burstein: We both agree that the Coma–Sculptor Cloud has a peculiar motion of 400 km sec^{-1} with respect to other regions. Our analysis has identified about 16 regions of size around 1500 km sec^{-1}, also moving with three–dimensional motions of 250–450 km sec^{-1}. How would this affect your determination of the Hubble Constant?

Tully: If we are sitting in the potential well of the great attractor, then the statistical probability is that measurement within the domain of the perturbation should lead to an underestimate of the Hubble Constant. Fortunately, the percentage effect appears to decrease with increasing scale. Within the local velocity anomaly, velocity perturbations are perhaps 200 km sec^{-1}, with observed velocities of 450 km sec^{-1}, or roughly a 30% effect. The Virgo Cluster and surroundings might influence us by 300 km sec^{-1} at an observed 1000 km sec^{-1}, for a 20% effect. The influence of the great attractor might be 400 km sec^{-1}, at 4000 km sec^{-1} for a 10% effect.

A Connection between the Pisces–Perseus Supercluster and the Abell 569 Cloud?

C. Balkowski, V. Cayatte, P. Chamaraux, and P. Fontanelli

Observatoire de Paris, Section de Meudon, France

The Pisces–Perseus supercluster and the cluster Abell 569 lie at low galactic latitude on each side of the obscuration zone, almost at the same redshift; thus, the question of a connection between them is naturally raised.

We have measured 41 redshifts of galaxies in the zone $3.^h5 \leq \alpha \leq 6.^h5$ and $40° \leq \delta \leq 50°$. Galaxies were collected from UGC (Nilson 1973), CGCG (Zwicky and Kowal 1968), Weinberger (1980) and from an examination of the Palomar Sky Survey charts. Observations were performed with the Nançay radiotelescope and with the 1.93m telescope at Haute Provence Observatory.

Figure 1 shows the sky distribution of the 286 redshifts available in a zone including the zone of the possible connection $3.^h5 \leq \alpha \leq 6.^h5$ and $40° \leq \delta \leq 55°$. Data taken from the literature are represented by crosses (Palumbo *et al.* , 1983; Giovanelli and Haynes 1982; Fanti *et al.* , 1982; Focardi *et al.* , 1986; Hauschildt 1987; Tifft, 1978). Data from the present study are represented by filled circles.

Figure 2 is a wedge diagram compressed in declination for galaxies in the zone $3^h \leq \alpha \leq 7.^h5$ and $38° \leq \delta \leq 52°$. It strongly suggests a connection between the Perseus supercluster at $3^h 15^m$ and A 569 at $7^h 15^m$, between 5000 and 6500 $\mathrm{km\,s^{-1}}$.

Figure 3 shows the distribution of galaxies in the same zone in the redshift interval 5000 to 7000 $\mathrm{km\,s^{-1}}$. The galaxies with $5000 < V < 7000\,\mathrm{km\,s^{-1}}$ lie mainly in the Perseus and A 569 clusters and between them, contrary to those with $V < 5000\,\mathrm{km\,s^{-1}}$.

The available data strongly suggest a connection between A 569 and the Pisces–Perseus supercluster. However, this conclusion should be confirmed by more observations in the zone 4^h to $4.^h5$ and 6^h to $6.^h5$ where few redshifts are presently available. If the connection is real, the dimensions of the Perseus–A 569 supercluster would be at least $130 \cdot h_{75}^{-1}$ Mpc, a very large value.

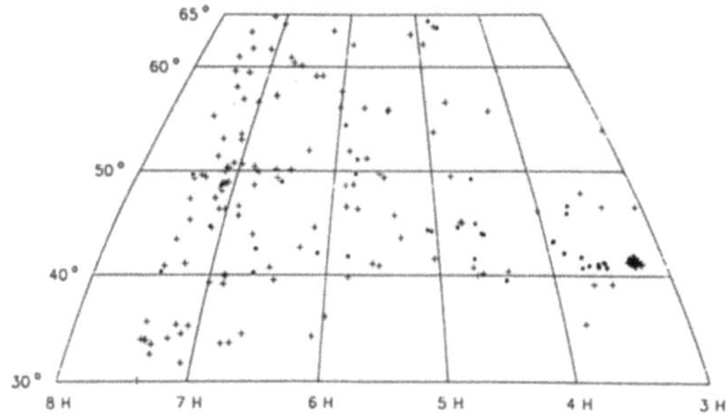

Figure 1: Distribution of the 286 galaxies with available redshifts in the zone of possible connection.

References

Fanti, C., Fanti, R., Feretti, L., Ficarra, A., Givia, I. M., Giovannini, G., Gregorini, L., Mantovani, F., Marano, B., Padrielli, L., Parma, P., Tomasi, P., and Vettolani, G. 1982, *Astron. Astrophys.* **105**, 200.

Focardi, P., Marano, B., and Vettolani, G. 1986, *Astron. Astrophys.* **105**, 200.

Giovanelli, R. and Haynes, M. P. 1982, *Astron. J.* **87**, 1355.

Hauschildt, M. 1987, *Astron. Astrophys.* **184**, 43.

Palumbo, G. G. C., Tanzella-Nitti, G., and Vettolani, G. 1983, *Catalogue of Radial Velocities of Galaxies* (New York: Gordon and Breach).

Tifft, W. G. 1978, *Astrophys. J.* **222**, 54.

Weinberger, R. 1980, *Astron. Astrophys. Suppl.* **40**, 123.

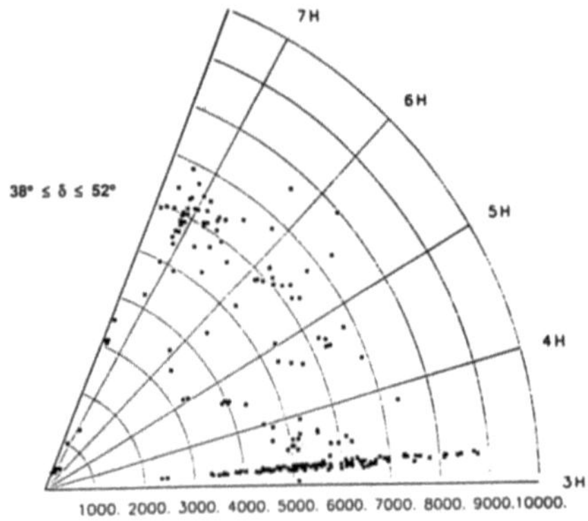

Figure 2: Wedge diagram in the zone $3^h \leq \alpha \leq 7.^h5$ and $38^\circ \leq \delta \leq 52^\circ$.

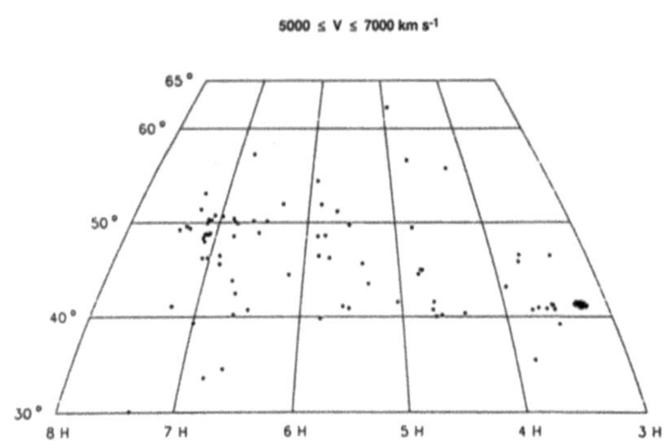

Figure 3: Distribution of the galaxies with a velocity between 5000 $\mathrm{km\,s^{-1}}$ and 7000 $\mathrm{km\,s^{-1}}$.

Automatic Algorithms for Grouping Nearby Galaxies

P. Fouqué[1], G. Paturel[2], P. Chamaraux[1], A. Fruscione[1,2], and J. F. Panis[2]

[1]Observatoire de Meudon, France
[2]Observatoire de Lyon, Saint–Genis Laval, France

Introduction

For several years, we have been studying groups of galaxies. The goal of our research is to detect groups of galaxies by means of automatic algorithms. We hope to achieve results similar to those of de Vaucouleurs (1975) in his famous paper "Nearby Groups of Galaxies" by using completely objective methods.

Historical Methods

The first automatic methods (*e.g.* Materne 1978, Paturel 1979) for grouping galaxies were based on the simultaneous use of position (right ascension and declination) and radial velocity to calculate the dispersion in three–dimensional space of each group. The optimal classification is obtained by minimizing the total dispersion. Two methods were used: by accreting new members (Materne) or by splitting a previous class (Paturel). The major problem with this "cluster analysis" comes from the fact that the radial velocity is a poor distance indicator for galaxies with low velocities.

Tully (1980) and Vennik (1984) used a method similar to the one used by Materne, but they introduced a force calculated from the mass divided by the square of the intergalactic distance. Again, the estimation of the distance between galaxies is the major problem with their method.

Yet another method was introduced by Huchra and Geller (1982). Companions with a given projected separation and a given velocity difference are searched for around every galaxy in a given sample. Those meeting the projected separation and velocity difference criteria are called a "group." The process is repeated until no further members can be added to the group. This method is very easy to implement on computers, and gives reasonable results. However, the problem of using velocity as a distance indicator remains, and there is still some subjectivity in selecting the separation criteria. This "HG" method has been used by Fruscione (1987) with some preliminary improvements.

Recent Improvements

Examining recent attempts to improve on the previous methods, we found that all try to use improved techniques of estimating distances from radial velocity. Recently, Chamaraux and Fouqué (1987) used a generalized HG method, where velocity and angular separation play symmetrical roles. Tully (1986) also improved his own method by calculating intergalactic separation in a more sophisticated way, in order to take into account the fact that for low velocity, the true separation between galaxies mainly depends on the angular separation, but not on the velocity. Tully applied his method to the sample in his *Nearby Galaxies Catalog* (Tully 1988). Finally, Panis and Paturel (1988) used a mean velocity for each galaxy instead of the observed one, this mean velocity being deduced from a pre–grouping analysis.

Conclusion

A preliminary comparison shows that the automated galaxy group detection methods all find essentially the same groups within 50 Mpc. Perhaps surprisingly, most of de Vaucouleurs's groups, found nearly a quarter of a century ago, using only the data in the first *Reference Catalogue* (de Vaucouleurs and de Vaucouleurs 1964), correspond closely to groups found by the recent methods.

References

Chamaraux, P. and Fouqué, P. 1988, in preparation.

Fruscione, A. 1987, DEA, Université de Paris.

Huchra, J. P. and Geller, M. J. 1982, *Astrophys. J.* **257**, 423.

Materne, J. 1978, *Astron. Astrophys.* **63**, 401.

Panis, J. F. and Paturel, G. 1988, in preparation.

Paturel, G. 1979, *Astron. Astrophys.* **71**, 106.

Tully, R. B. 1980, *Astrophys. J.* **237**, 390.

Tully, R. B. 1986, *Astrophys. J.* **303**, 25.

Tully, R. B. 1988, *Nearby Galaxies Catalog* (Cambridge: Cambridge Univ. Press).

de Vaucouleurs, G. 1975, "Nearby Groups of Galaxies," in *Galaxies and the Universe*, ed. A. Sandage, M. Sandage, and J. Kristian (Chicago: Univ. of Chicago Press), p. 557.
de Vaucouleurs, G. and de Vaucouleurs, A. 1964, *Reference Catalogue of Bright Galaxies* (Austin: Univ. of Texas Press).
Vennik, J. 1984, *Tartu Astron. Obs. Publ.* **73**, 1.

A Study of the Southern Supercluster

Shyamal Mitra

Astronomy Department, University of Texas, Austin, Texas, U.S.A.

The Southern Supercluster (de Vaucouleurs 1953) is a long chain of galaxies extending from Cetus to Dorado, through Fornax, Eridanus, and Horologium. The densest region of the supercluster lies between supergalactic longitudes 200° to 310°, and supergalactic latitudes −38° to −45°, which correspond to right ascensions of 2^h to 6^h hours and declinations of +5° to −69°.

From the contour maps made of galaxies from the *Second Reference Catalogue of Bright Galaxies* (de Vaucouleurs *et al.*, RC2), two regions [A ($180° \leq L \leq 320°$, $-30° \geq B \geq -50$) and B ($260° \leq L \leq 360°$, $0° \geq B \geq -30°$)] were delineated for detailed study. Galaxies from five catalogues were compiled and the basic parameters, diameters and axis ratios, were reduced to the RC2 system. Total magnitudes were measured for 105 galaxies, and the magnitudes for other galaxies were obtained from the RC3 (in preparation). The radial velocities of 23 galaxies were measured, but most of the radial velocities were culled from the literature.

A self-consistent distance scale was established using tertiary distance indicators: the luminosity index Λ_c, and the maximum rotation velocity $\log V_m$ obtained from H I line widths. The distance moduli were obtained using parameters $\log D_0$ and B_T^0. The distance moduli derived from the central velocity dispersion of early type galaxies, diameters of inner rings, and the method of "sosie" galaxies were obtained from other catalogues.

A database was assembled containing several parameters like $\log D$, B_T^0, μ, V_r, etc. A statistical study of their distribution was made and a measure of their completeness estimated. It was shown for galaxies brighter than $B_T^0 \sim 14.0$ that the probability that this areal distribution could have arisen by chance was less than 10^{-3}.

Several maps were constructed to investigate the structure of the supercluster in three dimensions using the positional coordinates and a distance estimate (either from radial velocity or distance modulus). The extent, size, and shape of the supercluster were determined. The supercluster lies within the velocity range 560 to 2240 $\mathrm{km\,sec^{-1}}$, and has a velocity dispersion of 345 $\mathrm{km\,sec^{-1}}$. It is 41 Mpc along its longest dimension and is at a mean distance of 20 Mpc.

To investigate the variation of velocity, distance modulus, and Hubble ratio, numerical maps were generated and departures from the uniform Hubble flow within the supercluster were determined. Fifteen groups belonging to the supercluster were identified. The mass, luminosity, and mass–to–light ratio of the groups were computed. The mean mass–to–light ratio is 100·h (h = H/100). The total luminosity of the supercluster is estimated to be $2.4 \times 10^{12}\ \mathrm{h}^{-2}\ L_{\odot}$ after making corrections for incompleteness. The total mass of the supercluster is estimated to be $2.4 \times 10^{14}\ \mathrm{h}^{-1} M_{\odot}$. The Southern Supercluster is less massive than the Local Supercluster, but is comparable to the Coma and Hercules Superclusters in terms of mass and luminosity.

Even though the Southern Supercluster is the nearest supercluster to the Local Supercluster, the two are not physically linked. However, there seems to be a tenuous connection between the Southern Supercluster and the Perseus Supercluster.

Acknowledgment: I should like to thank Prof. de Vaucouleurs for his guidance and encouragement during this project.

References

de Vaucouleurs, G. 1953, *Astron. J.* **58**, 30.

de Vaucouleurs, G., de Vaucouleurs, A., and Corwin, H. G. 1976, *Second Reference Catalogue of Bright Galaxies* (Austin: University of Texas Press).

Dwarf Galaxies, Voids, and the Topology of the Universe

Trinh X. Thuan

Astronomy Department, University of Virginia, Charlottesville, Virginia, U.S.A.
Institut d'Astrophysique de Paris,
and Service d'Astrophysique, CEN Saclay, France

Are we having a "biased" view of the universe because we have not surveyed the sky to a faint enough surface brightness level? Are the voids seen in the bright galaxy distribution filled with dwarf galaxies? In order to provide an answer to these questions, I have undertaken, in collaboration with S. E. Schneider, a 21–cm redshift survey of all galaxies classified by Nilson (1973) in the *Uppsala General Catalog* (UGC) as dwarf, magellanic irregular, or irregular galaxies, 1849 galaxies out of a total of ~13000 galaxies of all morphological types in the UGC. The redshift survey thus defined avoids the selection effect against dwarf galaxies present in previous redshift surveys: the UGC is a diameter–limited catalog (it lists all galaxies with a blue diameter larger than 1 arc–minute visible on the Palomar Sky Survey Prints and north of $\delta = -2°30'$) and thus does not discriminate against faint, low surface brightness dwarf galaxies, as is the case for magnitude–limited redshift surveys, such as the Center for Astrophysics (CfA) survey (de Lapparent *et al.* 1986).

The dwarf redshift survey was carried out using the Arecibo 305–m and Green Bank 100–m telescopes, with a velocity search range of −400 to 10000 km $\sec^{-1}$. The completed survey has a detection rate of ~85%, with only 283 non–detections. These vast amounts of data are being assembled in a catalog which will contain a total of ~1800 entries, giving for each dwarf galaxy an accurate H I flux (or upper limit), velocity and velocity width, with an estimated photographic magnitude. We found that the morphological selection criteria of "low surface brightness" and "little or no central concentration" which Nilson used to identify dwarf galaxies, have isolated a broad continuum of galaxies with absolute blue luminosities ranging from ~ -13 to ~ -21, from the "true" dwarf galaxies ($M_B \geq -16$) to more luminous and very low surface brightness (LSB) systems ($M_B < -16$).

As a first application of the redshift data, we have compared the spatial distribution of the dwarf and bright ($m_{pg} \leq 15.5$) galaxies in the CfA slice of the universe (de Lapparent *et al.* 1986). We find that dwarf and LSB galaxies also lie on the structures delineated by the high surface brightness galaxies, and are just as narrowly confined to the bright galaxy structures as the bright galaxies themselves (Figure 1a; see also Thuan *et al.* 1987). Gas-rich dwarf galaxies do not fill in the voids seen in the bright galaxy distribution. This conclusion is reinforced by a similar analysis for the contiguous 6° slice to the north (Figure 1b), and rules out a certain class of biased galaxy formation theories which predict explicitly that dwarfs should be present everywhere, including the voids. If biasing ideas are correct, the dark matter which is in the voids cannot be traced by dwarf galaxies with $M_B < -13$.

Another application is the analysis (in collaboration with J. R. Gott, S. E. Schneider, and J. Miller) of the topology of the universe using techniques developed by Gott *et al.* (1986), and by combining the dwarf sample with other available redshift samples. We found that the data are most consistent with a sponge-like topology (which imply Gaussian primordial density fluctuations) and a universe dominated by cold dark matter, with $\Omega = 1$ and $H = 50\,\mathrm{km\,sec^{-1}\,Mpc^{-1}}$.

References

Gott, J. R., Melott, A. L., and Dickinson, M. 1986, *Astrophys. J.* **306**, 341.

de Lapparent, V., Geller, M. J., and Huchra, J. P. 1986, *Astrophys. J. Lett.* **302**, L1.

Nilson, P. 1973, *Uppsala General Catalogue of Galaxies* (Uppsala: Royal Society of Sciences of Uppsala).

Thuan, T. X., Gott, J. R., and Schneider, S. E. 1987, *Astrophys. J. Lett.* **315**, L93.

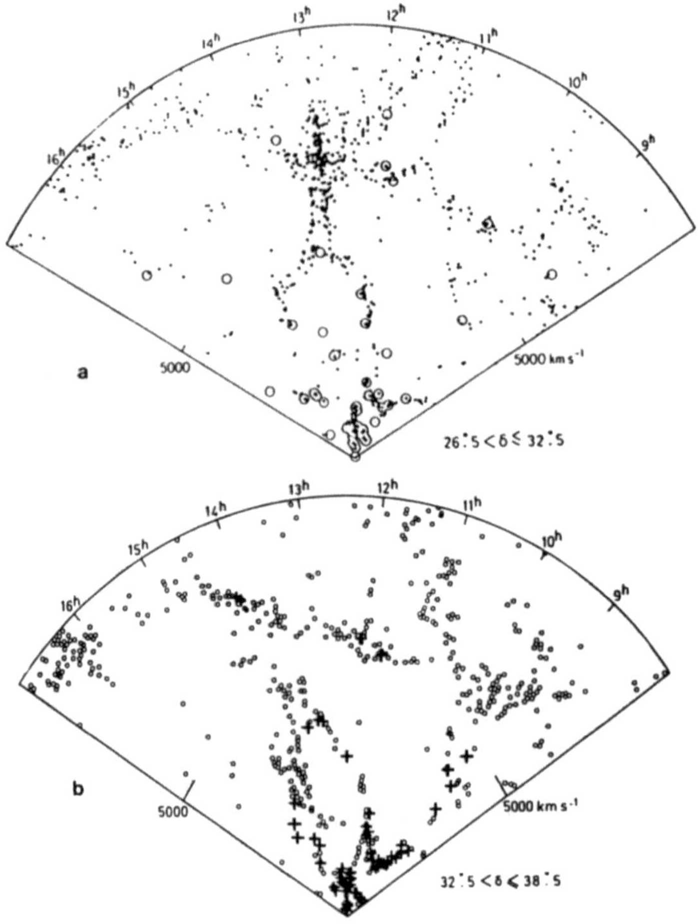

Figure 1: a) Comparison between the spatial distribution of dwarf and LSB galaxies (open circles) and that of bright ($m_{pg} \leq 15.5$) galaxies (crosses) in the CfA slice ($8^{\rm h} \leq \alpha \leq 17^{\rm h}$, $26°5 < \delta \leq 32°5$). There is no difference between the bright and dwarf galaxy distributions. b) Same as in Figure 1a, but for the 6° slice to the north ($8^{\rm h} \leq \alpha \leq 17^{\rm h}$, $32°5 < \delta \leq 38°5$). Bright galaxies are open circles and crosses are dwarf and LSB galaxies.

The Local Supercluster and Anisotropy of the Redshifts

Vera C. Rubin

Department of Terrestrial Magnetism, Carnegie Institution of Washington, Washington, D. C., U. S. A.

Introduction

This is a birthday party, so we must have a birthday card. Figure 1 is a birthday card for Gérard. It plots his published papers as a function of time, papers classified approximately according to subject matter. Gérard's output has been prodigious, and we anticipate continued valuable contributions at least until the year 2010. Happy Birthday!

Because this is a celebration, it is a good time to look back, to attempt to identify the steps which brought us to our present understanding. In order to deduce the existence of a Local Supercluster, two things are necessary: a knowledge of the structure and a knowledge of the motions within the nearby universe. I find it surprising that the machinery for deducing motions was at hand long before the corresponding machinery for deducing structure, although in some sense both go hand–in–hand, because distance determinations rest very much on measured redshifts. What follows is a very sketchy and very personal understanding of some of the early history.

Early Determination of Motions

Although Hooke, as early as 1700, suspected that the sun was moving, the first determination of a stellar proper motion came only in 1718, when Halley (1718) established that a few stars had altered their positions since the catalogue of Hipparchus over eighteen centuries earlier. In 1779, Lalande predicted that the motion of the sun would not be detectable until several centuries had passed and the sun had moved into a region of different stars. This prediction establishes

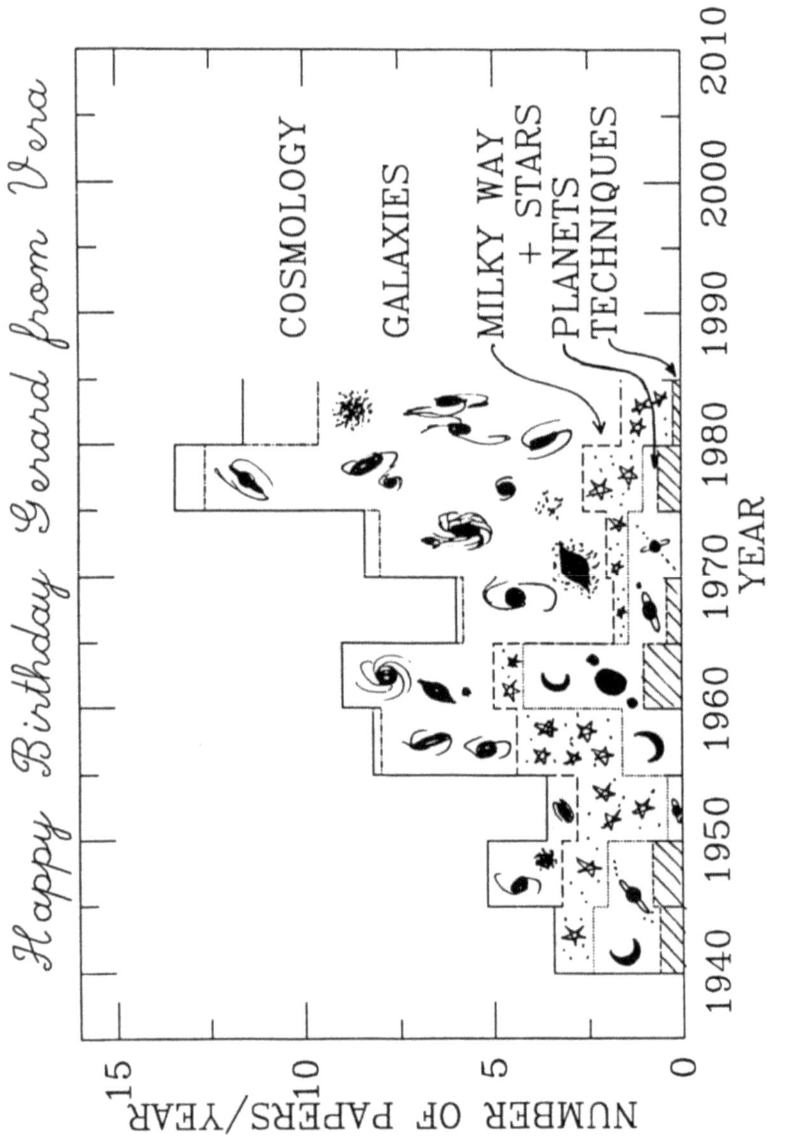

Figure 1: A birthday card for Gérard, showing the number of his published papers per year, subdivided by disciplines. Counts of papers are made to astronomical accuracy, with some popular accounts, book reviews and similar complications not included.

the "Lalande Principle," a warning to us all to avoid making solemn pronouncements. Only four years later, William Herschel (1783) determined the motion of the sun from proper motions of nearby stars. In fairness to Lalande, it is important to note that Herschel used a procedure very different (and far cleverer) than any envisioned by Lalande.

Two features of Herschel's work deserve special mention. First, Herschel understood that the sun could not be at rest: "there is not, in strictness of speaking, one fixed star in the heavens ..., when once it is known that some of them are in motion: for the change that must arise by such motions, in the value of a power which acts inversely as the squares of the distances, must be felt in all the neighboring stars; and if these be influenced by the motion of the former, they will again affect those that are next to them, and so on till all are in motion." Second, the Herschel manuscripts in the Royal Astronomical Society archives contain a sketch by Herschel showing the projections onto the equatorial plane of the proper motion components for stars tabulated by Meskelyne and Lalande. From these Herschel deduced the direction of the solar motion. The plot is especially interesting because the positions of the stars are reversed compared to those as seen from the earth (Hoskin 1980), suggesting that Herschel had plotted their positions and velocity vectors on a sphere. This procedure is of special interest to me, as I point out below.

The next major step in analyzing extraterrestrial motions came from the incorporation of new principles of physics into astronomical observations. The shift of spectral lines due to a line-of-sight velocity component, discovered by Doppler in 1842 and used by Huggins to obtain the radial velocity of a star in 1868, was applied in 1912 by Slipher to obtain the radial velocity of a galaxy, M31. Since that time, the number of known galaxy redshifts has increased to over 20,000 (Figure 2), as increased

light gathering power and more sensitive detectors in the visible and at 21-cm have speeded up the gathering process. But this early history indicates that by the first part of this century, knowledge existed that would make it possible to map radial velocities of galaxies all across the sky.

Early All-Sky Mapping

Obtaining radial velocities of selected galaxies is an easier task than obtaining an all-sky map of galaxies to a fixed limiting magnitude; understanding the biases and the systematic effects is especially complex for observations made separately from the northern and southern hemispheres. Halley (1715) was able to identify six "lucid spots among the fix't stars," of which only the Andromeda galaxy was extragalactic. By 1781, Messier (1781) had identified 103 nebulosities; 32 of these were extragalactic, and the great nebula in Andromeda became M31. Most of us know about the telescopic searches of the Herschels, William and son John, which culminated ultimately in the publication just one hundred

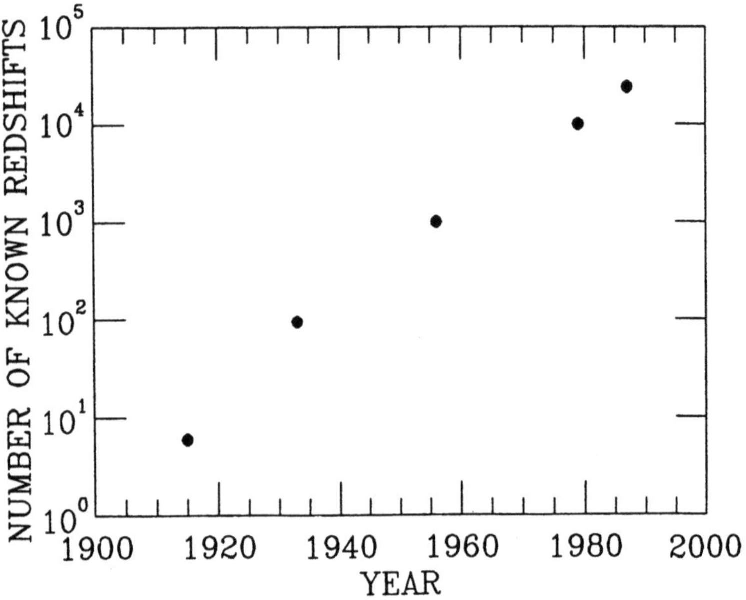

Figure 2: The number of known galaxy redshifts, as a function of time. We should know 10^6 by the year 2020!

years ago of the *New General Catalogue* by Dreyer (1888). Both Herschels were aware that the distribution of the spirals was distinct from that of the diffuse nebulosities. One of the most explicit statements of this comes from Alexander von Humboldt's (1849) *Cosmos*: "The stellar milky way, in the region of which, according to Argelander's admirable observations, the brightest stars of the firmament appear to be congregated, is almost at right angles with another milky way, composed of nebulae ... The milky way composed of nebulae does not belong to our starry stratum, but surrounds it at great distance without being physically connected with it, passing almost in the form of a large cross through the dense nebulae of Virgo, especially the northern wing, through Coma Berenicis, Ursa Major, Andromeda's girdle, and Pisces Boreales."

Less well-known now are the studies of the distribution of the nebulae, most of them published in the *Monthly Notices of the Royal Astronomical Society*, which followed the publication of the NGC. Especially attractive are the two color plots of Waters (1894). But almost 50 years were to pass before the Shapley–Ames Catalogue (1932) produced a major all-sky catalogue which included only extragalactic objects, and which attempted a fixed limiting magnitude. The difference in numbers of bright galaxies in the northern and southern galactic hemispheres (Figure 3) was the first major result from this catalogue.

In contrast to the 1888–1932 period in which there was little all-sky extra-

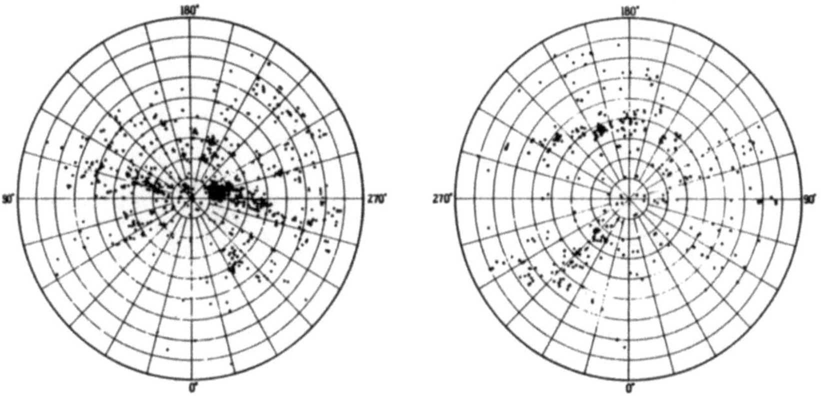

Figure 3: Distribution of 1025 galaxies brighter than 13th magnitude in the northern (left) and southern (right) galactic hemispheres, from the Shapley–Ames (1932) catalogue; reprinted with permission.

galactic mapping, the 1932–1984 was a period of enormous activity. Notable are the Palomar Sky Survey (Minkowski and Abell 1963) from which Abell (1958) extracted a catalogue of clusters of galaxies, (as well as its later companions from the Southern Hemisphere); the Shane–Wirtanen (1967) counts of northern galaxies to a faint limiting magnitude; the revisions to the Shapley–Ames catalogues by the de Vaucouleurs and their collaborators (RC1, 1964; RC2, 1976); the catalogues of Zwicky and collaborators (1961–1968b) and Vorontsov–Velyaminov (1962–1968); the Uppsala catalogue of Nilson (1973) for northern galaxies larger than 1′ and its southern extension (Lauberts 1982); and the remarkable infrared calalogue developed from the IRAS satellite observations (Neugebauer *et al.* 1984).

A figure copied from the RC2 (Figure 4) shows the clustering of galaxies to a plane, evidence cited by de Vaucouleurs for a Local Supercluster. A more resent map has been produced by Lahav (Lynden–Bell and Lahav 1988) based on a combination of the Upsala and Verontsov–Velyaminov catalogues. Galaxies larger than 1′ are plotted, and the concentration to a plane extending well across the sky is evident. Even to these greater distances, the sky projection of galaxies is not uniform.

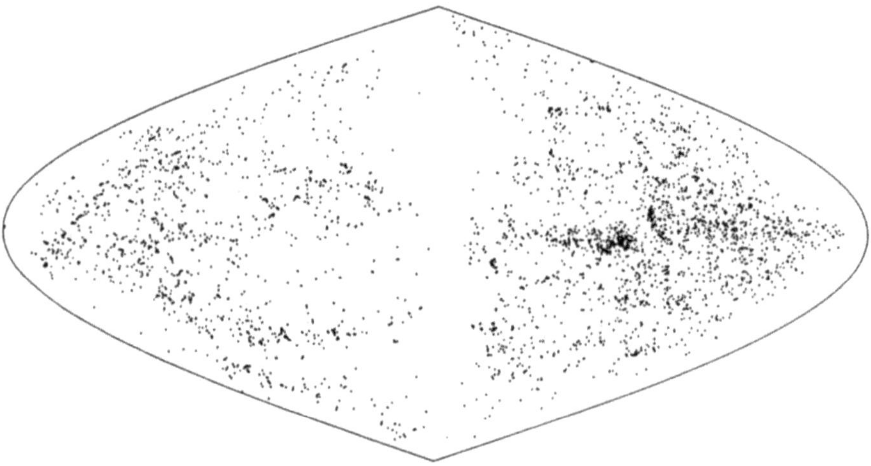

Figure 4: Distribution of 4364 galaxies from the RC2 (1976) in supergalactic coordinates. The central vertical void is due to obscuration from the galactic plane; the north galactic hemisphere is to the right. Reprinted with permission.

The Local Supercluster and Velocity Anisotropies

In view of this evidence, it is difficult to reconstruct the circumstances which prevented many astronomers from accepting the notion that we inhabited a flattened Local Supercluster. It had been hinted at for years by the early maps; some early history is described in Flin (1986) and more recently in Rubin (1988). The existence of the Supercluster was stressed in 1953 by de Vaucouleurs (1953). Elucidating its structure and the motions within it have occupied much of de Vaucouleurs's career, and he is almost solely responsible for convincing most of us of its reality. What may represent the dying gasp of the non–believers was published in the *Astrophysical Journal* in 1976, in a paper entitled, "Is the Local Supercluster a Physical Association?" The paper had at least one valuable result. It elicited an answer from de Vaucouleurs (1976) which contains the footnote, "A student of population statistics might just as well deny the existence of New York City by assigning a large enough radius to Greenwich Village!"

During the decades of the 1950's and 60's, numerous studies led to the general acceptance of a local supercluster, and attention turned next to the anisotropy of velocities induced by the irregular mass distribution. For this review, I will trace the history since 1950 by identifying several studies which have played significant roles in defining the structure of our nearby environment, and in examining systematic motions within this region. Even though I am not stressing the de Vaucouleurs papers, they make valuable reading as a set, and remind us once again how prescient were many of his ideas.

By late 1940's, radial velocities were available for about 100 galaxies; distances for these could best be estimated from total magnitudes, some of these of questionable accuracy. As a Master's degree student at Cornell University, I chose for my thesis an examination of a question which Gamow (1946) had independently raised. Do galaxies partake of a large-scale systematic motion in addition to the Hubble expansion? For each galaxy the Hubble velocity at the estimated distance was removed from the observed velocity; residuals were plotted on a globe. I then searched visually and found a great circle along which there were clumps of maxima and minima of mean residual velocities. To these I attempted to apply a first-order Oort-type analysis. The plane identified was virtually the present supergalactic plane; the velocity structure bears no resemblance to the velocity structure we identify today.

Although this work was presented at an AAS meeting at Haverford College in December 1950 (Rubin 1951), the paper was rejected for publication by both the *Ap. J.* and the *A. J.*, on the grounds that the statistics were poor and the analysis open to criticism. Both statements were true, but sometimes a glimmer of truth shows through nevertheless. At about the same time in the USSR, Ogorodnikov (1952) made a somewhat similar analysis. The significance of these works lies not in their conclusions, but in the support they offered to de Vaucouleurs (1953) a few years later when he initiated his more comprehensive studies, at a time when even weak support was welcome.

The direction which future work was to take can be deduced from a figure in an early de Vaucouleurs (1958) paper, reproduced here as Figure 5. For a sample of galaxies, the variation in mean radial velocity along the supergalactic plane is indicated for galaxies of increasing apparent magnitude, *i.e.* distance. The anisotropy in the velocity distribution along the supergalactic plane, evident as the minimum in the Virgo direction in this early work, has been the subject of numerous studies since then, both by de Vaucouleurs and others. Indeed, even today there is no certain agreement as to the magnitude of the mass excess in the direction of the Virgo cluster and how much this excess retards the expansion of the universe at the position of the Local Group.

The detection of the cosmic microwave radiation by Penzias and Wilson (1965) not only gave observational support for Big Bang cosmology, but it also established a fixed frame of reference (*i.e.* a sea of photons at 3°K) against which motions could be measured. Initial evidence from Partridge and Wilkinson (1967) indicated that the sun had a negligible motion with respect to this frame. But for my colleagues Kent Ford, Norbert Thonnard, Mort Roberts, and John Graham, and I, it suggested that the combination of advanced observing techniques and theoretical understanding made the time ripe for a fresh attack on the questions of deviations from a smooth Hubble flow.

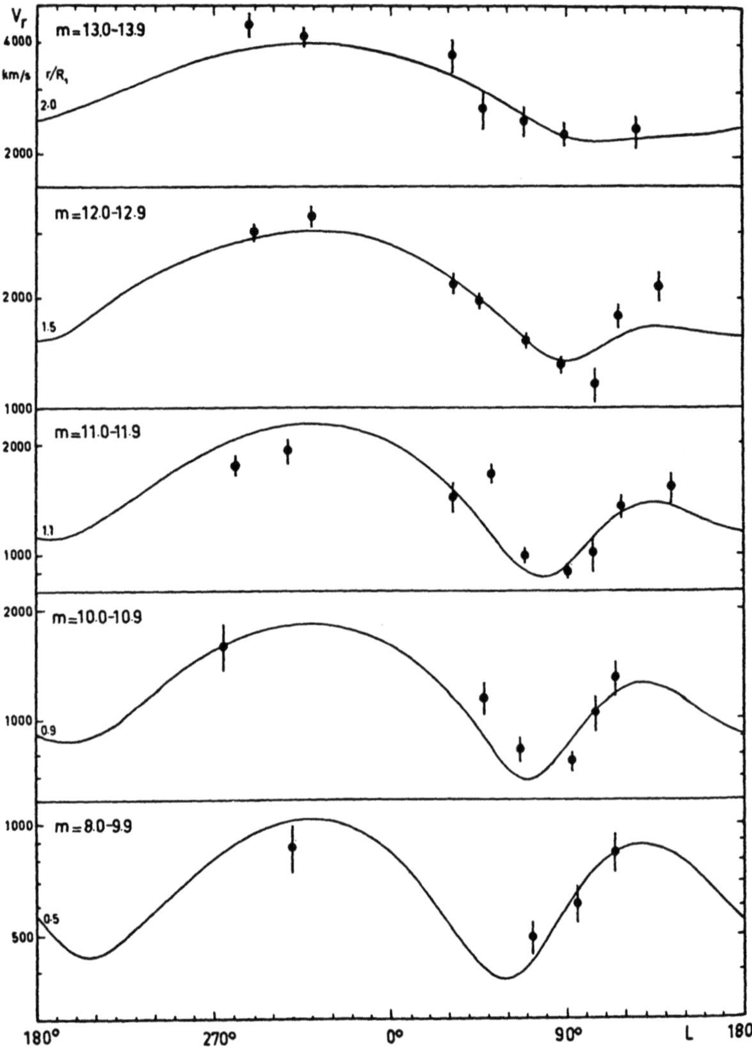

Figure 5: The mean radial velocities of galaxies in the supergalactic equatorial belt as a function of supergalactic longitude, for galaxies in five magnitude intervals (de Vaucouleurs 1958; reprinted with permission). The curves arise from the model containing both differential rotation and retardation of the expansion.

For an all–sky sample of over nearly 200 Sc I spirals chosen to be beyond the Virgo cluster, we obtained radial velocities (and magnitudes for the southern objects). The analysis of 96 galaxies with velocities between 3500 and 6500 $\mathrm{km\,s^{-1}}$ (Rubin *et al.* 1976a, 1976b) indicated an anisotropy across the sky, which could be reasonably well fit with a dipole variation. We interpreted this as a motion of the sun of $V_o = 600 \pm 125\,\mathrm{km\,s^{-1}}$, corresponding to a motion of the Local Group of $V_{LG} = 450 \pm 125\,\mathrm{km\,s^{-1}}$ toward $l = 163° \pm 15°$, $b = -11° \pm 15°$, more than 90° from the direction to the Virgo cluster. This result was generally met with the response that the velocity of the sun could not be this large. Within a year, however, a dipole anisotropy in the microwave background had been detected; the velocity was close to the present value (Fixsen *et al.* 1983; Lubin *et al.* 1985) for the Local Group of $V_{LG} = 540 \pm 50\,\mathrm{km\,s^{-1}}$ toward $l = 267° \pm 5°$, $b = +31° \pm 5°$, 44° from the direction to Virgo. Immediately the objection to the results of Rubin *et al.* changed. Now it was not the magnitude of the optical dipole, but the direction of the apex which was puzzling.

Unfortunately or perhaps fortunately, few other all–sky homogeneous data samples existed in the mid–70's, for which distances could be obtained independently of the Hubble velocities. Hence the Rubin *et al.* sample was subjected to numerous reanalyses during the next ten years (Schechter, 1977; Jackson, 1982; Peterson and Baumgart, 1986; Collins, Joseph, and Robertson, 1986). Improvements included more sophisticated statistical techniques, new photoelectric magnitudes, better values for foreground extinction, and IRAS infra–red galaxy magnitudes. Although it became clear that the original error was underestimated (200 $\mathrm{km\,s^{-1}}$ being a more likely value), the original direction and amplitude remained unaltered. Still, most astronomers were uncomfortable with the result, wished it would disappear, or chose to ignore it.

At about the same time, an examination of the observational consequences of an "infall" (actually a retardation in the expansion) to Virgo was carried out by Peebles (1976), followed by creative models and analysis of observational data by Schechter (1980) and by Tonry and Davis (1981). These studies, those of Aaronson, Huchra, and Mould (1979) plus the work of many others all contributed to produce an atmosphere in which an examination of large–scale motions was a legitimate, even trendy, field of research. As Malcolm Longair (1987) wrote in his summary of the I. A. U. symposium on *Observational Cosmology* held in Beijing in 1986, "Another hot topic was the observation of large–scale streaming velocities of galaxies. The history of these studies dates back to the much–discussed Rubin and Ford effect and is a good example of an observation, which many people wished would go away, suddenly becoming eminently respectable."

Current Studies of Large–Scale Motions

By the early 1980's, several observational studies were underway in an effort to obtain larger all–sky data sets. A few of these have altered our view of

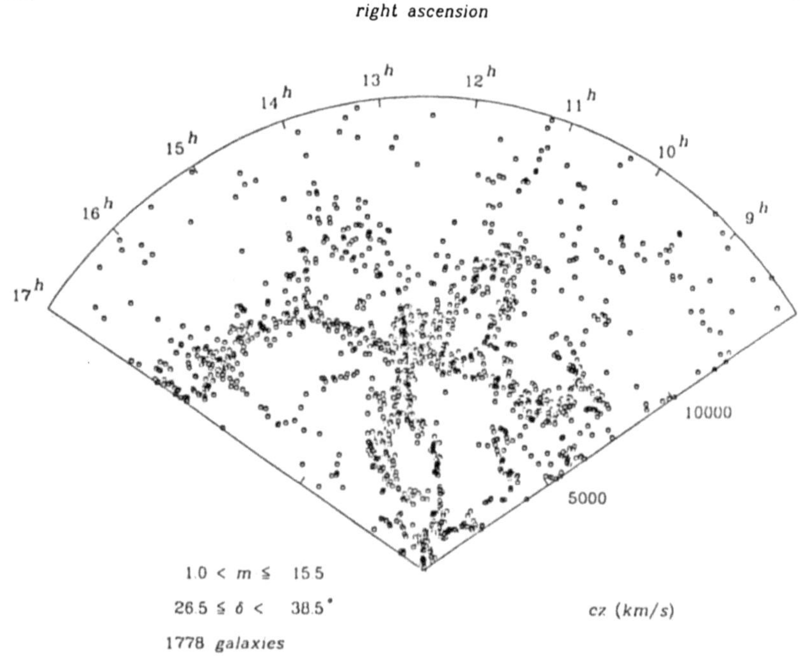

Figure 6: Observed velocity versus right ascension from the CfA survey for the strip 12° wide in declination and centered at 32.5°. From Geller and Huchra (1988); reprinted with permission. Note the large scales of the structures.

the large–scale structure and motions in the nearby universe: I mention here the CfA plots of the large–scale structure (de Lapparent, Geller and Huchra 1986; Geller and Huchra 1988), the elliptical data and analysis by Faber and colleagues (Faber and Burstein 1988), the gravitational field deduced from the IRAS infrared observations (Yahil 1988; Strauss and Davis 1988), and the cluster and supercluster studies of Bahcall (1988) and her colleagues.

Although evidence had been noted earlier, it was the plots of the Shane–Wirtanen galaxies made by Peebles and colleagues (Seldner *et al.* 1977), plus the studies of Jôeveer, Einasto, and Tago (1978) that shocked astronomers into noticing the large–scale irregularities in the distribution of galaxies. Galaxies were arranged in cells surrounding voids, with stringy filamentary connections between the cells. The discovery of the Boötes void (Kirshner *et al.* 1981) gave additional confirmation to this picture. But it was an initial plot of the distribution of galaxies from the Center for Astrophysics (CfA) redshift survey that showed us how widespread this phenomenon is. The CfA redshift survey is now complete for several more slices, and reemphasizes the spectacular views of voids and filaments, as shown in Figure 6. The voids, filaments, and general non–random pattern make a striking visual impression. Structures with scales as large as those covered by the observations force us to admit that we have not yet reached a domain where the distribution is smooth.

Given this non-uniform distribution of mass, at least of luminous mass, it is reasonable to expect departures from a smooth Hubble flow. Most recently, the velocity field within about $3000\,\mathrm{km\,s^{-1}}$ has been investigated by a group of seven (Faber and Burstein 1988) astronomers, using spectroscopic and photometric data for 400 elliptical galaxies distributed fairly well across the sky. Peculiar velocities relative to the cosmic microwave background indicate a complex pattern of motions: a bulk motion toward a Great Attractor in Centaurus, a small Virgocentric inflow, and a Local Velocity Anomaly. This model results from an examination of velocity residuals from a smooth Hubble flow; these vectors are plotted on the supergalactic plane in Figure 7.

A physical basis for understanding this complex velocity pattern is now emerging, based on work by Yahil (1988) and Strauss and Davis (1988), among others. To minimize the problem of obtaining an all–sky galaxy sample free of systematic biases, they chose as their sample a subset of those galaxies identified in the all–sky survey from the Infrared Astronomical Satellite (IRAS). Newly determined velocities are used initially to place each galaxy at its Hubble distance. On the assumption that luminosity traces mass (a big assumption), the density distribution is derived. The peculiar gravitational field is then calculated, and from it the peculiar motion for each galaxy, assuming that linear perturbation theory is valid. Each observed velocity is corrected for the predicted peculiar velocity, a new distance and hence density distribution is obtained, and the process repeated through several iterations. Convergence to an accuracy of $20\,\mathrm{km\,s^{-1}}$ produces the results shown in Figure 8.

Although some of the assumptions raise questions, and the sky coverage and radial velocity determinations are still incomplete, the resulting picture gives an amazing view of the universe out to $70h^{-1}$ Mpc. The gravitational field is dominated by two large concentrations. One, the Great Attractor, is located in the direction of Hydra–Centaurus; opposite it on the sky is the Perseus–Pisces complex. These cause the velocity field to bifurcate near the position of the Local Group. Overall, the gravitational field is one of enormous complexity, yet in nominal agreement with the model of Faber and Burstein. Note that different sky samples could give very different results. For example, a study of those galaxies with distances beyond $3500\,\mathrm{km\,s^{-1}}$ (Figure 9) will show a very different velocity pattern than a study with galaxies closer than $2000\,\mathrm{km\,s^{-1}}$.

So, for the first time, the circle is closed. We observe a non–random distribution of luminous galaxies, whose velocities indicate deviations from a smooth Hubble flow. In turn, the clumpy mass distribution implies velocity deviations which, in general, match those inferred from the observations. This is a remarkable result, which may initiate a new era for observational cosmology.

In closing, I wish to raise a question and present a result from recent work of my colleagues and myself (Rubin, Whitmore, and Ford 1988; Whitmore, Forbes, and Rubin 1988). Are we making a mistake when we assume that the properties of galaxies in clusters are similar to galaxies in the field? Are the kinematics

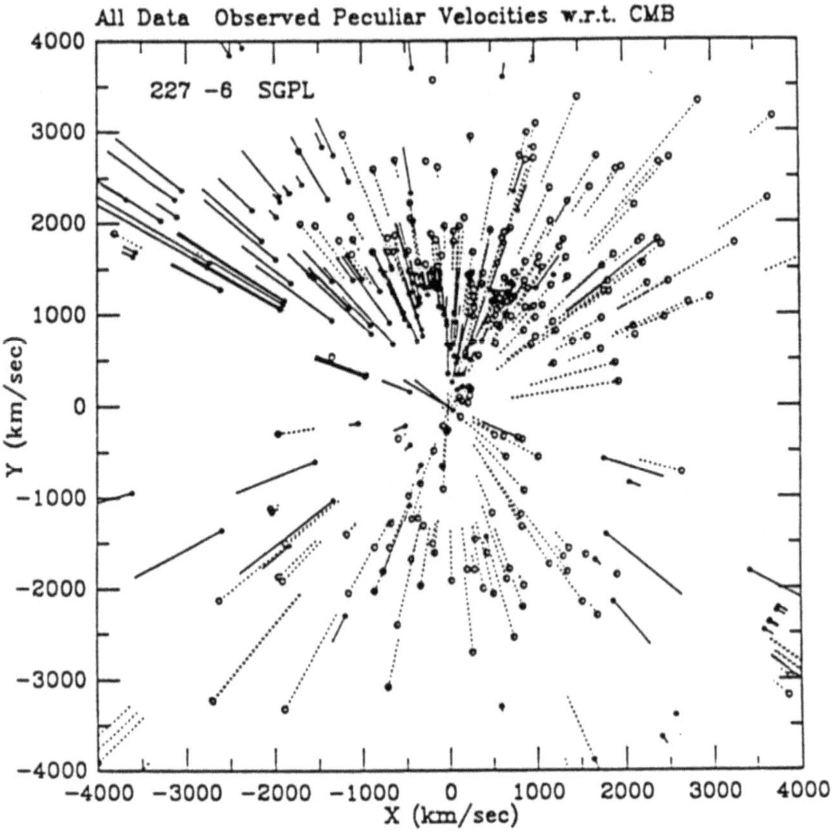

Figure 7: Velocity field map from Faber and Burstein (1988); reprinted with permission. Observed peculiar velocities for all ellipticals studied within $\pm 22.5°$ of the supergalactic plane, out to a distance of $4000\,\mathrm{km\,s^{-1}}$. Motions away from the Local Group are denoted by solid circles and solid lines; motions toward the Local Group are denoted by open circles and dotted lines.

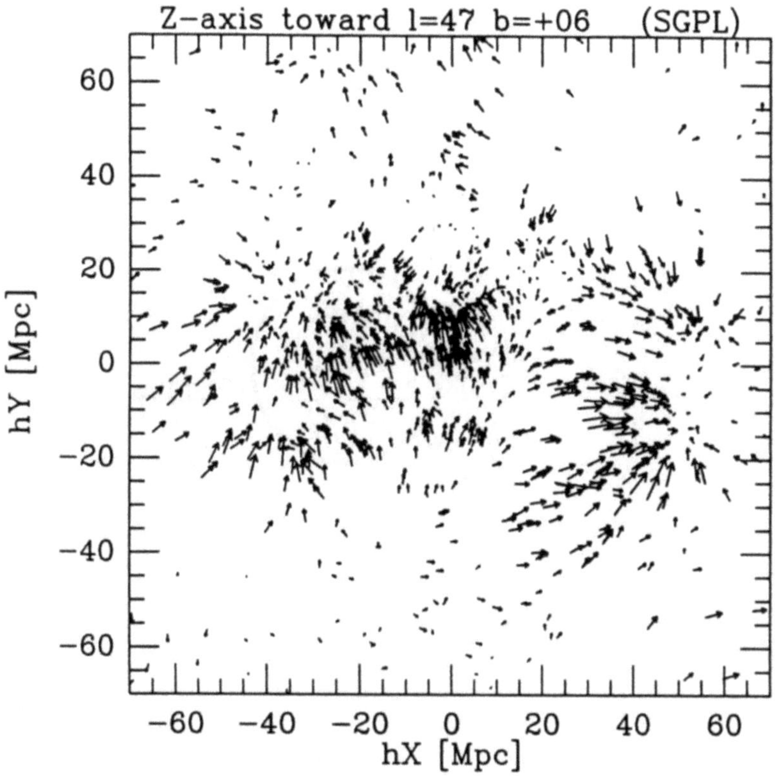

Figure 8: The gravitational field predicted by the IRAS galaxy distribution, from Yahil (1988); reprinted with permission. All galaxies within 22.5° of the supergalactic plane and with velocities within 7000 $km\,s^{-1}$ are shown as vectors whose lengths are proportional to the components of their peculiar gravities in the plane. Galaxies are located at the tails of the vectors; the relative vector lengths and directions correspond to the gravitational acceleration produced by all the other galaxies.

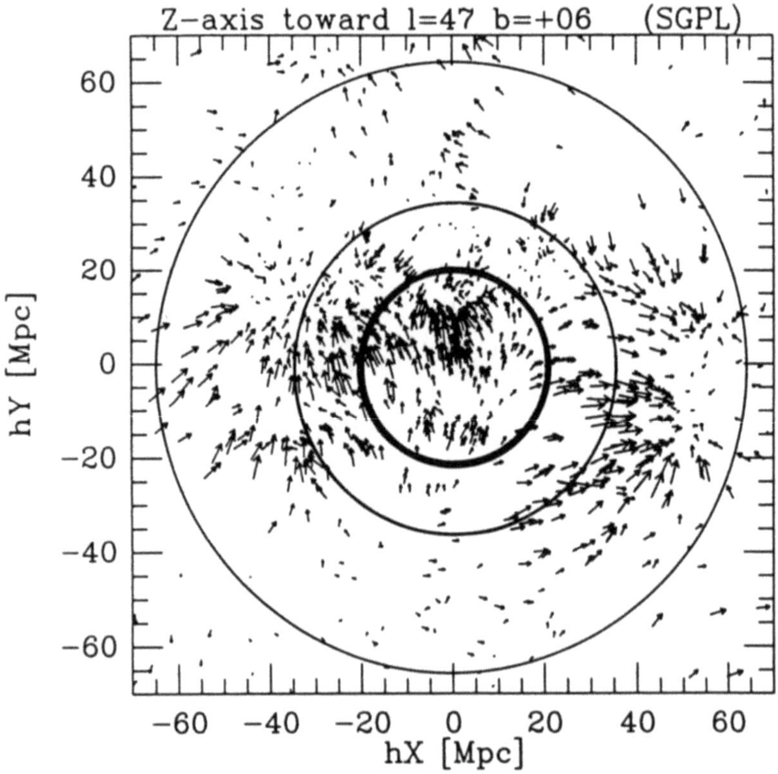

Figure 9: The same as Figure 8, with circles of constant distance superposed. The inner circle encloses a region within $2000\,\mathrm{km\,s^{-1}}$. The pull of the Virgo supercluster (located within the upper portion of the inner circle) is evident, although the detailed gravity is complex. The outer two circles enclose a ring (a shell in three dimensions) $3500\,\mathrm{km\,s^{-1}} < V < 6500\,\mathrm{km\,s^{-1}}$, the region of galaxies studied earlier by Rubin *et al.* (1976). Note how different will be the results which come from samples from different regions. Reprinted with permission.

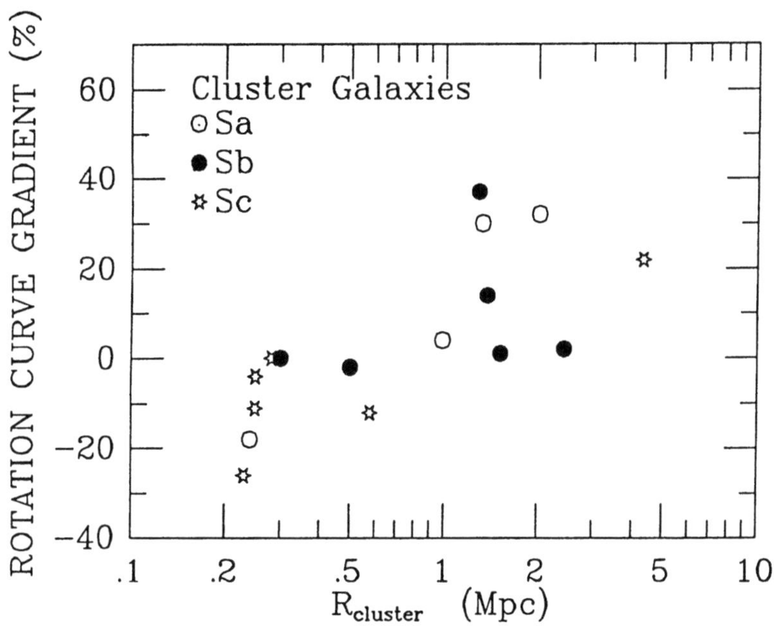

Figure 10: The value of the outer gradient of the rotation curve for cluster galaxies as a function of position of the galaxy in the cluster. The outer gradient is the difference in rotation velocity observed at 0.8 R_{25} minus the velocity at 0.4 R_{25}, normalized by V_{max}. Galaxies seen near the centers of clusters have flat or falling rotation curves, while those seen in the outer regions have flat or rising rotation velocities.

sufficiently similar that correlations like the Tully–Fisher relation have identical slopes and zero points for field and cluster galaxies? It is known that some parameters do differ, depending principally on the local density: neutral hydrogen content (Chamaraux *et al.* 1980), symmetry of the H I (Warmels 1985), clustering properties (Davis and Geller 1976), angular correlations (Haynes and Giovanelli 1988), luminosity functions (Sandage, Binggeli, and Tammann 1985), dwarf fraction (Sharp, Jones, and Jones 1988), among others.

In order to answer this question, we have observed extended rotation curves for 20 galaxies in four clusters — Cancer, Hercules, Peg I, and DC1842-63 — to see if the rotational properties of spirals are a function of galaxy environment. And, indeed, we find that clusters of galaxies contain near their cores a spiral population which has falling rotation curves. Such rotation curves have never been observed for field spirals. A plot of rotation curve gradient versus projected cluster distance (Figure 10) shows a good correlation: galaxies with flat or falling

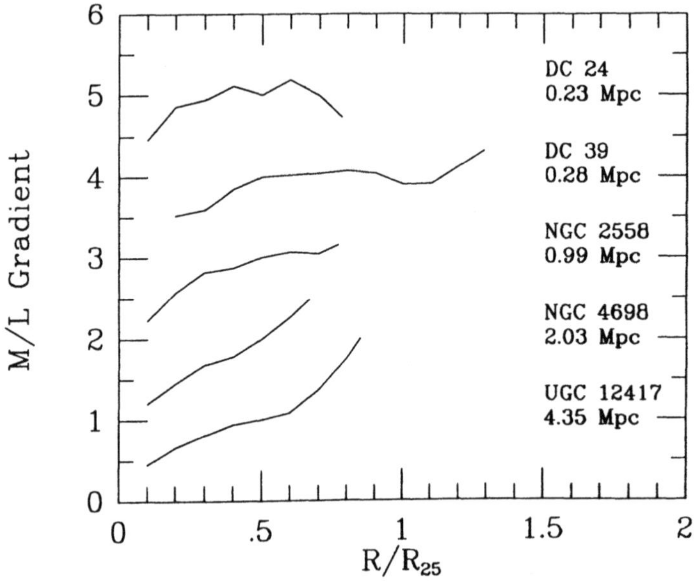

Figure 11: Mass–to–red light ratios (normalized to 1 at 0.5 R_{25}) versus the distance to the center of the galaxy; galaxies are arranged in order of increasing distances from the centers of the clusters.

rotation curves exist near cluster centers; galaxies with flat or rising rotation curves exist in the cluster outer regions and in the field.

Moreover, the gradient in mass–to–light ratio for a galaxy is also a function of clustocentric distance. Central galaxies have values of M/L nearly constant across the galaxy, while outer cluster galaxies and field galaxies show positive gradients in M/L (Figures 11, 12). These results suggest that in the dense cluster cores the mass distributions within the halos of spirals have been modified. Halos of spirals may have been stripped by gravitational interactions with other galaxies, or disrupted by mergers, or altered by interactions with the overall cluster gravitational field, or never have been permitted to form. These evolutionary effects which modify the galaxy dynamics will have to be taken properly into account for all investigations of galaxies in environments of differing densities.

Conclusions

Our knowledge of the structure and the velocities within the nearby universe continues to grow as observations reveal a greater complexity than had been anticipated. Once again we are reminded that most of cosmology is observa-

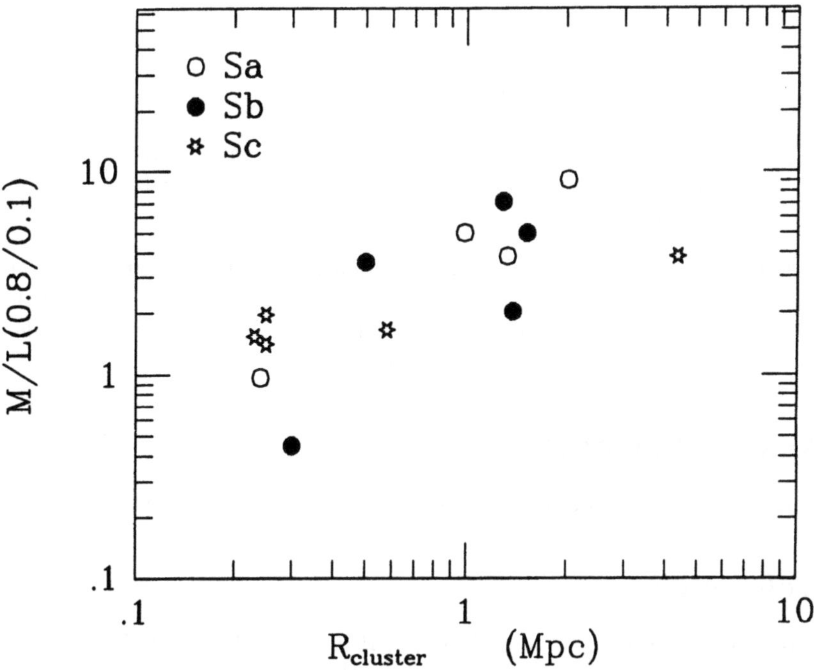

Figure 12: The gradient in the mass-to-light ratio (measured from 0.8 R_{25} to 0.1 R_{25}) as a function of clustocentric distance of the galaxy. Steeper M/L gradients, evidence of massive halos, are found in galaxies in the outer regions of the clusters.

tionally led. As befits an observational science, we can observe our progress by examining plots of a few recent results. Those shown here as Figures 6, 7, and 8 inspire us with their enormous potentials. I did not expect, during my lifetime, to see a plot of gravitationally induced velocity vectors for galaxies out to distances of 70 Mpc. It may not be a correct map, but it is a magnificently bold first step.

From the most recent studies of distributions and motions in the universe, I propose the following as conclusions.

1. Luminous matter in the presently observed universe is distributed in a non–random, clumpy manner.
2. This non–uniform distribution of matter gives rise to large–scale bulk motions. Scales of 10's of Mpc and 1000 $\mathrm{km\,s^{-1}}$ are involved.
3. Lacking evidence to the contrary, we assume that the distributions and motions in the nearby universe can be described by Newtonian gravitational theory.
4. Observations have raised many questions. Even though our answers may not yet be correct, I think we are asking the right questions. Many of these questions were first posed by Gérard de Vaucouleurs, and we acknowledge our scientific and personal debts to him.

Acknowledgements

I thank many colleagues for permission to reproduce figures from their work.

References

Aaronson, M., Huchra, J., and Mould, J. 1979, *Astrophys. J.* **229**, 1.

Abell, G. O. 1958, *Astrophys. J. Suppl.* **3**, 211.

Bahcall, N. 1988, *Ann. Rev. Astron. Astrophys.* **26**, 631.

Chamaraux, P., Balkowski, C., and Gerard, E. 1980, *Astron. Astrophys.* **83**, 38.

Collins, C. A., Joseph, R. D., and Robertson, N. A. 1986, in *Galaxy Distances and Deviations from Universal Expansion*, ed. B. F. Madore and R. B. Tully (Dordrecht: Reidel), p. 131.

Davis, M. and Geller, M. J. 1976, *Astrophys. J.* **208**, 13.

Dreyer, J. L. E. 1888, *Mem. Roy. Astron. Soc.* **49**, Part 1.

Faber, S. M. and Burstein, D. 1988, in *Large-Scale Motions in the Universe: A Vatican Study Week*, ed. V. C. Rubin and G. V. Coyne (Princeton: Princeton University Press), p. 115.

Fixsen, D. J., Cheng, E. S., Wilkinson, D. T. 1983, *Phys. Rev. Lett.* **50**, 620.

Flin, P. 1986, at the *10th Cracow Summer School of Cosmology*, preprint.
Gamow, G. 1946, *Nature* **158**, 549.
Geller, M. and Huchra, J. P. 1988, in *Large-Scale Motions in the Universe: A Vatican Study Week*, ed. V. C. Rubin and G. V. Coyne (Princeton: Princeton University Press), p. 3.
Haynes, M. P. and Giovanelli, R. 1988, in *Large-Scale Motions in the Universe: A Vatican Study Week*, ed. V. C. Rubin and G. V. Coyne (Princeton: Princeton University Press), p. 31.
Halley, E. 1715, *Phil. Trans. Roy. Soc. London* **29**, 39.
Halley, E. 1718, *Phil. Trans. Roy. Soc. London* **30**, 737.
Herschel, W. 1783, *Phil. Trans. Roy. Soc. London* **73**, 247.
Hoskin, M. 1980, *J. Hist. Astron.* **11**, 153.
von Humboldt, A. 1849, *A Sketch of the Physical Description of the Universe*, translated from the German by E. C. Otte (London: Henry G. Bohn), Vol. **1**, 141.
Jackson, P. 1982, Unpublished Ph. D. thesis, University of California, Santa Cruz.
Jôeveer, M., Einasto, J., and Tago, E. 1978, *Mon. Not. Roy. Astron. Soc.* **185**, 357.
Kirshner, R. P., Oemler, A., Jr., Schechter, P. L., and Shectman, S. A. 1981, *Astrophys. J. Lett.* **248**, L57.
de Lapparent, V., Geller, M. J., and Huchra, J. P. 1986, *Astrophys. J. Lett.* **202**, L1.
Lauberts, A. 1982, *The ESO/Uppsala Survey of the ESO (B) Atlas*, (Munich: European Southern Observatory).
Longair, M. 1987, in *Observational Cosmology*, ed. A. Hewitt, G. Burbidge, and L. Z. Fang (Dordrecht: Reidel), 823.
Lubin, P. M., Villela, G., Epstein, G. L., and Smoot, G. F. 1985, *Astrophys. J. Lett.* **298**, L1.
Lynden-Bell, D. and Lahav, O. 1988, in *Large-Scale Motions in the Universe: A Vatican Study Week*, ed. V. C. Rubin and G. V. Coyne (Princeton: Princeton University Press), p. 199.
Messier, C. 1781, in *Connoissance des temps pour l'annẽe bissextile 1784*, Paris, 227.
Minkowski, R. L. and Abell, G. O. 1963, in *Basic Astronomical Data*, ed. K. Aa. Strand (Chicago: University of Chicago Press), p. 481.
Neugebauer, G., Beichman, C. A., Soifer, B. T., Aumann, H. H., Chester, T. J., Gautier, T. N., Gillet, F. C., Hauser, M. G., Houck, J. R., Lonsdale, C. J., Low, F. J., and Young, E. T. 1984, *Science* **224**, 14.
Nilson, P. 1973, *Uppsala General Catalogue of Galaxies*, Uppsala Astron. Obs. Ann. **6**.
Ogorodnikov, K. F. 1952, *Problems of Cosmogony*, **1**, 150.
Partridge, R. B. and Wilkinson, D. T. 1967, *Phys. Rev. Lett.* **18**, 557.
Peebles, P. J. E. 1976, *Astrophys. J.* **205**, 318.

Penzias, A. A. and Wilson, R. W. 1965, *Astrophys. J.* **142**, 419.

Peterson, C. J. and Baumgart, W. 1986, *Astron. J.* **91**, 530.

Rubin, V. C. 1951, *Astron. J.* **56**, 47.

Rubin, V. C. 1988, in *Gérard and Antoinette de Vaucouleurs: A Life for Astronomy*, ed. M. Capaccioli and H. G. Corwin (Singapore: World Scientific), in press.

Rubin, V. C., Whitmore, B. C., and Ford, W. K. 1988, *Astrophys. J.* **333**, 522.

Rubin, V. C., Ford, W. K., Thonnard, N., and Roberts, M. 1976a, *Astron. J.* **81**, 719.

Rubin, V. C., Ford, W. K., Thonnard, N., Roberts, M., and Graham, J. A. 1976b, *Astron. J.* **81**, 687.

Sandage, A., Binggeli, B., and Tammann, G. A. 1985, *Astron. J.* **90**, 1759.

Schechter, P. L. 1977, *Astron. J.* **82**, 569.

Schechter, P. L. 1980, *Astron. J.* **85**, 801.

Seldner, M., Siebers, B., Groth, E. J., and Peebles, P. J. E. 1977, *Astron. J.* **82**, 249.

Shane, C. D. and Wirtanen, C. A. 1967, *Publ. Lick. Obs.* **22**, Part I.

Shapley, H. and Ames, A. 1932, *Ann. Harvard Coll. Obs.* **88**, No. 2.

Sharp, N. A., Jones, B. J. T., and Jones, J. E. 1988, *Mon. Not. Roy. Astron. Soc.* **185**, 457.

Strauss, M. A. and Davis, M., 1988, in *Large-Scale Motions in the Universe: A Vatican Study Week*, ed. V. C. Rubin and G. V. Coyne (Princeton: Princeton University Press), p. 255.

Tonry, J. L. and Davis, M. 1981, *Astrophys. J.* **246**, 680.

de Vaucouleurs, G. 1953, *Astron. J.* **58**, 30.

de Vaucouleurs, G. 1958, *Astron. J.* **63**, 253.

de Vaucouleurs, G. 1976, *Astrophys. J.* **203**, 33.

de Vaucouleurs, G. and de Vaucouleurs, A. 1964, *Reference Catalogue of Bright Galaxies* (RC1) (Austin: University of Texas Press).

de Vaucouleurs, G., de Vaucouleurs, A., and Corwin, H. G. 1976, *Second Reference Catalogue of Bright Galaxies* (RC2) (Austin: University of Texas Press).

Vorontsov–Velyaminov, B. A., Arkipova, V. P., and Krasnogorska, A. A. 1962, 1963, 1964, 1968, 1974, *Morphological Catalogue of Galaxies*, **I–V**, (Moscow: Moscow State University).

Warmels, R. H. 1985, in *The Virgo Cluster*, ed. O.–G. Richter and B. Binggeli (Munich: ESO), p. 51.

Waters, S. 1894, *Mon. Not. Roy. Astron. Soc.* **54**, 526.

Whitmore, B. C., Forbes, D. A., and Rubin, V. C. 1988, *Astrophys. J.* **333**, 542.

Yahil, A. 1988, in *Large-Scale Motions in the Universe: A Vatican Study Week*, ed. V. C. Rubin and G. V. Coyne (Princeton: Princeton University Press), p. 219.

Zwicky, F., Herzog, E., Kowal, C. T., Karpowicz, M., and Wild, P. 1961, 1963, 1965, 1966, 1968a, 1968b, *Catalogue of Galaxies and of Clusters of Galaxies*, **I–VI**, (Pasadena: California Inst. of Technology).

Discussion

W. T. Sullivan: Your evidence that spirals in cluster centers have falling rotation curves may be related to the lack of outer neutral hydrogen known for central spirals in the Virgo cluster (van Gorkom, Balkowski *et al.* ; Warmels) and in the Coma cluster (my own poster paper at this meeting). Do you find any correlations between H I deficiency and the nature of your rotation curves?

Rubin: Only eight of our galaxies have measured values for the H I deficiency, and there is a slight correlation between deficiency and "peculiarity" of the rotation curve. However, I am presently observing a large sample of Virgo cluster galaxies, in order to answer your question, which is an important one.

Sullivan: Two historical comments: (1) You mentioned that the full version of your early paper was not accepted by either *A. J.* or *Ap. J.*, but it should be noted that one can at least find the abstract of your A. A. S. talk of December 1950 in the 1951 *A. J.* (2) In your historical overview, I feel that Fritz Zwicky should also be included, for both (a) his studies of clusters of galaxies, extending back to the 1930's, and (b) his monumental *Catalog of Galaxies and of Clusters of Galaxies.*

Rubin: Correct on both counts.

Roger Davies: Does your plot of the slope of the outer part of the rotation curve *versus* distance from the cluster center imply that the underlying mass distribution in spirals differs with environment? Will the smaller extent and smaller amount of gas in the cluster galaxies cause a different part of the rotation curve to be sampled as a function of environment, perhaps contributing to the effect?

Rubin: Yes, we interpret our results to mean that the mass distribution in spirals is a function of the local environment. It is unlikely that the effect is enhanced by the region of the galaxy that is luminous, for our gradient in M/L is virtually independent of whether we measure the gradient from 0.1 R_{25} to 0.8 R_{25} or 0.4 R_{25} to 0.8 R_{25}. I think the gradient in M/L is the more fundamental parameter, rather than slope of the rotation curve, though of course both are responding to the mass distribution.

J.–C. Pecker: I was quite interested to see the Yahil diagram of the velocities derived from IRAS data. If I understand this correctly, it is a dynamical computation assuming that mass is proportional to IR luminosity. But it must assume also, as an initial condition, the actual velocity field as measured from radial velocities only. Is this sufficient? Is this correct?

Rubin: You are correct that the initial distances of the galaxies come from their Hubble distances, which distances are corrected on each iteration, as the peculiar motion is removed form the observed velocity. The solution is a linear approximation only, which may produce the major uncertainty.

K. Rudnicki: Zwicky and I have suggested that there is no real difference between a cluster and a large agglomeration of galaxies called a supercluster. There is a continuous transition between these two kinds of objects. Zwicky also stated that clusters can have different shapes. In addition, he wrote that we (our Galaxy) belongs to the Virgo cluster, which is just another way of stating that we belong to the Supergalaxy (the Local Supercluster).

Rubin: Thanks for the historical perspective.

F. J. Kerr: I was present at the A. A. S. meeting in late 1950 (this was my first A. A. S. meeting). I happened to hear by chance three prestigious council members discussing whether Vera Rubin's paper should be allowed. They seemed to be shocked by this brash young outsider (and female at that) making such an outrageous and impossible suggestion. Fortunately, they did not stop the presentation of the paper.

Local Supercluster Velocity Field from Unbiased B–band T–F Distances

L. Bottinelli[1], L. Gouguenheim[1], and P. Teerikorpi[2]

[1]Observatoire de Paris, Section de Meudon, and Université de Paris Sud, France
[2]Turku University Observatory, Piikkiö, Finland

We have shown in previous studies (Bottinelli *et al.* 1986, 1987, 1988a,b) the importance of the Malmquist bias and its consequences on the extragalactic distance scale; in particular, the local velocity field can be studied safely only from unbiased data. We have collected the largest available complete sample of B–band Tully–Fisher (hereafter TF) distances extracted from the Lyon–Meudon Observatories extragalactic data base (Paturel *et al.* 1988).

The unbiased distances determined from this sample allow us to study the peculiar motions in the local universe, and in particular (1) the comparison between a simple dipolar model and a Peebles spherical Virgocentric infall model, and (2) the determination of the Local Group peculiar velocity v_{LG}.

The dipolar model, adopting for the Local group apex $l = 125°$ and $b = 20°$ (de Vaucouleurs *et al.* 1981), gives a minimal r. m. s. dispersion of the individual values of the Hubble parameter H according to v_{LG}, which depends on the distance of the sample. The "best value" of v_{LG} increases continuously with the limiting distance of the sample (Figure 1b). On the contrary, a Peebles (1976) spherical infall model (with a Virgocentric infall velocity inversely proportional to the distance to Virgo) does not depend on the distance of the sample (Figure 1a) and leads to a lower dispersion; the best value of v_{LG} is in the range 200–300 km sec^{-1}.

The two different subsamples selected (1) in the Virgo cluster direction and (2) in the opposite one, give the best agreement for the H mean values when

v_{LG} is in the range 300–400 km sec^{-1} (Figure 2).

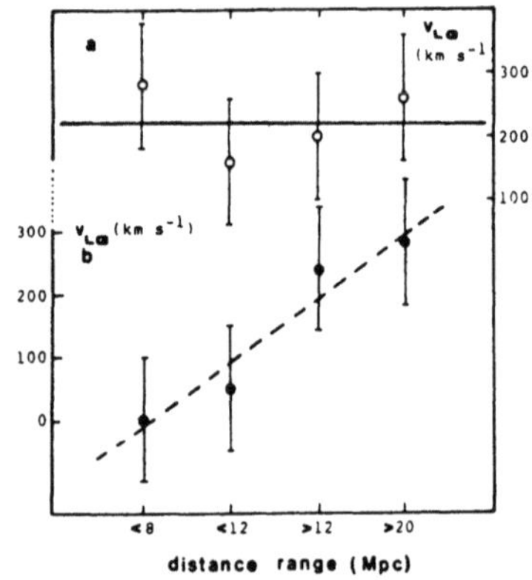

Figure 1. v_{LG} is plotted against the distance range of the sample for the Peebles model (open dots) and the dipolar model (black dots).

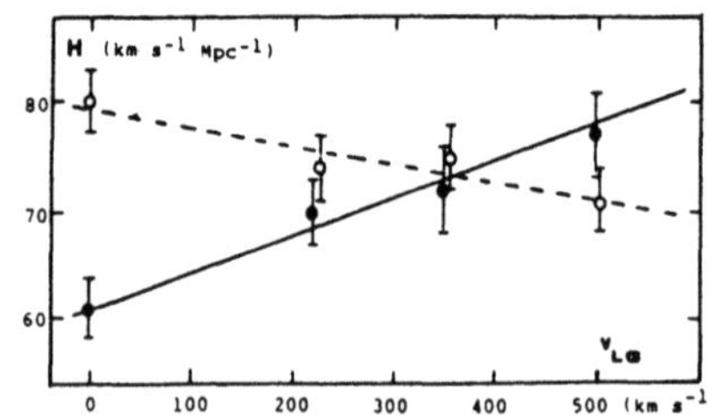

Figure 2. The mean H values obtained from galaxies either in the Virgo direction (black dots) or in the opposite direction (open dots) are plotted against the values adopted for v_{LG}.

From the adopted Peebles model, we get: $V_o - V_o^v d = v_{LG}(d - R)$ which can be written $Y = v_{LG}X$ with the following data: the radial velocity V_o of the galaxy, the mean radial velocity V_o^v (= 980 km sec^{-1}) of the Virgo cluster with respect to the Local Group, the galaxy distance d to the Local Group in units of Virgo distance (taken to be 16 Mpc), the angular distance θ between the center of Virgo and the galaxy, and

$$R = d(1 + d\cos\theta - 2\cos 2\theta)/(1 + d^2 - 2d\cos\theta).$$

v_{LG} is thus given by the slope of the Y vs. X linear regression, where X and Y are directly derived from the observations of unbiased TF distances and velocities. The least–squares fit forced through the origin gives v_{LG} = 320 km sec^{-1} (Figure 3).

Our conclusions obtained from unbiased distances are: (1) the Peebles spherical Virgocentric infall model gives a better description of the observed data than the simple dipolar model; and (2) the best value of v_{LG} is in the range 200–400 km sec^{-1}, with a "best choice" of 320 km sec^{-1}.

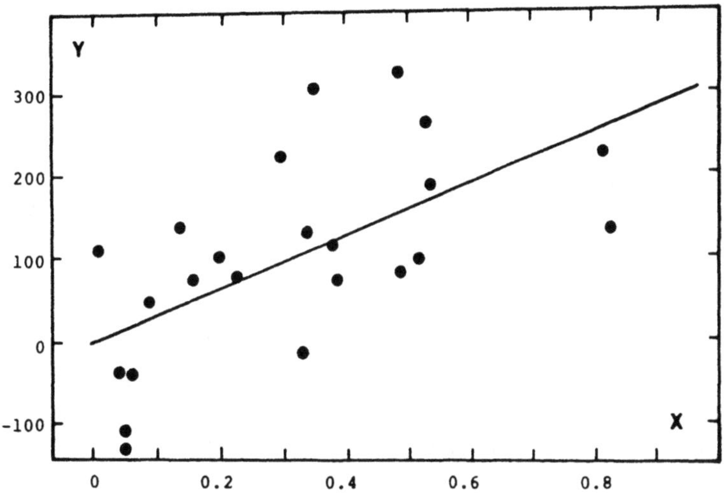

Figure 3: The slope of the best straight line corresponds to $v_{LG} = 320\,\mathrm{km\,sec^{-1}}$.

References

Bottinelli, L., Gouguenheim, L., Paturel, G., and Teerikorpi, T. 1986, *Astron. Astrophys.* **156**, 157.

Bottinelli, L., Fouqué, P., Gouguenheim, L., Paturel, G., and Teerikorpi, P. 1987, *Astron. Astrophys.* **181**, 1.

Bottinelli, L., Gouguenheim, L., and Teerikorpi, P. 1988a, *Astron. Astrophys.* **196**, 17.

Bottinelli, L., Gouguenheim, L., Paturel, G., and Teerikorpi, P. 1988b, *Astrophys. J.* **328**, 4.

Paturel, G., Bottinelli, L., Fouqué, P., and Gouguenheim, L. 1988, Workshop de Munich, in press.

Peebles, P. J. E. 1976, *Astrophys. J.* **205**, 318.

de Vaucouleurs, G., Peters, W. L., Bottinelli, L., Gouguenheim, L., and Paturel, G. 1981, *Astrophys. J.* **248**, 408.

Spiral Galaxies as Indicators of the Hubble Flow

R. D. Davies[1] and L. Staveley-Smith[2]

[1]University of Manchester, Nuffield Radio Astronomy Laboratories, Jodrell Bank, England, U.K.
[2]Anglo-Australian Observatory, Epping, N.S.W., Australia

We have chosen a sample of 307 Sb–Sc galaxies for an accurate study of the motions of local galaxies relative to the uniform Hubble flow. This work extends that of Hart and Davies (1982) who studied a sample of 100 Sbc galaxies. A deliberate effort was made to select for H I observation an isotropic distribution of galaxies over the sky. Northern hemisphere observations were made with the 76–m Lovell Telescope at Jodrell Bank while the southern hemisphere objects were observed with the Parkes 64–m telescope. Flux and velocity scales at the two observatories were carefully intercompared. For the sample as a whole, r. m. s. errors in H I integrated flux density were 11%, in velocity width were $10\,\mathrm{km\,sec^{-1}}$ and in systemic velocity were $5\,\mathrm{km\,sec^{-1}}$. Optical data on blue magnitudes, inclinations and diameters were taken from the literature. The basic observational data and an assessment of the intrinsic properties of Sb–Sc galaxies are given in Staveley-Smith and Davies (1987, 1988).

The galactic absorption corrections for the optical parameters to be used in subsequent studies are calculated from local Milky Way H I integrals in a 12 arcmin × 12 arcmin beam centred at the position of each galaxy. We believe that this is the best indicator currently available for these corrections. Our opinion is confirmed in a separate study of local H I and IRAS 100 micron brightness at intermediate and high latitudes; the correlation between the dust and gas is very tight. A further important consideration which required careful study was the inclination correction which it was necessary to apply to the three distance indicators, H I flux, blue luminosity, and blue diameter. The data showed no change of H I self-absorption with inclination. However, there was a significant change of optical diameter and blue luminosity with inclination; the corrections change significantly between Sb and Sc galaxies. For example, edge–on Sc galaxies have diameters 17±7% larger than face–on objects while their blue luminosities are 1.2±0.2 magnitudes fainter. Our data base enables us to make a more comprehensive study of these parameters than is available in the literature.

We have chosen the neutral hydrogen mass (M_H), blue luminosity (L_B), and diameter (D) as distance indicators. The third parameter, which is used to establish the absolute value of any of these three parameters for a galaxy, is the half-power velocity width. This velocity width is first corrected to take account of the galaxy inclination. Our approach is analogous to the Tully–Fisher method of distance determination, except that we use the three independent indicators M_H, L_B, and D. We estimate that the accuracy in determining the distance for a typical spiral galaxy is 40, 32, and 23% using M_H, L_B, and D, respectively, as distance indicators. The accuracy of H–band infrared luminosity for the same galaxy sample is 23%. These values can be compared with 26% for the Faber–Jackson method of estimating the distance of elliptical galaxies or 23% for the modified Faber–Jackson method used by Lynden–Bell *et al.* (1988).

Spiral galaxies provide a good data base for sampling the nearby Hubble flow; their volume density is 5–10 times greater than that of elliptical galaxies. Our sample enables us to make a thorough analysis of the flow pattern of luminous matter out to a redshift of 3–4,000 $\mathrm{km\,sec^{-1}}$. Our data set was used to estimate the Virgocentric infall velocity based on a model of the Virgo cluster potential. We find that this infall velocity is not well determined from the data. Our estimate of the Virgocentric infall at the distance of the Local Group of galaxies is $100 \pm 50\,\mathrm{km\,sec^{-1}}$.

The motion of the Local Group relative to the backdrop of more distant galaxies is obtained from a dipole fit to the velocity field. Our best estimate of the motion of the Local Group is $210\,\mathrm{km\,sec^{-1}}$ towards $l = 260°$, $b = 10°$ with respect to galaxies at a mean redshift of $2500\,\mathrm{km\,sec^{-1}}$. This result is fairly insensitive to small–scale streaming such as that induced by the Local Supercluster, since our sample has a very isotropic distribution in the sky. It is in reasonable agreement with earlier work by de Vaucouleurs and Peters (1981) and Aaronson *et al.* (1982). Solutions with respect to galaxies more distant than $3000\,\mathrm{km\,sec^{-1}}$ can be interpreted either as a bulk motion with respect to the Cosmic Microwave Background (CMB) or $\sim 1000\,\mathrm{km\,sec^{-1}}$ towards $l = 244°$, $b = -6°$, or else is indicative of either contamination by larger scale local streaming motions or sample biases at the high redshift end of our survey. We are making further optical and H I observations of spiral galaxies in the velocity range of 4–6000 $\mathrm{km\,sec^{-1}}$ which should help to resolve these important questions.

We have made estimates of the peculiar velocities in the foregrounds of two superclusters, namely Perseus–Pisces ($l = 135°$, $b = -25°$) and Hydra–Centaurus ($l = 286°$, $b = 24°$). Galaxies within 45° of the direction of the Perseus–Pisces supercluster appear to be at rest with respect to the CMB, with a mean velocity of $-45\pm\ 165\,\mathrm{km\,sec^{-1}}$. On the other hand, galaxies within 45° of the Hydra–Centaurus region have a component of streaming directed away from us of $372 \pm 99\,\mathrm{km\,sec^{-1}}$. This value is much smaller than the mean flow of 912 $\pm$ 181 km sec^{-1} reported by Lynden–Bell *et al.* (1988) for elliptical galaxies within 60° of the Great Attractor which lies only 25° away. As more data become available it should be possible to quantify these peculiar flows with more precision.

Another outcome of our multiwavelength study of Sb–Sc galaxies is a clarification of the relation between the parameters which describe spiral galaxies. A multivariate analysis shows that three dimensions are sufficient. The first is the size which is given by the blue luminosity, total mass, H I mass, and the diameter; these account for 69% of the variance. The second is the quiescent star formation which is given by the two colour indices $(B-V)$ or $(B-H)$. The third is embedded activity which is given by the far infrared colour measured between 60 and 100 microns and the existence of a bar. This third dimension has not been so clearly identified previously.

References

Aaronson, M., Huchra, J., Mould, J., Schechter, P. L., and Tully, R. B. 1982, *Astrophys. J.* **258**, 64.

de Vaucouleurs, G. and Peters, W. L. 1981, *Astrophys. J.* **248**, 395.

Hart, L. and Davies, R. D. 1982, *Nature* **297**, 191.

Lynden–Bell, D., Faber, S. M., Burstein, D., Davies, R. L., Dressler, A., Terlevich, R. J. and Wegner, G. 1988, *Astrophys. J.* **326**, 19.

Staveley–Smith, L. and Davies, R. D. 1987, *Mon. Not. Roy. Astron. Soc.* **224**, 953.

Staveley–Smith, L. and Davies, R. D. 1988, *Mon. Not. Roy. Astron. Soc.* **231**, 833.

Flat Edge–On Galaxies and Large–Scale Streamings

Igor Karachentsev

Special Astrophysical Observatory, USSR Academy of Sciences, U.S.S.R. and Arecibo Observatory, Cornell University, Ithaca, New York, U.S.A.

When the Tully–Fisher (T–F) relation is used to determine the distances to galaxies, the edge–on, spiral galaxies are usually excluded, due to large fluctuations of the observed extinction in these objects. Based on a sample of 324 galaxies observed at Arecibo, we demonstrate that this effect is less significant than it has been usually assumed. Moreover, we show that the scatter of galaxies in the T-F diagram ("H I profile width versus linear diameter") steadily decreases when the axial ratio, a/b, increases. For the flattest and the most inclined galaxies ($a/b > 7$), the scatter in the T–F diagram corresponds to the mean square error in distance estimates of $10^{\pm 0.09}$. Not only the H I line widths, but also the H I fluxes of flat galaxies are strongly correlated with their optical diameters: at a given flux, the r.m.s. scatter in diameter is only $10^{\pm 0.03}$.

At present, the effect of "latent parameters" on the T–F diagram cannot be properly estimated, due to the lack of reliable photometric data for a representative sample of flat galaxies. Using the photometric data of Watanabe (1983), we have obtained the r.m.s. deviation of $10^{\pm 0.048}$ with respect to the relation $W_{HI} \approx h^{0.70} \cdot I_0^{0.52}$ for 15 galaxies with $a/b > 5$ (W_{HI} is the 21–cm line width, h is the linear scale of an exponential disk and I_0 is its central surface brightness). This result slightly differs from the expected relationship for a Freeman disk ($V_m \approx h^{0.5} \cdot I_0^{0.5}$) and, surprisingly, resembles a similar relation, $\sigma_v \approx R_e^{0.72} \cdot I_e^{0.65}$, that has been derived for E-galaxies.

We have applied the criterion $a/b > 7$ to select 1547 flat galaxies from the UGC, ESO(B), and MCG catalogs. Their distribution over the sky has considerably less contrast than in the case of other galaxies; in particular, the distribution does not show a signature of the Local Supercluster.

Here, we give a brief description of the currently compiled, new Catalogue of Flat Galaxies (FGC), which will cover the whole sky with an effective depth $z_e \approx$

0.05 and will contain about 8,000 objects. The catalog has several characteristics that make it particularly suitable for the studies of large-scale streaming motions in the Universe: (i) the catalog members are disk-like, flat galaxies with a simple structure; (ii) the selection criterion used ($a/b > 7$) makes the sample morphologically homogeneous; (iii) the 21-cm line based detection rate of these galaxies is nearly 100%; and (iv) the flat galaxies avoid volumes occupied by groups and clusters so that their structure remains undisturbed, and they are not affected by large virial motions.

Reference

Watanabe, M. 1983, *Ann. Tokyo Astron. Obs.*, Second Ser., **19**, No. 2, 121.

On the de Vaucouleurs Density–Radius Relation and the Cellular Intermediate Large–Scale Structure of the Universe

Remo Ruffini

International Center for Relativistic Astrophysics, Department of Physics, University of Rome, Italy

Summary

The de Vaucouleurs mass density relation is interpreted in terms of a fractal and cellular structure of the Universe: the relation between this fractal and cellular structure and the usually adopted cosmological conditions of homogenity and isotropy, in three–dimensional space, are critically discussed.

Introduction

Among the solutions of the equations of general relativity, the ones obtained by A. A. Friedmann (1922) are, possibly, those with the greatest impact on physics and astronomy and, more generally, on Man's understanding of the Universe. The validity of such solutions can be tested in three different regimes:

I. By observations of galaxies, clusters, and superclusters, and by the verification of the Hubble law;

II. By observations of the infrared background radiation, which put very stringent limits on the anisotropies and inhomogeneities of the spatial geometry in the very early epochs of the Universe ($Z \approx 10^3$), as well as by testing the theory of cosmological nucleosynthesis.

III. By the analysis of the still unsolved problem of galaxy formation, a topic of active theoretical development today on the basis of Jeans instability, applied to the dark matter component of the Universe.

In the following, I will mainly address recent developments in the first regime, directly related to two classical and seminal papers written by Gérard de Vaucouleurs in 1970 and 1971.

The Friedmann solutions were obtained under the strong assumptions that the three–dimensional spatial geometry of the Universe be both homogeneous and isotropic. Under these strong symmetry conditions, it is possible to introduce a cosmological time; the three–dimensional metric is simply one in a space of constant curvature (positive, negative, or null) . The time dependence of the radius of curvature $a(t)$ is then given by the Einstein equations for a variety of equations of state describing the matter content of the Universe (see *e.g.* Landau and Lifsits 1970). At any given time t_0, for any point at a distance l — much smaller than the horizon size — from a given observer it is possible to define a velocity of recession (Landau and Lifsits 1970):

$$v_{rec} = \left(\frac{\dot{a}}{a}\right)_{t_0} l. \tag{1}$$

The observations made by Hubble in 1929 of a sample of nineteen "nebulae" and his interpretation of their redshifts as being due to a recession velocity lead to the well-known linear relationship of velocity with distance l of a galaxy:

$$v_{rec} = H\, l, \tag{2}$$

where H is the Hubble constant. It is interesting to see how much the estimates of the distances of those nineteen galaxies have changed in the intervening years (see *e.g.* Burbidge 1981). This lead Burbidge (1988) to remark how "good scientists are often also very lucky scientists."

Equations (1) and (2) gave the first evidence that the Universe followed, quite closely, the dynamical laws predicted by Friedmann on the grounds of Einstein's equations. This clear success in turn lead to a strong and almost emotional support of the idea that the Universe had to be spatially homogeneous and isotropic, even on scales of a few Mpc. It may be appropriate to recall here the words of Robertson (1955):

> "Observations which have been coming to a head over this period of time have given rise to the notion that the distribution and motion of matter in any sufficiently large spatial region of this Universe are, by and large, intrinsically much the same as those in any other similar region, regardless of its position and orientation. This presumed uniformity in the large implies a certain form of a principle of relativity, sometimes called, appropriately enough, the 'cosmological principle.' Clearly, any approximation to the actual Universe as

crude as this, must be one in which the observed agglomeration of matter into stars and nebulae, and even into clusters of nebulae, is to be replaced by a smeared–out substratum which preserves only the uniformities common to all regions — and thereby robs it of most of those individual characteristics which make the skies a delight and a challenge to a poet and astronomer alike!"

In the two papers of 1970 and 1971, de Vaucouleurs reviews the work done in the previous forty years pointing in a quite different direction: the observational and empirical evidence for a highly inhomogeneous and anisotropic distribution of galaxies and cluster of galaxies (*i.e.* matter), extending all the way to dimensions of hundreds of Mpc. After reviewing many contributions including the ones by Holmberg, Zwicky, Abell, Karachentsev, Kiang, and Kiang and Saslaw, de Vaucouleurs presents the distinct possibility of a Hierarclical Universe along the lines of Charlier's (1908, 1922a,b) earlier work. Of special interest to us has been, in de Vaucouleurs's 1970 paper, the establishment of a density–radius relation among a variety of cosmological objects, ranging from compact elliptical galaxies all the way to superclusters, following Carpenter's (1938 and references therein) classical density restriction. De Vaucouleurs takes an arbitrary spherical sample of radius R and moves it in space until he finds the largest mass contained in this particular volume, repeating the operation for different values of R. He then derives a maximum density–radius relation given by the simple power law:

$$\rho_{max}(R) = K\, R^{-\alpha}, \tag{3}$$

where $K = 1.55 \times 10^{15}\,\mathrm{g\,cm^{-1.3}}$ and $\alpha \approx 1.7$ (see also de Vaucouleurs 1974). For the definition of the density and the radius of systems not spherically symmetric and/or without a definite boundary, he assumes the effective radius R, corresponding to a sphere having the same volume as that within the equidensity surface which encloses half the total mass of a system. As a consequence, the effective mean density of the astrophysical object is given by:

$$\rho = \frac{3\,M}{8\pi\, R^3}. \tag{4}$$

He then applies this rule on all possible scales, going from planets to the deepest sample available (the Lick catalog), finding that the relation holds for scales from 0.02 Mpc all the way to dimensions of hundreds of Mpc.

In the intervening years, both of the above points of view have gained additional supporters. The point of view of those purporting an effective cosmological homogenity and isotropy from dimensions of the order of a few Mpc has found strong support in the contributions of Peebles and his collaborators; see *e.g.* Yu and Peebles (1969), Davis and Peebles (1983), Groth and Peebles (1977), Peebles (1980), and references therein. An alternative support for the homogeneity hypothesis has come from the work of Sandage, Tammann, and

Hardy (1972 and references therein) relating the surface density $A(m)$ to the spatial density $D(r)$ of the distribution of galaxies.

The second point of view has received a further clear observational and empirical support by the discovery of cosmological voids on a variety of extragalactic scales; see *e.g.* Chincarini and Martins (1975), Chincarini and Rood (1976), Chincarini (1978), Tarenghi *et al.* (1980), Jôeveer *et al.* (1978), Einasto *et al.* (1980), Kirshner *et al.* (1981), Bahcall and Soneira (1982), and de Lapparent *et al.* (1986).

A very important tool which will possibly enable us to discriminate between these two points of view may come from the theoretical interpretation of the analysis of the angular and spatial correlation functions of galaxies, clusters, and superclusters. This work pioneered by Totsuji and Kihara (1969) has been greatly developed by — among others — the work of Peebles (1980 and references therein), Peebles (1981), Davis and Peebles (1983), Bahcall (1986 and references therein), Klyping and Kopylov (1983), and Einasto *et al.* (1984, 1986). In the following discussion, we shall emphasize the analysis of spatial correlation functions and delay the analysis of the angular correlation functions to a later paper (Calzetti *et al.* 1988c).

The spatial correlation function can be defined as (see *e.g.* Calzetti *et al.* 1988a):

$$1+\xi(r)=\frac{<n(\vec{r}_1)n(\vec{r}_1+\vec{r})>}{<n>^2}, \tag{5}$$

where $n(\vec{r})$ is the particle density of our statistical sample of mean number density $<n>$ and the averages are made over all the points of $\vec{r}_1$ and on all the angles Ω_r of $\vec{r}$.

The correlation function $\xi(r)$ has been observed to have a very simple power law form

$$\xi_i(r)\approx A_i r^{-\gamma}, \tag{6}$$

with a common value of $\gamma \simeq 1.8$ and the amplitude A_i generally depending on the sample of galaxies, clusters, or superclusters considered. The law given by Equation (6) has been tested in the range from a few galactic radii all the way to hundreds of Mpc, the upper limit being given by the correlation functions of superclusters.

We have recently advanced a possible interpretation of these results (Calzetti *et al.* 1987a,b, 1988a, and Pietronero 1987) in terms of a simple fractal structure which is both hierarchical (in the de Vaucouleurs–Charlier sense) and observer homogeneous. It is also interesting that such a fractal structure, duly implemented by the existence of upper (L_c) and lower (l_c) cut–offs, can fulfill the very stringent requirements imposed by the observations of the cosmological black body radiation in the early phases of cosmology (Point II above). In Ruffini *et al.* (1988), we have advanced a model postulating the existence of a cellular structure in an "intermediate" large scale of the Universe (see Figure 1). Within each cell of dimension L_c, there is the previously mentioned fractal

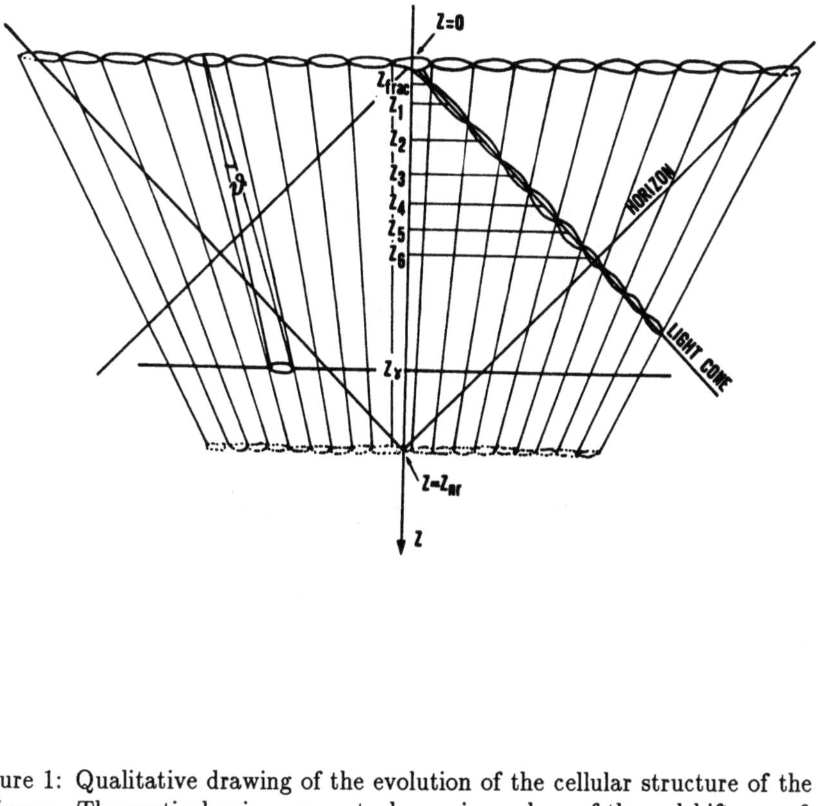

Figure 1: Qualitative drawing of the evolution of the cellular structure of the Universe. The vertical axis represents decreasing values of the redshift. $z = 0$ represents the cosmological observer today, embedded in his own "elementary cell"; z_1, z_2, ..., z_n are the centers of the successive "elementary cells" seen by this observer. $z = z_\gamma$ represents the surface of last scattering (decoupling) of the cosmological background radiation. θ is the angular scale subtended today by an "elementary cell" at the time of decoupling. Finally $z = z_{nr}$ is the time at which the cell formed via gravitational instability.

structure which could be formed by succesive fragmentations of the dark matter component of the Universe by Jeans instability (Calzetti *et al.* 1987a).

In some sense, we are proposing a hybrid solution between the two extreme viewpoints above: one purporting homogeneity and isotropy, and the other purporting a very non–isotropic and inhomogeneous hierarchical structure. The novelty is twofold: the introduction of an upper cut–off, L_c, in the fractal structure, and the proposal of a cellular structure in the global scale structure of the Universe. Looking at distances much larger than the cutoff L_c, or equivalently, at earlier cosmological eras, the traditional results of a Friedmann cosmology are recovered, while for smaller distances the typical features of a hierarchical or fractal structure are observed. One of the advantages of using the fractal structure and the concept of a cellular structure in the Universe, is that the traditional cosmological principle of homogeneity and isotropy is, in some sense, minimally violated, being substituted by the more observationally meaningful concept of observer homogeneity in the sense introduced by Mandelbrot (1985).

Our aim in the next section is to summarize two recent results (Calzetti *et al.* 1988a,b) showing how the de Vaucouleurs density–radius relation given in Equation (3) can be readily interpreted within the fractal model proposed here.

Fractal Structure and the de Vaucouleurs Density Relation

Calzetti *et al.* (1987b, 1988a) have considered a simple system both self–similar and observer homogeneous, namely one in which every particle of the ensemble sees the same average distribution. This idealized mathematical model has been constructed by a three–dimensional generalization of a Sierpinski gasket (see *e.g.* Sierpinski 1974), developing an observer homogeneous hierarchical structure out of the corners of an equilateral tetrahedron (see Figure 2). For such a "fractal" system, the correlation function can be expressed in a very simple analitical form (Calzetti *et al.* 1987b):

$$\xi(r) = \frac{n_d(r)}{< n >} - 1, \tag{7}$$

where $n_d(r) = \frac{1}{4\pi r^2}\frac{dN(r)}{dr}$ is the differential density, and $< n >$ is the average density of the sample contained in a radius R_S. The correlation function can then be written (Calzetti *et al.* 1988c) as

$$\xi(r) = A\, r^{-\gamma} - 1, \tag{8}$$

where

$$A = \frac{3-\gamma}{3} R_S^{\gamma}, \tag{9}$$

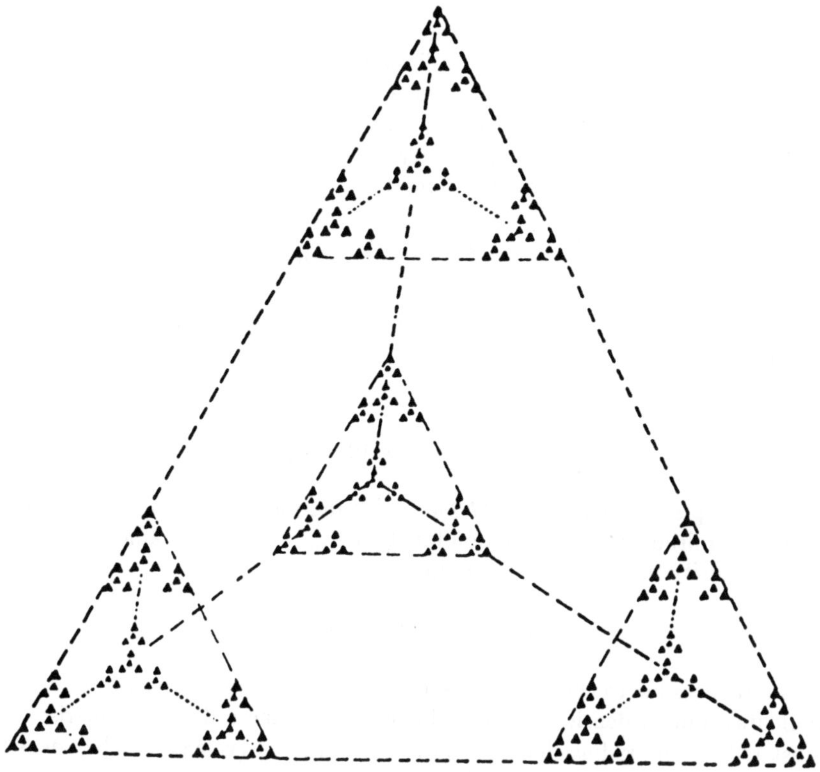

Figure 2: An example of a deterministic fractal structure built by a hierarchical tetrahedric construction, with a fractal dimension $D = \log 4/\log 3 = 1.2618\ldots$.

and $\gamma = 3 - D$; 3 is dimensionality of the space and D is the fractal dimension. It is important to stress that Equation (8) does not depend on any geometrical realization of the fractal structure, namely tetrahedrons, spheres, or other geometrical solids, but only on the intrinsic properties of the self-similar distribution, *e.g.* the space dimensions, the fractal dimension and the sample radius.

In order to compare the results given by Equation (8) with the observations, we must consider a stochastic fractal, that is a stochastic distribution of high density regions and a scale-free distribution of voids which, on average, fulfill the same equations of a geometrical fractal (see Calzetti *et al.* 1987b and references therein). All the considerations obtained for the deterministic fractal of Figure (1) and, clearly, also Equation (4) still apply to the case of a stochastic fractal. The dependence of the amplitude A in Equation (9) on the sample radius R_S has been verified by Calzetti *et al.* (1987a).

From Equation (8), we can now very simply estimate the average density of a structure of radius R within the fractal

$$n_V(R) = \frac{4\pi}{V}\int_0^R n_d(r) r^2\, dr = \frac{3}{3-\gamma} < n > A\, R^{-\gamma} = K'\, R^{-\gamma}. \tag{10}$$

Calzetti *et al.* (1988b) have estimated the numerical value of Equation (10) for the distribution of galaxies. Using the estimate of the average density from Groth and Peebles (1977)

$$n_V(R_S) \approx 5.84 \times 10^{-2}\, \mathrm{Mpc}^{-3} \tag{11}$$

with a sample radius of $R_S = 47.2\,\mathrm{Mpc}$, estimating from Equation (9) the expected amplitude of the correlation function A_{gal}, we have obtained a predicted galaxy density in a volume V of radius R

$$n_V(R) \approx 60\, R^{-1.8}\, \mathrm{Mpc}^{-3}. \tag{12}$$

It is now possible to compare this estimate with the one derived from de Vaucouleurs's radius–density relation. In order to make a direct comparison, we must reduce the radius–density relation to a radius–(galaxy–number–density) relation by dividing Equation (3) by the maximum mass of a galaxy in a radius of 20 Kpc (Calzetti *et al.* 1988b). If we assume that this mass is $M_{gal} = 10^{12}\, M_\odot$, we obtain

$$n_V(R) \approx 53\, R^{-1.7}\, \mathrm{Mpc}^{-3}, \tag{13}$$

quite close agreement with Equation (12). This result, on one hand, gives a theoretical explanation for the validity of the phenomenological relation found by de Vaucouleurs and, *vice versa*, also gives further confirmation, in addition to the ones based on the scaling of the amplitude of the correlation functions (Calzetti *et al.* 1987a), that galaxies are indeed members of self-similar and observer homogeneous structures (*i.e.* "fractal").

Conclusions

Clearly, all the above considerations apply only in the range of existence of fractal structures. It has been shown by Ruffini *et al.* (1988a) that such structures must have both a lower and upper cutoff. The existence of an upper cutoff is necessary (a) in order not to run into contradictions with the observed homogeneity of the cosmological background radiation and (b) in order to have the formation of these structures from a process of Jeans instabilities acting on the dark matter component of the Universe. Estimates of the masses and chemical potentials of the "–inos" composing the dark matter have been evaluated for selected values of this upper cut–off L_c for 100, 200 and 300 Mpc (see *e.g.* Ruffini *et al.* 1988a). The existence of a lower cut–off l_c follows, in the same scenario, from the formation of galactic halos and consequently of galaxies, which occur at a much later era than the one usually considered in traditional theories (at $Z \simeq 5 - 10$ (again, see Ruffini *et al.* 1988a, as well as Ruffini *et al.* 1988b). Therefore, we will not try to give a theoretical interpretation to the possible existence of a de Vaucouleurs density–radius relation at subgalactic scales.

There is, however, a further point in the 1970 de Vaucouleurs paper which acquires particular significance in light of the above results: this is the difficulty of defining, in the presence of a density–radius relation like the one given in Equation (12), a suitable cosmological density. This issue has been considered by Ruffini *et al.* (1988a), and it was argued that only measurements of densities performed on distances larger than the upper cutoff L_c, and suitably averaged over several fractal cells, can lead to a testable cosmological density distribution (see Figure (3)). As de Vaucouleurs correctly pointed out, the definition of a cosmological density from measurements within the cell is clearly meaningless.

All these matters have profound consequences for the determination of the Hubble constant: as I stressed in the introduction, the Hubble constant has a cosmological role only if it is measured in a spatially homogeneous and isotropic Universe. Only under these basic conditions can we make the identification

$$H = \frac{\dot{a}}{a} \tag{14}$$

with all its consequences (*e.g.* we can infer a meaningful determination of a cosmological critical density). In order to obtain a meaningful value of the Hubble constant, the Hubble law given in Equation (2) should be measured at distances larger than the upper cut–off, L_c, where homogeneity and isotropy, for measurements averaged over a few fractal cells, has been reached (see Ruffini *et al.* 1988a). All measurements at distances smaller than L_c will be dominated by the local dynamics of the inhomogeneous and anisotropic fractal structure, which is, clearly, expected to be very different from the one considered by Friedmann in Equation (1).

It is conceivable, as proposed by Mo and Ruffini (1988), that the amount of regularity of the fractal structure will allow us to define, for distances smaller

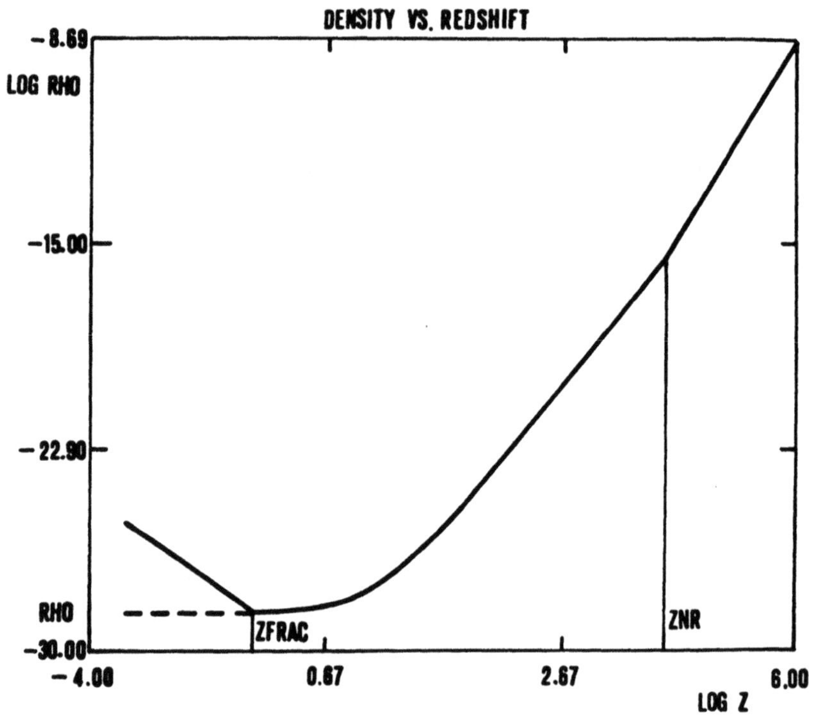

Figure 3: Log–log plot of the behaviour of the density of the Universe as a function of the cosmological redshift. The first zone, up to $z = z_{frac}$, represents the self–similar "elementary cell." For a larger value of z, the density increases as the cube of the cosmological redshift up to $z = z_{nr}$. For $z \geq z_{nr}$, the density follows the fourth power of the redshift as in the classical Friedmann Universe. The dashed line gives the effective average density of the Universe, extrapolated to the present as deduced from the value of $\rho(z)$ at $z = z_{frac}$.

then the upper cutoff, a "Hubble ratio" which is a function of the distance, and is clearly not directly related to a cosmological interpretation. Only the extrapolation of such a Hubble ratio to distances larger than the upper cutoff L_c would then lead to a determination of a cosmological meaningful value of the Hubble constant.

References

Bahcall, N. A. 1986, *Astrophys. J. Lett.* **302**, L41.

Bahcall, N. A. and Soneira, R. M. 1982, *Astrophys. J.* **262**, 419.

Burbidge, G. 1981, *Ann. N. Y. Acad. Sci.* **375**, 123.

Burbidge, G. 1988, "Modern Cosmology: The Harmonious and the Discordant Facts," *Int. J. Theor. Phys.*, to be published.

Calzetti, D., Einasto, J., Giavalisco, M., Ruffini, R., and Saar, E. 1987a, *Astrophys. Sp. Sci.* **137**, 101.

Calzetti, D., Giavalisco, M., Pietronero, L., and Ruffini, R. 1987b, *Phys. Rev. Lett.*, submitted.

Calzetti, D., Giavalisco, M., and Ruffini, R. 1988a, *Astron. Astroph.* **198**, 1.

Calzetti, D., Giavalisco, M., and Ruffini, R. 1988b, *Astron. Astrophys.*, submitted.

Calzetti, D., Giavalisco, M., and Ruffini, R. 1988c, in preparation.

Carpenter, E. F. 1938, *Astrophys. J.* **88**, 344.

Charlier, L. V. 1908, *Ann. Math. Astron. Phys.* **4**, No. 24.

Charlier, L. V. 1922a, *Ann. Math. Astron. Phys.* **16**, No. 22.

Charlier, L. V. 1922b, *Medd. Lund Obs.* No. **98**.

Chincarini, G. and Martins, D. 1975, *Astrophys. J.* **196**, 335.

Chincarini, G. and Rood, H. J. 1975, *Nature*, **257**, 294.

Chincarini, G. 1978, *Nature*, **272**, 515.

Davis, M. and Peebles, P. J. E. 1983, *Astrophys. J.* **267**, 465.

Einasto, J., Jôeveer, M., and Saar, E. 1980, *Mon. Not. Roy. Astron. Soc.* **193**, 353.

Einasto, J., Klypin, A. A., Shandarin, S. F., and Saar, E. 1984, *Mon. Not. Roy. Astron. Soc.* **206**, 529.

Einasto, J., Klypin, A. A., and Saar, E. 1986, *Mon. Not. Roy. Astron. Soc.* **219**, 457.

Friedmann, A. A. 1922, *Z. Phys.* **10**, 377.

Groth, E. J. and Peebles, P. J. E. 1977, *Astrophys. J.* **211**, 1.

Groth, E. J. and Peebles, P. J. E. 1977, *Astrophys. J.* **217**, 385.

Jôeveer, M., Einasto, J., and Tago, E. 1978, *Mon. Not. Roy. Astron. Soc.* **185**, 530.

Kirshner, R. P., Oemler, A., Schechter, P. L., and Schechtman, S. A. 1981, *Astrophys. J. Lett.* **248**, L57.

Klypin, A. A. and Kopylov, A. I. 1983, *Soviet Astr. Lett.* **9**, 41.

Landau, L. and Lifsits, E. M. 1970, *Theory des Champs* (Moscow: MIR).
de Lapparent, V., Geller, M., and Huchra, J. P. 1986, *Astrophys. J. Lett.* **302**, L1.
Mandelbrot, B. B. 1985, *The Fractal Geometry of Nature* (New York: W. H. Freeman).
Mo, H. J. and Ruffini, R. 1988, in preparation.
Peebles, P. J. E. 1980, *The Large Scale Structure of the Universe* (Princeton: Princeton Univ. Press).
Peebles, P. J. E. 1981, *Astrophys. J. Lett.* **243**, L119.
Pietronero, L. 1987, *Physica*, **144a**, 257.
Robertson, H. R. 1955, *Publ. Astron. Soc. Pac.* **67**, 82.
Ruffini, R., Song, D. J., and Taraglio, S. 1988a, *Astron. Astrophys.* **190**, 1.
Ruffini, R., Song, D. J., and Taraglio, S. 1988b, in preparation.
Sandage, A., Tammann, G. A., and Hardy, E. 1972, *Astrophys. J.* **172**, 253.
Sierpinski, W. 1974, *Oeuvres Choisies*, ed. S. Hartman (Warsaw: Edition Scientifiques).
Tarenghi, M., Chincarini, G., Rood, H. J., and Thompson, L. A. 1980, *Astrophys. J.* **235**, 724.
Totsuji, H. and Kihara, T. 1969, *Publ. Astron. Soc. Japan* **21**, 221.
de Vaucouleurs, G. 1970, *Science* **167**, 1203.
de Vaucouleurs, G. 1971, *Publ. Astron. Soc. Pac.* **83**, 113.
de Vaucouleurs, G. 1974, in *I. A. U. Symp. 58, The Formation and Dynamics of Galaxies*, ed. J. R. Shakeshaft (Dordrecht: Reidel), p. 1.
Yu, J. T. and Peebles, P. J. E. 1969, *Astrophys. J.* **158**, 103.

Analytical Models for Large–Scale Structure in the Universe

T. Buchert

Max–Planck–Institut für Astrophysik, Garching bei München, F.R.G.

Summary

I report briefly on some analytical models for the formation of large–scale structures valid at the early and late nonlinear stages in the language of gravitational instability theory. Some aspects of the applicability of those models to observed structures are pointed out.

Introduction

Large–scale structures observed today appear to be *organized*; in popular terms, they appear as cell or network structures to the observer (de Lapparent *et al.* 1988). Beside the observational problems of pattern recognition with high uncertainties, the structure has to be viewed as a snapshot of a *dynamical* formation process, *i.e.* the large–scale objects have to be understood in terms of *cosmological* scenarios.

The Models

One model which relates observed structures to early epochs characterized by small inhomogeneities superimposed on an isotropic and homogeneous state (*i.e.* the epoch of decoupling of matter and radiation) is provided by the **pancake** picture proposed by Zel'dovich (reviews by Shandarin *et al.* 1983, Shandarin and Zel'dovich 1988). This model is based on an approximation for the evolution of collisionless self–gravitating matter in an expanding Friedmann universe and relies on the assumption that some "dark matter" candidate (*e.g.* massive neutrinos) dominates the dynamical evolution of the structures under consideration, therefore attracting the baryonic matter component on the considered spatial scales.

The approximation scheme can be understood rigorously in terms of solutions for self–gravitating "dust" (*i.e.* pressureless matter) in the framework of Newtonian cosmology (Buchert 1988). The most striking feature of the model is the description of a highly anisotropic collapse of matter to surfaces of infinite density (caustics) approximating high density objects. The metamorphoses of these caustics can be subsequently studied, their qualitative aspects can be understood in terms of Lagrange singularity theory (Arnol'd 1982, Arnol'd *et al.* 1982). These metamorphoses develop multi–stream flows (systems of three–stream flows are called pancakes). Phenomenologically, the distribution of the density constrast in the vicinity of caustics is *hierarchical* in a dynamical sense: after the formation of a cellular structure, matter concentrates within cell walls to form filamentary structures, then filaments collapse to clumps of matter representing rich clusters.

Up to this stage, the pancake model gives qualitatively correct results in comparison with numerical simulations of the scenario, especially at the stage of formation of the cellular structure and shortly after that. Advantages of this analytical approach compared with numerical simulations concern the high *spatial resolution* of caustic metamorphoses, which is demonstrated by Buchert (1988); see Figure 1, which illustrates a two–dimensional calculation. This offers the possibility of studying the internal structure of high density objects as well as statistics of patterns, *e.g.* as a result of an initially Gaussian distribution of spectral amplitudes. However, concerning comparisons with observational structures, a morphological classification of large–scale patterns must take into account effects of projection of (three–dimensional) singular structures as seen by the observer. Buchert (1988) also presents an approximation which allows him to specify initial density contrasts and peculiar velocities separately in all Newtonian universes of Friedmann–Lemaître type, including models with "curvature" and a cosmological constant.

Drawbacks of these models are quantitative in nature. One shortcoming concerns the growth of the pancake thickness after the crossing of trajectories. Since the approximation neglects gravitational interaction of flows in multi–stream systems, the thickness growth is unlimited. In contrast to this behaviour, gravity acts as restraining force on particles (which leave high density regions) causing them to oscillate around the central plane of the pancake, as was shown in numerical simulations. A model which keeps some advantages of an analytical description but overcomes the problem of unlimited growth of the pancake thickness, was proposed by Gurbatov *et al.* (1985), Shandarin and Zel'dovich (1988), and Gurbatov *et al.* (1988). In a first approximation, the crossing of

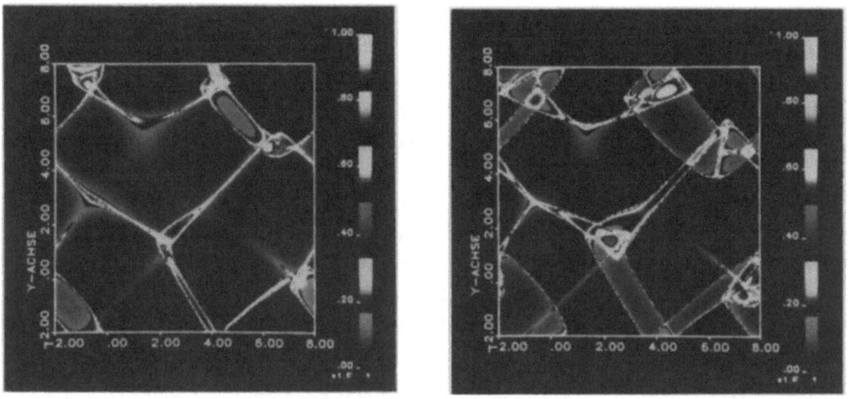

Figure 1: (left) A two-dimensional calculation based on the pancake model with Gaussian spectra for the peculiar velocity and gravitational potentials with cut-off (in total 50 fourier modes with high spatial resolution 100 Mio particles collected in 512^2 elements of an Eulerian plane) at $Z = 2$ in one box (100×100 Mpc). (right) Same as left hand figure, but at $Z = 1$. See Buchert (1988) for details.

trajectories is substituted by "sticking" of particles; this results in a realistic description of structure characteristics, such as masses, peculiar velocities and spatial distribution of density clumps, also at the late nonlinear stage of the development of the cell structure. The asymptotic behaviour of structure characteristics, in particular spatial scales, as well as of statistics of velocity and density fields, can be estimated.

Conclusion

All models mentioned above depend explicitely on the initial structure imprint, *i.e.* the spectrum of fluctuations given at the initial epoch. Therefore, characteristics of observed structures are determined by the parameters of the given spectrum. From the point of view of self-gravitating systems, the models are primarely *kinematical* in the sense that they imply a high specialization of nonlinearities of the underlying equations (see Buchert 1988). A satisfactory agreement with results of numerical calculations is explained by the argument that kinematical motion dominates effects of self-gravity up to the time of formation of the first pancakes. A further advantage of analytical models is that they offer the possibility of studying details of the structure without "step by step" numerical integration.

References

Arnol'd, V. I. 1982, *Trudy Sem. Petrovskogo* **8**, 21.

Arnol'd, V. I., Shandarin, S. F., and Zel'dovich, Ya. B. 1982, *Geophys. Astrophys. Fluid Dyn.* **20**, 111.

Buchert, T. 1988, *Astron. Astrophys.*, submitted.

de Lapparent, V., Geller, M. J., Huchra, J. P. 1988, *Astrophys. J.*, in press.

Gurbatov, S. N., Saichev, A. I., and Shandarin, S. F. 1985, *Dokl. Acad. Nauk* **285**, 323; and *Sov. Phys. Dokl.* **30**, 921.

Gurbatov, S. N., Saichev, A. I., Shandarin, S. F. 1988, *Mon. Not. Roy. Astron. Soc.*, submitted.

Shandarin, S. F., Doroshkevich, A. G., and Zel'dovich, Ya. B. 1983, *Usp. Fiz. Nauk* **139**, 83; and *Sov. Phys. Usp.* **26**, 46.

Shandarin, S. F. and Zel'dovich, Ya. B. 1988, *Rev. Mod. Phys.*, submitted.

The Distance-Redshift Relation in the Inhomogeneous Universe

Laurent Nottale
Observatoire de Paris, Meudon, France

Summary

The inhomogeneity and hierarchical distribution of matter in the Universe up to scales of several hundred Megaparsecs imply that the redshift–distance relation may deviate strongly from its mean for some lines–of–sight. We indicate ways to deal with this problem in the case of large scale density fluctuations.

Introduction

The most general way to study the redshift–distance relation is to work with the Optical Scalar Equation (OSE: Sachs 1961, Kantowski 1969) which describes the propagation of a light beam in a curved space–time for the geometrical optics approximation. It was recently written in a new form (Nottale and Chauvineau 1986, Nottale 1988):

$$d^2\lambda/d\Delta^2 + \omega^2(\Delta)\lambda + \xi^2\lambda^{-3} = 0,$$

where Δ is the optical distance (Nottale and Hammer 1984):

$$\Delta = q_0^{-2}[(q_0 z + 1 - q_0)(1 + 2q_0 z)^{-1/2} + q_0 - 1],$$

and where λ is related to the cross-sectional area of the beam A by the relation:

$$\lambda = (1+z)^2(1+2q_0 z)^{-1/2} A^{1/2}.$$

The light beam propagation depends on two terms; first, the "Ricci" matter term:

$$\omega^2 = 3q_0(1+2q_0 z)^2(1+z)^{-3}[(\rho/\rho_0)(1+z)^{-3} - 1],$$

and, second, the "Weyl" shear term, which, for example, can be written in the case of a Schwarzschild mass $m = GM/c^2$:

$$\xi = 3mL \int_{r_1}^{r} \frac{A(r)dr}{r^5(\frac{1-2m/r_m}{r_m^2} - \frac{1-2m/r}{r^2})}$$

where L is a constant.

Distance-Redshift Relations

In a completely homogeneous and isotropic Universe, one gets $\omega = 0$ and $\xi = 0$. The OSE reduces to $d^2\lambda/d\Delta^2 = 0$, and the solution is $\lambda = \Delta$, which is nothing but Mattig's relation in our notations. In an inhomogeneous Universe, one finds a generalized magnitude–redshift relation:

$$\log cq_0^{-2}\{q_0 z+(q_0-1)[-1+(1+2q_0 z)^{1/2}]\} - \log Amp^{1/2} - 0.2m = \log H_0 - 0.2M - 5.$$

The amplification Amp may always be written *exactly* in the form (Nottale, 1988):

$$Amp = [(1-\kappa)^2 - \gamma^2]^{-1},$$

where, to lowest order, κ denotes the "matter" contribution to the light beam convergence, and γ the shear contribution. This last term is well-known in the point mass lens case or for deflectors with simple density profiles, while analytical solutions lack for two or more lenses. However, it has been shown in several situations that the average amplification due to clumps is yielded by Ricci amplification of the same amount of matter smoothed out uniformly (Weinberg 1976, Young 1981, Peacock 1986, Nottale 1988). This is a favourable point, since in every realistic astrophysical situation, analytical solutions to the amplification by matter terms are now known.

When large density excesses with strong density gradients are encountered (galaxies, rich clusters of galaxies), the actual lens extensions are small enough so as to apply the thin lens approximation. The solution for one lens is well-known (see *e.g.* Nottale and Hammer 1984). The solution for any number of lenses on the line-of-sight from the observer to the source has been obtained by Nottale and Chauvineau (1986).

On the contrary, when one looks at very large scale ($\geq 1\,\mathrm{Gpc}$), only small density gradients are found. This is expressed in our notations by the condition $d\omega/d\Delta \ll \omega^2$. As already remarked elsewhere (Nottale 1988), the identity of the OSE with the one-dimensional Schrödinger equation allows us to apply the BKW approximation in this last case. This yields the following solution for a positive density fluctuation above the background Universe described by the function $\omega(\Delta)$:

$$Amp = \frac{\omega(0)\omega(\Delta)\Delta^2}{\sin^2 \int_0^{\Delta} \omega(\Delta)d\Delta}$$

This result allows us to generalize Mattig's relation in an inhomogeneous Universe.

References

Kantowski, R. 1969, *Astrophys. J.* **155**, 89.
Nottale, L. and Hammer, F. 1984, *Astron. Astorphys.* **141**, 144.
Nottale, L. and Chauvineau, B. 1986, *Astron. Astrophys.* **162**, 1.
Nottale, L. 1988, *Ann. Phys.* **13**, in press.
Peacock, J. A. 1986, *Mon. Not. Roy. Astron. Soc.* **223**, 113.
Sachs, P. K. 1961, *Proc. Roy. Soc. London* **A264**, 309.
Weinberg, S. 1976, *Astrophys. J. Lett.* **208**, L1.
Young, P. 1981, *Astrophys. J.* **244**, 756.

The Impact of Space Projects on Extragalactic Research

R. A. E. Fosbury

Astrophysics Division, Space Science Department, European Space Agency, Space Telescope — European Coordinating Facility, European Southern Observatory, Garching bei München, F. R. G.

Abstract

Observations in previously inaccessible regions of the spectrum, made possible by space–borne instruments, have produced fundamental changes in our perception of galaxies and their nuclei. I discuss some of these changes and point out the new possibilities to be offered by the next generation of observatory satellites.

Introduction

The placing of astronomical instruments into space, not as one–off experiments, but as fully–fledged observatories often operating for several years, has become almost routine. Most astronomers who handle observational data will be familiar with the use of one or other of these orbital facilities. In fact, the inclusion of satellite data in a research programme is ubiquitous, and so it is interesting to look back and make an assessment of some of the achievements in extragalactic astronomy deriving from their use. For the future, there is a broad range of missions planned or awaiting launch which will allow a mature exploitation of the pioneering discoveries made as a result of the first surveys in previously inaccessible wavebands.

In addition to obvious advantages of spectral coverage, there are other benefits of the space environment and the style of operation of the space observatories which are having significant impacts on the way we do astronomy. Even in familiar wavebands, the potential for high angular resolution observations outside the atmosphere coupled with fainter backgrounds, will soon allow the Hubble Space Telescope to probe much deeper into the universe, hopefully to the time of galaxy formation.

One of the consequences of the inherent stability of the space observatories is the availability of well-described and homogeneous data archives which are being exploited by many astronomers, not necessarily those involved with the instrument planning and operation, to produce new scientific results. Indeed, these large data sets often allow the selection of very clean samples of objects for further study: the IRAS results must be an outstanding example of this.

Having the facilities for studying a broad wavelength range is, for observations of variable objects, on its own insufficient; there has to be a level of organisation and cooperation to allow *simultaneous* observations with instruments which often have severe operational constraints. These problems would be greatly eased by dedicated broad-band observatories, preferably in a high orbit which would permit long, uninterrupted observations.

Past Extragalactic Achievements

With a history of experiments and mature space observatories already behind us, it is permissible, perhaps, to make some value judgements and identify some of the major achievements in extragalactic astronomy which have resulted from the data. Astronomers will always argue about the order in, and even the content of, such a list but mine is as follows:

The discovery of hot *coronal* ($T_e \sim 10^{6-8}$K) gas in clusters and individual ellipticals (EINSTEIN)

The new view of the infrared sky provided by IRAS:

- detection of ~25,000 galaxies
- very large infrared luminosities — starbursts and/or active nuclei, mergers
- new Seyferts — the obscured population
- space distribution of galaxies — tracing the mass?

The up-turn of elliptical galaxy energy distributions in the UV — K-correction, galaxy evolution (IUE)

Absorption in extragalactic sight lines — properties of galactic halos (IUE)

Discovery of very broad-band emission from AGN (IR, UV, X-ray, γ-ray)

High frequency energy distribution of AGN — the nature of BL Lac's, Blazars — soft X-ray excess — ionization of the gas

Variability monitoring of AGN, lines and continuum (IUE, EXOSAT)

Considering the first of these topics, the impact on our view of galaxies and their formation/evolution of the discovery of diffuse X–ray emission associated first with clusters of galaxies and then with individual ellipticals is profound. Future measurements of the properties of the hot, gaseous coronae, in particular the temperature, chemical composition and bulk motion, of which at present we have only rather crude indications, are eagerly awaited. The cluster X–ray sources have luminosities of $10^{42-45}\,\mathrm{erg\,s^{-1}}$, $T \sim 10^{7-8}\,\mathrm{K}$, $n \sim 10^{-4} - 10^{-2}\,\mathrm{cm^{-3}}$, and a gas mass comparable with the mass of stars in the galaxies in the cluster. X–ray lines of Si, S, O, Fe have been seen which give temperature and mass measurements. When the observations are interpreted in terms of the *cooling flow* picture, mass inflow rates up to 100's of $M_\odot\ \mathrm{yr^{-1}}$ are derived.

Of some 80 E and S0 galaxies observed by the EINSTEIN satellite, more than 50 were detected. In all but the lowest luminosity objects, it is thought that most of the X–ray emission is from a hot interstellar medium rather than from discrete sources. These galaxy coronae have $L \sim 10^{39-42}\,\mathrm{erg\,s^{-1}}$, $T \sim 10^{6-7}\,\mathrm{K}$, $n \sim 10^{-1} - 10^{-3}\,\mathrm{cm^{-3}}$, and a gas mass $\sim 5 \times 10^8 - 5 \times 10^{10}\,M_\odot$. It is found that $M(gas) \propto L_B^2$. Hydrostatic equilibrium calculations give an estimate of the mass of the galaxy, yielding values up to $\sim 10^{13}\,M_\odot$. The trouble is that the radial temperature profile is critical to the interpretation and this has not yet been measured satisfactorily in enough galaxies. The mass–to–light ratios are implied to be up to $\sim 100\,M_\odot/L_\odot$, and the cooling flow models give mass inflow rates up to a few $M_\odot\ \mathrm{yr^{-1}}$.

Where does this gas come from? Is it all accounted for by stellar mass loss processes or is there a primordial component? The problem of the angular momentum content of the gaseous coronae is crucial and one which has not been addressed by the X–ray observations, but rather relies on optical observations of gaseous filaments which are assumed to have cooled out of the hot phase. The scene is set for a rich harvest of results from the coming X–ray satellites such as ROSAT.

The view of the sky in the far infrared provided by the IRAS satellite is spectacular, and has provided a database of observations which not only provides a survey for future missions, but is sure to yield a continuing stream of new results for many years to come.

Some 96% of the sky was surveyed to a completeness of between 0.5 Jy and 1.5 Jy in the four bands at 12, 25, 60 and 100μm. Most of the galaxies detected are late–type spirals with ellipticals and S0's being rare. Individual luminosities range from $10^6 - 10^{13}\,L_\odot$. There is a tight correlation between radio (extended emission) and far IR fluxes, but the most luminous radio sources and the most luminous IRAS sources are *not* the same objects.

The ratio of the far IR continuum to CO line emission (representing the mass of molecular hydrogen) can be interpreted as a measure of the "star forming activity". L_{fir}/M_{H_2} ranges from 4–10 for normal galaxies, 20–40 for starburst galaxies, to 100–200 for extreme cases like Arp 220 and NGC 6240. When detected, ellipticals have a lower colour temperature than spirals. Is this dust heated by the hot gas rather than by the UV radiation field?

In the study of active galaxies as well, there have been many significant results. The most powerful radio galaxies are also strong emitters in the far IR, although the emission mechanism not clearly understood. Morphological peculiarities point to the importance of tidal interactions and merger processes which presumably trigger large bursts of star formation and could also provide the means for dumping fuel on an AGN.

For BL Lacs and Blazars, the IRAS measurements fit smoothly into the overall energy distributions, which suggests that the dominant emission processes are nonthermal. Many new Seyfert galaxies have been found as a result of the IRAS survey and they appear to have "warmer" energy distributions than normal galaxies. There are probably both thermal and nonthermal emission mechanisms at work here, although disentangling them from the low spatial resolution data seems to be a difficult problem. Not many quasars were seen by IRAS. Smooth radio–IR–Optical energy distributions for the flat–spectrum quasars suggest nonthermal processes; these are more luminous than the steep–spectrum or radio quiet objects whose emission meccanisms are not so clear.

The high luminosity infrared galaxies comprise a substantial fraction of all galaxies with bolometric luminosities in excess of $\sim 3 \times 10^{10} L_{\odot}$. This is possibly the same population as the (non–Seyfert) Markarian galaxies over the luminosity range up to $\sim 3 \times 10^{11} L_{\odot}$. Star formation is the predominant luminosity source. The most luminous sources show evidence for interaction/merging and also, perhaps, for shrouded nuclear activity.

Since the characteristic depth for galaxies of the IRAS survey is about 200 Mpc, it contains significant cosmological information. The number counts of 60μm sources can, after careful removal of the effects of Galactic "cirrus", be used to map the local gravitational field on the assumption that the detected galaxies trace the mass distribution. There appears to be good agreement between the number count anisotropy and the dipole component of the microwave background.

These are just some of the highlights of recent extragalactic research from space; there are many others which I am only too aware of having omitted. What is the future likely to offer?

Future missions

There are many new missions in various stages of planning and completion forming the scientific programmes of the major space agencies and collaborators. These plans extend up to and beyond the end of the century and it is neither appropriate nor possible in this talk to describe all the proposals in any detail. What I do below is to list, for those missions with an extragalactic relevance and have progressed beyond the "concept" stage, a few salient points which give an indication of the scope of the instruments. There is a summary given in Table 1.

ESA, in particular, has constructed a coherent plan for developments in space science intended to extend for the next 15 to 20 years. This plan, called *Horizon 2000*, consists of a mixed programme of large ("cornerstone"), medium, and small (quick reaction) missions, some of whose components are already launched [GIOTTO] or well-developed and awaiting launch [ULYSSES, HIPPARCOS, HST (with NASA)].

GRASP — An ESA mission for what would be the first γ-ray observatory with both high spatial and spectral resolution. It uses Ge and Na I detectors with a coded aperture mask for positioning.

GRO — NASA's Gamma-Ray Observatory designed for high sensitivity observations in the 0.05–30,000 MeV region. There are four instruments: EGRET, a γ-ray telescope; COMPTEL, an imaging Compton telescope (in collaboration with the MPI); a set of four collimated "phoswich" scintillation detectors (OSSE); and a burst detector with a time resolution of better than one millisecond and high sensitivity (BATSE).

GAMMA-1 — A Soviet/French collaboration for a low-orbit spcecraft with a pointing accuracy of half a degree. The main instrument is a 50–1000 MeV time-of-flight spark chamber with a 30 degree FOV. In addition, there is a 2–20 keV X-ray telescope and a 0.2–20 MeV Na I shielded telescope with a modulated "anti-collimator."

GRANAT — This is a Soviet spacecraft for γ- and X-ray astronomy planned for an orbit with an apogee of 200,000 km, placing it beyond the Earth's magnetosphere for periods of three days out of four. It contains a telescope (SIGMA, a collaboration with Toulouse and Saclay) to image in hard X- and soft γ-rays; instruments for imaging, energy spectra, and timing in the 3–100 keV range; and several detectors to investigate the cosmic γ-ray bursts.

Spectrum-X-γ — This is another large Soviet spacecraft planned for an apogee of 200,000 km, and to have two large grazing incidence X-ray telescopes with a one degree field and a resolution of about two minutes of arc. It may also have two smaller telescopes, but with higher angular resolution.

ROSAT — This is a large X-ray and XUV telescope built in a collaboration between the FRG (MPI), the UK (SERC), and the USA (NASA), and designed for the first all-sky X-ray survey using an imaging system, making it two orders of magnitude more sensitive than previous surveys. It will also perform imaging and spectroscopy of individual sources.

Table 1: A list of future space missions of relevance to extragalactic astronomy. Included are those which have progressed beyond the "concept" stage and stand some chance of being launched.

Mission	Agency	Launch	Status	Purpose
GRASP	ESA	Ariane	Phase A	γ–ray
GRO	NASA	Shuttle	1990?	γ–ray
GAMMA-1	USSR/France	Proton		γ–ray
GRANAT	USSR/France	Proton	1988	γ– hard X–ray
SPECTRA-2	USSR/FRG?		1994	γ–ray
Spectrum–X–γ	USSR/DK/FRG		1992/3	X–ray
ROSAT	FRG/NASA/SERC	Delta-2	1990	X–ray/EUV
EUVE	NASA	Rocket	1990	EUV
XTE	NASA	Rocket	Mid 1990's	X–ray timing
SAX	I/NL/ESA	Atl/Cen	1992/3	X–ray
AXAF	NASA	Shuttle	Late 1990's	X–ray
XMM	ESA	Ariane	Cornerstone	X–ray Spect
ASTRO-D	Japan		1992/3	X–ray Spect
SPECTRA-1	USSR/Denmark?		1992	X–ray
HST	NASA/ESA	Shuttle	1989	Opt/UV
Hipparcos	ESA	Ariane	mid-89	Astrometry
ASTRO/HUT	NASA	Shuttle		UV–spect
Lyman	ESA+(Can/NASA)?	Ariane	Phase A	UV
ISO	ESA	Ariane	1993	IR
COBE	NASA	Delta-2		Cosmic b/g
SIRTF	NASA	Shuttle	late 1990's	IR/FIR
FIRST	ESA	Ariane	Cornerstone	FIR/sub–mm
Quasat	ESA+Can?	Ariane	Phase A	VLBI
Radioastron–cm	USSR		1992–6	VLBI
Radioastron–mm	USSR		1996–2000	VLBI
Radioastron–ss	USSR		2001–3	VLBI

EUVE — The extreme ultraviolet explorer from NASA is designed to spend six months on a sky survey in three bands spanning 80–900 Å, and another six months for follow–up spectroscopy over the 70–760 Å range.

XTE — NASA's X–ray Timing Explorer. This consists of an array of large area proportional counters (LAPC), the high energy timing experiment (HEXTE), and a scanning shadow camera (SSC) for performing all–sky X–ray monitoring.

SAX — An Italian mission with collaboration from the Netherlands and ESA. It includes four imaging telescopes or concentrators with a resolution of about one minute of arc, equipped with sealed imaging gas proportional counters covering in total the energy range 0.15–10 keV.

AXAF — NASA's advanced X–ray astrophysics facility. This is a large observatory satellite with a telescope for imaging in the band 0.1–8.5 keV over a one degree field at an angular resolution of half an arcsecond. Five focal plane instruments have been studied including cameras, spectrometers and an X–ray calorimeter.

XMM — An ESA "cornerstone" for high spectral resolution studies of X–ray sources using a multimirror telescope with the largest possible collecting area in the band 0.2–10 keV. The highest resolution spectroscopy ($\lambda/\Delta\lambda \sim 1000$) will be concentrated around the oxygen and iron K lines. It is planned as an observatory satellite with a lifetime of about 10 years.

ASTRO–D — A Japanese X–ray satellite, possibly in collaboration with NASA, planned as a successor to *Ginga*. The instrument complement is still under study, but the prime purpose will be spectroscopy.

HST — The Hubble Space Telescope is a joint NASA–ESA mission for a 2.4–m aperture, diffraction limited, optical–ultraviolet telescope in low–Earth orbit. It's initial complement is six instruments for imaging, spectroscopy, high–speed photometry, and astrometry. To be launched by the Shuttle, it is possible to revisit it in orbit and to return it to the ground for refurbishment.

Hipparcos — This ESA mission is designed to obtain precise (2 milliarcsecond) positions, parallaxes, and annual proper motions for some 100,000 stars brighter than magnitude 13. The lifetime is two and a half years in a geostationary orbit.

ASTRO/HUT — The Hopkins Ultraviolet Telescope destined for a Shuttle launch. This is a one–metre telescope with a Rowland circle spectrograph with iridium and osmium coating respectively for good FUV and EUV performance.

Lyman — The primary objective of the Lyman mission is high–resolution spectroscopy with an 80–cm glancing incidence telescope in the 900–1200 Å region, with subsidiary aims for the FUV (1200–2000 Å) and EUV (100–300 Å). As a collaboration between ESA (and possibly Canada) and NASA it would be put into a 120,000 × 1000–km 48 hour orbit.

ISO — The ESA Infrared Space Observatory is the logical successor to IRAS and will have a cooled 60–cm telescope with instruments for imaging, photo–polarimetry, and spectroscopy in the band from 3–200 μm. It is planned for a 24–hour orbit, and the 2000 litres of liquid helium should give it an orbital lifetime of at least 18 months.

COBE — NASA's Cosmic Background Explorer will carry three instruments into orbit and will map the entire sky twice in its one–year lifetime. The differential microwave radiometer will investigate the microwave background isotropy at wavelengths of 3.3, 5.7, and 9.6 millimetres. The FIR absolute spectrophotometer is designed to map the spectrum of the microwave background, and the diffuse infrared background experiment is designed to search for the background in the band 1–300 μm.

SIRTF — NASA's Space Infrared Telescope Facility. The idea is for a one–metre class telescope producing diffraction limited images over the band 2–700 μm from a 900–km orbit. Planned focal plane instruments are an infrared array camera, a multiband imaging photometer system, and an infrared spectrometer.

FIRST — An ESA "cornerstone" designed to work in the FIR and sub–mm (50 μm — 1 mm) regions using heterodyne detectors. An eight–metre aperture dish is proposed.

Quasat — An ESA medium–size mission under Phase A study, and aimed at extending the worldwide VLBI network to obtain very high resolution images of compact sources. The antenna would be ten metres in aperture, and the apogee possibly 12,500–km.

Radioastron — This is a series of three missions planned by the Soviet Union for VLBI from space. The first two — cm and mm — will be on board spacecraft of the SPECTRUM series in orbits with an apogee of 77,000–km. The third — S–S — will consist of three 30–m dishes.

The information for this talk was gleaned from many articles and conversations. I thank all those who helped, but make no attempt at individual acknowledgments. The errors are mine.

Discussion

J. L. Masnou: I suggest you add to the list of space experiments in the γ-ray range the SIGMA experiment, a France/USSR collaboration. It is a high angular resolution imaging experiment, working from 30 keV up to 2 meV. It will be launched in December 1988 by a Russian launcher.

Radial Distribution of Radio Emitting Galaxies in Clusters

L. Feretti[1,2], G. Giovannini[1,2], and T. Venturi[1]

[1]Dipartimento di Astronomia, Bologna, Italy
[2]Istituto di Radioastronomia, Bologna, Italy

The radio properties of galaxies have been studied in the past by many authors, by means of radio observations of large, complete samples of objects (Auriemma *et al.* 1977, Hummel 1980). It is now well-established that the probability of having radio emission from a galaxy increases with its optical magnitude, for any optical morphological type. The difference of the radio behaviour between isolated and cluster galaxies has also been investigated: while the probability of radio emission in elliptical galaxies seems to be rather independent of the environment, for the spiral galaxies, discordant results have been found (Fanti *et al.* 1982, and references therein; Gavazzi and Jaffe 1986).

Here we investigate the dependence of radio emissivity of cluster galaxies on their position in the cluster. For this study, radio information on galaxies up to a large distance from the cluster center are needed. We use radio data on the clusters Coma and A 262, obtained with the Westerbork Synthesis Radio Telescope at 90–cm. The large field of view of the instrument at this wavelength allowed us to properly map the two clusters up to ~1.5–2 Abell radii.

Optical data on the galaxy morphology were taken from the literature (Kent and Gunn 1982, Righetti *et al.* 1988; and references therein). We then derived the radial profile of the number of galaxies per square degree as a function of the radial distance from the cluster center. The two clusters are quite different in morphological content: Coma shows a prevalence of ellipticals and lenticulars in the center (within ~30 arcmin) and no significant morphological segregation at the periphery, while A 262 is dominated in the whole field by spiral galaxies.

In order to have a homogeneous sample, we consider only galaxies with absolute optical photographic magnitude brighter than −18.5, out to 1.5 Abell radii from the cluster center. In this way, the number of surveyed galaxies is as follows:

	Coma	A 262
Ellipticals	62 (9)	12 (4)
Spirals	57 (15)	48 (5)
S0's	66 (1)	19 (0)

where in parentheses the number of detected galaxies is given.

For the whole sample of ellipticals and spirals, we computed the detection rate of objects lying in concentric rings of 0.5 Abell radius around the cluster center. The results are given in the following table:

Ellipticals

	Total	$-20 < M \leq -18.5$	$M \leq -20$
$R_A \leq 0.5$	7/41 (17%)	3/34 (9%)	4/7 (57%)
$0.5 < R_A \leq 1$	4/20 (20%)	1/15 (7%)	3/5 (60%)
$1 < R_A \leq 1.5$	2/13 (15%)	1/9 (11%)	1/4 (25%)

Spirals

	Total	$-20 < M \leq -18.5$	$M \leq -20$
$R_A \leq 0.5$	12/33 (36%)	8/27 (30%)	4/6 (67%)
$0.5 < R_A \leq 1$	5/45 (11%)	4/38 (11%)	1/7 (14%)
$1 < R_A \leq 1.5$	3/27 (11%)	1/23 (4%)	2/4 (50%)

For the elliptical galaxies, there is no indication that the probability of radio emission depends on the position within the cluster. For the spirals, instead, even allowing for the large errors, there is some evidence of a higher number of galaxies with radio emission near the cluster center, especially in the range of lower optical magnitudes.

In the table, we have simply given the detection rate, without taking into account the radio power of the detected object. To overcome this limitation, we have computed the bivariate radio luminosity function in the same magnitude intervals and ranges of distance from the cluster center. We have obtained a trend in full agreement with the result obtained above for the detection rate.

The radio segregation found here points at some strong effect of the cluster environment on the radio emissivity of spirals. In order to better understand the origin and nature of the effect it would be important to study separately the nuclear and disk emissivity in the cluster spirals. Unfortunately, this is not possible with the present radio data because of their insufficient angular resolution (~1 arcmin).

References

Auriemma, C., Perola, G. C., Ekers, R., Fanti, R., Lari, C., Jaffe, W. J., and Ulrich, M. H. 1977, *Astron. Astrophys.* **57**, 41.

Fanti, C., Fanti, R., Feretti, L., Ficarra, A., Gioia, I. M., Giovannini, G., Gregorini, L., Mantovani, F., Marano, B., Padrielli, L., Parma, P., Tomasi, P., and Vettolani, G. 1982, *Astron. Astrophys.* **105**, 200.

Gavazzi, G. and Jaffe, W. 1986, *Astrophys. J.* **310**, 53.

Hummel, E. 1980, Ph. D. Thesis, Groningen.

Kent, S. M. and Gunn, J. E. 1982, *Astron. J.* **87**, 945.

Righetti, G., Giovannini, G., and Feretti, L. 1988, *Astron. Astrophys. Suppl.* **73**, 173.

An Unusual Red Envelope Galaxy in an X–ray Selected Cluster

Isabella M. Gioia[1,2], B. Garilli[3], T. Maccacaro[1,2], D. Maccagni[3], G. Vettolani[2], and A. Wolter[1]

[1]Harvard–Smithsonian Center for Astrophysics, Cambridge, Mass., USA
[2]Istituto di Radioastronomia del CNR, Bologna, Italy
[3]Istituto di Fisica Cosmica del CNR, Milano, Italy

Summary

In the process of identifying X–ray sources from the Einstein Observatory Extended Medium Sensitivity Survey (Gioia *et al.* 1987), we have observed the field of 1E 1111.9−3754. A photometric and spectroscopic study of the optical counterpart of the X–ray source 1E 1111.9−3754 has revealed a poor cluster of galaxies dominated by a very luminous giant elliptical. The galaxy sits at the bottom of the potential well described by the X–ray emitting gas. This cluster is similar to the Albert/Morgan groups studied in the X–rays by Kriss *et al.* (1983). The presence of such a bright and large galaxy in this cluster can be explained by invoking cannibalism and stripping in this specific environment which may be at the origin of the large, red envelope detected in the Gunn i and r filters. We have given the name GREG (Giant Red Envelope Galaxy) to the brightest cluster member of 1E 1111.9−3754. A Hubble constant of 50 km s^{-1} Mpc^{-1} and $q_0 = 0$ have been used, implying a scale of 3.17 kpc arcsec^{-1} at GREG's redshift.

Observations

The X–ray source 1E 1111.9−3754 was found by analyzing the IPC field 4923 which has an exposure of 1600 sec. The source has only ∼100 net counts and

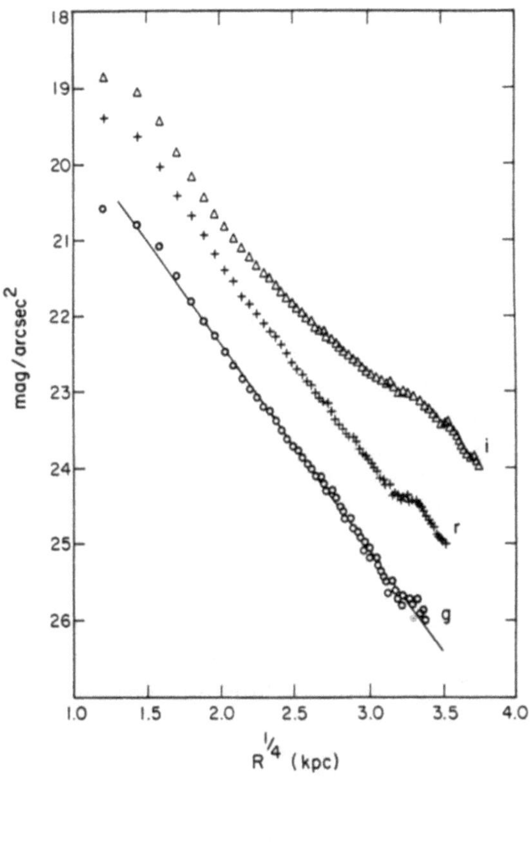

Figure 1: Surface Brightness Profiles of GREG.

appears elongated in the NE–SW direction. The assumption of a Raymond thermal spectrum with a temperature of ~4 keV combined with the hydrogen column density in the direction of 1E 1111.9–3754 ($N_H = 9 \times 10^{20}\,\mathrm{cm}^{-2}$) allows us to obtain a flux of $5.1 \times 10^{-12}\,\mathrm{erg\,cm^{-2}\,sec^{-1}}$ (0.3–3.5) keV. Spectra of 9 galaxies in the cluster were obtained with the ESO 3.6-m telescope with EFOSC in the MOS mode. The radial velocity of the cluster is $39340 \pm 153\,\mathrm{km\,sec^{-1}}$ with a spatial velocity dispersion of $1226\,(+297/-165)\,\mathrm{km\,sec^{-1}}$. GREG's spectrum clearly shows Ca II H and K, the G band, Hβ, Mg I and Na D absorption at z = 0.1312. Three pairs of 8 min images were taken in the Gunn g, r, and i filters. The surface brightness profiles of GREG are shown in Figure 1, where magnitudes arcsec^{-2} are plotted against $r^{1/4}$ (kpc). While in the g filter, the surface brightness profile can be represented by a de Vaucouleurs law out to a radius of about 150 kpc, at about 100 kpc radius in the *r* filter profile, and even from about 80 kpc radius in the *i* filter, an excess surface brightness with respect to de Vaucouleurs's law is clearly visible.

We have searched the IRAS data base for possible infrared emission at GREG's position. Only upper limits were derived of 100 mJy at 12μm, 200 mJy at 25μm, and 180 mJy at 60μm.

Results and Discussion

From the X–ray data, we derive a luminosity for the cluster of 3.6×10^{44} erg sec^{-1} in 0.3–3.5 keV. On the ESO/SRC J plates, the cluster appears as a rather poor cluster with a richness between 0 and −1 (within 1 Abell radius) and with a central galaxy density $N_0 \sim 10$ galaxies in 0.5 Mpc. GREG has a very high absolute magnitude $M_V = -25.23$, comparable to that of cD's in rich clusters, and an extremely large and red halo.

From the velocity dispersion, we expect N_0 a factor of 5 larger (Bahcall 1981). From the measured value of N_0, we expect X–ray luminosity a factor of 10 smaller (Bahcall 1980). Given the agreement between the X–ray luminosity and the velocity dispersion, (see Melnick and Quintana 1982), we should have a much higher central galaxy density. The low value of N_0 may indicate that dynamical evolution has abnormally advanced in this poor cluster and that the "missing galaxies" are to be looked for in the giant structure of GREG. Its extreme luminosity can be tied to the high velocity dispersion of the cluster, since the total accretion rate onto the central region is about twice that in poor clusters.

Numerical simulations (Carnevali *et al.* 1981) find that in poor compact groups extensive merging and the consequent formation of a central giant object is enhanced. The mechanism of cannibalism seems to work differently in poor and rich clusters. While in the latter, cannibalism should take place after cluster relaxation, with cannibalism of the low velocity dispersion objects, in poor clusters, cannibalism and formation of the central objects should have proceeded at the time of cluster collapse.

How did the envelope form? The X–ray luminosity testifies to the depth of the potential well and consequently of GREG's large mass. Its large envelope could be made up of the remnants of galaxy stripping sitting at the bottom of the potential well. The infrared excess could be produced by an intermediate age population of stars where cool AGB stars contribute a large fraction of the total luminosity (Renzini and Buzzoni 1986). Alternatively, a peculiar initial mass function could be invoked with a large excess of very low mass stars.

We plan to study further the history of this cluster by obtaining velocities of more galaxies, photometry of enough cluster members to derive a luminosity function, measurements of the velocity dispersion of GREG, and its *H* and *K* colors. A detailed discussion of GREG and its environment can be found in

Maccagni *et al.* (1988).

I. M. Gioia acknowledges financial support from the Smithsonian Institution under the "Research Opportunities Fund" program. Partial financial support for this work has come from the Smithsonian Scholarly Studies Program Grant SS88–3–87 and from NASA contract NAS8–30751.

References

Bahcall, N. A. 1980, *Astrophys. J. Lett.* **238**, L117.
Bahcall, N. A. 1981, *Astrophys. J.* **247**, 787.
Carnevali, P., Cavaliere, A., and Santangelo, P. 1981, *Astrophys. J.* **249**, 449.
Gioia, I. M., Maccacaro, T., and Wolter, A. 1987, in *Observational Cosmology*, ed. A. Hewitt, G. Burbidge, and L. Z. Fang (Dordrecht: Reidel), p. 593.
Kriss, G. A., Cioffi, D. F., and Canizares, C. R. 1983, *Astrophys. J.* **272**, 439.
Maccagni, D., Garilli, B., Gioia, I. M., Maccacaro, T., Vettolani, G., and Wolter, A. 1988, *Astrophys. J. Lett.*, submitted.
Melnick, J. and Quintana, H. 1982, *Astron. J.* **87**, 972.
Renzini, A. and Buzzoni, A. 1986, in *Spectral Evolution of Galaxies*, ed. C. Chiosi and A. Renzini (Dordrecht: Reidel), p. 195.

Blue Compact Dwarf Galaxies in the W Cloud of Virgo

George Helou[1], G. Lyle Hoffman[2], and E. E. Salpeter[3]

[1]IPAC, California Institute of Technology, Pasadena, California, U.S.A.
[2]Lafayette College, Easton, Pennsylvania, U.S.A.
[3]Cornell University, Ithaca, New York, U.S.A.

As part of our study of Blue Compact Dwarf (BCD) galaxies in the Virgo Cluster (Hoffman *et al.* 1988), we have examined how location in the Cluster affects the properties of these objects. For a source list, we used the Virgo Cluster Catalog by Binggeli, Sandage, and Tammann (1987), which contains 64 objects classified as "BCD" with varying degrees of uncertainty. A neutral hydrogen survey at Arecibo and a subsequent spectroscopic survey with the Palomar 5-meter reflector (in addition to other sources), have yielded 35 detections at heliocentric redshift $V_\odot < 3000\,\mathrm{km\,s^{-1}}$, and 26 in the background (Hoffman *et al.* 1987, Helou and Schombert 1988, Schulte-Ladbeck and Cardelli 1987), leaving only three objects with unknown redshifts.

The BCDs have an extremely patchy distribution in position-velocity space, as illustrated in Table 1. Three broad classes of objects are considered in the Table, all restricted to $12^h10^m <$ R.A. $< 12^h25^m$: "large spirals" meaning those found in the Revised Shapley-Ames Catalog (1981); BCD galaxies based on the morphological classification in Binggeli, Sandage, and Tammann (1987); and irregulars, also from classifications in the latter catalog. It is clear that the mix of these classes, and in particular the ratio of BCD/Irr varies radically with position in this space. The BCDs lie mainly in two clumps: a low $V_\odot$ clump within the central core, most likely the same physical association as the low velocity spirals thought to be falling through the Cluster core from behind; and a diffuse group centered on the W Cloud (de Vaucouleurs and Corwin 1986), behind and to the South, of the Cluster core. The Tully-Fisher relation for the latter group is consistent with this group being 1.7 times further away from us than the cluster core.

IRAS survey data have been co-added for all 64 BCDs (Helou *et al.* 1988). The detection rate is 6/9 for BCDs in the W Cloud, but only 7/26 for the remainder of the cluster members. In addition, the W Cloud BCDs are warmer in their 60μm/100μm colors the than the other BCDs, and warmer than spirals (see Figure 1). Thus, not only do BCDs occur preferentially in the W Cloud, but those found there are forming stars more actively than the others, judging by their enhanced far infrared luminosity.

We conjecture that BCDs "favor" regions such as the W Cloud at the present epoch, while other regions may contain only about one BCD for every five irregular galaxies, which is the average for the cluster as a whole. One reason may be that the present epoch is a "dynamically special" one for these clouds (Silk, Wyse, and Shields 1987), in the sense that they are on the point of falling into the cluster core. Singular dynamical epochs on the scale of clouds of galaxies may favor star formation activity in the irregulars in these clouds (evidently, spirals are not affected), possibly in a fashion similar to star bursts occuring in interacting pairs of galaxies.

This work was supported in part by NASA through the IRAS Extended Mission at the Infrared Processing and Analysis Center, JPL, Caltech.

Table 1. The position–redshift distribution of Virgo galaxies with $12^h10^m <$ R.A. $< 12^h\ 25^m$

Class	Dec	$V_\odot/100\,\mathrm{km\,s^{-1}}$ < 15	15–22	22–30	Dec	$V_\odot/100\,\mathrm{km\,s^{-1}}$ < 15	15–22	22–30
RSA	< 7°	1	7	5	7–14°	13	2	4
BCD	< 7°	2	9	0	7–14°	4	0	0
Irr	< 7°	4	8	2	7–14°	18	7	3

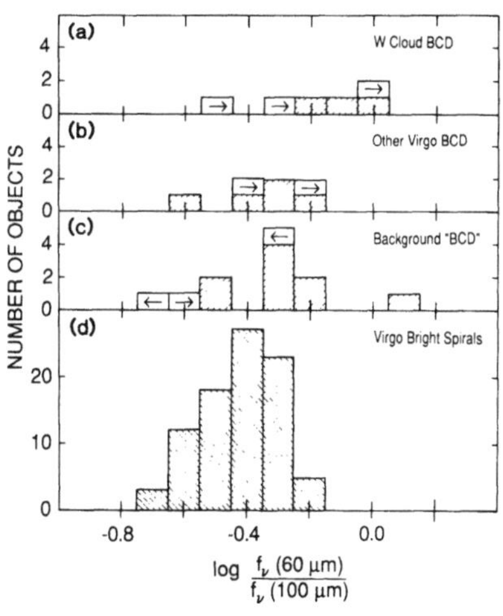

Figure 1. Histograms of the 60μm/100μm color ratios for several samples of galaxies. The top frame (a) is for BCDs in the W cloud, defined here by dec < 7°, R.A. < 12^h25^m, and 1500 < $V_\odot < 3000\,\mathrm{km\,s^{-1}}$. All other BCDs in Virgo ($V_\odot < 3000\,\mathrm{km\,s^{-1}}$) are shown in frame (b). Frame (c) contains data for all objects on our original source list whose redshifts place them beyond the Virgo Cluster. Frame (d) (lowest) contains data for spirals (type Sb and later) brighter than $B_0^T = 12.8$, or larger than $D_{25} = 2'\cdot 4$.

References

Binggeli, B., Sandage, A., and Tammann, G. A. 1987, *Astron. J.* **94**, 251.

Helou, G., Khan, I., Malek, L., and Boehmer, L. 1988, *Astrophys. J. Suppl.* **68**, in press.

Helou, G. and Schombert, J. 1988, in preparation.

Hoffman, G. L., Helou, G., Salpeter, E. E., Glosson, J., and Sandage, A. 1987, *Astrophys. J. Suppl.* **63**, 247.

Hoffman, G. L., Helou, G., Salpeter, E. E., and Lewis, B. M. 1988, *Astrophys. J.*, submitted.

Sandage, A. and Tammann, G. A. 1981, *A Revised Shapley-Ames Catalog of Bright Galaxies* (Washington, D.C.: Carnegie Institution of Washington).

Schulte–Ladbeck, R. E. and Cardelli, J. A. 1987, *Bul. Am. Astron. Soc.* **19**, 1061.

Silk, J., Wyse, R. F. G., and Shields, G. A. 1987, *Astrophys. J. Lett.* **322**, L59.

de Vaucouleurs, G. and Corwin, H. G. 1986, *Astron. J.* **92**, 722.

Intergalactic Dust as Seen in Emission

Bogdan Wszolek[1], Konrad Rudnicki[1], Paolo de Barnardis[2], and Silvia Masi[2]

[1]Jagiellonian University Observatory, Cracow, Poland
[2]Istituto di Fisica, Università degli Studii, Rome, Italy

Gérard and Antoinette de Vaucouleurs together with H. G. Corwin (1972), as well as Takase (1972), put forth the problem of intergalactic extinction within our Supergalaxy (sometimes called the Local Supercluster), and thus the problem of the existence of significant amounts of intergalactic dust in large agglomerations of galaxies. The problem is by no means an easy one. The very method of reducing photometric measurements can directly influence the results (see Guła *et al.* 1975 and Rudnicki 1986). On the other hand, it is not possible to correctly reduce the measurements without knowing intergalactic extinction parameters. This indeed seems to be a kind of vicious circle. Even if one does not consider it to be so, any data on intergalactic dust obtained by means other than investigation of extinction due to the dust, are of substantial importance for the problem.

We attempted to find whether the most distinct agglomerations of such dust, namely the four individual intergalactic obscuring clouds presently known (Rudnicki 1986) radiate any observable amounts of energy. We searched the IRAS data in its four bands. We found significant infrared radiation only for the largest (in the sense of angular size) cloud, the Okroy Cloud (Murawski 1983). The cloud is visible in IRAS data only in the 100 μm band (Figure 1), while it is not seen in 60, 25, and 12 μm. If we consider that the physical conditions of the dust are similar to those of interstellar dust, the upper limit of the cloud's temperature can be estimated as 23 K, which is somewhat lower than for the analogous interstellar clouds.

The coincidence of (1) the coordinates of the centre, (2) the shape, and (3) the size (of the central region), practically excludes the possibility of incidental superposition of the Okroy Cloud with some interstellar infrared feature within our Galaxy.

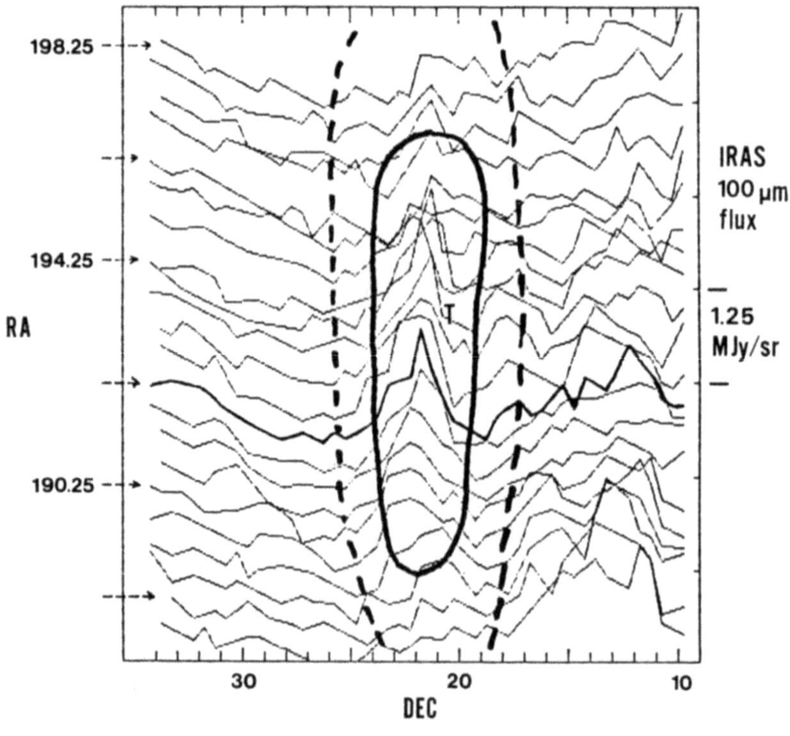

Figure 1: Emission after linear subtraction of the background, for $\lambda = 100\mu m$ (IRAS) in the region of the Okroy Cloud. The labeled contours (thin lines) correspond to right ascension (in degrees of arc). The continuous outline is an approximate boundary of the central region of the Okroy Cloud, within which the average number of galaxies per square degree is less than 1. The dashed line encloses the region of density lower than 4 galaxies per square degree. The thick line indicates the profile passing through the Cloud's centre, as determined from the galaxy counts. The coordinates of the centre of the Okroy Cloud, as determined from the galaxy counts, are $\alpha = 192.5°$, $\delta = 22°$. They are in good coincidence with the coordinates of the maximum of the $100\mu m$ emission.

At present, we know only that the Okroy Cloud lies some distance beyond the galactic halo (the stellar spherical component of the Galaxy), but the actual distance may be as large as 100 Mpc (Murawski 1983). It may be located far away in a void between the extragalactic filamentary structure, or it may be simply a dust satellite of our Galaxy.

The very difficult problem of determining the distance of the cloud, as well as the (perhaps easier) one of determining its actual temperature, should be handled before it can be stated whether the discovery of infrared radiation from the intergalactic cloud has any direct or only indirect significance for the problem of intergalactic dust in the supergalactic structure. In either case, we hope that it opens up new possibilities for approaching the problem initiated sixteen years ago by Gérard and Antoinette de Vaucouleurs and their co–workers.

Computations were made in Cracow on two IBM PC XT computers — the first on loan from Princeton University, and the second bought from funds provided by RPBR-I-11/i-05.

References

de Vaucouleurs, G., de Vaucouleurs, A., and Corwin, H. G. 1972, *Astron. J.* **77**, 285.

Guła, R., Rudnicki, K., and Tarraro, I. 1975, *Acta Cosmol.* **3**, 39.

Murawski, W. 1983, *Acta Cosmol.* **12**, 7.

Rudnicki, K. 1986, *Proc. Intern. School of Phys. "Enrico Fermi"* **86**, 480.

Takase, B. 1972, *Publ. Astron. Soc. Japan* **24**, 295.

Spectral Evolution of Galaxies

B. Rocca–Volmerange and B. Guiderdoni

Institut d'Astrophysique, Paris, France

We present a new model of spectrophotometric evolution for galaxies which allows us to compute evolving synthetic spectra with few assumptions about the star formation history of the galaxy. These synthetic spectra can be directly compared with observational spectra, or with apparent magnitudes and colors after convolution with the filter response curves. By coupling with Friedmann–Lemaître cosmological models, the model also gives photometric predictions for high-redshift galaxies taking into account the entangled effects of cosmology and evolution.

This model simulates the spectral evolution of a galaxy with time. At each time step, stars form from the gaseous component with the classical parameters (star formation rate and initial mass function). These stars are positioned on the ZAMS in the HR diagram. Stars evolve in the HR diagram along theoretical stellar tracks which are an important input. The convolution of the distribution of the stellar population with a library of stellar spectra spanning most of current spectral types from O to M stars, allows us to compute the synthetic spectrum of the galaxy at each time step. The principles of the model are presented in Rocca–Volmerange *et al.* (1981), and Guiderdoni and Rocca-Volmerange (1987). The atlas of evolving synthetic spectra is given in Rocca-Volmerange and Guiderdoni (1988), and is also available on request on magnetic tapes. Apparent magnitudes and colors for high-redshift galaxies, computed from this atlas, are proposed in Guiderdoni and Rocca–Volmerange (1988).

This model has a number of important improvements with respect to previous work (*e.g.* Bruzual 1981). (i) The theoretical stellar tracks include the last stages of stellar evolution (horizontal branch, asymptotic giant branch). (ii) The library of observational spectra has been obtained by connecting the UV spectra (1200 Å to 3710 Å) from the IUE atlas (Wu *et al.* 1983) to data in the catalogue by Gunn and Stryker (1983) (3170 Å to 10000 Å). (iii) The nebular emission (following Stasińska 1985) and an estimate of the internal extinction are taken into account. The resulting spectra give good fits of the observed UV and visible spectra of spiral and elliptical galaxies (see the other papers of this series cited herein).

Several papers describing applications of the model are currently in press or in preparation. The star formation in nearby and high–redshift elliptical galaxies is studied from the far–UV emission and the Lyα nebular line, in correlation with cooling flows or gas accretion (Rocca–Volmerange, 1988). These results are extended to the various cases of Lyα galaxies (Valls–Gabaud *et al.* 1988). The model is used to make predictions of faint galaxy counts (Guiderdoni and Rocca–Volmerange 1988). Finally, *BRK* colors of optical counterparts of high–redshift radio galaxies from the Parkes Selected Region sample are analyzed in terms of old stellar population and star formation (Dunlop *et al.* 1988).

References

Bruzual, G. 1981, Ph.D. Thesis, University of California, Berkeley.

Dunlop, J., Longair, M., Guiderdoni, B., and Rocca–Volmerange, B. 1988, in preparation.

Guiderdoni, B. and Rocca–Volmerange, B. 1987, *Astron. Astrophys.* **186**, 1.

Guiderdoni, B. and Rocca–Volmerange, B. 1988a, *Astron. Astrophys. Suppl.*, in press.

Guiderdoni, B. and Rocca–Volmerange, B. 1988b, in preparation.

Gunn, J. E. and Stryker, L. L. 1983, *Astrophys. J. Suppl.* **52**, 121.

Rocca–Volmerange, B., Lequeux, J., and Maucherat–Joubert, M. 1981, *Astron. Astrophys.* **104**, 177.

Rocca–Volmerange, B. and Guiderdoni, B. 1988, *Astron. Astrophys. Suppl.*, in press.

Rocca–Volmerange, B. 1988, *Mon. Not. Roy. Astron. Soc.*, in press.

Stasińska, G. 1985, private communication.

Valls–Gabaud, D., Rocca–Volmerange, B., and Guiderdoni, B. 1988, in preparation.

Wu, C.–C., Ake, T. B., Boggess, A., Bohlin, R. C., Imhoff, C. L., Holm, A. V., Levay, Z. G., Panek, R. J., Schiffer, F. H., and Turnrose, B. E. 1983, *IUE Ultraviolet Spectral Atlas*, NASA No. **22**.

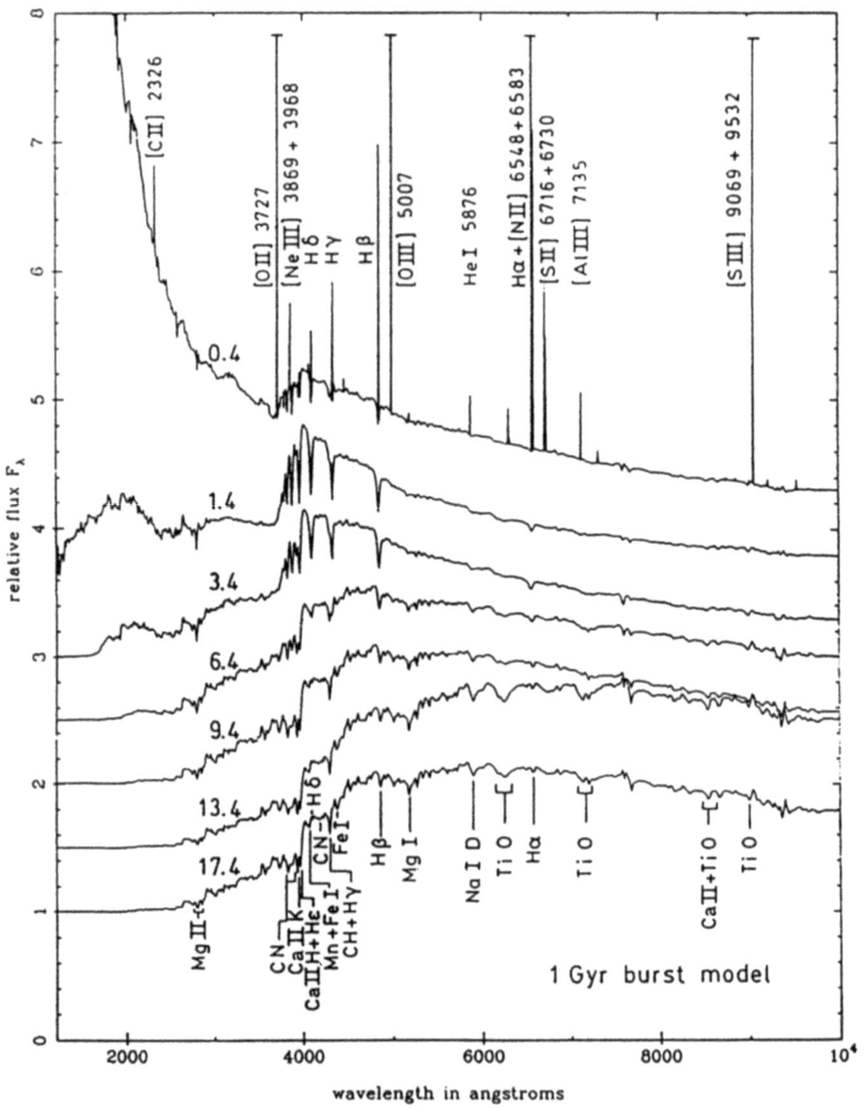

Figure 1: Example of evolution of a 1 Gyr burst model for a late-type galaxy

Distance Relationships

G. Paturel

Observatoire de Lyon, Saint-Genis Laval, France

Summary

From very general relationships between observable parameters, we present a method which uses all available information for estimating extragalactic distances.

Introduction

When reading de Vaucouleurs's papers devoted to the problem of the extragalactic distance scale, we see that he tries to use all available information to solve the problem. Is it possible to do this in a more systematic way? Is it possible to solve the problem using a single step method? Our aim in this paper is to answer these questions by discussing the general relationship between parameters attached to a galaxy, with the ambitious project of finding a generalized criterion for distance determination.

Preliminary Relationships

First, let us describe the data and tools we have for determining the distance of a given galaxy. Five properties are needed.

The most general data available on a galaxy is its brightness distribution. This distribution can be written as a very complicated function:

$$B = \mu(\alpha, \delta, \lambda, t) \tag{1}$$

This is the brightness at a location α, δ for the wavelength λ at a time t. All information is contained in this distribution. Very often the brightness can be considered as independent of time.

From this apparent brightness distribution it is possible to derive *apparent observable parameters*, p_i. From a mathematical point of view, such parameters are simply obtained by applying operator T_i on the function B.

$$p_i = T_i(B) \tag{2}$$

An example will make this clearer. Let an operator T be defined:

$$T = \int_A \int_{-\infty}^{+\infty} f(\lambda)\, d\lambda\, dA$$

where f is a function defining a bandpass and A is a circular aperture. T applied on B leads to an apparent observable parameter p, which is the apparent magnitude deduced from the aperture photometry. This can be written as

$$m(A) = T(B)$$

Apparent observable parameters p are related to *intrinsic observable parameters*, P, through the very simple relation

$$P = p \cdot r^n \tag{3}$$

where r is the distance of the galaxy, and n is an integer which can take only three values 0, 1, or 2 (in a non-relativistic domain). This amazing property is probably related to the fact that we are in a three–dimensional space. We will assume that corrections for other non–intrinsic effects (galactic extinction or inclination) have been applied.

Nature's laws and our confidence that they apply to remote galaxies suggest that a model reproducing the brightness distribution of a galaxy can be built on a basis of *physical parameters*, X_i (mass, angular momentum, age, etc.). This is written as

$$B' = \mu'(\alpha, \delta, \lambda) = F(X_i), \quad i = 1, N \tag{4}$$

where B' is the brightness distribution predicted by the model, using N physical parameters X_i. Obviously, such a model is still out of reach, and probably will be for a long time.

Unfortunately, the physical parameters X_i are not directly observable. The only hope we have is that they can be deduced from intrinsic observable parameters. Of course, we cannot prove that it is true. But, if it is not, extragalactic distances are out of reach. This last condition is written as

$$X_i = f_i(P_j),\ j = 1, M\ (M \geq N)\ for\ any\ i \tag{5}$$

The Σ' Surface

Using the five preliminary relations (1) to (5), it is possible to demonstrate the existence of a very general equation between intrinsic observable parameters P.

Using equations (5) and (4), we deduce that the predicted brightness distribution is a function of P

$$B' = G(P_j),\ j = 1, M \tag{6}$$

Then, an operator T_k, k being a member of $(1, M)$, is applied on B'. For a conventional distance $r = 1$ (the B' distribution can be calculated for any distance) we obtain from (2)

$$T_k(B') = P_k,\ k\ member\ of\ (1, M) \tag{7}$$

Finally, the following relation is obtained by combining equations (6) and (7):

$$P_k = H(P_j),\ j = 1, M,\ k\ member\ of\ (1, M)$$

This equation is the equation of a surface. It will be called Σ' because it has the same meaning as the Barbier-Chalonge-Divan Σ surface (1952). The final equation of Σ' will be written in the form:

$$F(P_j) = constant,\ j = 1, M \tag{8}$$

Some comments have to be made. A galaxy will have a representative point on the Σ' surface if the basis constituted by the P_j parameters is complete (in the mathematical sense); the better the basis, the closer the point. Of course, it is impossible to be sure that this completeness is fullfilled, thus, the thickness of the cloud of representative points will depend on the completeness of the basis. On the other hand, we hope to build the best basis by taking all known parameters P. However, these parameters must be independent to constitute a mathematical basis.

First Use of the Σ'–Surface

Let us assume that the equation of Σ' is linear in log scale. This is

$$\textstyle\sum_{i=1}^{M} q_i \cdot \log P_i = constant$$

where q_i are constants. Using equation (3) gives

$$\log r = \textstyle\sum_{i=1}^{M} -q_i \log p_i / \sum_{i=1}^{M} n_i q_i + constant$$

This equation shows that the distance can be calculated just by using apparent observable parameters. From this equation, it is possible to retrieve almost all distance criteria. Let us give two examples: with $M = 2$, $p_1 = D$ (apparent isophotal diameter) and $p_2 = m$ (apparent total magnitude), we retrieve exactly the diameter–luminosity relation (Heidmann, 1969); with $M = 2$, $p_1 = W$ (21–cm line width) and $p_2 = m$ (apparent total magnitude), we retrieve the Tully-Fisher (1977) relation.

The only condition which is required is that $\Sigma n_i q_i$ differs from zero. When the chosen parameters are not completely independent (in other words, when they do not constitute a good basis), $\Sigma n_i q_i$ is close to zero and the corresponding distance criterion is poor.

Generalized Method

When using all available parameters, the basis is as good as possible. However, this basis has first to be transformed in such a way that all parameters are distance–independent except one. This is done without loss of information simply by linear combination (see Paturel 1981 and Bottinelli *et al.* 1983). Hereafter, the parameter which is distance–dependent will be noted P', while the $M-1$ distance–independent remaining parameters will be noted $P_i^o (i=1, M-1)$.

With this transformation the Σ'–surface is represented by:

$$P' = F'(P_j^o),\ j = 1, M-1 \tag{9}$$

Because the intrinsic observable parameters, P_j^o, do not differ from the apparent observable parameters ($n_j = 0$), the distance can easily be written from Equations (9) and (3). Two ways of doing this will be given here.

"Sosie" (look alike) galaxies

If one selects all galaxies having the same P_j^o parameters as a calibrating galaxy, it can be seen from Equation (9) that they have the same P' parameter as the calibrator (say, the same absolute magnitude). Thus, the distance can be easily found. This method does not require any mathematical representation of the Σ'–surface because it just utilizes a small region of Σ' around a calibrator.

Regression on Principal Components

Using Principal Component Analysis (PCA), it is possible to transform the $M-1$ parameters P_j^o into only m parameters U_i ($i = 1, m;\ m < M-1$). The new basis (U_i) carries all significant information and is a true basis, because by definition the components are independent. P' can thus be represented by a general regression on U_i. For instance, P' could be written as a linear representation:

$$P' = \sum_{i=1}^{m} c_i \cdot U_i + constant.$$

We are not sure that such a representation is perfectly correct (because of Equation (5)), but it is the best that we can do.

Conclusion

We have devised a general method which uses all available information for estimating extragalactic distances. The best indication of the success of this method is probably that two of its techniques (sosies and principal component analysis) receive the approval of a great specialist: G. de Vaucouleurs. This result was not a foregone conclusion.

References

Barbier, D., Chalonge, D., and Divan, L. 1952, *Ann. d'Astrophys.* **15**, 201.
Bottinelli, L., Gouguenheim, L., and Paturel, G. 1983, *C. R. Acad. Sci. Paris* **296**, 1725.
Heidmann, J. 1969, *C. R. Acad. Sci. Paris* **268B**, 1782.
Paturel, G. 1981, Thesis, Université Lyon I.
Tully, R. B. and Fisher, R. 1977, *Astron. Astrophys.* **54**, 661.

Discussion

M. Capaccioli: Can you insert errors on parameters in your formulae?

Paturel: Not yet, but I think it would possible because in this kind of calculation we are using sums of parameters; thus, it would be possible to use weighted sums, the weight being a function of errors.

Local Calibrators and Globular Clusters

David A. Hanes

Astronomy Group, Physics Department, Queen's University, Kingston, Ontario, Canada

I. Introductory Remarks

We are here to commemorate the careers of Gérard and Antoinette de Vaucouleurs on the occasion of Gérard's seventieth birthday. Of course our pleasure at the event is muted by our awareness of Antoinette's all–too–recent loss: she would have taken great delight in these proceedings. Let us, however, honour both her memory and Gérard's unmistakable presence by considering how they have contributed to the subject of the extragalactic distance scale. Some of the problems in the determination of the asymptotic (far–field) value of the Hubble parameter will be discussed elsewhere in this volume. Meanwhile, as requested by the organizers, I will restrict myself to a discussion of the very local calibrators of the distance scale. In particular, I want to draw special atttention to the philosophical differences between Gérard's construction and the approaches taken by others in the last couple of decades, for I believe that these differences of philosophy provide the key to understanding the continuing dispute over the correctness of the competing distance scales.

Having mentioned the dispute, let me at once skirt the issue! In the spirit of cheerful celebration enjoyed here, I want to provoke no ill feeling from any deeply committed member of the audience by quoting my preferred value of the Hubble constant. In any event, since many of those in attendance are non–experts in the area of extragalactic distance determinations, it may prove just as profitable to speak purely qualitatively as to deal with specific values, and I propose to be quite conversational. Moreover — dare I say it? — it seems that there is now general agreement about the distances of the members of the Local Group, the nearest galaxies outside our own. This was the cheerful conclusion of the NATO Advanced Study Institute held two years ago in Kona (Madore and Tully 1986). Beyond there, of course, things break down.

II. Is the Game Worth the Candle?

Let me repeat a sensible question put to me in all seriousness by Keith Taylor: "Is H_0 really important?" One hears variants of this from time to time, usually with the deprecatory remark that "After all, agreement to a factor of two is pretty good in astronomy." Why should we not be content to adopt, say, $H_0 = 75 \pm 25$ ($\mathrm{km\,sec^{-1}Mpc^{-1}}$)?

I would argue that it really is important, for two reasons. One is largely philosophical: because we *can* discriminate, we *should*, partly as the rigorous completion of an academic exercise, but also because important science may result. [Suppose, for instance, that red supergiants are demonstrably standard candles of constant luminosity, irrespective of the total size or the mean metallicity of the populations from which they are drawn. This must tell us something about the evolution of massive stars (Humphreys 1987).] Some answers, and some methodologies, are better than others, and it is possible to evaluate them objectively. We have a scientific responsibility to identify those methods which have been misused or which have honestly misled us.

The second reason is more important. To resort to a cliché which I heartily dislike, it is simply that this question is at the "cutting edge" of astrophysics. To take an extreme example, suppose that stellar evolutionary codes implied ages of 7 billion years for the globular clusters. Then the present arguments over expansion ages spanning the range 10–20 billion years (as inferred from the Hubble parameter) would be regarded as less pressing. It is precisely because the globular cluster ages seem not to admit pure expansion models consistent with the "short" distance scale (VandenBerg 1988) that it is critical to resolve the controversy. If the "short" scale holds eventual sway, then changes will presumably be enforced in our understanding of stellar evolution or of cosmology itself. The argument is far from irrelevant.

III. A Brief History

To put Gérard de Vaucouleurs's contributions in perspective, let me describe in just a few lines the historical developments during this century. Every astronomer here will have at least conversational familiarity with the important events, except perhaps for the rapid-fire and confusing developments of the last decade.

Of course our story begins with the discovery of the period–luminosity law for Cepheid variables in the Magellanic Clouds (Leavitt 1908) and the subsequent identification of Cepheids in other nearby galaxies (Hubble 1925). Within a few years, Hubble (1929) had detected the universal expansion, using a methodology which persisted more or less unchanged —except for the details of its absolute calibration — until about 1976 (see Sandage and Tammann 1974a). This methodology calls first for the identification of Cepheids in as many nearby galaxies as possible; next, farther–reaching distance indicators, such as the luminosities of the brightest stars and the diameters of HII regions, are calibrated within the small sample of galaxies of known distance and applied beyond there. In this stepwise way, the construction proceeds outwards until a hierarchy is established which yields distances to a set of very remote objects (the prominent Sc I galaxies, in Sandage and Tammann's (1974a,b,c,d) construction) beyond any local irregularities in the expansion.

For several decades, any real progress in cosmological distance mapping centered upon a sharpening of the calibration (with, as we all know, net downward revisions in the deduced Hubble parameter). Some of the revisions had elements of high drama, such as the recognition of the confusion between two types of Cepheids as announced by Baade at the 1952 General Assembly of the IAU; others represented more gradual refinements. At almost every turn, the sundry revisions have been accompanied by optimistic statements implying that the particular technique has the virtue of the highest precision available. In retrospect, however, it seems clear that historical accidents rather than objective value judgements dictated the methodology: Sandage and his co–workers merely adopted wholesale the plan laid out by Hubble at a time when no other techniques were available. It is especially interesting to reconsider now their claims of high precision for what can be seen to be indicators of dubious or nil value [such as the brightest blue stars in galaxies (see Hanes 1982), or the diameters of HII regions, now admitted to be essentially worthless (Sandage and Tammann 1985)].

Questions of precision aside, there is a fundamental philosophical weakness in the methodology, which is simply this: it provides *one* avenue to the Hubble parameter, and if so much as one systematic error or invalid assumption intrudes at some stage, the entire chain of reasoning will be broken at and beyond that point. Paradoxically, therefore, we should not be too ready to welcome refinements in the distance scale [for example, refinements of the sort where a revised Hyades distance modulus implies an immediate proportional change in the Hubble parameter (Sandage and Tammann 1976)]. While we may hope and believe that the new scale is a closer approximation to reality, the direct impact of the revision is warning us that our approach is dangerously one–tined. What other dangers may lie hidden?

Of course, this weakness has long been apparent. Writing in 1969, van den Bergh (1975) drew a nice analogy to the sailors of antiquity who carried several marine chronometers on board their ships in the hope that the fair concordance among several of them would provide reassurances about their average timekeeping and would allow the identification of any erroneous clock(s). Likewise, a multiplicity of distance indicators may allow salutary checks upon the trustworthiness of any one of them.

It is exactly this robust multi–pronged approach that Gérard de Vaucouleurs has championed for many years. In my rather spotty chronology, let me introduce Gérard a bit belatedly (for he has surely been involved in the subject for several decades) by remembering the impact of his presentation (de Vaucouleurs 1977a) at IAU Colloquium 37, held here in Paris in 1976. In Figure 1, I reproduce his sketch of a "Tour Eiffel" of distance indicators: a secure construction, solidly supported by several completely independent local calibrators and consequently relatively insensitive to revisions in any one of them. His role has of course not simply been that of the critic: in the years following the Paris colloquium, he set about constructing exactly what he had called for: a cosmic distance scale securely founded on a multiplicity of indicators (de Vaucouleurs 1978a,b,c,d, 1979a,b; de Vaucouleurs and Bollinger 1979; see de Vaucouleurs 1980 for a summary and a comparison with the Sandage and Tammann methodology). Of course, as is well known, he found results completely at variance with those espoused by some other workers, Sandage and Tammann in particular.

The dozen years since 1976 have seen an explosion of work in the subject, largely through the introduction of completely new distance determination techniques (some of which spring from new technologies). Let me consider just one: the Tully–Fisher method (Tully and Fisher, 1977), now used in both the B and infrared bandpasses. This welcome new distance indicator is of great power in providing good *relative* distances; however, it remains contingent upon a secure calibration of the very local distance scale for its own absolute calibration. (Thus we need, at the very least, reliable distances to M 33 and M 31, the nearest spiral galaxies outside our own.) In other words, this is not a fundamental or primary distance indicator. As it happens, however, our armoury of fundamental calibrators has also been very much enriched in recent years. Let us now see how.

IV. Local Calibrators: the Various Flavours

In his series of papers, Gérard de Vaucouleurs has been admirably careful to distinguish between various kinds of distance indicator: those he calls "primary," for example, are of high precision, are visible in nearby galaxies, and can be calibrated directly in our own Galaxy by direct, fundamental methods such as trigonometric parallaxes. As an example, consider the RR Lyrae stars. Statistical or secular parallaxes provide the absolute local calibration, and their visibility in the Magellanic Clouds (and more recently in M 31; Pritchet 1988) affords a direct distance determination for those galaxies. Unfortunately, not everyone has been as scrupulous as Gérard in the terminology, and one finds the expression "primary distance indicator" used to mean one which is thought to be especially far–reaching or well–behaved, regardless of its position in the hierarchy of calibration.

Even Gérard's own construction contains an uncharacteristic slip: he refers to AB supergiants calibrated via Barbier–Chalonge–Divan spectrophotometry (Chalonge and Divan 1973) as primary indicators (of low quality). In fact, they

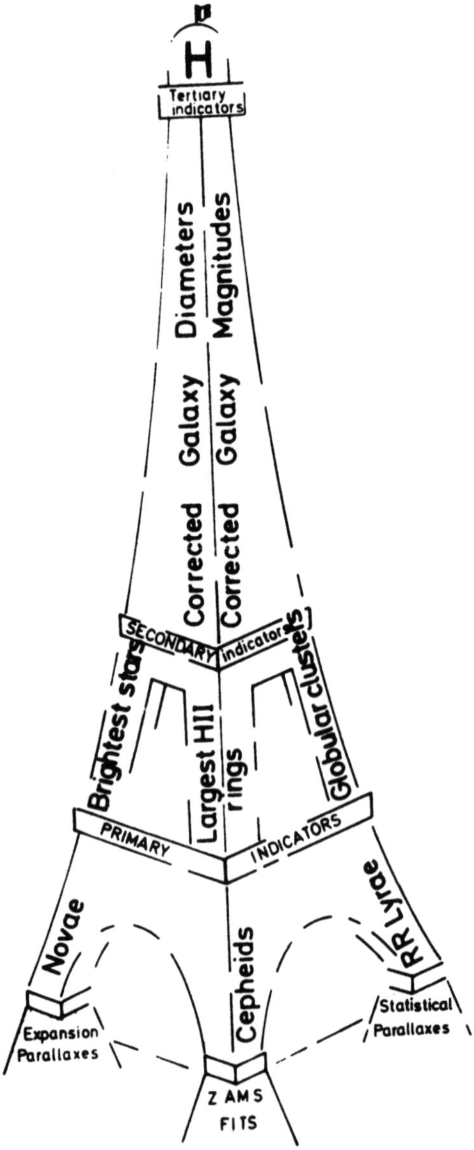

Figure 1: Une construction solide et durable pour atteindre H.

are secondary distance indicators in his terminology: their calibration relies upon the supergiants in the h and χ Persei double cluster, the distance of which is known from main sequence fitting with respect to the Hyades.

In what follows, buttressed by the knowledge that even Gérard can slip in these matters of definition, I will distinguish two kinds of local calibrator in a way which differs slightly from his. I shall first discuss those indicators which are visible in galaxies over a moderate range of distances but which can be calibrated in our own Milky Way Galaxy (perhaps through a couple of steps, as for the AB supergiants). Thus my first subset contains mostly primary, but also some secondary indicators, in the de Vaucouleurs sense.

Subsequently, I shall go on to describe a set of distance indicators which can also be calibrated locally (sometimes by pure thought, as I shall describe) but which can be applied to very large distances indeed. Few of these are persuasive — indeed, some of them are quite fanciful! — but it is at least entertaining to consider them.

V. The "Primary" Local Calibrators

V.1 Cepheid Variables

For obvious historical reasons, Cepheid variables are the first indicators which spring to mind when the cosmic distance scale is discussed. Indeed, such is their hold that the resolution of the distance scale controversy has been predicted to follow within days of the mere identification of Cepheids in sufficiently remote galaxies (Maddox 1985; I will return to this point in a moment.) Are the Cepheids really such uniquely promising distance indicators, or are there subtle traps for the unwary?

To turn first to the question of absolute calibration: de Vaucouleurs has emphasized that it is unnecessary, as well as unwise, to pin the whole calibration upon those few Galactic Cepheids found in open clusters and associations with distances inferred from main–sequence fits. Such is the approach taken by Sandage and Tammann, for example, who calibrate the zero points of the Cepheid Period–Luminosity (PL) and Period–Luminosity–Colour (PLC) relationships with reference to barely more than a dozen Cepheids. In this way, their entire scale is immediately dependent upon the adopted Hyades distance modulus. De Vaucouleurs prefers to include the zero–point calibrations afforded by statistical parallaxes and the modified Baade–Wesselink methods developed by Barnes and Evans (1976), with a consequently more robust calibration.

Cepheids have lately been much in the press, thanks largely to the extension of their study to long wavelengths; this stems from the Ph.D. thesis work of Welch (see Welch *et al.* 1984) and subsequent applications by Madore, Freedman, and others (as described in Madore's 1985 review; see also Feast and Walker 1987). At these longer wavelengths, the amplitudes are much reduced (so that random–phase observations show as little scatter in the PL relation as do the time–averaged B magnitudes) and interstellar obscuration is less a problem. In this way, new photoelectric distance moduli have been determined for many nearby galaxies, with large revisions required in a few cases (Freedman 1988). These studies, however, have not resolved the sticking point of the uncertainties in the zero–point calibration.

Moreover, even if the zero points were not contentious, and even given the internal precision in the infrared, there seems to me to be no guarantee of the continued preeminence of Cepheids as distance indicators, except in fields where they are already known to exist in some numbers. The reason is one of practicality: any use of the PL or PLC relationship requires the determination of periods for newly discovered Cepheids. Those found in remote galaxies are likely to be intrinsically bright and thus of long period (hundreds of days): inordinate amounts of telescope time will be required for the phasing of the variables before the photometry has any relevance at all. It is sobering to realize that we have studied more than one thousand Cepheids in each of the Large and the Small Magellanic Clouds, yet we know of only two Cepheids in each of M 81 (Freedman and Madore 1988) and M 101 (Cook *et al.* 1986).

In light of this, one is struck by the remarkable naiveté of Maddox's (1985) statement, in a *Nature* "News and Views" article, that the detection of Cepheid variables in galaxies of the Virgo cluster within a few days of the launch of the Hubble Space Telescope will resolve the controversy over the extragalactic distance scale. The problem is much more difficult than that, and it is not clear to me that a protracted search for remote Cepheids represents the most intelligent use of HST time given the demonstrable power of other distance indicators.

One often hears another reason stated for the appeal of the Cepheids. Many astronomers (*e.g.* Rowan–Robinson 1985) give them much higher weight in distance determinations on the grounds that we understand the astrophysics of the pulsation mechanism: in their view, this is preferable to an indicator of high empirical reliability, but mysterious operation. This may be slippery reasoning. Remember, for example, that an explanation of Kepler's laws, dating from about 1620, took more than half a century until the development of Newton's theory of universal gravitation; yet the empirical relationship was no less correct or precise before that theoretical development.

There is, moreover, an extra danger. Our persuasion that our theoretical understanding is near perfect may blind us to extrinsic effects which vitiate any practical application. No matter how well theory explains the Cepheid pulsation mechanism, for example, the fact remains that at present the zero–point calibration requires the study of a few variables of sometimes dubious cluster membership in crowded, differentially–obscured fields. Are the reddening laws sufficiently understood? Are the longest–period (youngest, most massive) Cepheids subject to circumstellar cocooning? Are multiple crossings of the instability strip important? It is for such reasons that the reality of the PLC relationship itself has been questioned (Stifft 1982), despite its apparently impressive underpinning in astrophysical theory (Sandage and Tammann 1969). (The security of that underpinning has in turn been questioned; see Clube *et al.* 1980.)

We must be careful not to lose sight of these dangers, but it is more important still that we remember de Vaucouleurs's appeal for a stable "Tour Eiffel," poised on several solid legs. Whatever the merits of the Cepheids, it would surely be short–sighted to concentrate on this indicator to the exclusion of others of comparable merit — even if they are imperfectly understood — in the foreseeable future.

V.2 Novae

The second leg of the de Vaucouleurs "Tour Eiffel" in its 1978 construction comes from the study of novae. The relative distance information comes from the well–known systematics of nova decay rates: intrinsically brighter novae decay more rapidly than their fainter brethren (McLaughlin, 1945). In de Vaucouleurs's (1978a) construction, the zero–point calibration depends upon the expansion parallaxes of Galactic novae.

Ford (1988) has recently reminded us of the serious problems introduced by the non–symmetric expansion of nova shells (the evidence for which is quite persuasive). This may impose systematic uncertainties in the calibration which ultimately limit the usefulness of novae as primary indicators; Cohen's (1988) work also bears upon this important question. Nonetheless, novae may continue as far–reaching relative distance indicators: at maximum light, they are brighter than the Cepheids and lie preferentially in the photometrically smoother galaxy haloes. The decay–rate relationship has been calibrated in M 31 (Rosino 1972), and novae have now been detected in the giant elliptical galaxy M 87 (Pritchet and van den Bergh 1987), which admits a single–stride distance determination from the Local Group to the Virgo cluster. [Indeed, van den Bergh (1986) has enthusiastically described this as part of "the Royal Road to the Hubble constant" and has even advocated the complete abandonment of the Cepheids, in striking contrast to his earlier multi–pronged approach (van den Bergh 1975).]

V.3 RR Lyrae Stars

The RR Lyrae variable stars have played an important historical role in this subject: it was the failure of the then–new Hale 5–meter telescope to detect them in M 31 that confirmed the need for a large revision in the extragalactic distance scale in 1952. Calibration is provided by statistical parallaxes and through the fitting of local subdwarf main sequences to globular cluster colour–magnitude diagrams. (Unfortunately, the RR Lyrae stars are remote enough that even the HIPPARCOS satellite will yield no precise trigonometric parallaxes, except for RR Lyrae itself; Monet 1988.) As is well known, however, there is a continuing dispute over the metallicity dependence (or not) of RR Lyrae intrinsic luminosity. This important lingering uncertainty limits the precision with which distances may be derived. On the positive side, the short periods of RR Lyrae stars and their characteristic light curves permit their ready identification in nearby galaxies without the need for extremely long-term photometric study.

Although less luminous than the Cepheids, the RR Lyrae stars can still be studied in the Local Group galaxies, including M 31: indeed, here is one simple indication of the progress made in the four decades since the construction of the 5-metre telescope. Sensitive panoramic detectors, such as CCDs, and superb sites, such as Mauna Kea, allow this important observation (Pritchet 1988) with a telescope of only 3.6 metres aperture (the CFHT). Even with the Hubble Space Telescope, however, the RR Lyrae variables will be out of reach much outside the Local Group.

V.4 Mira Variables

The long-period red supergiant pulsators of which Mira is the prototype have long disappointed us because of their failure to obey a sensible period- luminosity relationship in the usual *UBV* bandpasses (van den Bergh 1975). Recent work, however, has shown that at near-infrared wavelengths, the Miras in the Magellanic Clouds obey a PL relationship which is comparable in its internal scatter to that displayed by the Cepheids (Glass and Lloyd Evans 1981). In this early treatment of the new-found relationship, those authors carried out the "obvious" step of calibrating the relationship through the adoption of the Magellanic Cloud Cepheid distance modulus. Of course, this plausible procedure fails to take advantage of the potential of the Miras as completely independent distance indicators: they can be calibrated within the Milky Way by statistical parallaxes (Robertson and Feast 1981) or (secondarily) by the study of Miras within globular clusters of known distance. Such a procedure affords an independent check of the Cepheid moduli for the Magellanic Clouds; see Glass and Lloyd Evans (1981) for such analyses.

The Mira variables may now provide us with an important fourth "primary" leg for the "Tour Eiffel" construction appealed for by de Vaucouleurs; they may be as far-reaching as the Cepheids, but present the same problem of the time-consuming determination of periods. They have already been used in an independent determination of the distance to the Galactic centre (Glass and Feast 1982).

V.5 AB Supergiant Stars

I have already referred to the method known as BCD (Barbier, Chalonge, and Divan) spectrophotometry, most recently described in detail by Chalonge and Divan (1973). Empirically, it has long been known that the AB supergiants of the highest luminosity manifest their preeminence spectroscopically. Early in this century, for example, Petrie used a photographic Hγ index based upon observed hydrogen line emission reversals as a luminosity indicator. In the most recent modern study, emission reversals at Hα and Hβ have been explored photoelectrically and correlated with intrinsic luminosity by Tully and Nakashima (1986). BCD spectrophotometry itself relies upon more than just the hydrogen line emission reversals: it provides quantitative measurements of the Balmer discontinuity and so forth. The method has the power to be very far-reaching, given the extreme luminosity of the AB supergiants, and the method surely warrants further development and calibration with modern instrumentation. (In its classical form it is carried out photographically.) De Vaucouleurs (1978a) recognizes both the power and the present limitations of the method in referring to it as a primary indicator of lower precision; that may change with time. It is worth remembering, of course, that AB supergiants will be resolved in Virgo galaxies by the HST, but will present none of the difficulties of long-term photometric monitoring required for Cepheid work.

V.6 Eclipsing Binaries

The other primary indicator of low precision referred to by de Vaucouleurs (1978a) is that of the study of eclipsing variables in the Magellanic Clouds by Dwořák (1974; see also Gaposchkin 1938, 1940). The technique is conceptually simple: eclipse timings yield stellar radii as fractions of the binary separation; the spectra or colours yield effective temperatures; and one adopts a mass–luminosity law, bolometric corrections, Kepler's laws and the Stefan–Boltzmann law to deduce, eventually, intrinsic luminosities and (therefore) the distance.

A recent reconsideration of the limitations of the method has been provided by Davidge in his Ph.D. thesis and subsequent publications (Davidge 1987, 1988). Not surprisingly, there are difficulties. Since most eclipsing systems are close binaries, one needs to worry about mass transfer, the non-applicability of the adopted mass–luminosity law (especially if one or the other star is evolved), and so forth. Davidge also points out that the $(B-V)$ colour is a poor indicator of effective temperature, and concludes that the method is of marginal power at best.

V.7 Direct Main Sequence Fitting

Until the development of detectors which combined great sensitivity with panoramic digital imaging (allowing the photometric decomposition of stars in crowded fields through the use of software such as DAOPHOT; Stetson 1987), it was unrealistic to dream of calibrating stellar main sequences in galaxies outside our own. Now, however, CCDs on large southern telescopes have made that dream a reality in studies of the Magellanic Clouds (see Olszewski 1988). Such studies provide immediate intercomparisons with two of the classical primary indicators: the Cepheids and the RR Lyrae pulsators. Such intercomparisons in the Magellanic Clouds are especially valuable because, as a consequence of the lower mean metallicities of the Large and Small Clouds, they allow the further exploration of the metallicity dependence of the various indicators (Feast 1988).

The Hubble Space Telescope holds great promise in this respect for the other galaxies of the Local Group: in M 31, for example, its expected photometric limit near $V = 28$ mag will permit main sequence photometry to the M = 3.5 magnitude level — adequate for the Population I component at least, although the turnoff of the halo population will still be tantalizingly out of reach.

V.8 The W Virginis Variables

Little progress has been reported on the use of W Vir stars as distance indicators since van den Bergh's (1977) summary. As he noted then, the application of Breger and Bregman's (1975) PLC relationship for W Vir stars to five such variables in M 31 leads to distance moduli at variance with the usually accepted values, perhaps because of some intrinsic curvature in the PLC relationship.

Do the W Vir stars represent a particularly valuable line of future work for distance determinations? My answer would be a qualified no: qualified, because every extra tool is welcome (and especially when it provides a Pop II distance scale to compare with the Pop I Cepheid construction); but no, because of a familiar problem: their long periods demand extensive photometric monitoring; worse still, they are fainter than the classical Cepheids. Nonetheless, their study in at least the Local Group galaxies is of continuing interest (although perhaps not entirely in the context of distance determinations).

VI. Globular Clusters as Primary Indicators

The globular clusters warrant a separate subsection in this presentation by virtue of my commitment to them, if for no other reason! Given my interests, it is perhaps forgiveable if I spend what some may consider a disproportionate amount of time discussing them.

VI.1 Two Methods Contrasted

Let me point out at the outset that the globular clusters have been used by me and by Gérard de Vaucouleurs in two quite different ways (with similar results, however). Gérard treats them as secondary or tertiary indicators (de Vaucouleurs 1977b, 1978c). He uses a set of galaxies with known distance, and a set of Virgo cluster galaxies at a common distance, to explore and calibrate the dependence of the luminosity of the brightest cluster upon parent galaxy luminosity. By good fortune, the distance–independent difference in magnitude of these two is a well–behaved non–degenerate function of parent galaxy luminosity, so that an observation of that difference in brightness yields an iterative estimate of parent galaxy intrinsic luminosity (and hence a distance modulus). We should applaud again the de Vaucouleurs empiricism: nearly two decades ago (de Vaucouleurs 1970), he had already developed this straightforward method despite our ignorance of the underlying luminosity functions and population scalings (which came from Hanes 1977 in the first instance). Subsequent developments (de Vaucouleurs 1977b, 1978c) have been merely the sharpening of the tool to a finer point.

In my approach, the globular clusters are calibrated *within* the Milky Way (and thus earn the title of "primary" indicator in the loose sense I am permitting here). Our sample of more than one hundred globulars permits the delineation of the luminosity function (LF) to sufficient precision that a simple intercomparison of LFs between the Milky Way and, say, the richer LFs observed in Virgo cluster ellipticals permits a one-step estimate of this cosmologically important distance (Hanes 1979).

To my mind, this is the outstandingly attractive appeal of the globular clusters: the brightest of them are near $M = -10$, and the bright tail of the LF can be readily explored over a depth of several magnitudes to very large distances indeed (Harris 1988). Moreover, the globulars lie in the photometrically smooth galaxy halos, are non-variable (we trust!) so that no time-consuming monitoring is required, and are extraordinarily numerous in many galaxies.

The de Vaucouleurs method has the extra virtue that it requires only the identification of the few brightest clusters, but it does depend upon some simple scaling of total cluster content with parent galaxy luminosity; this seems to be violated here and there (as, for example, in the super-abundant M 87 cluster population) in a way which may be a function of environment (Hanes and Harris 1986). My method makes no such assumption, but does require that the LF be the same in all galaxies, regardless of the total population. Some may see this as improbable (Tammann and Sandage, 1983), but there is so far no contrary evidence in a handful of admittedly weak tests (Pritchet and van den Bergh 1985; see Hanes and Whittaker 1987).

VI.2 The Zero-Point Calibration

The determination of distances for globular clusters internal to the Milky Way is a problem with a long history and some profound implications (Shapley 1918). In its most precise form (Harris 1980), it has until lately relied upon horizontal branch (HB) stars — including the RR Lyrae pulsators — observed within the clusters, but calibrated in the field. Recent developments have permitted two more props in this construction.

The first springs again from the development of panoramic digital detectors and powerful data reduction software: for the first time we can do really precise subdwarf main-sequence fitting to the globular clusters; see, for example, Hesser *et al.* 's (1987) exemplary work in 47 Tuc. (We should not forget, however, to acknowledge Sandage's (1970) Herculean efforts based on photographic photometry in four globular clusters.)

A second, purely geometrical method has lately reached fruition largely through the efforts of Kyle Cudworth (see Cudworth and Peterson, 1988). The dispersions in proper motion (measured so far in the nearest half-dozen globulars) can be equated to the radial velocity dispersions to yield distances to an advertised precision of 10%. The method is attractive in that it is entirely immune to the effects of interstellar obscuration and thus allows an important check on the luminosity-dependent indicators. Moreover, there is a clear upward path here: measurements of proper motion can only get better as our temporal baseline lengthens, and the precision must always improve!

It is also worth remembering that extreme precision is not needed for individual clusters in the calibrating sample for a subsequent use outside the Milky Way. Uncertainties in the shape of the intrinsic globular cluster LF are more dominated by the sparseness of the calibrating sample than by any imprecisions in individual determinations of cluster distance and luminosity. We must, however, be alert to systematic scale errors occasioned by inappropriate models for the obscuration or the (contentious) metallicity dependence of HB luminosity.

VI.3 The Power of the Method

Until lately, the use of globular cluster LFs has been problematic in the view of many because in remote systems the "turnover" in the function has not been clearly reached. I have argued elsewhere (Hanes 1983, Hanes and Whittaker 1987) that this is not critical because the LF is not self-similar in a logarithmic representation: in other words, distance information can be unravelled from mere population scalings (which would not be the case, for example, if the LF were a pure power law). In Hanes and Whittaker (1987), we have argued that reaching the level of the turnover (which requires photometry to fainter than 24th magnitude in the Virgo cluster, for example) is an unnecessary luxury: the form of the LF is entirely adequate to constrain the distance estimate even if we work to some rather brighter limit. Looking further out, this means that useful distance estimates are possible to very remote systems where the turnover is well beyond our reach. (Remember that globular cluster systems have been detected in the galaxies of the Coma cluster: Harris, 1987; Thompson and Valdes, 1987.)

There is one important proviso: an accurate distance estimate requires a good calibration of the intrinsic *shape* of the (assumed universal) globular cluster LF. This has been the major stumbling block: the statistical uncertainties imposed by our sparse local calibrating sample permit a fairly broad range of distance moduli in any real application. The way around this is to sample a very rich population deeply enough to determine the shape of the LF and then to use the local calibrating sample to set the zero point. (Of course, this is entirely analagous to our use of Cepheids: the form of the PLC is determined in the Magellanic Clouds, while the zero point is determined in the Milky Way.) The globular-cluster-rich Virgo galaxies are a natural target for this critical improvement, and much work has gone into this endeavour.

A first detection of the turnover in the Virgo galaxy M 87 was claimed by van den Bergh, Pritchet and Grillmair (1985), but it is clear from a close scrutiny of their data and figures (see the discussion by Hanes and Whittaker 1987) that the detection is marginal at best, and provides little real constraint on the form of the function. But now Harris (1988) has convincingly reached well beyond the turnover in no fewer than four Virgo cluster ellipticals. Encouragingly, they agree in magnitude to within the expected errors (taking into account the line–of–sight depth of the cluster itself) and confirm the predictions of the Hanes and Whittaker formalism (when the appropriate value for the shape parameter is adopted).

VI.4 The Future Promise

It is customary to remark that the Hubble Space Telescope will resolve almost every outstanding problem in astronomy. Here I feel that I can honestly claim to have found a natural target for HST! It will resolve the globular clusters in Virgo, thus winnowing out field objects, and refine our knowledge of the intrinsic LF to high precision. It will permit the study of the less–rich globular cluster systems in the spiral galaxies of the Virgo cluster and allow a test of the universality of that function among galaxies of various types. And it will permit the study of globular cluster systems in very remote clusters — to Coma and beyond — and yield their distances in essentially a single stride from the Local Group. Do you wonder at my enthusiasm for these beautiful stellar systems?

VII. The Farther–Reaching Indicators

Owing to the exigencies of time (in this talk) and space (in the written version), I regret that I can only comment briefly upon my second class of "local" calibrators. I described these earlier as being extraordinary far–reaching but, at least in some cases, rather fanciful in their calibration or practicality.

VII.1 Supernovae

As lately nicely summarized by Branch (1988), supernovae can be used as distance indicators in several ways. The first use is exemplified by Tammann and Sandage (1983), and relies upon determining distances stepwise out to one or more galaxies which have contained well–studied supernovae in this century. This permits a calibration of the typical maximum luminosity, and if all supernovae are alike, they are now standard candles useful at yet larger distances. In this application, of course, they are not primary, but we should remember that a modern Galactic supernova would permit a direct calibration by expansion parallax or light echo studies. Indeed, Supernova 1987A in the LMC might have provided such a calibration on the lower rungs of the hierarchy had it not been so atypical. (We should remember that its atypicality is worrying news for those who would argue that supernovae of a given class are all much alike.)

A second approach involves a variant of the Baade–Wesselink method. Here, one integrates the velocity curve of the expanding photosphere to determine the size of the shell at any epoch. The broadband colours imply a temperature, and the Stefan–Boltzmann law yields an intrinsic luminosity — ergo, a distance. The sticking point here is the correctness of the effective temperature (for surely the expanding photosphere is far from blackbody); see Branch (1988) for details of how this is accommodated.

Finally, we may calibrate supernova explosions by "pure thought" — or rather with computer models, which may not be quite the same thing! For example, Craig Wheeler will tell you just how bright at maximum light a Type I supernova will ever get as it undergoes carbon deflagration. The impressive feature of these models is just how beautifully they reproduce the observed spectral evolution (Branch 1988).

VII.2 Neo–Copernican Arguments

De Vaucouleurs has lately made use of the concept of astronomical "sosies," a French word which means "look–alikes" (Bottinelli *et al.* 1985). He points out, in a way which recalls Copernicus and Shapley, that we should be reluctant to believe in the atypicality of our own Milky Way galaxy, since we believe ourselves to be in no preferred position. The scale factors of our own Galaxy are moderately well–determined, and the requirement that they match some suitable average for galaxies of a similar type imposes a preferred distance scale.

VII.3 The Sunyaev-Zeldovich Effect

Sunyaev and Zeldovich (1972) pointed out that inverse Compton scattering of microwave background photons off relativistic electrons in hot intracluster gas would yield an observed decrement in the cosmic background in the direction of the cluster. Moreover, the particular form of the decrement depends upon the gas temperature, geometry, etc., in a way which tells us about the distance to the cluster (and thus the Hubble constant): see Chase *et al.* (1987) for a discussion. At various times successes in the detection of the decrement have been claimed, but the technique is certainly not without problems, not the least of which is in the requisite assumptions about the gas distribution. No one, I think, would pretend that this is a particularly valuable tool for distance determinations.

VII.4 Gravitational Lenses

Similar remarks apply in the use of the optics of gravitational lenses as indicative of distance. In principle, the desired information is implicit in the trajectories followed by the light from remote lensed sources (Alcock and Anderson 1985), but in practice so many assumptions have to made about the distribution of scattering centres that the exercise is moot. Indeed, it makes more sense to apply the argument the other way: if the distances are known, the observed lensing provides a tracer of the scattering centres.

VII.5 Superluminal Sources

Donald Lynden–Bell (1977) suggested at one time that one should expect angular expansion velocities to correspond to apparent linear velocities of only just twice the speed of light (with rare larger values) in simple models of superluminally expanding sources. Of course, this allows one to set a bound on the distance of the sources (and, in fact, his original discussion was supportive of the "short" distance scale), but he has subsequently concluded that the argument was too simplistic in its assumptions: he does not now consider this a useful distance indicator.

VII.6 Trigonometric Parallaxes

Let me give the last word to Donald Lynden–Bell with a return to the classical technique of trigonometric parallax measurements. Speaking at the Woolley Symposium, he pointed out (Lynden–Bell 1972) that to carry out a trigonometric parallax measurement of stars within the Andromeda galaxy all we really need do is park a space probe a mere parsec from the sun — not yet as far away as the nearest star. While this is at present an unrealistic aspiration, it would be a brave man who would rule out such a line of inquiry in the relatively near future. Even posting one end of the baseline at the orbit of Pluto, for example, would yield an improvement of nearly two orders of magnitude in the precision of trigonometric parallaxes. Such a purely geometrical determination would be invaluable in nailing down the calibration of a host of other indicators and would provide a check on the importance of interstellar obscuration in a way which could not be gainsaid.

VIII. Concluding Remarks

My intention in this conversational review has been twofold: first, to demonstrate the array of tools which we have at our disposal, even in the very local calibration of the cosmological distance scale; and second, to remind you all — unnecessarily, I feel sure! — of the important role played in this field by Gérard de Vaucouleurs in particular. Naturally enough, these aspects go hand in glove: Gérard has encouraged us at every turn to think of supporting our constructions in as many independent ways as possible, and has himself developed the successful implementation of many. His research has been marked by attention to detail, by rigour, by pragmatism, and, perhaps most obviously, by a love of astronomy which is revealed in his hard work and enthusiasm. I am pleased to have had the opportunity to remember his contributions in these pleasant circumstances.

References

Alcock, C. and Anderson, N. 1985, *Astrophys. J. Lett.* **291**, L29.

Barnes, T. G. and Evans, D. S. 1976, *Mon. Not. Roy. Astron. Soc.* **174**, 489.

van den Bergh, S. 1975, in *Galaxies and the Universe*, ed. A. Sandage, M. Sandage, and J. Kristian (Chicago: Univ. of Chicago Press), p. 505.

van den Bergh, S. 1977, in *I. A. U. Colloq. 37, Décalages vers le Rouge et Expansion de l'Univers*, ed. C. Balkowski and B. E. Westerlund (Paris: CNRS), p. 13.

van den Bergh, S. 1986, in *Galaxy Distances and Deviations from Universal Expansion*, ed. B. F. Madore and R. B. Tully, (Dordrecht: Reidel), p. 41.

van den Bergh, S., Pritchet, C. J., and Grillmair, C. 1985, *Astron. J.* **90**, 595.

Bottinelli, L., Gouguenheim, L., Paturel, G., and de Vaucouleurs, G. 1985, *Astrophys. J. Suppl.* **59**, 293.

Branch, D. 1988, in *The Extragalactic Distance Scale*, ed. S. van den Bergh and C. J. Pritchet, (San Francisco: Astron. Soc. Pac.), in press.

Breger, M. and Bregman, J. N. 1975, *Astrophys. J.* **200**, 343.

Chalonge, D. and Divan, L. 1973, *Astron. Astrophys.* **23**, 69.

Chase, S. T., Joseph, R. D., Robertson, N. A., and Ade, P. A. R. 1987, *Mon. Not. Roy. Astron. Soc.* **225**, 171.

Clube, S. V. M. and Dawe, J. A. 1980, *Mon. Not. Roy. Astron. Soc.* **190**, 591.

Cohen, J. 1988, in *The Extragalactic Distance Scale*, ed. S. van den Bergh and C. J. Pritchet, (San Francisco: Astron. Soc. Pac.), in press.

Cook, K. H., Aaronson, M., and Illingworth, G. 1986, *Astrophys. J. Lett.* **301**, L45.

Cudworth, K. M. and Peterson, R. C. 1988, in *The Extragalactic Distance Scale*, ed. S. van den Bergh and C. J. Pritchet, (San Francisco: Astron. Soc. Pac.), in press.

Davidge, T. J. 1987, *Astron. J.* **94**, 1169.

Davidge, T. J. 1988, *Astron. J.* **95**, 731.

Dwořák, T. Z. 1974, *Acta Cosmo.* **2**, 13.

Feast, M. W. 1988, in *The Extragalactic Distance Scale*, ed. S. van den Bergh and C. J. Pritchet, (San Francisco: Astron. Soc. Pac.), in press.

Feast, M. W. and Walker, A. R. 1987, *Ann. Rev. Astron. Astrophys.* **25**, 345.

Ford, H. C. 1988, in *The Extragalactic Distance Scale*, ed. S. van den Bergh and C. J. Pritchet, (San Francisco: Astron. Soc. Pac.), in press.

Freedman, W. L. 1988, in *The Extragalactic Distance Scale*, ed. S. van den Bergh and C. J. Pritchet, (San Francisco: Astron. Soc. Pac.), in press.

Freedman, W. L. and Madore, B. F. 1988, in *The Extragalactic Distance Scale*, ed. S. van den Bergh and C. J. Pritchet, (San Francisco: Astron. Soc. Pac.), in press.
Gaposchkin, S. 1938, *Harvard Obs. Report* No. 151.
Gaposchkin, S. 1940, *Harvard Obs. Report* No. 201.
Glass, I. S. and Feast, M. W. 1982, *Mon. Not. Roy. Astron. Soc.* **198**, 199.
Glass, I. S. and Lloyd Evans, T. 1981, *Nature* **291**, 303.
Hanes, D. A. 1977, *Mem. Roy. Astron. Soc.* **84**, 45.
Hanes, D. A. 1979, *Mon. Not. Roy. Astron. Soc.* **188**, 901.
Hanes, D. A. 1982, *Mon. Not. Roy. Astron. Soc.* **201**, 145.
Hanes, D. A. 1983, *Highlights of Astron.* **6**, ed. R.M. West (Dordrecht: Reidel), p. 227.
Hanes, D. A. and Harris, W. E. 1986, *Astrophys. J.* **309**, 564.
Hanes, D. A. and Whittaker, D. G. 1987, *Astron. J.* **94**, 906.
Harris, W. E. 1980, in *I. A. U. Symp. 85, Star Clusters*, ed. J. E. Hesser (Dordrecht: Reidel), p. 81.
Harris, W. E. 1987, *Astrophys. J. Lett.* **315**, L29.
Harris, W. E. 1988, in *The Extragalactic Distance Scale*, ed. S. van den Bergh and C. J. Pritchet, (San Francisco: Astron. Soc. Pac.), in press.
Hesser, J. E., Harris, W. E., VandenBerg, D. A., Allwright, J. W. B., Shott, P., and Stetson, P. B. 1987, *Publ. Astron. Soc. Pac.* **99**, 739.
Hubble, E. 1925, *Observatory* **48**, 140.
Hubble, E. 1929, *Proc. Nat. Acad. Sci.* **15**, 168.
Humphreys, R. 1987, *Publ. Astron. Soc. Pac.* **99**, 5.
Jacoby, G. 1988, in *The Extragalactic Distance Scale*, ed. S. van den Bergh and C. J. Pritchet, (San Francisco: Astron. Soc. Pac.), in press.
Leavitt, H. S. 1908, *Harvard Col. Obs. Ann.* **60**, 87.
Liller, W. and Alcaino, G. 1983a, *Astrophys. J.* **264**, 53.
Liller, W. and Alcaino, G. 1983b, *Astrophys. J.* **265**, 166.
Lynden-Bell, D. 1972, *Quar. J. Roy. Astron. Soc.* **13**, 133.
Lynden-Bell, D. 1977, *Nature* **270**, 396.
Maddox, J. 1985, *Nature* **313**, 347.
Madore, B. F. 1985, in *I. A. U. Colloq. 82, Cepheids: Theory and Observations*, ed. B. F. Madore (Cambridge: Cambridge Univ. Press), p. 166.
Madore, B. F. and Tully, R. B. 1986, eds. *Galaxy Distances and Deviations from Universal Expansion*, (Dordrecht: Reidel).
McLaughlin, D. B. 1945, *Publ. Astron. Soc. Pac.* **57**, 69.
Monet, D. 1988, in *The Extragalactic Distance Scale*, ed. S. van den Bergh and C. J. Pritchet, (San Francisco: Astron. Soc. Pac.), in press.
Olszewski, E. W. 1988, in *I. A. U. Symp. 126, Globular Cluster Systems in Galaxies*, ed. J. E. Grindlay and A. G. Davis Philip (Dordrecht: Kluwer Academic), p. 159.
Pritchet, C. J. 1988, in *The Extragalactic Distance Scale*, ed. S. van den Bergh and C. J. Pritchet, (San Francisco: Astron. Soc. Pac.), in press.

Pritchet, C. J. and van den Bergh, S. 1985, *Astron. J.* **90**, 2027.
Pritchet, C. J. and van den Bergh, S. 1987, *Astrophys. J.* **318**, 507.
Robertson, B. S. C. and Feast, M. W. 1981, *Mon. Not. Roy. Astron. Soc.* **196**, 111.
Rosino, L. 1972, *Astron. Astrophys. Suppl.* **9**, 347.
Rowan-Robinson, M. 1985, *The Cosmological Distance Ladder*, (Cambridge: Cambridge Univ. Press).
Sandage, A. R. 1970, *Astrophys. J.* **162**, 841.
Sandage, A. R. and Tammann, G. A. 1969, *Astrophys. J.* **157**, 683.
Sandage, A. R. and Tammann, G. A. 1974a, *Astrophys. J.* **190**, 525.
Sandage, A. R. and Tammann, G. A. 1974b, *Astrophys. J.* **191**, 603.
Sandage, A. R. and Tammann, G. A. 1974c, *Astrophys. J.* **194**, 223.
Sandage, A. R. and Tammann, G. A. 1974d, *Astrophys. J.* **194**, 559.
Sandage, A. R. and Tammann, G. A. 1976, *Astrophys. J.* **210**, 7.
Sandage, A. R. and Tammann, G. A. 1985, in *Supernovae as Distance Indicators*, ed. N. Bartel, (Heidelberg: Springer–Verlag), p. 1.
Shapley, H. 1918, *Publ. Astron. Soc. Pac.* **30**, 42.
Stetson, P. B. 1987, *Publ. Astron. Soc. Pac.* **99**, 191.
Stifft, M. J. 1982, *Astron. Astrophys.* **112**, 149.
Sunyaev, R. A. and Zeldovich, Ya. B. 1972, *Comments Astro. Sp. Sci.* **4**, 173.
Tammann, G. A. and Sandage, A. R. 1983, in *Highlights of Astronomy*, **6**, ed. R. M. West, (Dordrecht: Reidel).
Thompson, L. A. and Valdes, F. 1987, *Astrophys. J. Lett.* **315**, L35.
Tully, R. B. and Fisher, J. R. 1977, *Astron. Astrophys.* **54**, 661.
Tully, R. B. and Nakashima, J. M. 1986, in *Galaxy Distances and Deviations from Universal Expansion*, ed. B. F. Madore and R. B. Tully, (Dordrecht: Reidel), p. 25.
VandenBerg, D. 1988, in *I. A. U. Symposium 126, Globular Cluster Systems in Galaxies*, ed. J. E. Grindlay and A. G. Davis Philip (Dordrecht: Kluwer Academic), p.107.
de Vaucouleurs, G. 1970, *Astrophys. J.* **159**, 435.
de Vaucouleurs, G. 1977a, in *I. A. U. Colloq. 37, Décalages vers le Rouge et Expansion de l'Univers*, ed. C. Balkowski and B. E. Westerlund (Paris: CNRS), p. 301.
de Vaucouleurs, G. 1977b, *Nature* **266**, 126.
de Vaucouleurs, G. 1978a, *Astrophys. J.* **223**, 351.
de Vaucouleurs, G. 1978b, *Astrophys. J.* **223**, 730.
de Vaucouleurs, G. 1978c, *Astrophys. J.* **224**, 14.
de Vaucouleurs, G. 1978d, *Astrophys. J.* **224**, 710.
de Vaucouleurs, G. 1979a, *Astrophys. J.* **227**, 380.
de Vaucouleurs, G. 1979b, *Astrophys. J.* **227**, 729.
de Vaucouleurs, G. 1980, in *Some Strangeness in the Proportion (An Einstein Centenary Celebration)*, ed. H. Woolf (Reading, MA: Addison-Wesley), p. 416.

de Vaucouleurs, G. and Bollinger, G. 1979, *Astrophys. J.* **233**, 433.
Welch, D. L., Weiland, F., McLaren, R., McAlary, C. W., Madore, B. F., and Neugebauer, G. 1984, *Astrophys. J. Suppl.* **54**, 647.

Discussion

R. Buta: Can you comment on the number of Local Group galaxies for which infrared photometry of Cepheids has been obtained?

Hanes: For this I should defer to Barry Madore, who will know the latest count. First, however, let me point out that there is a distinction to be drawn between true infrared photometry and an extension to, say, the *I* bandpass. Working at *I* gives you the benefits of going to the true infrared in that interstellar obscuration is less and the Cepheid amplitudes are down, but at *I* there are many panoramic detectors available (CCDs) which permit the unscrambling of crowded fields in remote galaxies. (In the true infrared, we are only starting to see good area detectors.) At *I*, Cepheid photometry has been done in many Local Group galaxies (LMC, SMC, M 31, M 33, IC 1613, NGC 6822 ...) as well as out to NGC 2403 and, recently, M 81 (Freedman and Madore 1988).

Roger Davies: Would you comment on the promise of using the planetary nebula luminosity function to provide a zero point for the distance scale based on ellipticals (as Ford and collaborators are pursuing)?

Hanes: The PN luminosity function seems a remarkably robust relative distance indicator (see, for example, Jacoby 1988), but it cannot be considered a primary calibrator — which is why I did not include the method here — since there is no obvious way to calibrate it internal to our own Galaxy. The function is well–determined, however, in the nuclear parts of M 31 and M 81 and in NGC 5128, and should be observable in the Virgo cluster ellipticals, so at this next level in the hierarchy the method is very promising. According to Jacoby, the PN luminosity function is apparently insensitive to metallicity, and, of course, one expects to find them in all populations.

L. Gouguenheim: I should like to make a comment on the Cepheid distance criterion. Madore and his collaborators have obtained beautiful results in the infrared, but, as you said, what is needed now is a good zero–point calibration of this relationship from Galactic Cepheids. The ESA satellite HIPPARCOS is expected to measure Cepheid parallaxes: this will be very important — it is the first step!

Hanes: I certainly agree that these will be welcome results. I remain less optimistic about the practicality — and therefore the wisdom — of concentrating too much on Cepheid work in remote galaxies, however.

G. Alcaino: Even in spirals similar to our own within distances of 3 Mpc, such as NGC 55 and NGC 253 in the Sculptor Group, it is immensely difficult with current ground–based telescopes to discriminate the globular cluster population. It is even more difficult to learn if the luminosity function mimics that in our own Galaxy so that it can be used as a stable distance indicator.

Hanes: I could not agree more. You and Bill Liller (Liller and Alcaino 1983a,b) succeeded in identifying likely globular cluster candidates in these galaxies, I know, and I myself have hunted for them, using AAT plates and an automated scanning device. Paradoxically, what hurts us is the proximity of these galaxies! They are in fact not quite like the Milky Way, being less luminous, but they subtend relatively large solid angles. So, one expects to see perhaps 50 globulars at 20th magnitude (on average) scattered over a degree of sky. They may be incipiently resolved, it is true, but it is still really a needle–in–a–haystack search through the rich population of foreground stars.

To test the universality of the luminosity functions, it is much easier to work to fainter levels in a more remote globular–cluster–rich spiral such as the Sombrero (NGC 4594 = M 104). I have CFHT time for just this exercise. In fact, I fly there tomorrow!

P. Brosche: The use of globular clusters as distance indicators may be difficult because we have found already that those in our own Galaxy constitute a three–dimensional manifold (to compare with the two significant dimensions of the galaxies).

Hanes: I approach this as an empiricist. We do not yet understand how globular clusters form and how they are shredded in subsequent dynamical events; worse, we do not understand fully how both these phenomena depend, if they do at all, on galaxy morphology and environment. Certainly globular cluster systems are not all alike (see Hanes and Harris 1986), but so far the simple–minded working assumption of the universality of the luminosity function cannot be excluded and indeed yields quite sensible results. Like Kepler, I will continue to use the tool — always on the lookout for problems, of course, and testing the validity in every way possible — until the case is proven one way or the other.

W. T. Sullivan: A further distance–determining method of the future (AD 2000? 2010?) that should be mentioned is the possibility of radio interferometry on the scale of the inner solar system. If one considers a baseline of a few AU at a wavelength of 1–cm, then it turns out that the entire observable universe lies within the near field (that is, the Fresnel region) of this interferometer! In other words, one will be able, in principle at least, to directly measure the distance to *any* radio source of sufficiently small size and large intensity by measuring the curvature of the wavefront as it passes through the solar system. There will be problems with dispersion by the interplanetary, interstellar, and intergalactic plasma, but these can probably be overcome.

Distances to the Galaxies M 81 and NGC 2403 from CCD I–Band Photometry of Cepheids

Wendy L. Freedman[1] and Barry F. Madore[2]

[1]Mt. Wilson and Las Campanas Observatories, Pasadena, California, U. S. A.
[2]California Institute of Technology, Pasadena, California, U. S. A.

M 81 and NGC 2403 are two of the closest northern hemisphere galaxies outside of the Local Group for which Cepheid variables have been detected and periods accurately determined. I–band CCD photometry has been obtained for the only two known Cepheids in M 81, and for eight of the known Cepheids in NGC 2403. For M 81, an apparent I-band distance modulus and formal fitting error of $(m-M)_I = 27.64 \pm 0.09$ mag are derived, while for NGC 2403 $(m-M)_I = 27.57 \pm 0.14$ mag. Taking a true distance modulus of 18.50 mag for the Large Magellanic Cloud (and $E_{B-V}) = 0.07$), a true modulus of 27.59 mag $\pm$ 0.31 is obtained for M 81, and 27.51 mag $\pm$ 0.24 for NGC 2403 (foreground reddenings of $A_I = 0.05$ and 0.06 mag being adopted for M 81 and NGC 2403, respectively, from Burstein and Heiles 1984). These moduli correspond to distances of 3.3 $\pm$ 0.5 and 3.2 $\pm$ 0.4 Mpc, respectively, but they are still upper limits since we have not yet independently determined the reddening internal to the parent galaxies.

At present, the largest factors contributing to the uncertainty in the distances to M 81 and NGC 2403 are: (1) the number of Cepheids so far observed, (2) the distance to the Large Magellanic Cloud (*i.e.* the zero point of the Cepheid P–L relation) and finally (3) the internal reddening affecting each of the Cepheids in the individual program galaxies. No correction for reddening internal to each of the program galaxies has been applied in this study, but the uncertainty in the provisional true distance modulus has been increased. The total reddening can be assessed using high signal–to–noise multi–color observations, and will be dealt with in a forthcoming paper (Freedman and Madore, in preparation). Surveys for more Cepheids in M 81 are underway, but are several years from completion. With phased observations of all of the known Cepheids

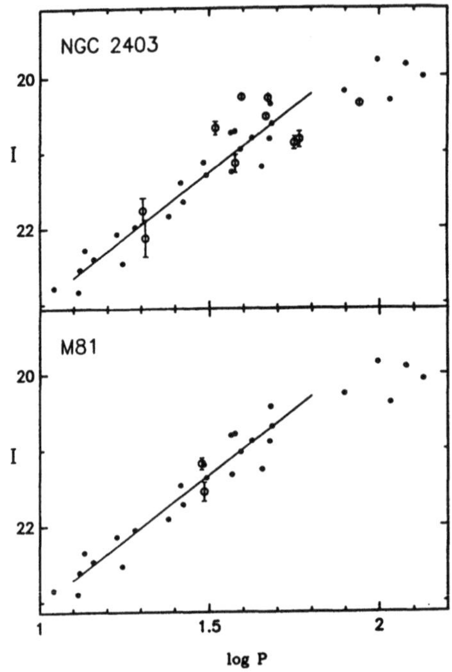

Figure 1: Period–Luminosity relations. *Upper Panel*: NGC 2403. *I*–band P–L relation for the averaged observations of ten Cepheids observed in NGC 2403 (open circles). *Lower Panel:* M 81. Open circles mark the periods and averaged *I*–band magnitudes for the two Cepheids (V 2 and V 30) with known periods in M 81. Filled circles in both panels give the time- -averaged photoelectric data for the Large Magellanic Cloud Cepheids, shifted in apparent modulus until minimum scatter was achieved for each galaxy.

in NGC 2403, the sample–size error for this galaxy could be further decreased by a factor of two.

The results presented here are based on observations made using prime focus CCD detectors at the Canada–France–Hawaii telescope (CFHT) and the KPNO 4.0–m; details will presented in the *Astrophysical Journal.*

Reference

Burstein, D. and Heiles, C. 1984, *Astrophys. J. Suppl.* **54**, 33.

Distances From H I Line Widths in Disk Galaxies

L. Gouguenheim

Observatoire de Paris, Meudon, and Université Paris Sud, Paris, France

I. Introduction

An accurate determination of the extragalactic distance scale is needed in order to obtain (1) a good determination of the expansion rate of the universe, and (2) information on deviations from a uniform Hubble flow. The accuracy is needed for two reasons. First, we need a good linearity of the distance scale; this is particularly important for the study of the velocity field. Second, we need a good calibration, which determines the accuracy of the Hubble constant H_0.

The extragalactic distance scale is built in two steps, using the so-called primary and secondary distance indicators.

The primary calibration is still uncertain. It depends on a small number of nearby galaxies in which different classes of individual stars whose absolute magnitudes are known are recognized (mainly Cepheids, novae, and RR Lyrae stars). The major remaining problems are the extinction of blue light coming from these stars, and the calibration of the absolute magnitude of galactic Cepheids. Promising results on IR observations of Cepheids are being obtained by B. Madore and collaborators (see for example McGonegal *et al.* 1982).

These primary indicators generally have a sound physical basis related to our knowledge of stellar properties and evolution. However, they fail at large distances (except for the supernovae) where individual stars are no longer observable.

A second kind of indicator is thus needed. These involve global properties of galaxies.[1] They are calibrated from the properties of the previous sample whose distances have been obtained from primary calibrators. Their physical

[1] These are called *tertiary* indicators by de Vaucouleurs in his many papers on the distance scale. — Ed.

bases are not as well understood — in part due to our limited knowledge of galaxy evolution — but they have a larger range, up to ~ 100 Mpc.

During the past dozen years, there have been two different approaches to the extragalactic distance scale problem. Sandage and Tammann (1982 and references therein) have chosen, at each step, what seems to them to be the *best* indicator, "putting all the eggs in the same basket," while de Vaucouleurs (1982 and references therein) has preferred to "spread the risks," using at each step as large a number of independent indicators as possible.

The following symbolic relations are examples of secondary indicators:

$$-M = a \log V_m + b \tag{1}$$

$$-M = a \log \sigma_v + b \tag{2}$$

$$-M = a \Lambda_c + b \tag{3}$$

They give the absolute magnitude of a galaxy as a function of (1) the maximum circular velocity, V_m in disk galaxies [the Tully–Fisher (1977) relation, Equation (1)]; of the internal velocity dispersion σ_v [the Faber–Jackson (1976) relation, Equation (2)] for spheroidal galaxies; or de Vaucouleurs's (1979) luminosity index Λ_c [Equation (3)], which again applies to disk galaxies. The first two relations are the expression of a total mass to total luminosity relation, and all of them have the general shape

$$-M = ap + b \tag{4}$$

where p is a directly observable parameter.

It is generally considered that the Tully–Fisher (hereafter TF) relation is the most accurate of these three, with a dispersion of 0.4 – 0.5 mag at given $p = \log V_m$, and the following discussion will concentrate on it.

II. Calibration of the TF Relationship

The maximum circular velocity V_m is deduced from the 21–cm line of neutral hydrogen, observed in a galaxy seen as a point source by the radiotelescope, after correcting for inclination effects. Non–circular motions must also be taken into account, because they tend to widen the 21–cm line. Bottinelli *et al.* (1983) have applied a linear summation of the random motions, described by a gaussian function with $\sigma_x = \sigma_y = 1.5\sigma_z = 10\text{–}15\,\mathrm{km\,sec^{-1}}$ (the z–axis being perpendicular to the disk); the rotation components V_m have then been deduced from the different widths of the 21–cm line measured respectively at 20, 40 and 50% of maximum intensity. They show good agreement with V_m measured directly from rotation curves. Lewis (1984) has found $\sigma_x = \sigma_y = \sigma_z = 12\,\mathrm{km\,sec^{-1}}$, and Tully and Fouqué (1985) have confirmed the validity of a linear summation for non–dwarf galaxies.

Not applying this correction leads to a steeper slope of the TF relation, and will add some extra noise to the distance determinations; however, no *systematic* effect is expected if all the line widths, including those of the calibrators, are corrected in the same way.

Two main systems of magnitudes have been used: the B_T^o system of RC2 (de Vaucouleurs *et al.* 1976), and the $H_{-.5}$ system of Aaronson *et al.* (1979). The B_T^o magnitudes have the advantage of being total, but they are sensitive to extinction effects. On the contrary, the $H_{-.5}$ magnitudes are relatively insensitive to extinction effects, but they refer to an aperture which is one–third of the *blue* photometric diameter a_{25}. This has two consequences: first, the H magnitudes do not measure the same fraction of the total light in different galaxies, depending of the bulge–to–disk ratio; and second, they involve B–band diameters, which are subject to extinction.

The calibration of the TF relation needs the determination of two parameters: the slope a, using local calibrators, cluster data, or kinematic distances; and the zero point, b, which is determined from local calibrators, and is thus affected by the uncertainty of the primary calibration.

The value of the slope depends on the system of magnitudes, and to a smaller extent, as discussed above, on the system of line withs, corrected or not for non–circular motions. The slope is in the range 5–6 at B (Bottinelli *et al.* 1983) and 10–12 at H (Aaronson *et al.* 1979; Aaronson and Mould 1983).

The variation of the slope with the wavelength of the magnitude can be easily understood, keeping in mind that the TF relation relies on a total mass to total luminosity relation. B–band and H–band magnitudes do not measure the same fraction of the total luminosity for blue or red galaxies. The bluest are expected to have the smallest $\log V_m$, and thus $M_B - M_H$ is expected to increase with $\log V_m$; this effect is reinforced by the limited aperture of the $H_{-.5}$ magnitude system: the luminosity of the blue, late–type galaxies, with small $\log V_m$ and small bulge–to–disk ratio, is more underestimated than the luminosity of red, large $\log V_m$, and large bulge–to–disk ratio early–type galaxies. For both of these reasons, one expects a systematic increase of $M_B - M_H$ with $\log V_m$.

The steeper slope in the red thus reflects more widely on M_H than on M_B the errors in $\log V_m$ (including both line widths and inclination measurements). This point has been derived from arguments related to the observed dispersion of the TF relation; this dispersion, in fact, is not easy to determine, owing to the statistical problems discussed below.

A possible type effect has also been discussed: either the slope a, or the zero point b, or possibly both, could be functions of a second parameter, the morphological type. For example, Giraud (1985) shows a segregation between late–type Scd–Sd galaxies and early–type Sab–Sb ones, when plotting the Hubble parameter H *versus* the recession velocity; at a given velocity, the late–type galaxies have systematically larger values of H than the early–type ones. Giraud suggests introducing a shift of 0.5 mag in the zero point of the TF relation between Scd–Sd and Sab–Sb galaxies. This effect is curiously not present in the

H-band data, which could be interpreted as an argument favouring the $H_{-.5}$ system over the B_T^0 system. These so-called "type effects" are in fact related to the Malmquist bias, as discussed below.

III. Possible Interpretation of the TF Relationship

The general idea is that the TF relation comes from a relationship between total mass and bolometric luminosity among galaxies; this interpretation holds if the maximum rotational velocity is a good indicator of the total mass.

Interesting results have been obtained by Lafon (1976) who has developed dynamical models of self-consistent thin disks, from microscopic distribution functions with no energy truncature. His main results are the following:

1. flat rotation curves are quite normal,
2. these flat rotation curves neither require nor exclude galactic halos, massive or not,
3. the rotation curves depend on two parameters, the total mass of the galaxy and the kinematic temperatures of the interior regions,
4. the maximum rotational velocity is essentially sensitive to the total mass of the system.

IV. Malmquist Bias in the TF Relation

1. The Method

The bias arising when determining distances from a magnitude-limited sample using equation (4) has been studied by Teerikorpi (1984). Its main properties can be understood from Figure 1.

We consider first a class of galaxies characterized by the same value of $p = \log V_m$. Through the TF relation, their mean absolute magnitude M_p is known ($-M_p = ap+b$), and the individual magnitudes are distributed around M_p. We assume this distribution to be gaussian, with dispersion σ_{M_p}. If the sample is cut at the apparent limiting magnitude m_l, an absolute limiting magnitude at distance d, $M_l(d) = m_l - 5\log d - 25$ results. When d increases, $M_l(d)$ becomes brighter, the mean absolute magnitude $< M >$ of the sample selected becomes more luminous, and the bias $\Delta M_d = -(< M > - M(d))$ increases. For a larger m_l, the limiting absolute magnitude is less luminous and the bias is smaller.

If two different samples of galaxies, characterized by p and $p' < p$ are considered, it is also easily seen that the bias is stronger for the p' class, which is the less luminous. Thus:

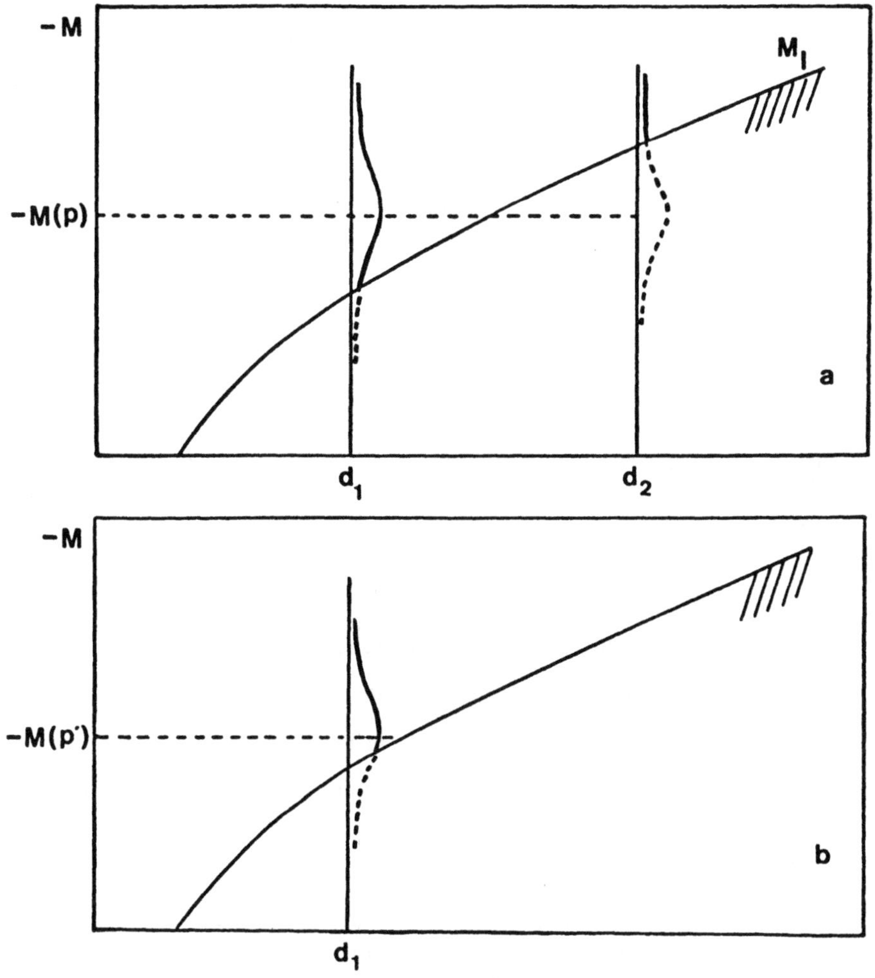

Figure 1: (a) only the galaxies more luminous than the absolute limiting magnitude M_l, *i.e.* above the limiting curve, are observed; the more distant the galaxy, the more severe the cut-off. For a given value of p, the luminosity function is more severely cut at the larger distance; (b) for a class of lower luminosity galaxies, characterized by $p' < p$, the bias at the same distance is larger.

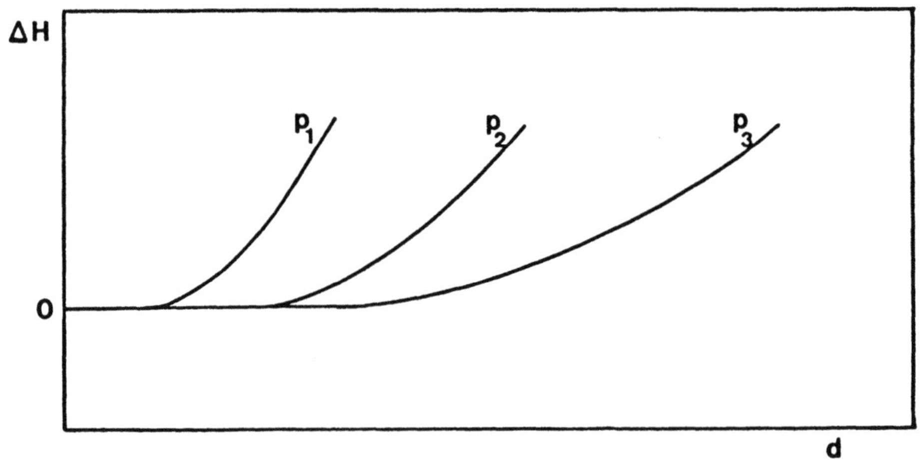

Figure 2: Three bias curves, obtained for three different values of the parameter p, with $p_1 < p_2 < p_3$, are represented; the other parameters, m_l and σ_{M_p}, are the same.

- *at given m_l and given p, the bias increases with the distance;*
- *at given m_l and distance, the bias is stronger for smaller p;*
- *at given p and distance, the bias is smaller for larger m_l.*

The shapes of the bias curves are shown in Figure 2.

- For a given class of galaxies p, the bias is negligible up to a threshold and then increases with distance.
- The threshold depends on p: it is larger for larger p.
- When considering a sample of galaxies with all p, it gives a cloud of points, distributed around the various curves, and the remaining plateau is the smallest one. As a consequence, the bias is not conspicuous.

In order to overcome these problems, Bottinelli *et al.* (1986a) have introduced the concept of ***normalized distance***:

$$d' = d \times \text{dex}[-0.2a(2.7 - logV_m)]$$

where d is the kinematic distance, which is an unbiased estimate of the distance. At the same d', all galaxies with different p suffer from the same amount of bias, provided that all these subsamples are characterizd by the same m_l and σM_P. All the plateau data are also selected.

2. Results

This method has been applied to B–band TF distances, with line widths corrected for non–circular motions (Bottinelli *et al.* 1986a) and to the H–band TF relation with uncorrected line widths (Bottinelli *et al.* 1988a). The main results are the following:

- both samples are affected by a strong bias (Figure 3) and
- the plateau data give a mean value of the Hubble constant, in de Vaucouleurs's local scale, $H_0 = 72 \pm 3\,\mathrm{km\,sec^{-1}}$ $\mathrm{Mpc^{-1}}$ in both B–band and H–band, which is significantly different from the values of H_0 previously determined from similar samples (de Vaucouleurs *et al.* 1981; Aaronson and Mould 1983).

3. Discussion

The "type effect" mentioned at the end of Section II is actually expected from the bias, because late–type galaxies are, in the mean, less luminous than early–type ones, and are thus expected to suffer from a larger bias (Bottinelli 1986b). Moreover, this so–called "type effect" is not conspicuous in H–band data; it should, however, not be concluded that this has something to do with a "better" quality of the H–band data. This comes essentially from the constitution of the samples (Figure 4) where the limiting magnitude is larger for low luminosity galaxies (small p). Thus, the two different effects, of m_l and p, respectively, on the resulting bias, compensate for one another.

Some authors (*e.g.* Tammann 1986; Giraud 1986) have considered the bias arising in the whole sample (including all p) without any normalization; it has been seen previously that the bias is very difficult to find, and that the unbiased data are not so easily recognized. Moreover, Giraud has tried to compute the expected bias; he comes to the conclusion that there remains an intrinsinc increase of H with kinematic distance after correcting the bias, which should explain only one–third of the effect. In fact, this method is a step backward in comparison to the method using the normalized distance, and relies on a strong assumption concerning the luminosity function. The separate curves of Figure 2 indicate how galaxies in different $\log V_m$ intervals populate the diagram. In order to calculate the average dependence of the biased H on d, one must know how populated each curve is by galaxies. The expected bias depends on the mean absolute magnitude and σM of the global luminosity function of galaxies, which is assumed to be gaussian. When dealing with a subsample of galaxies with same value of $\log V_m$, the mean unbiased absolute magnitude of the sample is known from the TF relation. This is not true in the present case, where M is determined from apparent magnitudes and (biased!) distances. Moreover, the choice of the limiting magnitude, on which the bias is strongly dependent,

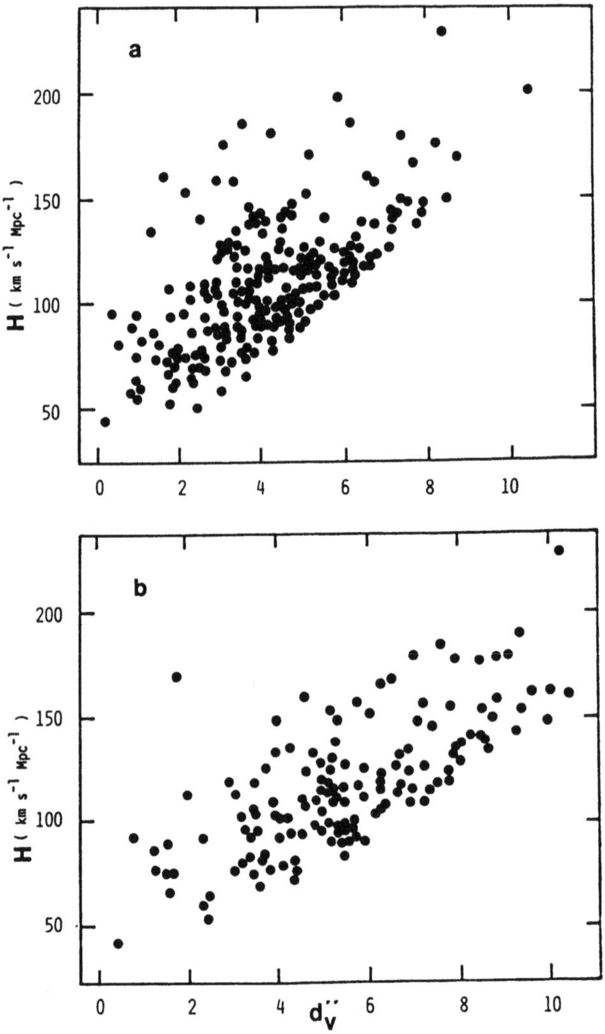

Figure 3: H *versus* normalized distance for (a) the B–band sample, cut at the limiting magnitude $B_T^o = 12$, and (b) the H–band sample, cut at the limiting magnitude $H_{-.5} = 10$.

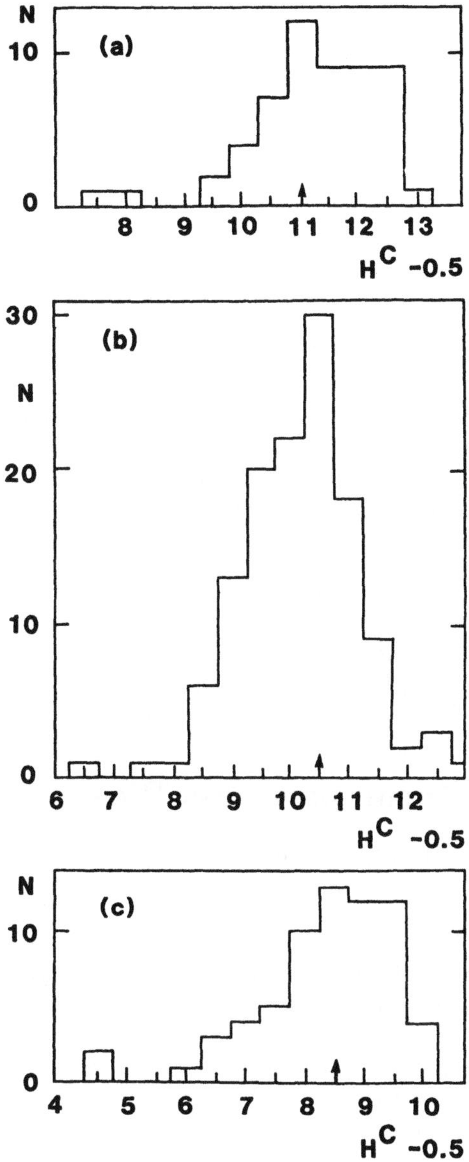

Figure 4: Histograms of $H_{-.5}$ magnitudes in three different $\log V_m$ ranges: (a) $1.9 \leq \log V_m < 2.1$; (b) $2.1 \leq \log V_m < 2.3$; (c) $\log V_m \geq 2.3$. The arrows indicate the magnitude up to which the sample can be considered as being complete.

has not been discussed. For all these reasons, it is difficult credit some the conclusions of previous studies.

4. Conclusions

The main conclusions of this discussion of Malmquist bias in the TF relation are the following:

1. The bias is the predominant effect. It is much more important than the accuracies of B_T^0 or $H_{-.5}$ magnitudes, or the effects of non–circular motions.

2. The bias does not depend only, or even predominantly, on the observed dispersion of the relation. It does, however, depend strongly on the limiting magnitude of the sample. This leads to the result that — contrary to statements commonly seen — when comparing different distance criteria, the best one is not necessarily characterized by the smallest scatter; and the bias at a given distance is not necessarily stronger for a criterion with larger dispersion.

3. Is the determination of $H_0 = 72 \pm 3$ using de Vaucouleurs's primary calibration only local (because of possible local motions) or global? There are two ways to answer this question. The first is to increase the sample and, therefore, m_l and the plateau threshold. The second relies on cluster data, which we discuss in the next section.

V. Cluster Incompleteness Bias

Contrary to general belief, a bias is also expected within a cluster.

1. Method

The bias arising when determining distances from a magnitude–limited sample of galaxies within a cluster using Equation 4 has been studied by Teerikorpi (1987). The bias is illustrated in Figure 5. The bias expected at small p is larger because the luminosity function of galaxies with same value of p is more severely cut. A bias decreasing with increasing p and negligible at large p (plateau region) is thus expected.

2. Results

The bias arising in a sample of 10 clusters with velocities ranging from 4000 to 11000 $\mathrm{km\,sec^{-1}}$ plus the Virgo cluster, has been studied both in B–band and in H–band light (Bottinelli *et al.* 1987, 1988a). A normalized $\log V_m$ taking into account the clusters distances and limiting magnitudes has been used in order

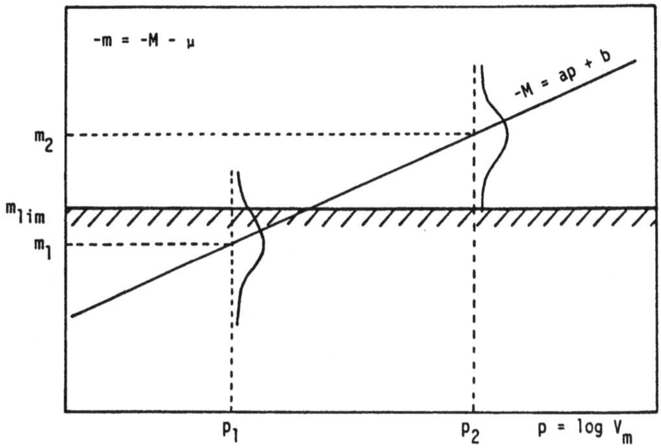

Figure 5: Apparent magnitude m *versus* $p = \log V_m$ diagram in a cluster having a distance modulus μ and observed to a limiting magnitude m_l (hatched line). The straight line stands for the TF relation. It is seen that the luminosity function is more severely cut at p_1 than at p_2 ($p_1 < p_2$).

to put all the cluster data together. The data, plotted in Figure 6, clearly show the trend expected. The plateau data in the B–band lead to $H_o = 73 \pm 4$, again using de Vaucouleurs's primary calibration and adopting an infall velocity of $200\,\mathrm{km\,sec^{-1}}$ for the Local group. This result is in remarkable agreement with the value obtained from the field sample.

The limiting magnitude of the $H_{-.5}$ sample is bright, leading to a very small number of plateau data. Using a sample of 19 galaxies including biased data near the threshold, and an iterative method for computing the bias, a similar value of H_o, in the range 70 – 75, is obtained.

VI. Conclusions

1. The effect of the Malmquist bias and the cluster population incompleteness bias have been strongly underestimated or even ignored (Bottinelli *et al.* 1988b). We desperately need large samples of galaxies complete to faint limiting magnitudes. The B_T^o system is presently probably the best magnitude system in which to work.

2. As a general comment, any physical property obtained statistically from a biased sample must be treated with great care. In the present context, examples include:

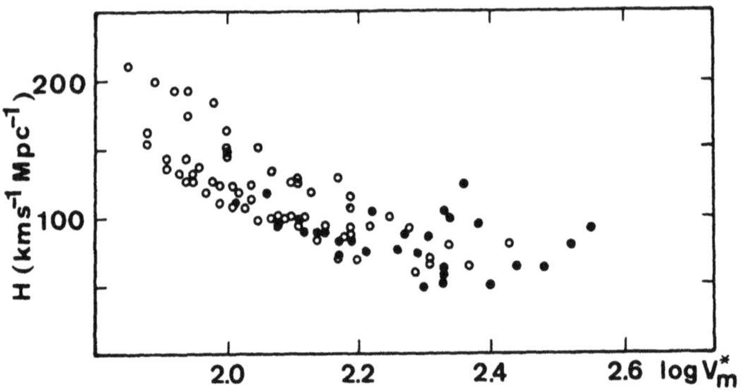

Figure 6: *H versus* normalized $\log V_m$ for 10 clusters. The Virgo cluster points (•) have been added, adopting an infall velocity of the Local Group equal to 220 km sec^{-1}.

- The type effect in the TF relation (this is, in fact, a differential Malmquist bias)
- The particular physical properties of cluster galaxies giving a different slope of the TF relation (the cluster population incompleteness bias predicts a shallower slope)
- All the kinematic studies performed with biased distances.

References

Aaronson, M. and Mould, J. 1983, *Astrophys. J.* **265**, 17.

Aaronson, M., Huchra, J., and Mould, J. 1979, *Astrophys. J.* **229**, 1.

Bottinelli, L., Gouguenheim, L., and Teerikorpi, P. 1988a, *Astron. Astrophys.* **196**, 17.

Bottinelli, L., Gouguenheim, L., Paturel, G., and Teerikorpi, P. 1986a, *Astron. Astrophys.* **156**, 157.

Bottinelli, L., Gouguenheim, L., Paturel, G., and Teerikorpi, P. 1986b, *Astron. Astrophys.* **166**, 393.

Bottinelli, L., Gouguenheim, L., Paturel, G., and Teerikorpi, P. 1988b, *Astrophys. J.* **328**, 4.

Bottinelli, L., Gouguenheim, L., Paturel, G., and de Vaucouleurs, G. 1983, *Astron. Astrophys.* **118**, 4.

Bottinelli, L., Fouqué, P., Gouguenheim, L., Paturel, G., and Teerikorpi, P. 1987, *Astron. Astrophys.* **181**, 1.

Faber, S. M. and Jackson, R. E. 1976, *Astrophys. J.* **204**, 668.
Giraud, E. 1985, *Astron. Astrophys.* **153**, 125.
Giraud, E. 1986, *Astron. Astrophys.* **174**, 23.
Lafon, J. P. 1976, *Astron. Astrophys.* **46**, 461.
Lewis, B. M. 1984, *Astrophys. J.* **285**, 483.
McGonegal, R., McLaren, R. A., McAlary, C. W., and Madore, B. F. 1982, *Astrophys. J. Lett.* **257**, L33.
Sandage, A. and Tammann, G. A. 1982, *Astrophys. J.* **256**, 339.
Tammann, G. A. 1986, *I. A. U. Symp. No. 124, Observational Cosmology*, ed. A. Hewitt, G. Burbidge, and L. Z. Fang (Dordrecht: Reidel), p. 151.
Teerikorpi, P. 1984, *Astron. Astrophys.* **141**, 407.
Teerikorpi, P. 1987, *Astron. Astrophys.* **173**, 39.
Tully, R. B. and Fisher, J. R. 1977, *Astron. Astrophys.* **54**, 661.
Tully, R. B. and Fouqué, P. 1985, *Astrophys. J. Suppl.* **58**, 67.
de Vaucouleurs, G. 1979, *Astrophys. J.* **227**, 380.
de Vaucouleurs, G. 1982, *The Cosmic Distance Scale and the Hubble Constant* (Canberra: Mt. Stromlo and Siding Spring Obs., Australian Nat. Univ.).
de Vaucouleurs, G. and Peters, W. L. 1986, *Astrophys. J.* **303**, 19.
de Vaucouleurs, G., de Vaucouleurs, A., and Corwin, H. G. 1976, *Second Reference Catalog of Bright Galaxies* (Austin: Univ. of Texas Press).
de Vaucouleurs, G., Peters, W. L., Bottinelli, L., Gouguenheim, L., and Paturel, G. 1981, *Astrophys. J.* **248**, 408.

Discussion

R. B. Tully: Two comments. (1) As clarification, the difference between the value of H° of 72 described here and my value of around 90 is *not* due to a calibration difference, but is due to whether or not all of the upturn in the value of H_0 with distance is due to Malmquist bias. I agree that there can be Malmquist bias, but it is mixed with the competing effect of the "local velocity anomaly." (2) The amplitude and even the sign of Malmquist bias depends on the *slope* of the distance estimator relationship. With the slope used here, there must be considerable bias. However, it *is possible* to use a recipe that avoids bias.

Gouguenheim: Concerning your first point, I stress again that the "Local velocity anomaly", if it exists, can be studied only using unbiased distances. We do not find in our recent sample of about 80 unbiased distances any clue favouring this hypothesis: galaxies which are members neither of the Leo Spur nor of the Coma–Sculptor Cloud give exactly the same low mean H value. Secondly, concerning the slope of the relation: we determined it from the unbiased sample, using the direct regression ($-M$ *versus* $\log V_m$), thus dealing with the completeness in magnitude, and then with the Malmquist bias. Schechter first and then Teerikorpi (for the TF relation) have discussed the case of the inverse regression, and have shown from analytical considerations that distances determined from this inverse regression not affected by the Malmquist bias. But it is also assumed when applying this inverse regression that, at a given apparent magnitude, the sample is complete in $p = \log V_m$, which has to be verified. It appears from the study made by Tammann and collaborators, and also from our own unpublished work, that another bias occurs in this relation. It seems to me that this "recipe" has to be tested, before being sure of its results.

B. F. Madore: If, as you suggest, magnitude cut–offs are producing Malmquist bias effects on the Tully–Fisher relations for galaxies or clusters, how do you account for the remarkably constant *slopes* of the observed relations, and for the remarkably similar dispersions of the relations for clusters at different distances?

Gougenheim: The cluster bias is not expected to modify in a strongly different manner the slope a of the relation as it is observed in two different clusters, whose distance moduli differ by $\Delta\mu$. It can be shown that the same amount of bias is expected to occur at two different values of $p = \log V_m$ differing by $\Delta p = \Delta\mu/a$ (assuming that the limiting magnitudes are the same). As long as the *modified* slope can be considered as linear, this modified slope is almost the same in different clusters.

Nevertheless, a different situation appears when Δp is large enough, so that the non–linear part of the biased slope is observed. This was the case for the Hercules Supercluster studied by Buta and Corwin, where a shallower slope was found.

The observed dispersion is a complex combination of the intrinsic scatter of the intrinsic scatter of the relation, the observational errors, the luminosity function, and the incompleteness function, so no conclusion can be drawn from it only. As was the case for the "type effect" observed in the B–band and not in the H–band, different factors can contribute in opposite ways.

R. J. Buta: Using line widths from Bothun, and from Giovanelli, Chincarini, and Haynes, Corwin and I found a flat Tully–Fisher relation in the Hercules Supercluster from B–band magnitudes. You would interpret this as being due to a cluster incompleteness bias, but it is our belief that errors in the line widths caused most of this effect. Can you comment on this possibility?

Gougenheim: The errors in the 21–cm line width (including measurement errors and inclination uncertainties) are included in the width of the luminosity function at given p. The larger the errors are, the larger is σ_M^p and, consequently, the bias. This, in turn, lowers the slope.

The D_n – Log σ Relation for Elliptical Galaxies: Present Status and Future Work

David Burstein

Department of Physics, Arizona State University, Tempe, Arizona, U. S. A.

Foreword

As this is a meeting honoring the life's work of Gérard and Antoinette de Vaucouleurs, I feel it is appropriate that I place into print the sense of the comments I made at the time of my talk.

In many ways, I consider Gérard to be one of my teachers, although I never took a class from him and, indeed, did not directly meet him until after I obtained my Ph. D. degree. I learned from him in the way most astronomers learn from each other — by reading his papers. I was fortunate to be at a major observatory (Lick), which had the journals and observatory publications that permitted me to trace the evolution of Gérard's thinking on extragalactic astronomy, from his earliest work on the photographic process up to the present. It is from these papers that I learned about the pitfalls and problems involved in photographic surface photometry in particular, and photometry of galaxies in general. It is also from these papers that I learned many of the techniques required to overcome these obstacles in order to obtain reliable data.

I personally have been fortunate to have had many fine teachers who influenced my science in a direct way. I believe that we in astronomy are fortunate to have had Gérard and Antoinette de Vaucouleurs, who have influenced the science of many astronomers in the way science should work, by example.

Abstract

The present status of the D_n – Log σ relationship of elliptical galaxies as a means of estimating distances to galaxies is evaluated using the data assembled by the

"Seven Samurai," and the "Great Attractor" velocity field model that has been developed with these data. Specific tests are made to detect possible systematic errors in the estimated distances of ellipticals that might be due to environment, Galactic extinction, and stellar population differences. No systematic effects can be detected. Future projects to map peculiar motions elsewhere in the Universe using the D_n – Log σ relationship are described, with emphasis on the project in which this author is directly involved: a program to search for evidence of large-scale motions in other regions of space.

1. Introduction

Faber and Jackson (1976), expanding on an original suggestion by Fish (1964), showed that the absolute magnitudes and central velocity dispersions of elliptical galaxies were correlated in an apparent power-law relationship, $L \propto \sigma^n$, where n is in the range 3–5. This relationship was further explored by a number of individuals and groups (Schechter and Gunn 1976; Schechter 1979; Whitmore, Kirschner, and Schechter 1979; Tonry and Davis 1981a, b; Tonry 1981; Dressler 1984).

In the paper that initiated the collaboration that has come to be known as the "Seven Samurai" (this author, R. L. Davies, A. Dresser, S. M. Faber, D. Lynden-Bell, R. Terlevich, and G. A. Wegner), Terlevich *et al.* (1981) further investigated the Faber-Jackson relationship. They showed that, if one assumes distances predicted from a smooth Hubble flow, the scatter of central velocity dispersion with absolute magnitude was correlated with the scatter of the Mg_2 stellar population parameter with absolute magnitude: the so-called $\Delta - \Delta$ relationship, or "second parameter" problem for ellipticals.

The referee of the Terlevich *et al.* paper, Ken Freeman, proposed that the "second parameter" was the mass density of the galaxy. In Freeman's words, "A consistent interpretation of the $\Delta - \Delta$ correlation is that the more metal-rich systems at a given luminosity have higher surface density. This makes sense within conventional enrichment pictures: surface density is M/R^2, which is also the characteristic gravitational acceleration in the system ..." De Vaucouleurs and Olson (1982) independently proposed surface brightness as an additional parameter in the L – Log σ relationship.

In the 20–20 vision of hindsight, were these astronomers correct? Well, both yes and no. The answer is "Yes" if one asks, "Does the L – Log σ relationship have intrinsic scatter that is correlated with the surface brightness of the galaxies?" This interpretation was confirmed independently by both our group (Dressler *et al.* 1987) and by Djorgovski and Davis (1987), and has led to the adoption of the D_n – Log σ relation as the standard distance estimator for ellipticals (see further discussion of the D_n parameter below).

The answer is "No" if one asks, "Is the source of the $\Delta - \Delta$ relationship found by Terlevich *et al.* the fact that there is intrinsic scatter in the L – Log

σ relationship due to surface brightness?" The principal source of the $\Delta - \Delta$ relationship is errors in estimated distances. This was partly due to the fact that we used a smooth Hubble flow as our measure of distance, and partly because several of the galaxies in that sample have relatively large peculiar motions. That distance errors are the true source of the $\Delta - \Delta$ relationship is consistent with the null (or minimal) $\Delta - \Delta$ relationships found for ellipticals in clusters (Dressler 1984; Dressler *et al.* 1987).

Faber *et al.* (1987) have shown that the existence of surface brightness as a second parameter in the L – Log σ relation can be incorporated into a more direct, distance–dependent relationship; namely that between the D_n parameter and Log σ . The D_n parameter is both a difficult parameter to physically understand, as well as a much–misunderstood parameter in the existing literature. It is therefore helpful to state what the D_n parameter is *not*: D_n is *neither* an isophotal diameter, *nor* the diameter corresponding to a certain magnitude. D_n *is* the diameter *within* which an elliptical galaxy has a fiducial *mean* surface brightness (L/R^2), taken in our published studies as 20.75 B mag arc–sec^{-2}. In the Appendix to Dressler *et al.* (1987), it is shown that, if all elliptical galaxies have the same form of luminosity distribution, then a D_n–like parameter has a unique relationship to the combination of effective radius ($\propto$ luminosity) and effective surface brightness. The existence of this relationship is shown explicitly in Burstein *et al.* (1987).

One purpose of the present contribution is to use the available data on the D_n – Log σ relation to investigate the possible effects of known sources of systematic errors in distance estimates: Galactic extinction, enviromental dependence of galaxy properties, and differences in stellar populations among ellipticals. The other purpose is to outline how the D_n – Log σ relationship is being used to further investigate large–scale motions in the Universe. In this latter capacity, this paper will naturally detail the project in which the author is involved, namely a search for the existence of large–scale motions in other, more distance regions of space.

2. Tests for Systematic Effects in the D_n – Log σ Relationship

a. Methodology

Here, as in all papers from our group, distances will be measured in units of km sec^{-1}, relative to the cosmic microwave background (CMB) reference frame. To obtain a distance in Mpc, one simply divides the quoted distance by one's favorite value for the Hubble constant. The absolute scale to this system is provided by the distance to the Coma Cluster, which itself is permitted to have a peculiar velocity relative to the CMB.

Distances for galaxies can also be estimated from a velocity field model,

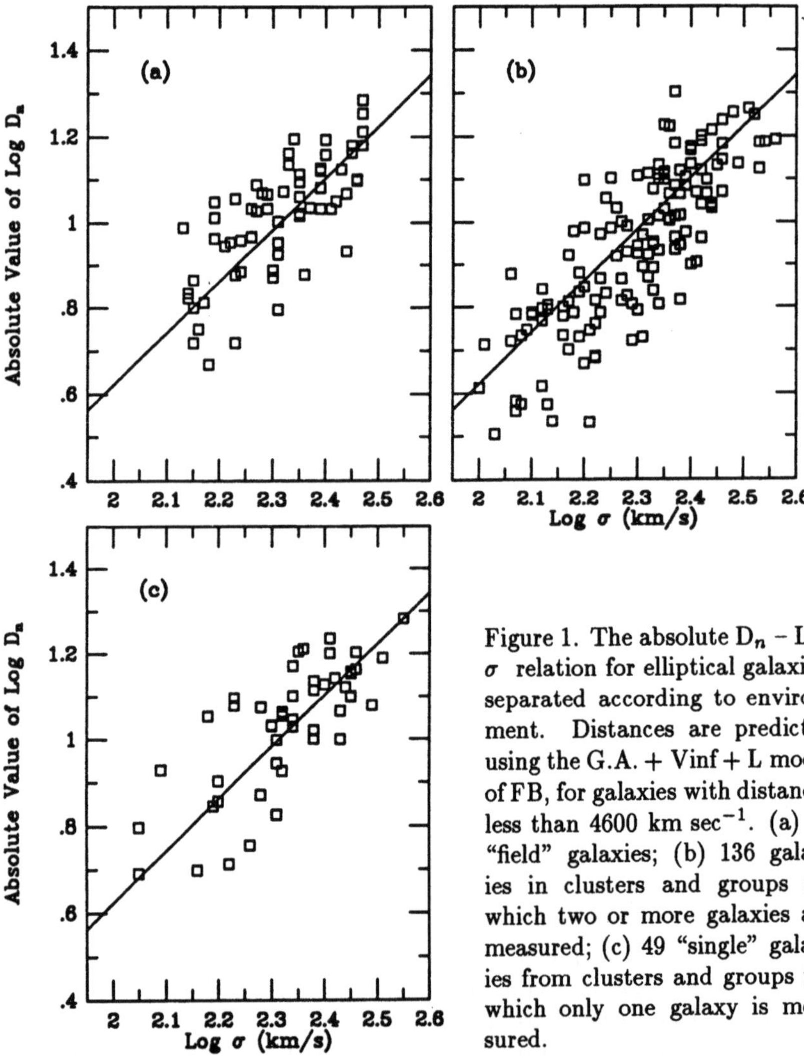

Figure 1. The absolute D_n – Log σ relation for elliptical galaxies, separated according to environment. Distances are predicted using the G.A. + Vinf + L model of FB, for galaxies with distances less than 4600 km sec^{-1}. (a) 64 "field" galaxies; (b) 136 galaxies in clusters and groups for which two or more galaxies are measured; (c) 49 "single" galaxies from clusters and groups for which only one galaxy is measured.

here taken to be the "Great Attractor + Virgo Infall + Local Anomaly" model adopted by Faber and Burstein (1988; FB), expanding on an original analysis by Lynden–Bell *et al.* (1988). In brief, this velocity field model describes the CMB motion of the Local Group and surrounding galaxies as due to the sum of three separately–generated motions:

1) The "Great Attractor" (G.A.). The velocity field within a distance of $3000\,\text{km}\,\text{sec}^{-1}$ of the Local Group is dominated by the gravitational pull of a centralized mass whose center lies in the direction $l = 310°$, $b = 18°$, at a distance of $\sim 4200\,\text{km}\,\text{sec}^{-1}$. This gravitational field produces a peculiar motion at the position of the Local Group of $535 \pm 60\,\text{km}\,\text{sec}^{-1}$. The current mathematical model of the G.A. allows only for a spherically–symmetric, centrally–concentrated, power–law velocity field with a core radius of $\sim 1300\,\text{km}\,\text{sec}^{-1}$.

2) Virgocentric Infall (Vinf): The peculiar motions of galaxies within 700 km sec^{-1} of the center of the Virgo cluster are influenced by the gravitational field of the Virgo cluster. At the position of the Local Group ($1350\,\text{km}\,\text{sec}^{-1}$ distant from the center of Virgo), motion due to Virgocentric infall is now estimated to be less than $100\,\text{km}\,\text{sec}^{-1}$ (Peebles 1988; FB). Previously larger estimates for Virgocentric infall were strongly biassed by the velocity field generated by the G.A. (FB).

3) The Local Anomaly (L): The peculiar motions of galaxies within 750 km sec^{-1} of the Local Group are essentially the same relative to the CMB (Peebles 1988; FB). These galaxies are distributed in a very flattened plane, termed the Coma–Sculptor cloud by Tully and Fisher (1987). This cloud appears to have an internal velocity disperison of less than 100 km sec^{-1} as seen from our vantage point. If the motions due to the G.A. and Vinf are taken into account, the remaining motion of the Local Anomaly is $360\,\text{km}\,\text{sec}^{-1}$ towards $l = 199°$, $b = 3°$. Lynden–Bell and Lahav (1988) point out that this direction is nearly opposite to the Local Void identified by Tully and Fisher (1987), and that this Local Anomaly motion could be the result of a lack of gravitational attraction from the void. FB show that the Coma Sculptor cloud is typical of other regions of space in terms of coherent size ($\sim 1500\,\text{km}\,\text{sec}^{-1}$) and a three–dimensional motion of $360\,\text{km}\,\text{sec}^{-1}$ ($= 200\,\text{km}\,\text{sec}^{-1}$ in one–dimension) relative to the G.A.

b. Is there a Dependence of Predicted Distance on Environment?

The elliptical galaxy sample can be divided into three subgoups, depending both on environment and how that environment was sampled (cf. Lynden–Bell *et al.* 1988): 1) "field" galaxies that belong to no readily indentifiable group or cluster; 2) "cluster" galaxies that belong to a group or cluster, for which we have predicted distances of two or more ellipticals; and 3) "single" galaxies that belong to a group or cluster, for which we have the predicted distance of only one galaxy.

The absolute D_n - Log σ relationship can be constructed for these galaxies using the G.A. + Vinf + L velocity field model, in a hybrid manner described in FB. Figure 1 plots the absolute value of D_n versus Log σ for galaxies with predicted distances less than 4600 km sec^{-1}, according to whether these are field galaxies (Figure 1a), cluster galaxies (Figure 1b), or "single" galaxies (Figure 1c). The line drawn in each figure represents the adopted relationship between Log σ and D_n , $\sigma^{1.2}/D_n$ = constant. The fact that the adopted D_n - Log σ relation does not appear to be the best fit to the fainter cluster galaxy data is due to selection effects in the cluster data relative to the field data, as discussed in an appendix to Lynden-Bell *et al.* (1988).

Figure 1 shows that the D_n - Log σ relationship and, hence, the predicted peculiar velocities for these ellipticals, have no apparent dependence on environment. An analogous test for spiral galaxies (using their HI deficiency parameter as a measure of environmental density) yields a similar null result (Burstein et al. 1988b). Moreover, the predicted distances for late-type spirals, early-type spirals, and elliptical galaxies in the same regions of space agree very well (FB). Thus, one concludes that the velocity field within 4000 km sec^{-1} that is obtained from distance measures to these galaxies is not detectably biassed by the environment of the galaxies.

c. Do Errors in Galactic Extinction lead to Systematic Errors in the Velocity Field?

A mis-estimation of Galactic extinction results in an error in D_n of size $\Delta D_n \cong 0.33 \Delta A_B$, where $A_B = 4 \times E_{B-V}$ (Dressler *et al.* 1987). An erroneous velocity field could result if Galactic extinction were systematically mis-estimated over the sky. An error in Galactic extinction is independent of the distance of a galaxy, so that the resulting error in distance must be proportional to the distance of the galaxy. As a measure of this possible error, we take $\log(R/V)$, with R being the predicted distance of the galaxy, and V the radial velocity of the galaxy (relative to the CMB).

An upper limit to possible errors in Galactic extinction can be taken from the scatter about the relationship between the reddening-free aborption-line index Mg_2 and reddening-dependent $(B-V)$ color (Burstein *et al.* 1988a), defined as $\Delta(B-V) = (B-V)_O - 1.12 Mg_2 + 0.62$. The quantity $(B-V)_O$ is corrected both for reddening and K-effect. Figure 2a plots $\log(R/V)$ versus $\Delta(B-V)$ for 325 elliptical galaxies (at all distances). As with the previous test for environmental dependence, no systematic correlation of predicted distances with errors in Galactic extinction can be detected with the present data.

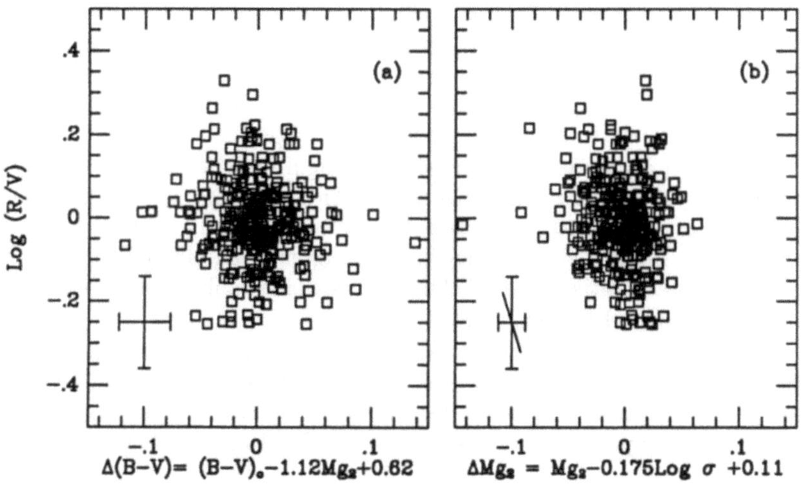

Figure 2: Two tests for potential systematic effects in predicted distances of elliptical galaxies, using the D_n – Log σ relation for 325 galaxies. Vertical axis is the logarithm of the ratio of predicted distance (R) to radial velocity (V) measured w.r.t. the CMB. (a) A test of extinction–related errors, using the residual of $(B-V)_0$ color with Mg_2 absorption line strength, $\Delta(B-V)$. (b) A test of stellar population–related errors, using the residual of Mg_2 as a function of central velocity dispersion, σ. Since both R and ΔMg_2 are functions of σ, the errors in σ translate into correlated errors, as shown. These tests show no evidence of systematic errors in predicted distances with either extinction errors or differences in stellar population.

d. How Do Differences in Stellar Population Affect Predicted Distances?

Both $(B - V)$ color and the absorption–line index Mg_2 are well–established as measures of the average stellar population in old stellar systems (cf. Burstein *et al.* 1984). The predicted errors in distances due to differences in stellar population are the same as that of errors in Galactic extinction, and will lead to errors that scale with distance.

Burstein *et al.* (1988a) show that there is an intrinsic scatter of both Mg_2 and $(B - V)$ color with the size of ellipticals, as measured by Log σ . Figure 2b plots $\log(R/V)$ versus the scatter of Mg_2 with Log σ , defined as $\Delta Mg_2 = Mg_2 - 0.175 \mathrm{Log}\sigma + 0.11$, for the same sample of 325 elliptical galaxies as in Figure 2a. Since both R and ΔMg_2 are functions of Log σ , the errors in both axes are correlated in the manner shown in the figure.

The extent of correlation observed in this figure can be accounted for by the known error in central velocity dispersion. Any remaining systematic effect as a function of stellar population is not detectable with the present data. Apparently, the integrated colors and line–strengths of an old stellar population are much more sensitive to small changes in stellar population than is the total stellar luminosity of the population.

Spiral galaxies also exhibit an intrinsic range of stellar population at given absolute size (Burstein, Condon, and Yin 1987; Burstein 1988). However, it is shown in Burstein *et al.* (1988b) that the existing data do not find that this scatter in stellar population influences the predicted distances for these spirals.

e. Predicted Distances and the Intrinsic Physical Properties of Galaxies

The intrinsic relationships between mass and luminosity that permit us to predict the distances of both ellipticals and spirals appear to be relatively insensitive to variations in either environment or in stellar population. This latter null result is mildly surprising, as with these same data one detects an intrinsic range of stellar population among galaxies of similar size and Hubble type.

Is it possible that the stellar population of a galaxy is directly correlated with the mass and/or mass density of the galaxy? Such a general correlation was suggested by Freeman (Section I, above), and it would be tempting to draw such a conclusion here. However, a proper evaluation of this question is both beyond the scope of the present paper and the accuracy of the present data. It is conceivable that systematic effects resulting from either stellar population or environment are present, but simply smaller than the current observational errors can reveal. If so, more accurate surveys of distances to both spiral and elliptical galaxies will provide valuable data for understanding the relationship of mass and luminosity within galaxies.

3. Future Work with the D_n – Log σ Relationship

By the middle of 1986, it was clear to the seven of us in the "Seven Samurai" collaboration that future work in measuring distances to galaxies was best pursued by us if we divided into smaller groups. The viable use of ellliptical galaxies as accurate distance estimators had been established, and we saw that there are many projects that could be done by applying these methods. Dividing into smaller groups also permitted us to each bring in new collaborators into the studies.

As of this writing, Dressler (1988) has nearly completed a redshift survey of all galaxies in the European Southern Observatory Catalogue (Lauberts 1982) that lie within a solid angle of a ~ steradian around the Great Attractor. The area being surveyed is nearly bisected by the Galactic plane. Dressler's study has found a substantial overdensity of galaxies in the radial velocity range 3500–6000 $\mathrm{km\,sec^{-1}}$ over this whole region, indicating that the Great Attractor is seen both in numbers of galaxies as well as in the velocity field.

Dressler and Faber are investigating the peculiar motions of galaxies on the "back" side of the Great Attractor. As explained in Faber *et al.* (1989), the original sample of elliptical galaxies for the Seven Samurai program was affected by a classfication error in the ESO Catalog, causing us to exclude ellipticals that would have been on the other side of the Great Attractor. Dressler and Faber have corrected for this bias, and are obtaining spectroscopic and photometric data for 100 galaxies, most of which are predicted to be falling into the Great Attractor region from the other side. It is hoped that these data will have been reduced and analyzed by the time this article is published.

The British-based contingent (Lynden-Bell and Terlevich) have joined with John Lucey (Durham), Ofer Lahav (Cambridge), and Jorge Melnick (ESO/Santiago) to intensively study the motions of galaxies in the regions around the Great Attractor. Faber and Dressler are investigating the motions of both elliptical and spiral galaxies to distances of ~ 6000 $\mathrm{km\,sec^{-1}}$ over the whole sky. They are also determining the accuracy of the D_n – Log σ relation for the bulges of S0 galaxies (Dressler 1987), in order to obtain more galaxies to map the nearby velocity field.

All of the above-mentioned programs will better define the velocity field within a distance approximately equal to the distance to the Coma Cluster (~ 7000 $\mathrm{km\,sec^{-1}}$). It is possible that the motions of galaxies in this region are either dominated, or substantially affected, by the presence of the mass concentration we call the Great Attractor. The analysis of FB has shown that velocity coherence due to the Great Attractor velocity field appears to extend to at least 4000 $\mathrm{km\,sec^{-1}}$ from the Local Group in some directions. The aforementioned new studies should define the total region of velocity coherence due to the Great Attractor, which may be larger.

It would be strange if the nearby Great Attractor were the only such object in the observable Universe. To search for the existence of other "Great Attractors," one must, by necessity, measure the peculiar velocities of galaxies at distances further than the distance to the Coma Cluster. Such a survey program was initiated in late 1986 by Davies, Wegner, and myself, focussing on two specific regions of the sky. In the year and a half since we began this survey, we have added three astronomers to our team: Matthew Colless (Durham), Robert McMahan (Center for Astrophysics), and Edmund Bertschinger (MIT). The remainder of this paper is devoted to detailing how this survey was assembled, and its status as of this writing.

4. A Search for Large–Scale Motions Elsewhere in the Universe

a. Why Use Elliptical Galaxies?

What would one expect to see if a Great Attractor–type velocity field is typical of other regions of space? Depending on viewing geometry, one should observe a general flow of $\sim$ 500-1000 km sec^{-1} over a region $\sim$ 4000 km sec^{-1} in size, culminating in an intrinsically very "noisy" region having a velocity dispersion of 1000–1500 km sec^{-1}. The region of coherent flow should have an intrinsic one–dimensional noise of about 150 km sec^{-1}, attributable to the motions of regions of galaxies that are $\sim$ 1500 km sec^{-1} in size. Thus, an observational error of 300 km sec^{-1} would be sufficient to detect the $\sim$ 1000 km sec^{-1} flow nearest the center of this more distant "GA." An observational error of 150 km sec^{-1} is required to detect the net flow across the whole volume of space.

The increase of observational error with distance is not the only problem in dealing with motions in other regions of space. In the near field, peculiar motions can be resolved both in terms of spatial coordinates and in terms of peculiar velocities. At distances greater than 8000 km sec^{-1}, spatial coordinates merge along the line–of–sight and observational errors in predicted distances dominate expected motions on the smallest scales. Moreover, the systematic effects of observational errors become very important at these distances: if the accuracy of predicting distances is 20%, the random observational error in peculiar velocity for a galaxy at a distance of 10,000 km sec^{-1} is 2,000 km sec^{-1}. Uncertainty in the amount of Malmquist bias correction (*i.e.* is the galaxy in a dense region of space, or a sparse region?) can lead to systematic errors of 500 km sec^{-1} or more at these distances.

Elliptical galaxies are better suited than late–type spirals for handling these problems in two ways. First, clusters and groups of ellipticals are readily identifiable at large distances (see below), and are well–localized in space. The accuracies of predicted distances for such groups are limited only by the number of galaxies within the group that can be measured. Second, elliptical–dominated groups tend to cluster together to form the "backbone" of large–scale structures (cf. Giovanelli and Haynes 1985). Thus, measurements of the peculiar motions of groups and clusters of elliptical galaxies should provide the best opportunity for measuring large–scale motions within other regions of space.

b. A Survey of Distances to Ellipticals with Radial Velocities of 8,000–14,000 km sec^{-1}

The limits of a such a survey are set by practical considerations: regions that appear to have coherent structure in two–dimensions are needed (much like the Great Attractor region would appear from a distance of 10,000 km sec^{-1}). These regions must be more distant than 7,000 km sec^{-1} to isolate them from the local flow. They cannot be more distant than 14,000 km sec^{-1}, due to the growing effects of observational errors.

The Struble and Rood (1987) compilation of 568 Abell cluster (1958) redshifts yields over 50 clusters with redshifts $0.027 < z < 0.045$. In addition, Robert Jackson, in an unpublished Ph.D. thesis (U.C. Santa Cruz, 1982) has identified 96 elliptical–dominated groups, randomly distributed on the sky. The compilation by Jackson adds another 20 aggregates in this redshift range to those in the Abell catalog.

Two regions of the sky are particularly well–populated in the desired redshift range: the Hercules–Corona Borealis (H–C–B) Supercluster Complex in the region +10° to +40° declination and $14^h - 18^h$ R.A., and the Pisces–Cetus (P–C) Supercluster Complex in the region -10° to +50° and $23^h - 4^h$ (nomenclature taken from Tully and Fisher 1987). Moreover, previous analyses have suggested that large–scale motions exist in both regions (H–C–B: Bahcall and Soniera 1983; Bahcall *et al.* 1986; Bahcall 1988; our original survey of elliptical galaxies. P–C: Batuski and Burns 1985a,b), and both regions can be observed by the Northern hemisphere observatories to which our team has ready access.

The Abell cluster radial velocities and Jackson data yield 40 elliptical–rich clusters in these two regions. With the positions of these clusters as our guide, we searched 80 plates in the Kitt Peak collection of Palomar Sky Survey plates for elliptical–rich associations in the region of each previously–known cluster or group. After some calibration, we were able to pick out elliptical–rich clusters in the right redshift range by using the characteristic size of the brightest galaxies as our guide. Altogether we found 83 such groups and clusters.

The positions of the selected groups on the sky are shown in Figure 3 in Galactic coordinates (note the offset of the center position in longitude in this diagram). The histogram of the number of prospective ellipticals in each group, as classified by eye, is given in Figure 4 in two parts: H–C–B (38 clusters, 279 prospective galaxies, northern Galactic latitudes), P–C (45 clusters, 319 galaxies, southern Galactic latitudes), and all 83 clusters (598 galaxies). The median number of prospective ellipticals per group is 7. From the CCD imaging data obtained so far, we expect the median number of usuable elliptical galaxies

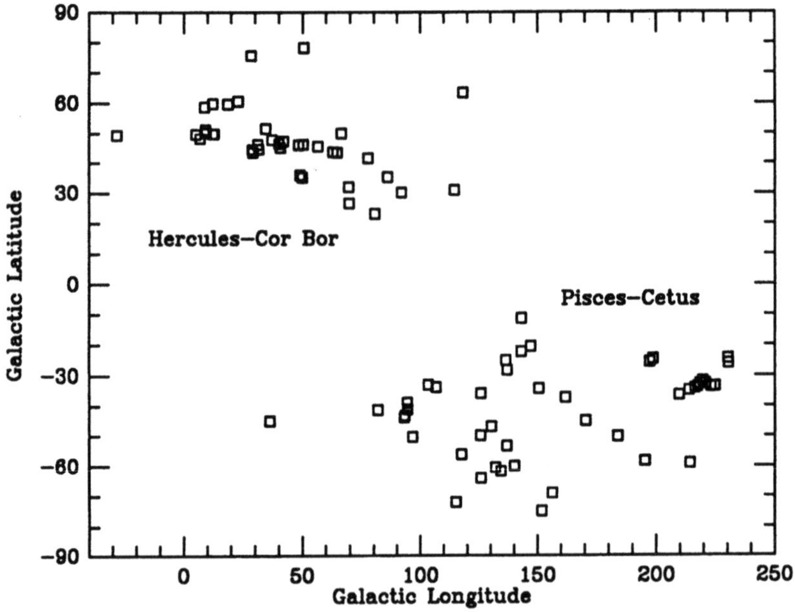

Figure 3: The distribution on the sky of the 83 elliptical–dominated groups and clusters at distances of 8,000 – 14,000 $\mathrm{km\,sec^{-1}}$. Galactic coordinates are used; note the offset of the center of the figure in longitude. The Hercules–Corona Borealis (positive Galactic latitude) and Pisces–Cetus (negative Galactic latitude) regions are noted.

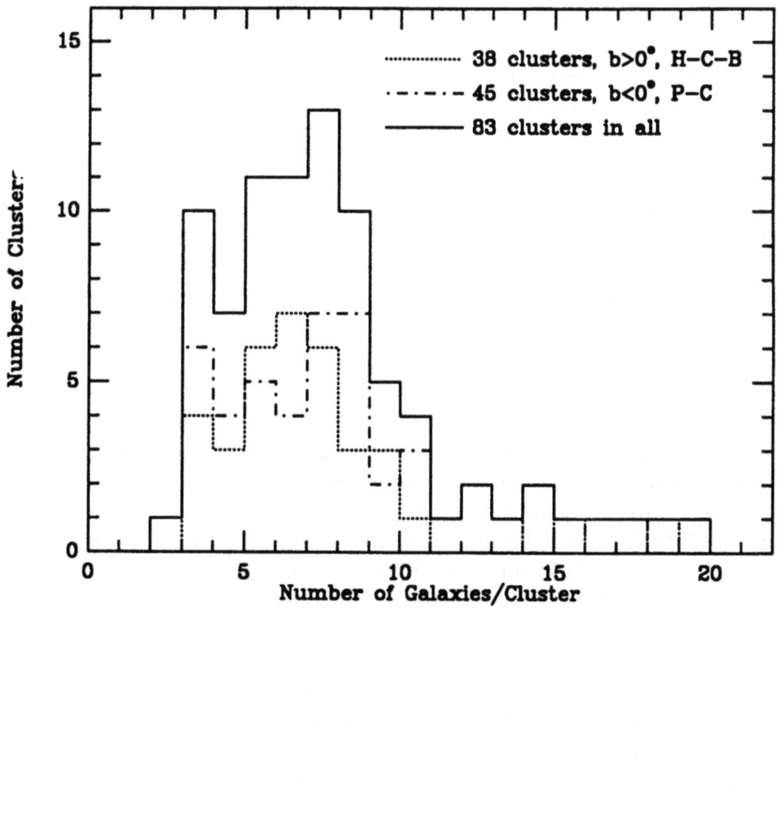

Figure 4: The histogram distributions of the number of clusters containing N prospective elliptical galaxies, divided into three parts: i) all clusters (solid line); ii) H–C–B clusters (dotted line); and iii) P–C clusters (dot–dash line). The median cluster contains 7 prospective ellipticals, of which 4 – 5 are expected to be usable for this survey.

for this program will be 4–5 per group.

c. What Peculiar Velocities Can Be Measured at these Distances, and over What Size Scale?

We expect that the average observational error in distance will be 20% for an individual galaxy. For a group of 4 galaxies, the error in peculiar velocity will be 1000 km sec^{-1}. It will take, on average, 10 such groups to sample the peculiar motions in a particular volume of space to an accuracy of 300 km sec^{-1}; 30 such groups to an accuracy of 175 km sec^{-1}. We expect there will be sufficient numbers of galaxies and clusters in both the H–C–B and the P–C regions (cf. Figure 3) to determine the total net motion of these regions to an accuracy of better than 150 km sec^{-1}. Subregions within these volumes (also apparent in Figure 3) are typically defined by 10–15 groups, implying that the motions of these subregions can be determined to an accuracy of $\sim$ 300 km sec^{-1}.

This survey is only feasible if the observational error for each galaxy can be kept to a minimum. Obtaining accurate data for these more distant galaxies is challenging, but feasible with CCD–based spectrographic and imaging systems. In the Seven Samurai survey, we found that values of D_n are more sensitive to errors in the zero point of the photometric system than to random errors (Burstein *et al.* 1987). The size of the seeing disk is also important to determining values of D_n , since the galaxies appear small. Problems with seeing can be handled both by accurate two–dimensional modelling and by choosing a fainter mean surface brightness for the D_n parameter.

Numerical experiments run by RLD and MC indicate that values of D_n are reliable obtained only for diameters that are larger than six times the seeing radius. Moreover, a 1% accuracy in sky background level (typical for these CCD images) means that the galaxy luminosity profile is accurate to better than 10% only at isopotal levels brighter than 10% of sky. In this context, we note that the surface brightness value of 20.75 B mag arc–sec^{-2} used for the D_n parameter in the Seven Samurai survey was an arbitrarily– chosen quantity. Figure 5 shows that the D_n value defined for a mean surface brightness of 21.75 B mag arc–sec^{-2} is well–correlated with the 20.75 B mag arc–sec^{-2} D_n parameter for 329 ellipticals from the Seven Samurai survey. We thus expect to be able to use a D_n parameter in the R passband that will be suitable for these more distant galaxies.

The accuracy of the D_n – Log σ relation is also dependent on the degree to which the luminosity profiles of these galaxies are the same within the D_n radii. Figure 6 shows the normalized growth curves for 120 elliptical galaxies from CCD R–band images obtained for the present survey. The growth curve for each galaxy has been normalized by a de Vaucouleurs (1959) $r^{1/4}$ law growth curve, and is plotted only for radii that are larger than 6 times the seeing disk and smaller than the isophotal radius corresponding to 10% of the sky brightness.

Due to seeing, the present data cannot measure the luminosity profiles of

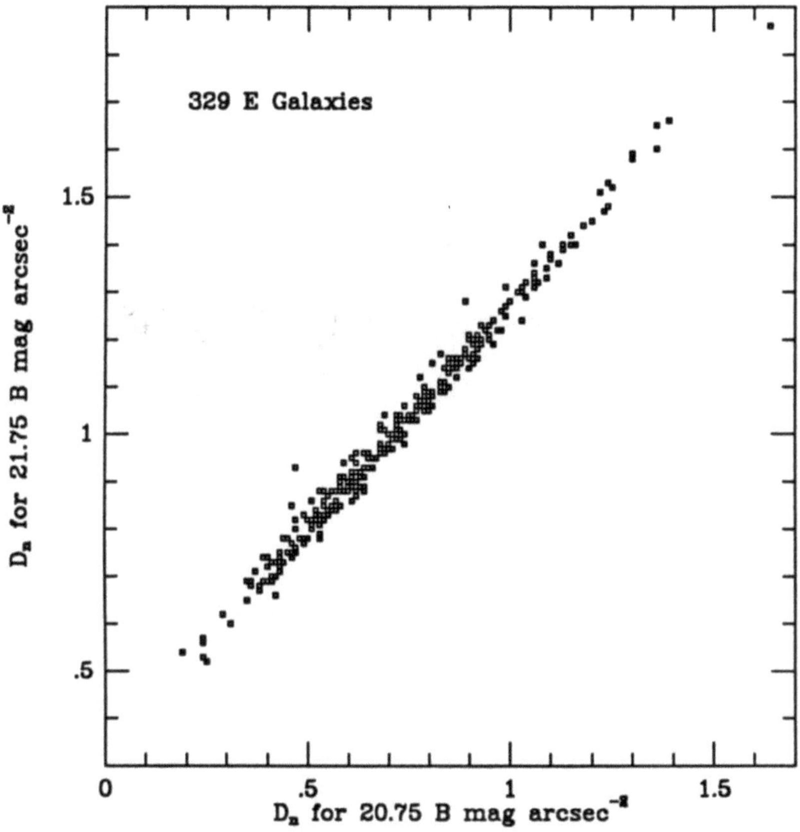

Figure 5: The D_n radius (in units of 0·1) measured for a mean surface brightness of 21.75 B mag arc–sec^{-2}, plotted versus the D_n radius measured for a mean surface brightness of 20.75 B mag arc–sec^{-2} (the original definition of D_n). The data for 329 elliptical galaxies with "good" photometry from the Seven Samurai survey are used (Burstein *et al.* 1987).

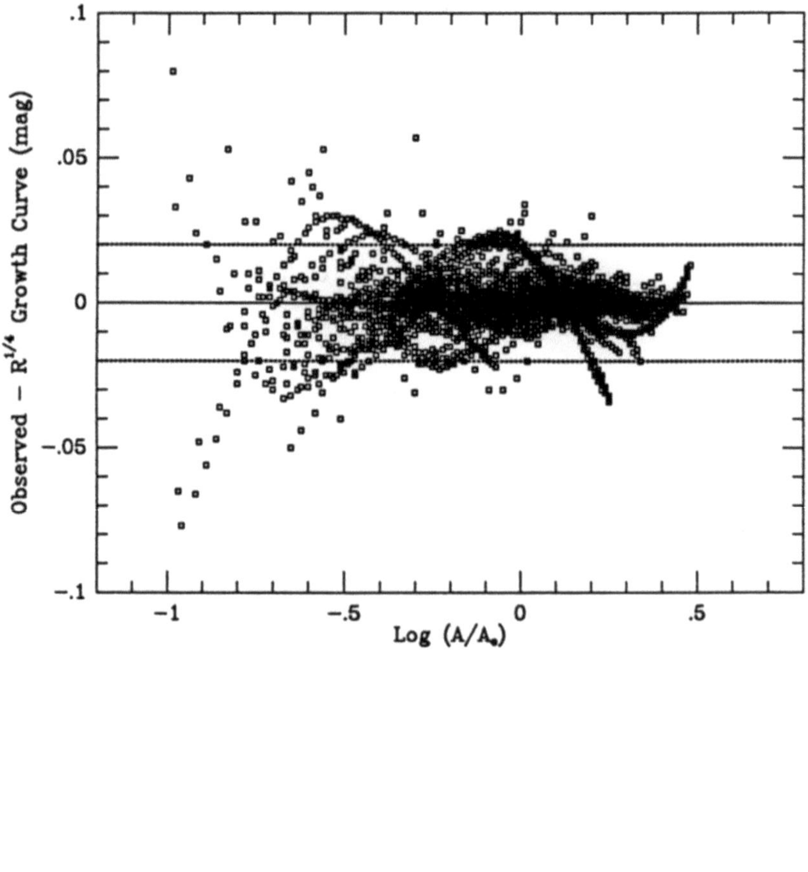

Figure 6: Residuals from a fit to an $r^{1/4}$ law for the CCD growth curves of 120 elliptical galaxies, plotted versus the logarithm of the ratio of the observed radius to the effective circular aperture, A_e. Each profile was fit with the $r^{1/4}$ law from an inner radius of six times the seeing disk radius, to an outer radius corresponding to an isophotal luminosity 10% above sky. The dotted horizontal lines are drawn at intervals are drawn at intervals of ± 0.02 mag in difference between the model profile and the observed profiles. Most galaxies have growth curves that agree with the standard model to better than 0.01 mag.

these galaxies near their centers and, as such, is insensitive to the probable differences in core properties (*e.g.* Lauer 1985). However, these data are ideal for measuring the luminosity profiles of ellipticals outside the core, since the sky background is obtained on the same CCD frame. It is therefore significant that the growth curves of these elliptical galaxies follow a de Vaucouleurs law to a very high degree of accuracy; the average deviation from an $r^{1/4}$ growth curve is less than 0.01 mag.

Of a random sampling of growth curves derived from 250 CCD R–band images of 190 galaxies, we find that 65% are elliptical galaxies whose $r^{1/4}$ laws are well–resolved with current CCD images, 23% are probably ellipticals for which better seeing images are required, and 12% are galaxies that are either too small or are not ellipticals. It is reasonable to expect an overall success rate of between 65–75% in obtaining data sufficent to accurately predict distances, yielding a final sample of more than 400 elliptical galaxies.

d. Random Errors and Systematic Errors: Malmquist Bias

The random effects of observational errors are well–known and relatively easy to quantitatively estimate. Such is not true for the errors due to Malmquist bias, which depend on the true distribution of galaxies along the line–of–sight. As extensively discussed in FB, Lynden–Bell *et al.* (1988) and elsewhere, the effect of Malmquist bias errors is fundamentally non–Gaussian in a clumpy medium. Gaussian techniques are viable in the near field only because the total observational error in peculiar velocities is relatively small. Gaussian techniques are not the methods of choice for investigations of peculiar motions in more distant regions.

We intend to do full–scale Monte Carlo modelling of the peculiar motions that we observe. One of our collaborators, Ed Bertschinger, was brought into the project primarily for his expertise in such modelling procedures. Bertschinger points out four advantages of using the Monte Carlo method, in addition to the obvious one of not having to assume Gaussian probability distributions: 1) Monte Carlo techniques can model the Malmquist–bias errors of peculiar motions for galaxies in a sparsely–sampled survey, such as the present survey. 2) It can directly examine the effects of systematic problems such as survey incompleteness, Galactic extinction, etc. 3) It can be used to find the optimal strategies for reducing the data to minimize errors. 4) Perhaps most importantly, theories can be processed just like the observations, starting from N–body simulations rather than telescopes.

5. Closing Thoughts

The study of large–scale motions in the universe is a fast–growing subject that is branching–out in several directions at once. Distance predictions accurate to better than 20% are now available for spiral galaxies with well-determined photometry, diameters and axial ratios (*e.g.* FB; Pierce and Tully 1988). CCD-based data should improve the observational error for ellipticals, perhaps to as small as 15%.

New surveys of peculiar motions are in progress that will more accurately define the nearby velocity field, define the extent of gravitational influence of the Great Attractor, and search for evidence of other Great Attractors elsewhere in the Universe. The use of elliptical galaxies and spiral galaxies in these programs are, in many ways, complimentary: clusters of elliptials are numerous and readily localized in space; spirals tend to be found less clustered and more generally distributed. There are many more spirals than ellipticals available for distance estimates. Spirals suitable for distance determinations using the Tully-Fisher (1977) method outnumber ellipticals usable in the D_n – Log σ relation by 5:1. Ellipticals are found in the densest regions of space; late-type spirals are found in the sparsest regions. Our survey is intended to measure distances of ellipticals. In the end analysis, however, both ellipticals and spirals will be used to map peculiar motions in other regions of space, as they have mapped the motions of galaxies in the nearby regions.

6. Acknowledgments

A review such as this would not be possible without all of the hard work of the people mentioned herein. All of the Seven Samurai, and all of the group of us sampling the distances to more distant galaxies have contributed substantially to the science reported here. As always, we thank the many Academic Institutions and Observatories that have, and are, supporting our research, including the organizers of this meeting. The research of this author has been supported in part by a Faculty-Grant-in-Aid from Arizona State University.

References

Abell, G. O. 1958, *Astrophys. J. Suppl.* **3**, 211.

Bahcall, N. A. 1988, in *Large-Scale Motions in the Universe*, proceedings of Vatican Study Week No. **27**, ed. G. Coyne and V. C. Rubin (Princeton: Princeton University Press) in press.

Bahcall, N. A., Soniera, R. and Burgett, W. S. 1986, *Astrophys. J.* **311**, 15.

Bahcall, N. A. and Soniera, R. 1983, *Astrophys. J.* **270**, 20.

Batuski, D. J. and Burns, J. O. 1985a, *Astron. J.* **90**, 1413.

Batuski, D. J. and Burns, J. O. 1985b, *Astrophys. J.* **299**, 5.

Burstein, D., 1988, in *Towards Understanding Galaxies at High Redshift*, ed. R. G. Kron and A. Renzini, (Boston: Kluwer) p. 93.

Burstein, D., Davies, R. L., Dressler, A., Faber, S. M., Stone, R. P. S., Lynden-Bell, D., Terlevich, R. J., and Wegner, G. 1987, *Astrophys. J.* **64**, 601.

Burstein, D., Davies, R. L., Dressler, A., Faber, S. M., Lynden–Bell, D., Terlevich, R. J., and Wegner, G. 1988a, in *Towards Understanding Galaxies at High Redshift*, ed. R. G. Kron and A. Renzini, (Boston: Kluwer) p. 17.

Burstein, D., Davies, R. L., Dressler, A., Faber, S. M., Lynden–Bell, D., Terlevich, R. J., and Wegner, G. 1988b, in *Large-Scale Structure and Motions in the Universe*, ed. G. Giuricin, F. Mardirossian, M. Maxxetti and M. Ramella, (Dordrecht: Reidel), in press.

Burstein, D., Davies, R. L., Dressler, A., Faber, S. M., Stone, R. P. S., Lynden–Bell, D., Terlevich, R. J., and Wegner, G. 1987, *Astrophys. J.* **64**, 601.

Burstein, D., Condon, J. J. and Yin, Q. F. 1987, *Astrophys. J. Lett.* **315**, L99.

Burstein, D., Faber, S. M., Gaskell, C. M., and Krumm, N. 1984, *Astrophys. J.* **287**, 586.

Djorgovski, S. and Davis, M. 1987, *Astrophys. J.* **313**, 59.

Dressler, A. 1984, *Astrophys. J.* **281**, 512.

Dressler, A. 1987, *Astrophys. J.* **317**, 1.

Dressler, A. 1988, *Astrophys. J.*, in press.

Dressler, A., Lynden–Bell, D., Burstein, D., Davies, R. L., Faber, S. M., Terlevich, R. J., and Wegner, G. 1987, *Astrophys. J.* **313**, 42.

Faber, S. M. and Burstein, D. 1988, in *Large-Scale Motions in the Universe*, proceedings of Vatican Study Week No. **27**, ed. G. Coyne and V. C. Rubin (Princeton: Princeton University Press), in press. (FB)

Faber, S. M. and Jackson, R. E. 1976, *Astrophys. J.* **204**, 668.

Faber, S. M., Dressler, A., Davies, R. L., Burstein, D., Lynden–Bell, D., Terlevich, R. L., and G. A. Wegner. 1987, in *Nearly Normal Galaxies: From the Planck Time to the Present*, ed. S. M. Faber, (Heidelberg: Springer–Verlag) p. 175.

Faber, S. M., Wegner, G. A., Burstein, D., Davies, R. L., Dressler, A., Lynden–Bell, D. and Terlevich, R. J. 1989, *Astrophys. J.* submitted.

Fish, R. A. 1964, *Astrophys. J.* **139**, 284.

Giovanelli, R. and Haynes, M. P. 1983, *Astron. J.* **88**, 881.

Lauberts, A. 1982, *ESO/Uppsala Survey of the ESO(B) Atlas*, (Garching bei München: ESO).

Lauer, T. 1985, *Astrophys. J.* **292**, 104.

Lynden–Bell, D. and Lahav, O. 1988 in *Large-Scale Motions in the Universe*, proceedings of Vatican Study Week No. **27**, ed. G. Coyne and V. C. Rubin (Princeton: Princeton University Press), in press.

Lynden–Bell, D., Faber, S. M., Burstein, D., Davies, R. L., Dressler, A., Terlevich, R. J., and Wegner, G. 1988, *Astrophys. J.* **326**, 19. (LFBDDTW).

Peebles, P. J. E. 1988, preprint.

Pierce, M. and Tully, R. B. 1988, preprint.

Schechter, P. L. and Gunn, J. E. 1978, *Astron. J.* **83**, 1360.

Schechter, P. L. 1979, *Astrophys. J.* **229**, 472.

Terlevich, R. J., Davies, R. L., Faber, S. M., and Burstein, D. 1981, *Mon. Not. Roy. Astron. Soc.* **196**, 381.

Struble, M. S. and Rood, H. J. 1987, *Astrophys. J. Suppl.* **63**, 555.
Tonry, J. L. 1981, *Astrophys. J. Lett.* **251**, L1.
Tonry, J. L. and Davis, M. 1981a, *Astrophys. J.* **246**, 666.
Tonry, J. L. and Davis, M. 1981b, *Astrophys. J.* **246**, 680.
Tully, R. B. and Fisher, J. R. 1987, *Nearby Galaxies Atlas*, (Cambridge: Cambridge University Press).
Tully, R. B. and Fisher, J. R. 1977, *Astron. Astrophys.* **54**, 661.
de Vaucouleurs, G. and Olson, D. W. 1982, *Astrophys. J.* **256**, 346.
Whitmore, B. C., Kirshner, R. P. and Schechter, P. L. 1979, *Astrophys. J.* **234**, 68.

Concluding Remarks

Frank Bash

Astronomy Department, University of Texas at Austin, U. S. A.

On behalf of the Honor Committee, Scientific Organizing Committee, and all the participants, I thank the Local Organizing Committee for a superbly organized meeting.

Unlike most meetings which concentrate on an area of research, which discuss the most important current problems, and in which the reviewer suggests the most important future problems, this meeting celebrated a career, the contributions of one person, Gérard de Vaucouleurs. It astonishes me that this celebration and the list of Gérard's accomplishments is also a listing of the foundations of the modern study of galaxies and of observational cosmology. It is very rare in modern science for one person (really two people) to have done so much work that was also so important. Gérard's work isn't done. (Jim Gunn at the end of the conference on Nearly Normal Galaxies suggests that two of the most important future problems include determining the Hubble Constant to 10% and doing a larger objective galaxy catalog with good photometry and radial velocities.)

In the last few years, Gérard has suffered personal blows, and he and astronomy lost a very important contributer, Antoinette de Vaucouleurs. I ask us all to help Gérard to keep going. We must all realize that now we are the only family he has.

Finally, I state the obvious: Gérard is unique in modern times. His contributions are nearly superhuman. He has published more than 360 research papers, 100 popular articles, and 20 books. He may be the last of the great astronomers. But I want to point to one special attribute which Gérard has and which, in this age of astrophysics, deserves special attention.

These days we are, or easily can be, buried in data. The rawest graduate student can go to the observatory and bring back 10 reels of magnetic tape and, in one month's time, by a mysterious process called data reduction, can turn those 10 reels into 20 reels of tape. The data are no longer sacred. The shrine to the data — the plate vault — is of historical interest only.

So we must operate as physicists. We must ask specific questions, motivated by physical theory, and limit our data taking to those data which bear on the question. But we must be careful. Our knowledge of Physics is still very imperfect, and especially in questions of galaxies and cosmology where Physics nears its limits or gives no specific answers, we see aesthetic criteria used to select favored theories. This reminds me, with some horror, of the failures of the Pythagoreans in their notion of the harmony of the spheres and of the failed cosmology of Plato and Aristotle, all based on "self-evident" aesthetic principles.

What Gérard especially has contributed is an ability to look at the empirical data — a lot of data — to listen to them, to think about them, and suggest descriptive theories which summarize them. This is how he found the Local Supercluster and the $r^{1/4}$ law. While we must operate as physicists, we must also continue to look beyond the boundaries of hypotheses which our current Physics can suggest, and really try to see what is actually there — not what we wish to be there. Gérard can do this. Maybe if 100 of us work together we can equal him.

Corrigenda

The Editors apologize to the authors and to the readers for the following problems found in the camera–ready text after it was sent to the publisher.

Page	Line	In Place of ...	Read ...
9	last	ignor	ignore
12	20	1985	1988
13	5th up	*á*	*à*
33	1	exposive	explosive
47	10	Principle	Principal
51	4	p-tupels	p-tuples
51	21	characteristica	characteristics
53	4	Récherche	Recherche
53	11	photographical	photographic
55	10	eccentricity	ellipticity
55	Last	d'Études	d'Etudes
65	caption	heirarchy	hierarchy
71	6	quiet	quite
109	15	ammounts	amounts
168	4th up	obseravtions	observations
168	2nd up	bifurication	bifurcation
169	6	bifuricated	bifurcated
174	11	interegrated	integrated
174	12	continumm	continuum
181	Figure	S	N
181	4	Court'es	Courtès
188	5	geometerical	geometrical
188	12	resusitation	resuscitation
189	14	Ray Soneira I	Ray Soneira and I
221	3	step 0.2	step 0.02
227	13	elliptical	ellipticals
227	3rd up	"mistery"	"mystery"
235	7th up	$\sim$ 5 arcmin	$\sim$ 3.5 arcmin
244	18	of dust lane	dust lane
256	9th up	isothermallity	isothermality
281	2nd up	Vaucoulerus	Vaucouleurs
312	14	explainations	explanations
312	4th up	occurance	occurence
313	3	nuclues	nucleus
313	7th up	The Galaxy	the Galaxy
313	2nd up	existance	existence
317	4th up	found have	found to have

320	8	chnages	changes
320	11	has be	has been
321	8	calulation	calculation
323	24	identfied	identified
324	26	teritiary	tertiary
329	20	anomolous	anomalous
351	9	discover	discovery
352	4th up	standard deviations	errors
354	7th up	verify	confirm
372	9th up	rlated	related
433	27 and 28		[Join these lines]
435	5th up	resent	recent
462	14 and 16	Lifsits	Lifshitz
463	24	$\mathrm{g\,cm^{-1.3}}$	$\mathrm{g\,cm^{-3}}$
463	8th up	0.02 Mpc	0.002 Mpc
463	5th up	homogenity	homogeneity
466	9th up	analitical	analytical
472	1	Lifsits	Lifshitz
499	3rd up	incidental	accidental
546	25	of the intrinsic scatter	[Delete]
555	20	classfication	classification
564	9	complimentary	complementary
564	9	elliptials	ellipticals

MIX
Papier aus verantwortungsvollen Quellen
Paper from responsible sources
FSC® C105338

FSC
www.fsc.org

If you have any concerns about our products,
you can contact us on
ProductSafety@springernature.com

In case Publisher is established outside the EU,
the EU authorized representative is:
Springer Nature Customer Service Center GmbH
Europaplatz 3, 69115 Heidelberg, Germany

Printed by Libri Plureos GmbH
in Hamburg, Germany